REFRIGERATION SYSTEMS AND APPLICATIONS

REFRIGERATION SYSTEMS AND APPLICATIONS

Second Edition

İbrahim Dinçer
Faculty of Engineering and Applied Science
University of Ontario Institute of Technology (UOIT)

Mehmet Kanoğlu
Department of Mechanical Engineering
University of Gaziantep

A John Wiley and Sons, Ltd., Publication

This edition first published 2010

First Edition published in 2003

Registered office
John Wiley & Sons Ltd, The Atrium, Southern Gate, Chichester, West Sussex, PO19 8SQ, United Kingdom

For details of our global editorial offices, for customer services and for information about how to apply for permission to reuse the copyright material in this book please see our website at www.wiley.com.

Library of Congress Cataloging-in-Publication Data

Dinçer, İbrahim, 1964-
Refrigeration systems and applications / İbrahim Dinçer, Mehmet Kanoğlu. – 2nd ed.
p. cm.
Includes bibliographical references and index.
ISBN 978-0-470-74740-7 (cloth)
1. Cold storage. 2. Frozen foods. 3. Refrigeration and refrigerating machinery. I. Kanoğlu, Mehmet. II. Title.
TP372.2.D56 2010
621.5′6 – dc22

2009051239

A catalogue record for this book is available from the British Library.

ISBN: 978-0-470-74740-7

Set in 9/11 Times by Laserwords Private Limited, Chennai, India

Contents

About the Authors

İbrahim Dinçer is a full professor of mechanical engineering in the faculty of engineering and applied science at University of Ontario Institute of Technology (UOIT). Renowned for his pioneering works in the area of sustainable energy technologies, he has authored and co-authored numerous books and book chapters, more than 500 refereed journal and conference papers, and many technical reports. He has chaired many national and international conferences, symposia, workshops, and technical meetings. He has delivered more than 150 keynote and invited lectures. He is an active member of various international scientific organizations and societies, and serves as editor-in-chief (for *International Journal of Energy Research* by Wiley and *International Journal of Exergy* and *International Journal of Global Warming* by Inderscience), associate editor, regional editor, and editorial board member on various prestigious international journals. He is a recipient of several research, teaching, and service awards, including a Premier's Research Excellence award in Ontario, Canada, in 2004. He has made innovative contributions to the understanding and development of sustainable energy technologies and their implementation, particularly through exergy. He has been working actively in the areas of hydrogen and fuel cell technologies, and his group has developed various novel technologies or methods.

Mehmet Kanoğlu is professor of mechanical engineering at University of Gaziantep. He received his B.S. in mechanical engineering from Istanbul Technical University and his M.S. (1996) and Ph.D. (1999) in mechanical engineering from University of Nevada, Reno, under the supervision of Professor Yunus A. Çengel, to whom he will be forever grateful. He spent the 2006–2007 academic year as a visiting professor at University of Ontario Institute of Technology, where he taught courses and was involved in research. Some of his research areas are refrigeration systems, gas liquefaction, hydrogen production and liquefaction by renewable energy sources, geothermal energy, energy efficiency, and cogeneration. He is the author or co-author of dozens of journal and conference papers. Professor Kanoğlu is the co-author of the book *Solutions Manual to Accompany: Introduction to Thermodynamics and Heat Transfer*, McGraw-Hill Inc., New York, 1997. Professor Kanoğlu has taught courses on thermal sciences at University of Nevada, Reno, University of Ontario Institute of Technology, and University of Gaziantep. He has consistently received excellent evaluations of his teaching.

Preface

Refrigeration is an amazing area where science and engineering meet for solving the humankind's cooling and refrigeration needs in an extensive range of applications, ranging from the cooling of electronic devices to food cooling, and has a multidisciplinary character, involving a combination of several disciplines, including mechanical engineering, chemical engineering, chemistry, food engineering, civil engineering and many more. The refrigeration industry has drastically expanded during the past two decades to play a significant role in societies and their economies. Therefore, the economic impact of refrigeration technology throughout the world has become more impressive and will continue to become even more impressive in the future because of the increasing demand for refrigeration systems and applications. Of course, this technology serves to improve living conditions in countless ways.

This second edition of the book has improved and enhanced contents in several topics, particularly in advanced refrigeration systems. It now includes study problems and questions at the end of each chapter, which make the book appropriate as a textbook for students and researchers in academia. More importantly, it now has comprehensive energy and exergy analyses presented in several chapters for better and performance improvement of refrigeration systems and applications, which make it even more suitable for industry. Coverage of the material is extensive, and the amount of information and data presented is sufficient for several courses, if studied in detail. It is strongly believed that the book will be of interest to students, refrigeration engineers, practitioners, and producers, as well as people and institutions that are interested in refrigeration systems and applications, and that it is also a valuable and readable reference text and source for anyone who wishes to learn more about refrigeration systems and applications and their and analysis.

Chapter 1 addresses general concepts, fundamental principles, and basic aspects of thermodynamics, psychrometrics, fluid flow and heat transfer with a broad coverage to furnish the reader with background information that is relevant to the analysis of refrigeration systems and applications. Chapter 2 provides useful information on several types of refrigerants and their environmental impact, as well as their thermodynamic properties. Chapter 3 delves into the specifics of refrigeration system components and their operating and technical aspects, analysis details, utilization perspectives and so on, before getting into refrigeration cycles and systems. Chapter 4 presents a comprehensive coverage on refrigeration cycles and systems for various applications, along with their energy and exergy analyses. Chapter 5 as a new chapter provides enormous material on advanced refrigeration cycles and systems for numerous applications with operational and technical details. There are also illustrative examples on system analyses through energy and exergy, which make it unique in this book. Chapter 6 deals with a number of technical aspects related to heat pump systems and applications, energy and exergy analyses and performance evaluation of heat pump systems, new heat pump applications and their utilization in industry, and ground source heat pump systems and applications. Chapter 7 is about heat pipes and their micro- and macro-scale applications, technical, design, manufacturing, and operational aspects of heat pipes, heat pipe utilization in HVAC applications, and their performance evaluation.

Incorporated through this book are many wide-ranging examples which provide useful information for practical applications. Conversion factors and thermophysical properties of various materials, as well as a large number of food refrigeration data, are listed in the appendices in the International System of Units (SI). Complete references are included with each chapter to direct the curious and interested reader to further information.

İbrahim Dinçer
Mehmet Kanoğlu

Acknowledgements

We are particularly thankful to various companies and agencies which contributed documents and illustrations for use in the first edition of this book. These valuable materials helped cover the most recent information available with a high degree of industrial relevance and practicality. We still keep most of them in the second edition as long-lasting materials.

We are grateful to some of our colleagues, friends and graduate students for their feedback and assistance for the first and the current editions of this book.

We acknowledge the support provided by our former and current institutions.

Also, we sincerely appreciate the exemplary support provided by Nicky Skinner and Debbie Cox of John Wiley & Sons in the development of this second edition, in many countless ways, from the review phase to the final product.

Last, but not least, we would like to take this opportunity to thank our families who have been a great source of support and motivation, and for their patience and understanding throughout the preparation of this second edition.

İbrahim Dinçer
Mehmet Kanoğlu

1

General Aspects of Thermodynamics, Fluid Flow and Heat Transfer

1.1 Introduction

Refrigeration is a diverse field and covers a large number of processes ranging from cooling to air conditioning and from food refrigeration to human comfort. Refrigeration as a whole, therefore, appears complicated because of the fact that thermodynamics, fluid mechanics, and heat transfer are always encountered in every refrigeration process or application. For a good understanding of the operation of the refrigeration systems and applications, an extensive knowledge of such topics is indispensable.

When an engineer or an engineering student undertakes the analysis of a refrigeration system and/or its application, he or she should deal with several basic aspects first, depending upon the type of the problem being studied, which may be of thermodynamics, fluid mechanics, or heat transfer. In conjunction with this, there is a need to introduce several definitions and concepts before moving into refrigeration systems and applications in depth. Furthermore, the units are of importance in the analysis of such systems and applications. One should make sure that the units used are consistent to reach the correct result. This means that there are several introductory factors to be taken into consideration to avoid getting lost inside. While the information in some situations is limited, it is desirable that the reader comprehends these processes. Despite assuming that the reader, if he or she is a student, has completed necessary courses in thermodynamics, fluid mechanics, and heat transfer, there is still a need for him or her to review, and for those who are practicing refrigeration engineers, the need is much stronger to understand the physical phenomena and practical aspects, along with a knowledge of the basic laws, principles, governing equations, and related boundary conditions. In addition, this introductory chapter reviews the essentials of such principles, laws, and so on, discusses the relationships between different aspects, and provides some key examples for better understanding.

We now begin with a summary of the fundamental definitions, physical quantities and their units, dimensions, and interrelations. We then proceed directly to the consideration of fundamental topics of thermodynamics, fluid mechanics, and heat transfer.

1.1.1 Systems of Units

Units are accepted as the currency of science. There are two systems: the *International System of Units* (*Le Système International d'Unitès*), which is always referred to as the SI units, and the *English System of Units* (the *English Engineering System*). The SI units are most widely used throughout the world, although the English System is the traditional system of North America. In this book, the SI units are primarily employed.

1.2 Thermodynamic Properties

1.2.1 Mass, Length and Force

Mass is defined as a quantity of matter forming a body of indefinite shape and size. The fundamental unit of mass is the kilogram (kg) in the SI and its unit in the English System is the pound mass (lbm). The basic unit of time for both unit systems is the second (s). The following relationships exist between the two unit systems:

$1\,\text{kg} = 2.2046\,\text{lbm}$ or $1\,\text{lbm} = 0.4536\,\text{kg}$

$1\,\text{kg/s} = 7936.6\,\text{lbm/h} = 2.2046\,\text{lbm/s}$

$1\,\text{lbm/h} = 0.000126\,\text{kg/s}$

$1\,\text{lbm/s} = 0.4536\,\text{kg/s}$

In thermodynamics the unit *mole* (mol) is commonly used and defined as a certain amount of substance containing all the components. The related equation is

$$n = \frac{m}{M} \tag{1.1}$$

if m and M are given in grams and gram/mol, we get n in mol. If the units are in kilogram and kilogram/kilomole, n is given in kilomole (kmol). For example, 1 mol of water, having a molecular weight of 18 (compared to 12 for carbon-12), has a mass of 0.018 kg and for 1 kmol, it becomes 18 kg.

The basic unit of length is the meter (m) in the SI and the foot (ft) in the English System, which additionally includes the inch (in.) in the English System and the centimeter (cm) in the SI. The interrelations are

$1\,\text{m} = 3.2808\,\text{ft} = 39.370\,\text{in.}$

$1\,\text{ft} = 0.3048\,\text{m}$

$1\,\text{in.} = 2.54\,\text{cm} = 0.0254\,\text{m}$

Force is a kind of action that brings a body to rest or changes the direction of motion (e.g., a push or a pull). The fundamental unit of force is the newton (N).

$1\,\text{N} = 0.22481\,\text{lbf}$ or $1\,\text{lbf} = 4.448\,\text{N}$

The four aspects (i.e., mass, time, length, and force) are interrelated by Newton's second law of motion, which states that the force acting on a body is proportional to the mass and the acceleration in the direction of the force, as given in Equation 1.2:

$$F = ma \tag{1.2}$$

Equation 1.2 shows the force required to accelerate a mass of 1 kg at a rate of 1 m/s^2 as $1\,\text{N} = 1\,\text{kg m/s}^2$.

It is important to note that the value of the earth's gravitational acceleration is 9.80665 m/s^2 in the SI system and 32.174 ft/s^2 in the English System, and it indicates that a body falling freely toward the surface of the earth is subject to the action of gravity alone.

1.2.2 *Specific Volume and Density*

Specific volume is the volume per unit mass of a substance, usually expressed in cubic meters per kilogram (m^3/kg) in the SI system and in cubic feet per pound (ft^3/lbm) in the English System. The *density* of a substance is defined as the mass per unit volume and is therefore the inverse of the specific volume:

$$\rho = \frac{1}{v} \tag{1.3}$$

and its units are kg/m^3 in the SI system and lbm/ft^3 in the English System. Specific volume is also defined as the volume per unit mass, and density as the mass per unit volume, that is

$$v = \frac{V}{m} \tag{1.4}$$

$$\rho = \frac{m}{V} \tag{1.5}$$

Both specific volume and density are intensive properties and are affected by temperature and pressure. The related interconversions are

$1\,\text{kg/m}^3 = 0.06243\,\text{lbm/ft}^3$ or $1\,\text{lbm/ft}^3 = 16.018\,\text{kg/m}^3$

$1\,\text{slug/ft}^3 = 515.379\,\text{kg/m}^3$

1.2.3 *Mass and Volumetric Flow Rates*

Mass flow rate is defined as the mass flowing per unit time (kg/s in the SI system and lbm/s in the English system). Volumetric flow rates are given in m^3/s in the SI system and ft^3/s in the English system. The following expressions can be written for the flow rates in terms of mass, specific volume, and density:

$$\dot{m} = \dot{V}\rho = \frac{\dot{V}}{v} \tag{1.6}$$

$$\dot{V} = \dot{m}v = \frac{\dot{m}}{\rho} \tag{1.7}$$

1.2.4 *Pressure*

When we deal with liquids and gases, pressure becomes one of the most important components. Pressure is the force exerted on a surface per unit area and is expressed in bar or Pascal (Pa). 1 bar is equal to 10^5 Pa. The related expression is

$$P = \frac{F}{A} \tag{1.8}$$

The unit for pressure in the SI denotes the force of 1 N acting on 1 m^2 area (so-called *Pascal*) as follows:

$1\,\text{Pascal (Pa)} = 1\,\text{N/m}^2$

The unit for pressure in the English System is pounds force per square foot, lbf/ft^2. The following are some of the pressure conversions:

1 Pa = 0.020886 lbf/ft^2 = 1.4504×10^{-4} lbf/in.^{2} = 4.015×10^{-3} in water = 2.953×10^{-4} in Hg

1 lbf/ft^2 = 47.88 Pa

1 lbf/in.^{2} = 1 psi = 6894.8 Pa

1 bar = 1×10^5 Pa

Here, we introduce the basic pressure definitions, and a summary of basic pressure measurement relationships is shown in Figure 1.1.

1.2.4.1 Atmospheric Pressure

The atmosphere that surrounds the earth can be considered a reservoir of low-pressure air. Its weight exerts a pressure which varies with temperature, humidity, and altitude. Atmospheric pressure also varies from time to time at a single location, because of the movement of weather patterns. While these changes in barometric pressure are usually less than one-half inch of mercury, they need to be taken into account when precise measurements are essential.

1 standard atmosphere = 1.0133 bar = 1.0133×10^5 Pa = 101.33 kPa = 0.10133 MPa
= 14.7 psi = 29.92 in Hg = 760 mmHg = 760 Torr.

1.2.4.2 Gauge Pressure

The *gauge pressure* is any pressure for which the base for measurement is atmospheric pressure expressed as kPa as gauge. Atmospheric pressure serves as reference level for other types of pressure measurements, for example, gauge pressure. As shown in Figure 1.1, the gauge pressure is either positive or negative, depending on its level above or below the atmospheric pressure level. At the level of atmospheric pressure, the gauge pressure becomes zero.

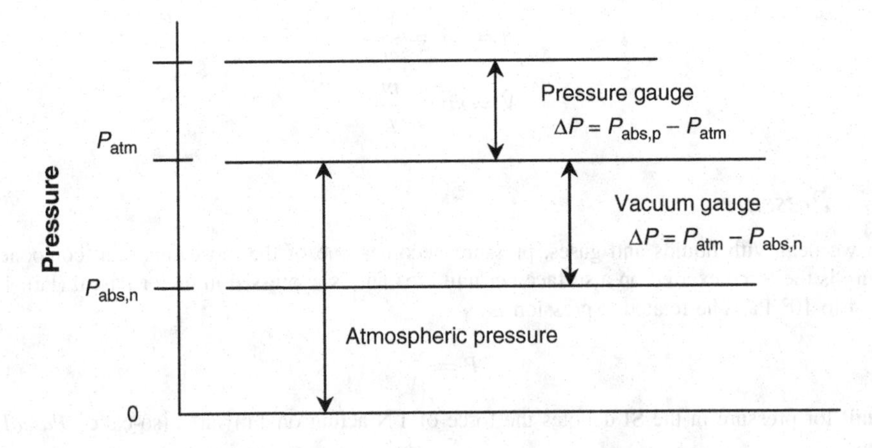

Figure 1.1 Illustration of pressures for measurement.

1.2.4.3 Absolute Pressure

A different reference level is utilized to obtain a value for absolute pressure. The absolute pressure can be any pressure for which the base for measurement is full vacuum, being expressed in kPa as absolute. In fact, it is composed of the sum of the gauge pressure (positive or negative) and the atmospheric pressure as follows:

$$\text{kPa (gauge)} + \text{atmospheric pressure} = \text{kPa (absolute)} \tag{1.9}$$

For example, to obtain the absolute pressure, we simply add the value of atmospheric pressure of 101.33 kPa at sea level. The absolute pressure is the most common one used in thermodynamic calculations despite the pressure difference between the absolute pressure and the atmospheric pressure existing in the gauge being read by most pressure gauges and indicators.

1.2.4.4 Vacuum

A vacuum is a pressure lower than the atmospheric one and occurs only in closed systems, except in outer space. It is also called the *negative gauge pressure*. As a matter of fact, vacuum is the pressure differential produced by evacuating air from the closed system. Vacuum is usually divided into four levels: (i) low vacuum representing pressures above 1 Torr absolute (a large number of mechanical pumps in industry are used for this purpose; flow is viscous), (ii) medium vacuum varying between 1 and 10^{-3} Torr absolute (most pumps serving in this range are mechanical; fluid is in transition between viscous and molecular), (iii) high vacuum ranging between 10^{-3} and 10^{-6} Torr absolute (nonmechanical ejector or cryogenic pumps are used; flow is molecular or Newtonian), and (iv) very high vacuum representing absolute pressure below 10^{-6} Torr (primarily for laboratory applications and space simulation).

A number of devices are available to measure fluid (gaseous or liquid) pressure and vacuum values in a closed system and require the fluid pressure to be steady for a reasonable length of time. In practice, the most common types of such gauges are the following:

- **Absolute pressure gauge.** This is used to measure the pressure above a theoretical perfect vacuum condition and the pressure value is equal to ($P_{abs,p} - P_{atm}$) in Figure 1.1. The most basic type of such gauges is the barometer. Another type of gauge used for vacuum measurements is the U-shaped gauge. The pressure value read is equal to ($P_{atm} - P_{abs,n}$) in Figure 1.1.
- **Mercury U-tube manometer.** These manometers use a column of liquid to measure the difference between two pressures. If one is atmospheric pressure, the result is a direct reading of positive or negative gauge pressure.
- **Plunger gauge.** This gauge consists of a plunger connected to system pressure, a bias spring, and a calibrated indicator. An auto tire gauge would be an example.
- **Bourdon gauge.** This is the most widely utilized instrument for measuring positive pressure and vacuum. Measurements are based on the determination of an elastic element (a curved tube) by the pressure being measured. The radius of curvature increases with increasing positive pressure and decreases with increasing vacuum. The resulting deflection is indicated by a pointer on a calibrated dial through a ratchet linkage. Similar gauges may be based on the deformation of diaphragms or other flexible barriers.
- **McLeod gauge.** This is the most widely used vacuum-measuring device, particularly for extremely accurate measurements of high vacuums.

Among these devices, two principal types of measuring devices for refrigeration applications are manometers and Bourdon gauges. However, in many cases manometers are not preferred because of the excessive length of tube needed, inconvenience at pressures much in excess of 1 atm, and less accuracy.

There are also pressure transducers available, based on the effects of capacitance, rates of change of strain, voltage effects in a piezoelectric crystal, and magnetic properties (Marquand and Croft, 1997). All have to be calibrated and the only calibration possible is against a manometer under steady conditions, even though they are most likely to be used under dynamic conditions.

It is important to note at another additional level that the *saturation pressure* is the pressure of a liquid or vapor at saturation conditions.

1.2.5 Temperature

Temperature is an indication of the thermal energy stored in a substance. In other words, we can identify hotness and coldness with the concept of temperature. The temperature of a substance may be expressed in either relative or absolute units. The two most common temperature scales are the Celsius (°C) and the Fahrenheit (°F). As a matter of fact, the Celsius scale is used with the SI unit system and the Fahrenheit scale with the English Engineering system of units. There are also two more scales: the Kelvin scale (K) and the Rankine scale (R) that is sometimes employed in thermodynamic applications. The relations between these scales are summarized as follows:

$$T_{(°C)} = \frac{T_{(°F)} - 32}{1.8} \tag{1.10}$$

$$T_{(K)} = T_{(°C)} + 273.15 = \frac{T_{(R)}}{1.8} = \frac{T_{(°F)} + 459.67}{1.8} \tag{1.11}$$

$$T_{(°F)} = 1.8T_{(°C)} + 32 = 1.8(T_{(K)} - 273.15) + 32 \tag{1.12}$$

$$T_{(R)} = 1.8T_{(K)} = T_{(°F)} + 459.67 \tag{1.13}$$

Furthermore, the temperature differences result in

$$1\,\text{K} = 1\,°\text{C} = 1.8\,\text{R} = 1.8\,°\text{F}$$

$$1\,\text{R} = 1\,°\text{F} = 1\,\text{K}/1.8 = 1\,°\text{C}/1.8$$

Kelvin is a unit of temperature measurement; zero Kelvin (0 K) is absolute zero and is equal to −273.15 °C. The K and °C are equal increments of temperature. For instance, when the temperature of a product is decreased to −273.15 °C (or 0 K), known as *absolute zero*, the substance contains no heat energy and supposedly all molecular movement stops. The saturation temperature is the temperature of a liquid or vapor at saturation conditions.

Temperature can be measured in many ways by devices. In general, the following devices are in common use:

- **Liquid-in-glass thermometers.** It is known that in these thermometers the fluid expands when subjected to heat, thereby raising its temperature. It is important to note that in practice all thermometers including mercury ones only work over a certain range of temperature. For example, mercury becomes solid at −38.8 °C and its properties change dramatically.
- **Resistance thermometers.** A resistance thermometer (or detector) is made of resistance wire wound on a suitable former. The wire used has to be of known, repeatable, electrical characteristics so that the relationship between the temperature and resistance value can be predicted precisely. The measured value of the resistance of the detector can then be used to determine the value of an unknown temperature. Among metallic conductors, pure metals exhibit the greatest change of resistance with temperature. For applications requiring higher accuracy,

especially where the temperature measurement is between −200 and +800 °C, the majority of such thermometers are made of platinum. In industry, in addition to platinum, nickel (−60 to +180 °C), and copper (−30 to +220 °C) are frequently used to manufacture resistance thermometers. Resistance thermometers can be provided with 2, 3, or 4 wire connections and for higher accuracy at least 3 wires are required.

- **Averaging thermometers.** An averaging thermometer is designed to measure the average temperature of bulk stored liquids. The sheath contains a number of elements of different lengths, all starting from the bottom of the sheath. The longest element which is fully immersed is connected to the measuring circuit to allow a true average temperature to be obtained. There are some significant parameters, namely, sheath material (stainless steel for the temperature range from −50 to +200 °C or nylon for the temperature range from −50 to +90 °C), sheath length (to suit the application), termination (flying leads or terminal box), element length, element calibration (to copper or platinum curves), and operating temperature ranges. In many applications where a multielement thermometer is not required, such as in air ducts, cooling water, and gas outlets, a single element thermometer stretched across the duct or pipework will provide a true average temperature reading. Despite the working range from 0 to 100 °C, the maximum temperature may reach 200 °C. To keep high accuracy these units are normally supplied with 3-wire connections. However, up to 10 elements can be mounted in the averaging bulb fittings and they can be made of platinum, nickel or copper, and fixed at any required position.
- **Thermocouples.** A thermocouple consists of two electrical conductors of different materials connected together at one end (so-called *measuring junction*). The two free ends are connected to a measuring instrument, for example, an indicator, a controller, or a signal conditioner, by a reference junction (so-called *cold junction*). The thermoelectric voltage appearing at the indicator depends on the materials of which the thermocouple wires are made and on the temperature difference between the measuring junction and the reference junction. For accurate measurements, the temperature of the reference junction must be kept constant. Modern instruments usually incorporate a cold junction reference circuit and are supplied ready for operation in a protective sheath, to prevent damage to the thermocouple by any mechanical or chemical means. Table 1.1 gives several types of thermocouples along with their maximum absolute temperature ranges. As can be seen in Table 1.1, copper–constantan thermocouples have an accuracy of ±1 °C and are often employed for control systems in refrigeration and food-processing applications. The iron–constantan thermocouple with its maximum temperature of 850 °C is used in applications in

Table 1.1 Some of the most common thermocouples.

Type	Common Names	Temperature Range (°C)
T	Copper–Constantan (C/C)	−250 to 400
J	Iron–Constantan (I/C)	−200–850
E	Nickel Chromium–Constantan or Chromel–Constantan	−200–850
K	Nickel Chromium–Nickel Aluminum or Chromel–Alumel (C/A)	−180–1100
-	Nickel 18% Molybdenum–Nickel	0–1300
N	Nicrosil–Nisil	0–1300
S	Platinum 10% Rhodium–Platinum	0–1500
R	Platinum 13% Rhodium–Platinum	0–1500
B	Platinum 30% Rhodium–Platinum 6% Rhodium	0 to 1600

the plastics industry. The chromel–alumel type thermocouples, with a maximum temperature of about 1100 °C, are suitable for combustion applications in ovens and furnaces. In addition, it is possible to reach about 1600 or 1700 °C using platinum and rhodium–platinum thermocouples, particularly in steel manufacture. It is worth noting that one advantage thermocouples have over most other temperature sensors is that they have a small thermal capacity and thus a prompt response to temperature changes. Furthermore, their small thermal capacity rarely affects the temperature of the body under examination.

- **Thermistors.** These devices are semiconductors and act as thermal resistors with a high (usually negative) temperature coefficient. Thermistors are either self-heated or externally heated. Self-heated units employ the heating effect of the current flowing through them to raise and control their temperature and thus their resistance. This operating mode is useful in such devices as voltage regulators, microwave power meters, gas analyzers, flow meters, and automatic volume and power level controls. Externally heated thermistors are well suited for precision temperature measurement, temperature control, and temperature compensation due to large changes in resistance versus temperature. These are generally used for applications in the range -100 to $+300$ °C. Despite early thermistors having tolerances of ±20% or ±10%, modern precision thermistors are of higher accuracy, for example, ±0.1 °C (less than ±1%).
- **Digital display thermometers.** A wide range of digital display thermometers, for example, handheld battery powered displays and panel mounted mains or battery units, are available in the market. Figure 1.2 shows a handheld digital thermometer with protective boot (with a high accuracy, e.g., ±0.3% reading ± 1.0 °C). Displays can be provided for use with all standard thermocouples or BS/DIN platinum resistance thermometers with several digits and 0.1 °C resolution.

It is very important to emphasize that before temperature can be controlled, it must be sensed and measured accurately. For temperature measurement devices, there are several potential sources of error such as sensor properties, contamination effects, lead lengths, immersion, heat transfer, and controller interfacing. In temperature control there are many sources of error which can be

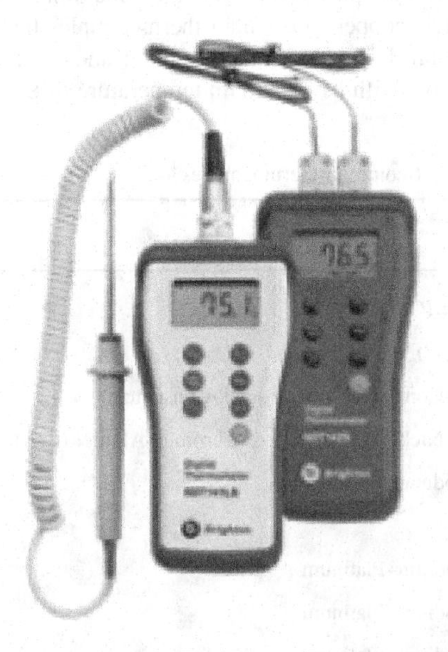

Figure 1.2 Handheld digital thermometers (*Courtesy of* Brighton Electronics, Inc.).

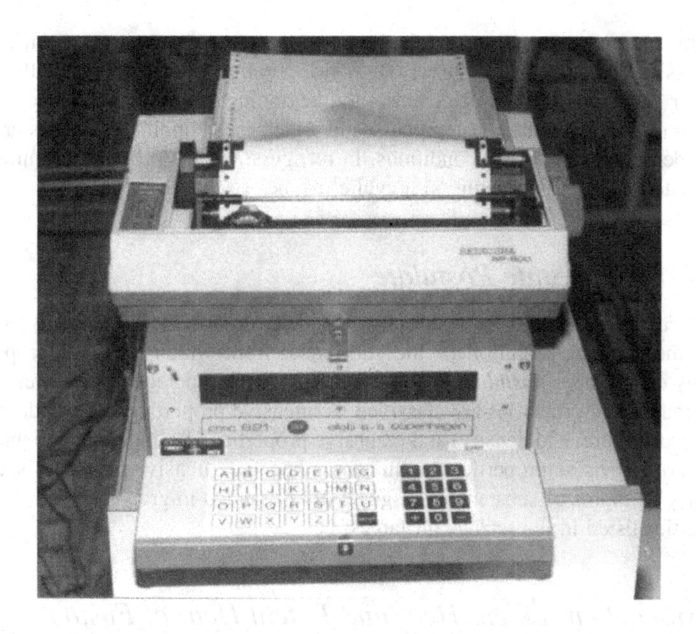

Figure 1.3 A data acquisition system for temperature measurements in cooling.

minimized by careful consideration of the type of sensor, its working environment, the sheath or housing, extension leads, and the instrumentation. An awareness of potential errors is vital in the applications dealt with. Selection of temperature measurement devices is a complex task and has been discussed briefly here. It is extremely important to remember to "choose the right tool for the right job." Data acquisition devices are commonly preferred for experimental measurements. Figure 1.3 shows a data acquisition system set-up for measuring temperatures during heating and cooling applications.

1.2.6 Thermodynamic Systems

These are devices or combination of devices that contain a certain quantity of matter being studied. It is important to carefully define the term *system* as that portion of all matter under consideration. There are three systems that we can define as follows:

- **Closed system.** This is defined as a system across the boundaries of which no material crosses. In other words, it is a system that has a fixed quantity of matter, so that no mass can escape or enter. In some books, it is also called a *control mass*.
- **Open system.** This is defined as a system in which material (mass) is allowed to cross its boundaries. It is also called a *control volume*.
- **Isolated system.** This is a closed system that is not affected by the surroundings at all, in which no mass, heat, or work crosses its boundary.

1.2.7 Process and Cycle

A process is a physical or chemical change in the properties of matter or the conversion of energy from one form to another. Several processes are described by the fact that one property remains constant. The prefix iso- is employed to describe a process, such as an isothermal process (a constant-temperature process), an isobaric process (a constant-pressure process), and an isochoric process (a constant-volume process). A refrigeration process is generally expressed by the conditions or properties of the refrigerant at the beginning and end of the process.

A cycle is a series of thermodynamic processes in which the endpoint conditions or properties of the matter are identical to the initial conditions. In refrigeration, the processes required to produce a cooling effect are arranged to operate in a cyclic manner so that the refrigerant can be reused.

1.2.8 Property and State Postulate

This is a physical characteristic of a substance used to describe its state. Any two properties usually define the state or condition of the substance, from which all other properties can be derived. This is called *state postulate*. Some examples are temperature, pressure, enthalpy, and entropy. Thermodynamic properties are classified as intensive properties (independent of the mass, e.g., pressure, temperature, and density) and extensive properties (dependent on the mass, e.g., mass and total volume). Extensive properties per unit mass become intensive properties such as specific volume. Property diagrams of substances are generally presented in graphical form and summarize the main properties listed in the refrigerant tables.

1.2.9 Sensible Heat, Latent Heat and Latent Heat of Fusion

It is known that all substances can hold a certain amount of heat; this property is their thermal capacity. When a liquid is heated, the temperature of the liquid rises to the boiling point. This is the highest temperature that the liquid can reach at the measured pressure. The heat absorbed by the liquid in raising the temperature to the boiling point is called *sensible heat*. The heat required to convert the liquid to vapor at the same temperature and pressure is called *latent heat*. In fact, it is the change in enthalpy during a state change (the amount of heat absorbed or rejected at constant temperature at any pressure, or the difference in enthalpies of a pure condensable fluid between its dry saturated state and its saturated liquid state at the same pressure).

Fusion is the melting of a material. For most pure substances there is a specific melting/freezing temperature, relatively independent of the pressure. For example, ice begins to melt at 0 °C. The amount of heat required to melt 1 kg of ice at 0 °C to 1 kg of water at 0 °C is called the *latent heat of fusion of water* and equals 334.92 kJ/kg. The removal of the same amount of heat from 1 kg of water at 0 °C changes it back to ice.

1.2.10 Vapor States

A vapor is a gas at or near equilibrium with the liquid phase – a gas under the saturation curve or only slightly beyond the saturated vapor line. *Vapor quality* is theoretically assumed; that is, when vapor leaves the surface of a liquid it is pure and saturated at the particular temperature and pressure. In actuality, tiny liquid droplets escape with the vapor. When a mixture of liquid and vapor exists, the ratio of the mass of the liquid to the total mass of the liquid and vapor mixture is called the *quality* and is expressed as a percentage or decimal fraction.

Superheated vapor is the saturated vapor to which additional heat has been added, raising the temperature above the boiling point. Let us consider a mass (m) with a quality (x). The volume is the sum of those of the liquid and the vapor as defined below:

$$V = V_{liq} + V_{vap} \tag{1.14}$$

It can also be written in terms of specific volumes as

$$mv = m_{\text{liq}}\, v_{\text{liq}} + m_{\text{vap}}\, v_{\text{vap}} \quad (1.15)$$

Dividing all terms by the total mass results in

$$v = (1 - x)v_{\text{liq}} + x v_{\text{vap}} \quad (1.16)$$

and

$$v = v_{\text{liq}} + x v_{\text{liq,vap}} \quad (1.17)$$

where $v_{\text{liq,vap}} = v_{\text{vap}} - v_{\text{liq}}$.

1.2.11 Thermodynamic Tables

Thermodynamic tables were first published in 1936 as steam tables by Keenan and Keyes, and later were revised and republished in 1969 and 1978. The use of thermodynamic tables of many substances ranging from water to several refrigerants is very common in process design calculations. In literature they are also called either steam tables or vapor tables. In this book, we refer to them as thermodynamic tables. These tables are normally given in different distinct phases (parts), for example, four different parts for water such as saturated water, superheated vapor water, compressed liquid water, and saturated solid–saturated vapor water; and two distinct parts for R-134a such as saturated and superheated. Each table is listed according to the values of temperature and pressure and the rest contains the values of four other thermodynamic parameters such as specific volume, internal energy, enthalpy, and entropy. When we normally have two variables, we may obtain the other data from the respective table. In learning how to use these tables, the most important point is to specify the state by any two of the parameters. In some design calculations, if we do not have the exact values of the parameters, we should make an interpolation to find the necessary values. Some people find this disturbing. However, further practice will provide sufficient confidence to do so. Beside these thermodynamic tables, recently, much attention has been paid to the computerized tables for such design calculations. Of course, despite the fact that this eliminates several reading problems, the students may not well understand the concepts and comprehend the subject. That is why in thermodynamics courses it is a must for the students to know how to obtain the thermodynamic data from the respective thermodynamic tables. The *Handbook of Thermodynamic Tables* by Raznjevic (1995) is one of the most valuable sources for several solids, liquids, and gaseous substances.

1.2.12 State and Change of State

The state of a system or substance is defined as the condition of the system or substance characterized by certain observable macroscopic values of its properties such as temperature and pressure. The term *state* is often used interchangeably with the term *phase*, for example, solid phase or gaseous phase of a substance. Each of the properties of a substance in a given state has only one definite value, regardless of how the substance reached the state. For example, when sufficient heat is added or removed, most substances undergo a state change. The temperature remains constant until the state change is complete. This can be from solid to liquid, liquid to vapor, or vice versa. Figure 1.4 shows the typical examples of ice melting and water boiling.

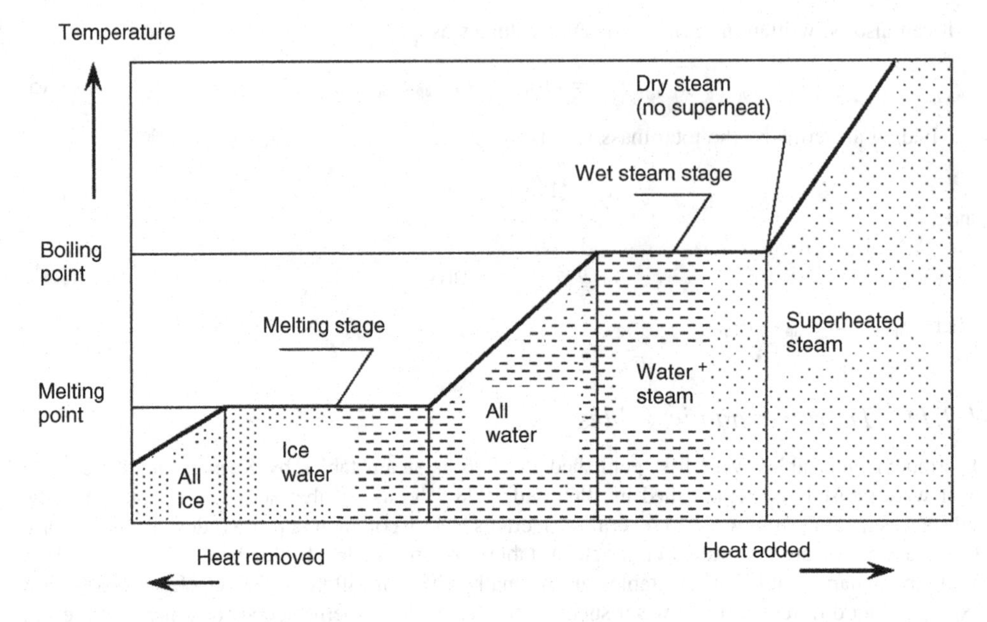

Figure 1.4 The state-change diagram of water.

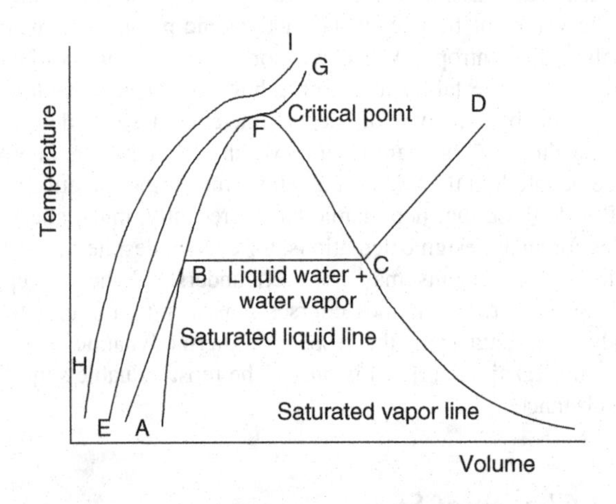

Figure 1.5 Temperature–volume diagram for the phase change of water.

A clearer presentation of solid, liquid, and vapor phases of water is exhibited on a temperature–volume ($T-v$) diagram in Figure 1.5. The constant-pressure line ABCD represents the states which water passes through as follows:

- **A–B.** This represents the process where water is heated from the initial temperature to the saturation temperature (liquid) at constant pressure. At point B it is fully saturated liquid water with a quality $x = 0$, with zero quantity of water vapor.
- **B–C.** This is the constant-temperature vaporization process in which there is only phase change from saturated liquid to saturated vapor, referring to the fact that the quality varies from 0 to

100%. Within this zone, the water is a mixture of liquid water and water vapor. At point C it is completely saturated vapor and the quality is 100%.

- **C–D.** This represents the constant-pressure process in which the saturated water vapor is superheated with increasing temperature.
- **E–F–G.** In this line there is no constant-temperature vaporization process. The point F is called the *critical point* where the saturated liquid and saturated vapor states are identical. The thermodynamic properties at this point are called *critical thermodynamic properties*, for example, critical temperature, critical pressure, and critical specific volume.
- **H–I.** This is a constant-pressure heating process in which there is no phase change from one phase to another (only one is present); however, there is a continuous change in density.

The other process which may occur during melting of water is *sublimation* in which the ice directly passes from the solid phase to vapor phase. Another important point that needs to be emphasized is that the solid, liquid, and vapor phases of water may be present together in equilibrium, leading to the *triple point*.

1.2.13 Pure Substance

This is defined as a substance which has a homogeneous and invariable chemical composition. Despite having the same chemical composition, it may be in more than one phase, namely, liquid water, a mixture of liquid water and water vapor (steam), and a mixture of ice and liquid water. Each one has the same chemical composition. However, a mixture of liquid air and gaseous air cannot be considered a pure substance because of the fact that the composition of each phase differs. A thorough understanding of the pure substance is of significance, particularly for air-conditioning applications. Thermodynamic properties of water and steam can be taken from tables and charts, in almost all thermodynamic books, based on the experimental data or real-gas equations of state through computer calculations. It is important to note that the properties of low-pressure water are of great significance in air conditioning, since water vapor existing in the atmosphere typically exerts a pressure less than 1 psi (6.9 kPa). At such low pressures, it is known that water vapor shows ideal gas behavior.

1.2.14 Specific Heats

The energy required to change (to raise or to drop) the temperature of a unit mass of a substance by a unit temperature difference is called the *specific heat* c. Its unit is kJ/kg · K or kJ/kg · °C. The specific heat is called the constant-pressure specific heat (c_p) if the process takes place at constant pressure (e.g., heating or cooling a gas in a piston-cylinder device). It is called the constant-volume specific heat (c_v) if the process takes place at constant volume (e.g., heating or cooling a gas in a rigid tank).

1.2.15 Specific Internal Energy

This represents the molecular state type of energy and is a measure of the energy of a simple system in equilibrium as a function of $c_v\,dT$. In fact, for many thermodynamic processes in closed systems the only significant energy changes are internal energy changes, and the significant work done by the system in the absence of friction is the work of pressure–volume expansion such as in a piston–cylinder mechanism. The specific internal energy of a mixture of liquid and vapor can be written in a form similar to Equations 1.16 and 1.17:

$$u = (1 - x)u_{\text{liq}} + xu_{\text{vap}} \tag{1.18}$$

and

$$u = u_{\text{liq}} + xu_{\text{liq,vap}} \tag{1.19}$$

where $u_{\text{liq,vap}} = u_{\text{vap}} - u_{\text{liq}}$.

1.2.16 Specific Enthalpy

This is a measure of the heat energy per unit mass of a substance, usually expressed in kJ/kg, as a function of $c_p\, dT$. Since enthalpy is a state function, it is necessary to measure it relative to some reference state. The usual practice is to determine the reference values which are called the *standard enthalpy of formation* (or the heat of formation), particularly in combustion thermodynamics. The specific enthalpy of a mixture of liquid and vapor components can be written as

$$h = (1 - x)h_{\text{liq}} + xh_{\text{vap}} \tag{1.20}$$

and

$$h = h_{\text{liq}} + xh_{\text{liq,vap}} \tag{1.21}$$

where $h_{\text{liq,vap}} = h_{\text{vap}} - h_{\text{liq}}$.

1.2.17 Specific Entropy

Entropy is a property resulting from the second law of thermodynamics (SLT). This is the ratio of the heat added to a substance to the absolute temperature at which it was added and is a measure of the molecular disorder of a substance at a given state. The unit of entropy is kJ/K and the unit of specific entropy is kJ/kg · K.

The entropy change of a pure substance between the states 1 and 2 is expressed as

$$\Delta s = s_2 - s_1 \tag{1.22}$$

The specific entropy of a mixture of liquid and vapor components can be written as

$$s = (1 - x)s_{\text{liq}} + xs_{\text{vap}} \tag{1.23}$$

and

$$s = s_{\text{liq}} + xs_{\text{liq,vap}} \tag{1.24}$$

where $s_{\text{liq,vap}} = s_{\text{vap}} - s_{\text{liq}}$.

The entropy change of an incompressible substance (solids and liquids) is given by

$$s_2 - s_1 = c \ln \frac{T_2}{T_1} \tag{1.25}$$

where c is the average specific heat of the substance.

An isentropic (i.e., constant entropy) process is defined as a reversible and adiabatic process.

$$s_2 = s_1 \tag{1.26}$$

1.3 Ideal Gases

In many practical thermodynamic calculations, gases such as air and hydrogen can often be treated as ideal gases, particularly for temperatures much higher than their critical temperatures and for pressures much lower than their saturation pressures at given temperatures. Such an ideal gas can be described in terms of three parameters, the volume that it occupies, the pressure that it exerts, and its temperature. As a matter of fact, all gases or vapors, including water vapor, at very low pressures show ideal gas behavior. The practical advantage of taking real gases to be ideal is that a simple equation of state with only one constant can be applied in the following form:

$$Pv = RT \tag{1.27}$$

and

$$PV = mRT \tag{1.28}$$

The ideal gas equation of state was originally established from the experimental observations and is also called the $P-v-T$ relationship for gases. It is generally considered as a concept rather than a reality. It only requires a few data to define a particular gas over a wide range of its possible thermodynamic equilibrium states.

The gas constant (R) is different for each gas depending on its molecular weight (M):

$$R = \frac{\overline{R}}{M} \tag{1.29}$$

where $\overline{R} = 8.314\,\text{kJ/kmol}\cdot\text{K}$ is the universal gas constant.

Equations 1.27 and 1.28 may be written in a mole-basis form as follows:

$$P\overline{v} = RT \tag{1.30}$$

and

$$PV = n\overline{R}T \tag{1.31}$$

The other simplification is that, if it is assumed that the constant-pressure and constant-volume specific heats are constant, changes in the specific internal energy and the specific enthalpy can be simply calculated without referring to the thermodynamic tables and graphs from the following expressions:

$$\Delta u = (u_2 - u_1) = c_v(T_2 - T_1) \tag{1.32}$$

$$\Delta h = (h_2 - h_1) = c_p(T_2 - T_1) \tag{1.33}$$

The following is another useful expression for ideal gases, obtained from the expression $h = u + Pv = u + RT$:

$$c_p - c_v = R \tag{1.34}$$

For the entire range of states, the ideal gas model may be found unsatisfactory. Therefore, the compressibility factor (Z) is introduced to measure the deviation of a real gas from the ideal gas equation of state, which is defined by the following relation:

$$Pv = ZRT \quad \text{or} \quad Z = \frac{Pv}{RT} \tag{1.35}$$

Figure 1.6 shows a generalized compressibility chart for simple substances. In the chart, we have two important parameters: reduced temperature ($T_r = T/T_c$) and reduced pressure ($P_r = P/P_c$).

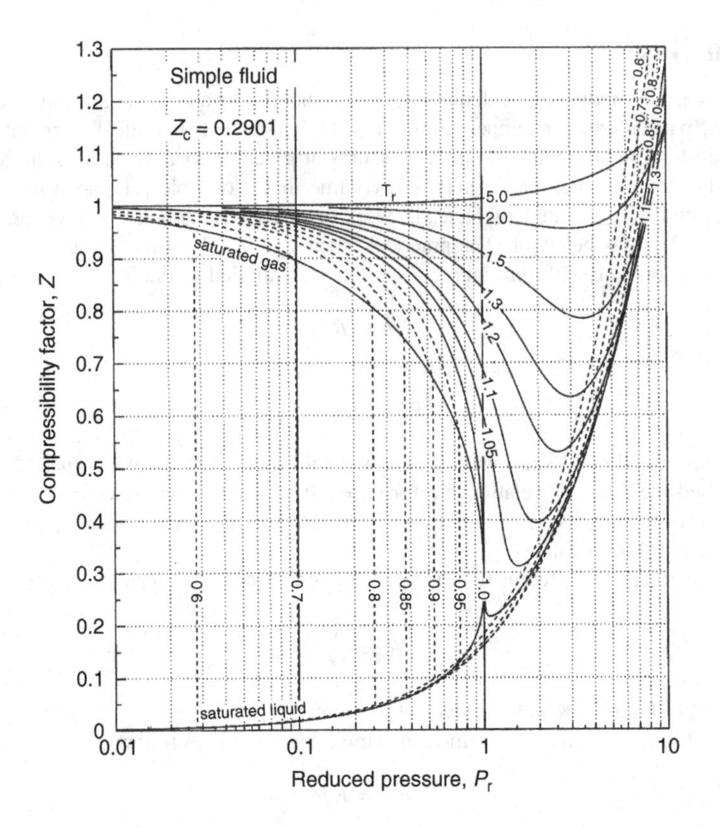

Figure 1.6 Generalized compressibility chart for simple substances (Borgnakke, and Sonntag, 2008).

Therefore, in order to calculate the compressibility factor the values of T_r and P_r should be calculated using the critical temperature and pressure values of the respective substance which can easily be taken from thermodynamics books. As can be seen in Figure 1.6, at all temperatures $Z \rightarrow 1$ as $P_r \rightarrow 0$. This means that the behavior of the actual gas closely approaches the ideal gas behavior, as the pressure approaches zero. For real gases, Z takes values between 0 and 1. If $Z = 1$, Equation 1.35 becomes Equation 1.27. In the literature, there are also several equations of state for accurately representing the $P-v-T$ behavior of a gas over the entire superheated vapor region, namely the Benedict–Webb–Rubin equation, van der Waals equation, and Redlich and Kwong equation. However, some of these equations of state are complicated because of the number of empirical constants, and require computer software to get the results.

There are some special cases where P, v, or T is constant. At a fixed temperature, the volume of a given quantity of ideal gas varies inversely with the pressure exerted on it (in some books this is called *Boyle's law*), describing compression as

$$P_1 V_1 = P_2 V_2 \tag{1.36}$$

where the subscripts refer to the initial and final states.

Equation 1.36 is employed by designers in a variety of situations: when selecting an air compressor, for calculating the consumption of compressed air in reciprocating air cylinders, and for determining the length of time required for storing air. Nevertheless, it may not always be practical because of temperature changes. If temperature increases with compression at a constant pressure, the volume of a gas varies directly with its absolute temperature in K as

$$\frac{V_1}{T_1} = \frac{V_2}{T_2} \tag{1.37}$$

If temperature increases at a constant volume, the pressure of a gas this time varies directly with its absolute temperature in K as

$$\frac{P_1}{T_1} = \frac{P_2}{T_2} \tag{1.38}$$

Equations 1.37 and 1.38 are known as Charles' law. If both temperature and pressure change at the same time, the combined ideal gas equation can be written as follows:

$$\frac{P_1 V_1}{T_1} = \frac{P_2 V_2}{T_2} \tag{1.39}$$

For a given mass, since c_{v0} is constant, the internal energy of an ideal gas can be written as a function of temperature:

$$dU = mc_{v0}\, dT \tag{1.40}$$

and the specific internal energy becomes

$$du = c_{v0}\, dT \tag{1.41}$$

The enthalpy equation for an ideal gas, based on $h = u + Pv$, can be written as

$$dH = mc_{p0}\, dT \tag{1.42}$$

and the specific enthalpy then becomes

$$dh = c_{p0}\, dT \tag{1.43}$$

The entropy change of an ideal gas, based on the general entropy equation in terms of $T\,ds = du + P\,dv$ and $T\,ds = dh - v\,dP$ as well as the ideal gas equation $Pv = RT$, can be obtained in two ways by substituting Equations 1.41 and 1.43:

$$s_2 - s_1 = c_{v0} \ln \frac{T_2}{T_1} + R \ln \frac{v_2}{v_1} \tag{1.44}$$

$$s_2 - s_1 = c_{p0} \ln \frac{T_2}{T_1} - R \ln \frac{P_2}{P_1} \tag{1.45}$$

For a reversible adiabatic (i.e., isentropic) process the ideal gas equation in terms of the initial and final states under $Pv^k = \text{constant}$ is

$$Pv^k = P_1 v_1^k = P_2 v_2^k \tag{1.46}$$

where k stands for the adiabatic exponent (so-called *specific heat ratio*) as a function of temperature:

$$k = \frac{c_{p0}}{c_{v0}} \tag{1.47}$$

Based on Equation 1.46 and the ideal gas equation, the following expressions can be obtained:

$$\left(\frac{T_2}{T_1}\right) = \left(\frac{v_1}{v_2}\right)^{k-1} \tag{1.48}$$

$$\left(\frac{T_2}{T_1}\right) = \left(\frac{P_2}{P_1}\right)^{(k-1)/k} \tag{1.49}$$

$$\left(\frac{P_2}{P_1}\right) = \left(\frac{v_1}{v_2}\right)^{k} \tag{1.50}$$

Note that these equations are obtained under the assumption of constant specific heats.

Let us consider a closed system with ideal gas, undergoing an adiabatic reversible process with a constant specific heat. The work can be derived from the first law of thermodynamics (FLT) equation as follows:

$$W_{1-2} = \frac{mR(T_2 - T_1)}{1-k} = \frac{P_2V_2 - P_1V_1}{1-k} \tag{1.51}$$

Equation 1.51 can also be derived from the general work relation, $W = \int P\,dV$.

For a reversible polytropic process, the only difference is the polytropic exponent (n) which shows the deviation from a log P and log V diagram, leading to the slope. Therefore, Equations 1.46, 1.48–1.51 can be rewritten with the polytropic exponent under Pv^n = constant as

$$Pv^n = P_1v_1^n = P_2v_2^n \tag{1.52}$$

$$\frac{P_2}{P_1} = \left(\frac{T_2}{T_1}\right)^{n/(n-1)} = \left(\frac{v_1}{v_2}\right)^{n} = \left(\frac{V_1}{V_2}\right)^{n} \tag{1.53}$$

$$W_{1-2} = \frac{mR(T_2 - T_1)}{1-n} = \frac{P_2V_2 - P_1V_1}{1-n} \tag{1.54}$$

In order to give a clear idea it is important to show the values of n for four different types of polytropic processes for ideal gases (Figure 1.7) as follows:

- $n = 0$ for isobaric process ($P = 0$),
- $n = 1$ for isothermal process ($T = 0$),
- $n = k$ for isentropic process ($s = 0$),
- $n = \infty$ for isochoric process ($v = 0$).

As is obvious from Figure 1.7, there are two quadrants where n varies from 0 to ∞ and where it has a positive value. The slope of any curve drawn is an important consideration when a reciprocating engine or compressor cycle is under consideration.

In thermodynamics a number of problems involve mixtures of different pure substances (i.e., ideal gases). In this regard, it is of importance to understand the related aspects accordingly. Table 1.2 gives a summary of the relevant expressions and two ideal gas models: the Dalton model and Amagat model. In fact, in the analysis it is assumed that each gas is unaffected by the presence of other gases, and each one is treated as an ideal gas. With regard to entropy, it is important to note that increase in entropy is dependent only upon the number of moles of ideal gases and is independent of its chemical composition. Of course, whenever the gases in the mixture are distinguished, the entropy increases.

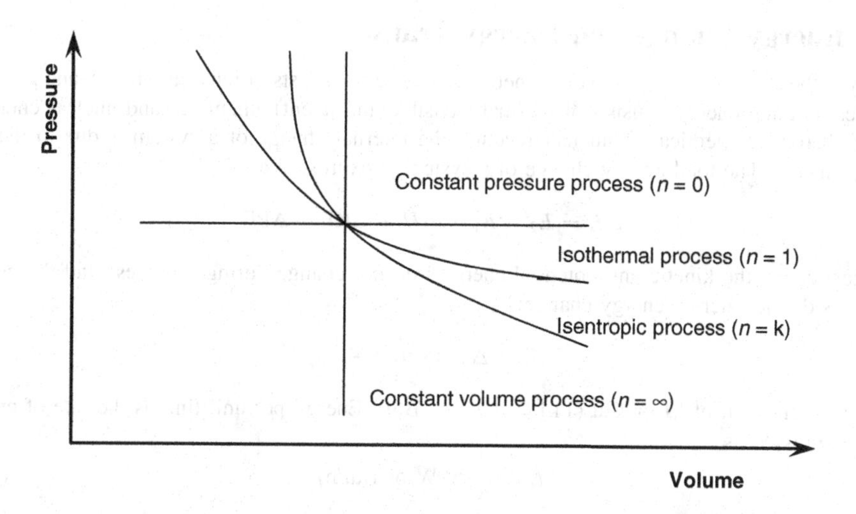

Figure 1.7 Representation of four different polytropic processes on a pressure–volume diagram.

Table 1.2 Equations for gas and gas mixtures and relevant models.

Definition	Dalton Model	Amagat Model
Total mass of a mixture of N components	$m_{tot} = m_1 + m_2 + \cdots + m_N = \sum m_i$	
Total number of moles of a mixture of N components	$n_{tot} = n_1 + n_2 + \cdots + n_N = \sum n_i$	
Mass fraction for each component	$c_i = m_i / m_{tot}$	
Mole fraction for each component	$y_i = n_i / n_{tot} = P_i / P_{tot} = V_i / V_{tot}$	
Molecular weight for the mixture	$M_{mixi} = m_{tot}/n_{tot} = \sum n_i M_i / n_{tot} = \sum y_i M_i$	
Internal energy for the mixture	$U_{mix} = n_1 \overline{U_1} + n_2 \overline{U_2} + \cdots + n_N \overline{U_N} = \sum n_i \overline{U_i}$	
Enthalpy for the mixture	$H_{mix} = n_1 \overline{H_1} + n_2 \overline{H_2} + \cdots + n_N \overline{H_N} = \sum n_i \overline{H_i}$	
Entropy for the mixture	$S_{mix} = n_1 \overline{S_1} + n_2 \overline{S_2} + \cdots + n_N \overline{S_N} = \sum n_i \overline{S_i}$	
Entropy difference for the mixture	$S_2 - S_1 = -\overline{R}(n_1 \ln y_1 + n_2 \ln y_2 + \cdots + n_N \ln y_N)$	
P, V, T for the mixture	T and V are constant. $P_{tot} = P = P_1 + P_2 + \cdots + P_N$	T and P are constant. $V_{tot} = V = V_1 + V_2 + \cdots + V_N$
Ideal gas equation for the mixture	$PV = n\overline{R}T$	
Ideal gas equations for the components	$P_1 V = n_1 \overline{R}T$ $P_2 V = n_2 \overline{R}T$ $\vdots$ $P_N V = n_N \overline{R}T$	$PV_1 = n_1 \overline{R}T$ $PV_2 = n_2 \overline{R}T$ $\vdots$ $PV_N = n_N \overline{R}T$

1.4 Energy Change and Energy Transfer

Energy is the capacity for doing work. Energy of a system consists of internal, kinetic, and potential energies. Internal energy consists of thermal (sensible and latent), chemical, and nuclear energies. Unless there is a chemical or nuclear reaction the internal change of a system is due to thermal energy change. The total energy change of a system is expressed as

$$\Delta E = E_2 - E_1 = \Delta U + \Delta \mathrm{KE} + \Delta \mathrm{PE} \tag{1.55}$$

For most cases, the kinetic and potential energies do not change during a process and the energy change is due to internal energy change:

$$\Delta E = \Delta U = m(u_2 - u_1) \tag{1.56}$$

Energy has the unit of kJ or Btu (1 kJ = 0.94782 Btu). Energy per unit time is the rate of energy and is expressed as

$$\dot{E} = \frac{E}{\Delta t} \text{ (kW or Btu/h)} \tag{1.57}$$

The unit of energy rate is kJ/s, which is equivalent to kW or Btu/h (1 kW = 3412.14 Btu/h). Energy per unit mass is called *specific energy*; it has the unit of kJ/kg or Btu/lbm (1 kJ/kg = 0.430 Btu/lbm).

$$e = \frac{E}{m} \text{ (kJ/kg or Btu/lbm)} \tag{1.58}$$

Energy can be transferred to or from a system in three forms: mass, heat, and work. They are briefly described in the following sections.

1.4.1 Mass Transfer

The mass entering a system carries energy with it and the energy of the system increases. The mass leaving a system decreases the energy content of the system. When a fluid flows into a system at a mass flow rate of $\dot{m}$ (kg/s), the rate of energy entering is equal to mass times enthalpy $\dot{m}h$ (kW).

1.4.2 Heat Transfer

The definitive experiment which showed that heat is a form of energy convertible into other forms was carried out by the Scottish physicist James Joule. Heat is the thermal form of energy and heat transfer takes place when a temperature difference exists within a medium or between different media. Heat always requires a difference in temperature for its transfer. Higher temperature differences provide higher heat-transfer rates.

Heat transfer has the same unit as energy. The symbol for heat transfer is Q (kJ). Heat transfer per unit mass is denoted by q (kJ/kg). Heat transfer per unit time is the rate of heat transfer $\dot{Q}$ (kW). If there is no heat transfer involved in a process, it is called an *adiabatic process*.

1.4.3 Work

Work is the energy that is transferred by a difference in pressure or force of any kind and is subdivided into shaft work and flow work. Shaft work is mechanical energy used to drive a mechanism such as a pump, compressor, or turbine. Flow work is the energy transferred into a system by fluid flowing into, or out of, the system. The rate of work transfer per unit time is called *power*. Work has the same unit as energy. Work is denoted by W. The direction of heat and work interactions can be expressed by sign conventions or using subscripts such as "in" and "out" (Cengel and Boles, 2008).

1.5 The First Law of Thermodynamics

It is simply known that thermodynamics is the science of energy and entropy and that the basis of thermodynamics is experimental observation. In thermodynamics, such observations were formed into four basic laws of thermodynamics called the zeroth, first, second, and third laws of thermodynamics. The first and second laws of thermodynamics are the most common tools in practice, because of the fact that transfers and conversions of energy are governed by these two laws, and in this chapter we focus on these two laws.

The first law of thermodynamics (FLT) can be defined as the law of conservation of energy, and it states that energy can be neither created nor destroyed. It can be expressed for a general system as the net change in the total energy of a system during a process is equal to the difference between the total energy entering and the total energy leaving the system:

$$E_{in} - E_{out} = \Delta E_{system} \tag{1.59}$$

In rate form,

$$\dot{E}_{in} - \dot{E}_{out} = \Delta \dot{E}_{system} \tag{1.60}$$

For a closed system undergoing a process between initial and final states involving heat and work interactions with the surroundings (Figure 1.8),

$$\begin{gathered} E_{in} - E_{out} = \Delta E_{system} \\ (Q_{in} + W_{in}) - (Q_{out} + W_{out}) = \Delta U + \Delta \mathrm{KE} + \Delta \mathrm{PE} \end{gathered} \tag{1.61}$$

If there is no change in kinetic and potential energies,

$$(Q_{in} + W_{in}) - (Q_{out} + W_{out}) = \Delta U = m(u_2 - u_1) \tag{1.62}$$

Let us consider a control volume involving a steady-flow process. Mass is entering and leaving the system and there is heat and work interactions with the surroundings (Figure 1.9). During a

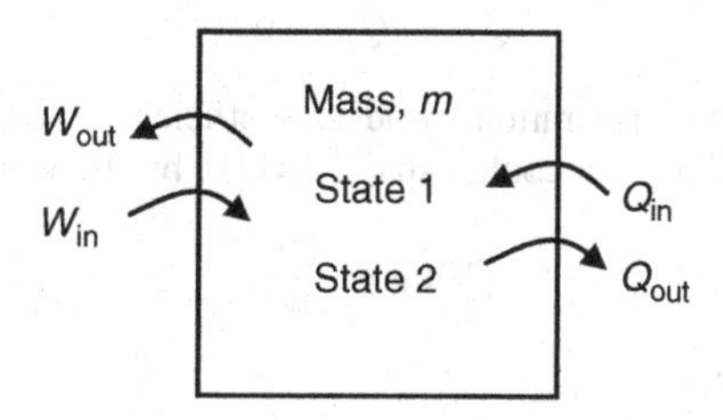

Figure 1.8 A general closed system with heat and work interactions.

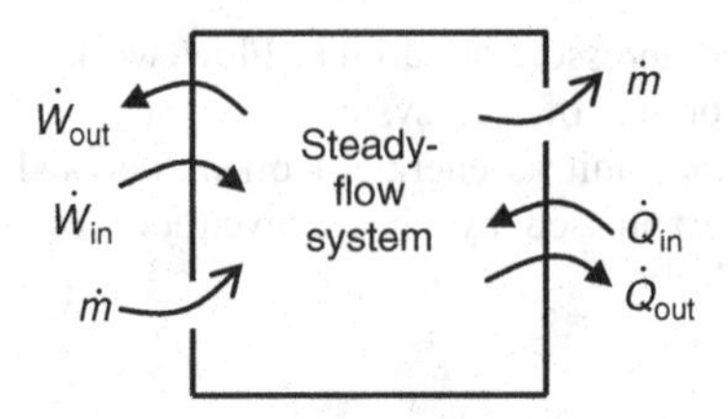

Figure 1.9 A general steady-flow control volume with mass, heat and work interactions.

steady-flow process, the total energy content of the control volume remains constant, and thus the total energy change of the system is zero. Then the FLT can be expressed as

$$\dot{E}_{in} - \dot{E}_{out} = \Delta \dot{E}_{system} = 0$$

$$\dot{E}_{in} = \dot{E}_{out} \tag{1.63}$$

$$\dot{Q}_{in} + \dot{W}_{in} + \dot{m}h_{in} = \dot{Q}_{out} + \dot{W}_{out} + \dot{m}h_{out}$$

Here, the kinetic and potential energies are neglected.

An important consequence of the first law is that the internal energy change resulting from some process will be independent of the thermodynamic path followed by the system, and of the paths followed by the processes, for example, heat transfer and work. In turn, the rate at which the internal energy content of the system changes is dependent only on the rates at which heat is added and work is done.

1.6 Refrigerators and Heat Pumps

A refrigerator is a device used to transfer heat from a low- to a high-temperature medium. They are cyclic devices. Figure 1.10a shows the schematic of a vapor-compression refrigeration cycle (the most common type). A working fluid (called *refrigerant*) enters the compressor as a vapor and is compressed to the condenser pressure. The high-temperature refrigerant cools in the condenser by rejecting heat to a high-temperature medium (at T_H). The refrigerant enters the expansion valve as liquid. It is expanded in an expansion valve and its pressure and temperature drop. The refrigerant is a mixture of vapor and liquid at the inlet of the evaporator. It absorbs heat from a low-temperature medium (at T_L) as it flows in the evaporator. The cycle is completed when the refrigerant leaves the evaporator as a vapor and enters the compressor. The cycle is demonstrated in a simplified form in Figure 1.10b.

An energy balance for a refrigeration cycle, based on the FLT, gives

$$Q_H = Q_L + W \tag{1.64}$$

The efficiency indicator for a refrigeration cycle is coefficient of performance (COP), which is defined as the heat absorbed from the cooled space divided by the work input in the compressor:

$$\text{COP}_\text{R} = \frac{Q_L}{W} \tag{1.65}$$

This can also be expressed as

$$\text{COP}_\text{R} = \frac{Q_L}{Q_H - Q_L} = \frac{1}{Q_H/Q_L - 1} \tag{1.66}$$

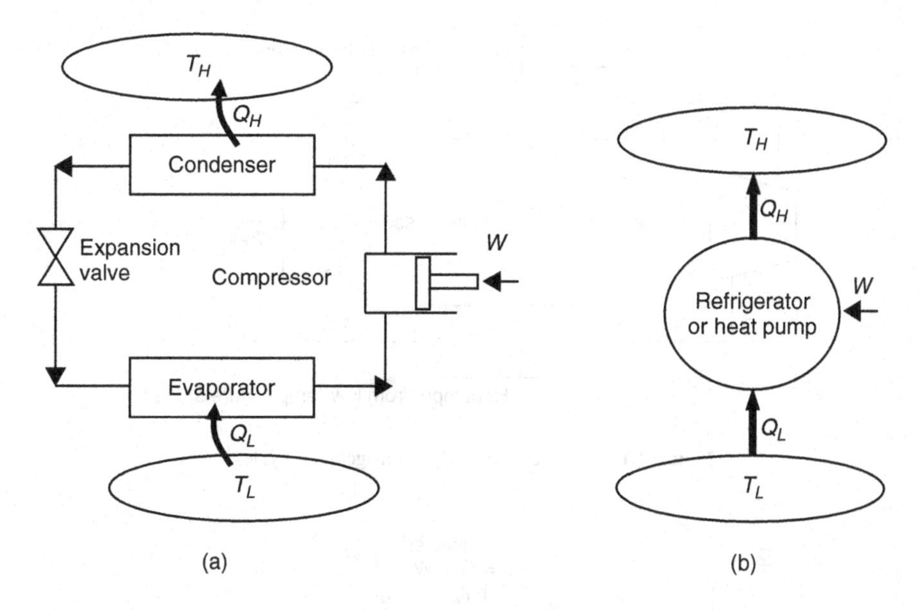

Figure 1.10 (a) The vapor-compression refrigeration cycle. (b) Simplified schematic of refrigeration cycle.

A heat pump is basically the same device as evaporator. The difference is their purpose. The purpose of a refrigerator is to absorb heat from a cooled space to keep it at a desired low temperature (T_L). The purpose of a heat pump is to transfer heat to a heated space to keep it at a desired high temperature (T_H). Thus, the COP of a heat pump is defined as

$$\mathrm{COP_{HP}} = \frac{Q_H}{W} \tag{1.67}$$

This can also be expressed as

$$\mathrm{COP_{HP}} = \frac{Q_H}{Q_H - Q_L} = \frac{1}{1 - Q_L/Q_H} \tag{1.68}$$

It can be easily shown that for given values Q_L and Q_H the COPs of a refrigerator and a heat pump are related to each other by

$$\mathrm{COP_{HP}} = \mathrm{COP_R} + 1 \tag{1.69}$$

This shows that the COP of a heat pump is greater than 1. The COP of a refrigerator can be less than or greater than 1.

1.7 The Carnot Refrigeration Cycle

The Carnot cycle is a theoretical model that is useful for understanding a refrigeration cycle. As known from thermodynamics, the Carnot cycle is a model cycle for a *heat engine* where the addition of heat energy to the engine produces work. In some applications, the Carnot refrigeration cycle is known as the *reversed Carnot cycle* (Figure 1.11). The maximum theoretical performance can be calculated, establishing criteria against which real refrigeration cycles can be compared.

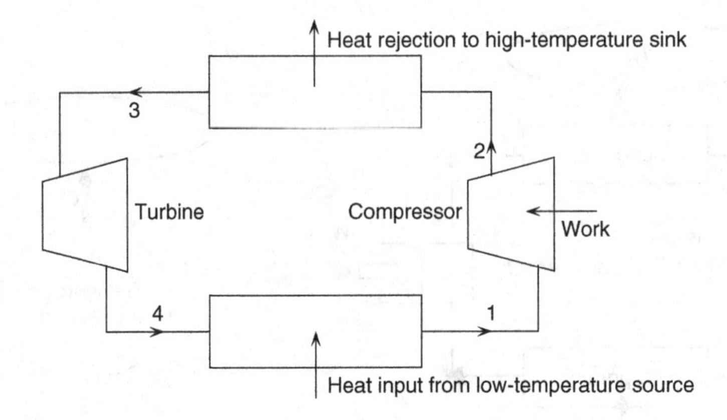

Figure 1.11 The reversed Carnot refrigeration cycle.

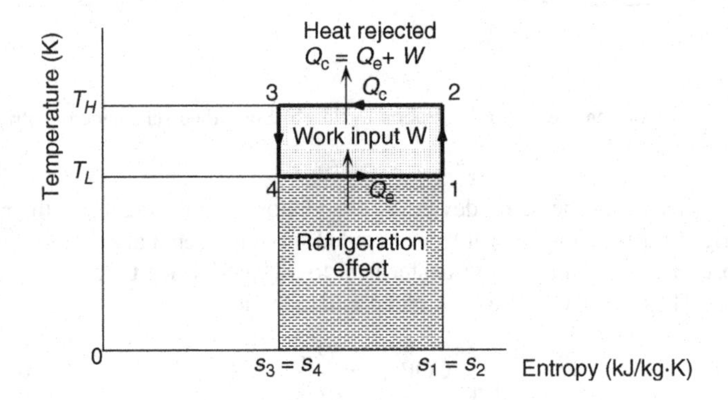

Figure 1.12 $T-s$ diagram of the Carnot refrigeration cycle.

The following processes take place in the Carnot refrigeration cycle as shown on a temperature–entropy diagram in Figure 1.12:

- (1–2) is the ideal compression at constant entropy, and work input is required. The temperature of the refrigerant increases.
- (2–3) is the rejection of heat in the condenser at a constant condensation temperature, T_H.
- (3–4) is the ideal expansion at constant entropy. The temperature of the refrigerant decreases.
- (4–1) is the absorption of heat in the evaporator at a constant evaporation temperature, T_L.

The refrigeration effect is represented as the area under the process line 4-1, as follows:

$$Q_L = T_L(s_1 - s_4) \tag{1.70}$$

The theoretical work input (e.g., compressor work) for the cycle is represented as the area within the cycle line 1–2–3–4–1, as follows:

$$W = (T_H - T_L)(s_1 - s_4) \tag{1.71}$$

After inserting Equations 1.70 and 1.71 into Equation 1.65, we find the following equation, which is dependent on the process temperatures:

$$\mathrm{COP}_{\mathrm{R,rev}} = \frac{Q_L}{W} = \frac{Q_L}{Q_H - Q_L} = \frac{T_L}{T_H - T_L} \tag{1.72}$$

It can also be expressed as

$$\mathrm{COP}_{\mathrm{R,rev}} = \frac{1}{Q_H/Q_L - 1} = \frac{1}{T_H/T_L - 1} \tag{1.73}$$

For a reversible heat pump, the following relations apply:

$$\mathrm{COP}_{\mathrm{HP,rev}} = \frac{Q_H}{W} = \frac{Q_H}{Q_H - Q_L} = \frac{T_H}{T_H - T_L} \tag{1.74}$$

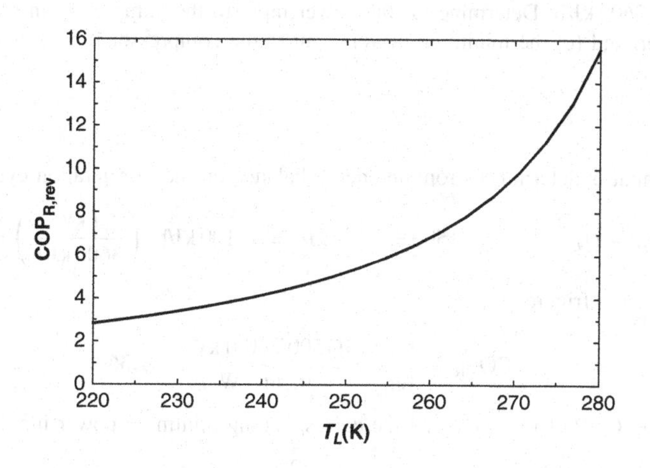

Figure 1.13 The COP of a reversible refrigerator as a function of $T_L \cdot T_H$ is taken as 298 K.

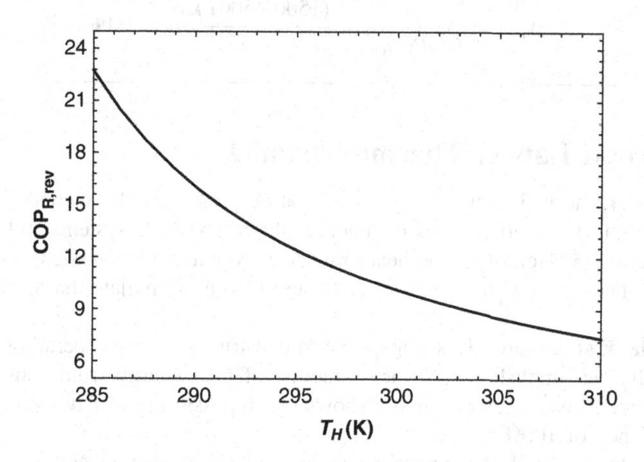

Figure 1.14 The COP of a reversible refrigerator as a function of $T_H \cdot T_L$ is taken as 273 K.

or

$$COP_{HP,rev} = \frac{1}{1 - Q_L/Q_H} = \frac{1}{1 - T_L/T_H} \tag{1.75}$$

The above relations provide the maximum COPs for a refrigerator or a heat pump operating between the temperature limits of T_L and T_H. Actual refrigerators and heat pumps involve inefficiencies and thus they will have lower COPs. The COP of a Carnot refrigeration cycle can be increased by either (i) increasing T_L or (ii) decreasing T_H. Figures 1.13 and 1.14 show that the COP of a reversible refrigerator increases with increasing T_L and decreasing T_H.

Example 1.1

A refrigeration cycle is used to keep a food department at −15 °C in an environment at 25 °C. The total heat gain to the food department is estimated to be 1500 kJ/h and the heat rejection in the condenser is 2600 kJ/h. Determine (a) the power input to the compressor in kW, (b) the COP of the refrigerator, and (c) the minimum power input to the compressor if a reversible refrigerator was used.

Solution

(a) The power input is determined from an energy balance on the refrigeration cycle:

$$\dot{W}_{in} = \dot{Q}_H - \dot{Q}_L = 2600 - 1500 = 1100\,\text{kJ/h} = (1100\,\text{kJ/h})\left(\frac{1\,\text{kW}}{3600\,\text{kJ/h}}\right) = \mathbf{0.306\,kW}$$

(b) The COP of the refrigerator is

$$COP_R = \frac{\dot{Q}_L}{\dot{W}_{in}} = \frac{(1500/3600)\,\text{kW}}{0.306\,\text{kW}} = \mathbf{1.36}$$

(c) The maximum COP of the cycle and the corresponding minimum power input are

$$COP_{R,\,rev} = \frac{T_L}{T_H - T_L} = \frac{258}{298 - 258} = 6.45$$

$$\dot{W}_{min} = \frac{\dot{Q}_L}{COP_{R,\,rev}} = \frac{(1500/3600)\,\text{kW}}{6.45} = \mathbf{0.065\,kW}$$

1.8 The Second Law of Thermodynamics

As mentioned earlier, the FLT is the energy-conservation principle. The second law of thermodynamics (SLT) refers to the inefficiencies of practical thermodynamic systems and indicates that it is impossible to have 100% efficiency in heat to work conversion. The classical statements such as the Kelvin–Plank statement and the Clausius statement help us formulate the SLT:

- **The Kelvin–Plank statement.** It is impossible to construct a device, operating in a cycle (e.g., heat engine), that accomplishes only the extraction of heat energy from some source and its complete conversion to work. This simply shows the impossibility of having a heat engine with a thermal efficiency of 100%.
- **The Clausius statement.** It is impossible to construct a device, operating in a cycle (e.g., refrigerator and heat pump), that transfers heat from the low-temperature side (cooler) to the high-temperature side (hotter).

A very easy way to show the implication of both the FLT and the SLT is a desktop game that consists of several pendulums (made of metal balls) in contact with each other. When you raise the first of the balls, you give energy to the system, potential energy. Upon release, this ball gains kinetic energy at the expense of potential energy. When this ball hits the second ball, small elastic deformations transform the kinetic energy again into another form of potential energy. The energy is transferred from one ball to the other. The last one gains kinetic energy to go up again. The cycle continues but every time lower, until it finally stops. The FLT explains why the balls keep moving, but the SLT explains why they do not do it forever. In this game the energy is lost in sound and heat and is no longer useful in keeping the balls in motion.

The SLT also states that the entropy in the universe is increasing. As mentioned before, entropy is the degree of disorder and every process happening in the universe is a transformation from a lower entropy to a higher entropy. Therefore, the entropy of a state of a system is proportional to (depends on) its probability, which gives us opportunity to define the SLT in a broader manner as "the entropy of a system increases in any heat transfer or conversion of energy within a closed system." That is why all energy transfers or conversions are irreversible. From the entropy perspective, the basis of the SLT is the statement that the sum of the entropy changes of a system and that of its surroundings must be always positive. Recently, much effort has been spent in minimizing the entropy generation (irreversibility) in thermodynamic systems and applications.

Moran and Shapiro (2007) noted that the SLT and deductions from it are useful because they provide means for

- predicting the direction of processes,
- establishing conditions for equilibrium,
- determining the best performance of thermodynamic systems and applications,
- evaluating quantitatively the factors that preclude the attainment of the best theoretical performance level,
- defining a temperature scale, independent of the properties of any thermometric substance, and
- developing tools for evaluating some thermodynamic properties, for example, internal energy and enthalpy using the experimental data available.

Consequently, the SLT is the linkage between entropy and usefulness of energy. The SLT analysis has found applications in a large variety of disciplines, for example, chemistry, economics, ecology, environment, and sociology far removed from engineering thermodynamics applications.

1.9 Exergy

The science of thermodynamics is built primarily on two fundamental natural laws, known as the first and the second laws. The FLT is simply an expression of the conservation of energy principle. It asserts that *energy* is a thermodynamic property, and that during an interaction, energy can change from one form to another but the total amount of energy remains constant. The SLT asserts that energy has quality as well as quantity, and actual processes occur in the direction of decreasing quality of energy. The high-temperature thermal energy is degraded as it is transferred to a lower temperature body. The attempts to quantify the quality or "work potential" of energy in the light of the SLT has resulted in the definition of the property named exergy.

Exergy analysis is a thermodynamic analysis technique based on the SLT, which provides an alternative and illuminating means of assessing and comparing processes and systems rationally and meaningfully. In particular, exergy analysis yields efficiencies which provide a true measure of how nearly actual performance approaches the ideal, and identifies more clearly than energy analysis the causes and locations of thermodynamic losses and the impact of the built environment on the natural environment. Consequently, exergy analysis can assist in improving and optimizing designs.

Performance of energy conversion systems and processes is essentially measured by efficiency, except that it becomes coefficient of performance for refrigeration and heat pump systems. There are two thermodynamic efficiencies, namely energy and exergy efficiencies. Although energy efficiency is commonly used by many for performance assessment, exergy efficiency is more beneficial, since it considers irreversibilities, and presents the actual performance of the systems. By considering both of these efficiencies, the quality and quantity of the energy used to achieve a given objective is considered and the degree to which efficient and effective use of energy resources is achieved can be understood. Improving efficiencies of energy systems is an important challenge for meeting energy policy objectives. Reductions in energy use can assist in attaining energy security objectives. Also, efficient energy utilization and the introduction of renewable energy technologies can significantly help solve environmental issues. Increased energy efficiency benefits the environment by avoiding energy use and the corresponding resource consumption and pollution generation. From an economic as well as an environmental perspective, improved energy efficiency has great potential (Dincer and Rosen, 2005).

An engineer designing a system is often expected to aim for the highest reasonable technical efficiency at the lowest cost under the prevailing technical, economic, and legal conditions and with regard to ethical, ecological, and social consequences. Exergy methods can assist in such activities and offer unique insights into possible improvements with special emphasis on environment and sustainability. Exergy analysis is a useful tool for addressing the environmental impact of energy resource utilization and for furthering the goal of more efficient energy resource use, for it enables the locations, types and true magnitudes of losses to be determined. Also, exergy analysis reveals whether and by how much it is possible to design more efficient energy systems by reducing inefficiencies. We present exergy as key tool for systems/processes analysis, design, and performance improvement.

1.9.1 What is Exergy?

The useful work potential of a given amount of energy at a specified state is called *exergy*. It is also called the availability or available energy. The work potential of the energy contained in a system at a specified state, relative to a reference (dead) state, is simply the maximum useful work that can be obtained from the system (Dincer, 2002; 2003).

A system is said to be in the dead state when it is in thermodynamic equilibrium with its environment. At the dead state, a system is at the temperature and pressure of its environment (in thermal and mechanical equilibrium); it has no kinetic or potential energy relative to the environment (zero velocity and zero elevation above a reference level); and it does not react with the environment (chemically inert). Also, there are no unbalanced magnetic, electrical, and surface tension effects between the system and its surroundings, if these are relevant to the situation at hand. The properties of a system at the dead state are denoted by subscript zero, for example, P_0, T_0, h_0, u_0, and s_0. Unless specified otherwise, the dead-state temperature and pressure are taken to be $T_0 = 25\,°C$ (77 °F) and $P_0 = 1$ atm (101.325 kPa or 14.7 psia). A system has zero exergy at the dead state.

The notion that a system must go to the dead state at the end of the process to maximize the work output can be explained as follows: if the system temperature at the final state is greater than (or less than) the temperature of the environment it is in, we can always produce additional work by running a heat engine between these two temperature levels. If the final pressure is greater than (or less than) the pressure of the environment, we can still obtain work by letting the system expand to the pressure of the environment. If the final velocity of the system is not zero, we can catch that extra kinetic energy by a turbine and convert it to rotating shaft work, and so on. No work can be produced from a system that is initially at the dead state. The atmosphere around us contains a tremendous amount of energy. However, the atmosphere is in the dead state, and the energy it contains has no work potential.

Therefore, we conclude that a system delivers the maximum possible work as it undergoes a reversible process from the specified initial state to the state of its environment, that is, the dead state. It is important to realize that exergy does not represent the amount of work that a work-producing device will actually deliver upon installation. Rather, it represents the upper limit on the amount of work a device can deliver without violating any thermodynamic laws. There will always be a difference, large or small, between exergy and the actual work delivered by a device. This difference represents the available room that engineers have for improvement, especially for greener buildings and more sustainable buildings per ASHRAE's Sustainability Roadmap.

Note that the exergy of a system at a specified state depends on the conditions of the environment (the dead state) as well as the properties of the system. Therefore, exergy is a property of the system–environment combination and not of the system alone. Altering the environment is another way of increasing exergy, but it is definitely not an easy alternative.

The work potential or exergy of the kinetic energy of a system is equal to the kinetic energy itself since it can be converted to work entirely. Similarly, exergy of potential energy is equal to the potential energy itself. On the other hand, the internal energy and enthalpy of a system are not entirely available for work, and only part of thermal energy of a system can be converted to work. In other words, exergy of thermal energy is less than the magnitude of thermal energy.

1.9.2 Reversibility and Irreversibility

These two concepts are highly important to thermodynamic processes and systems. The *reversibility* is defined as the statement that both the system and its surroundings can be returned to their initial states, just leading to the theoretical one. The irreversibility shows the destruction of availability and states that both the system and its surroundings cannot be returned to their initial states due to the irreversibilities occurring, for example, friction, heat rejection, and electrical and mechanical effects. For instance, as an actual system provides an amount of work that is less than the ideal reversible work, the difference between these two values gives the irreversibility of that system. In real applications, there are always such differences, and therefore, real cycles are always irreversible. For example, the entropy of the heat given off in the condenser is always greater than that of the heat taken up in the evaporator, referring to the fact that the entropy is always increased by the operation of an actual refrigeration system.

1.9.3 Reversible Work and Exergy Destruction

The reversible work W_{rev} is defined as *the maximum amount of useful work output or the minimum work input for a system undergoing a process between the specified initial and final states* in a totally reversible manner.

Any difference between the reversible work W_{rev} and the actual work W_u is due to the irreversibilities present during the process, and this difference is called *irreversibility* or *exergy destroyed*. It is expressed as

$$Ex_{destroyed} = W_{rev,out} - W_{out} \quad \text{or} \quad Ex_{destroyed} = W_{in} - W_{rev,in} \quad \text{or} \quad Ex_{destroyed} = W_{in} - W_{rev,in} \tag{1.76}$$

Irreversibility is a *positive quantity* for all actual (irreversible) processes since $W_{rev} \geq W$ for work-producing devices and $W_{rev} \leq W$ for work-consuming devices.

Irreversibility can be viewed as the *wasted work potential* or the *lost opportunity* to do useful work. It represents the energy that could have been converted to work but was not. It is important to note that lost opportunities manifest themselves in environmental degradation and avoidable emissions. The smaller the irreversibility associated with a process, the greater the work that is produced (or the smaller the work that is consumed). The performance of a system can be improved by minimizing the irreversibility associated with it.

A heat engine (an engine that converts heat to work output, e.g., a steam power plant) that operates on the reversible Carnot cycle is called a *Carnot heat engine*. The thermal efficiency of a Carnot heat engine, as well as other reversible heat engines, is given by

$$\eta_{\text{th,rev}} = 1 - \frac{T_L}{T_H} \tag{1.77}$$

where T_H is the source temperature and T_L is the sink temperature where heat is rejected (i.e., lake, ambient, and air). This is the maximum efficiency of a heat engine operating between two reservoirs at T_H and T_L.

A refrigerator or heat pump operating on reversed Carnot cycle would supply maximum cooling (in the case of refrigerator) and maximum heating (in the case of heat pump) and the COP of such reversible cycles are

$$\text{COP}_{\text{R,rev}} = \frac{1}{T_H/T_L - 1} \tag{1.78}$$

$$\text{COP}_{\text{HP,rev}} = \frac{1}{1 - T_L/T_H} \tag{1.79}$$

1.9.4 Exergy Balance

For a thermodynamic system undergoing any process, mass, energy, and entropy balances can be expressed as (see Cengel and Boles, 2008)

$$m_{\text{in}} - m_{\text{out}} = \Delta m_{\text{system}} \tag{1.80}$$

$$E_{\text{in}} - E_{\text{out}} = \Delta E_{\text{system}} \tag{1.81}$$

$$S_{\text{in}} - S_{\text{out}} + S_{\text{gen}} = \Delta S_{\text{system}} \tag{1.82}$$

In an actual process, mass and energy are conserved while entropy is generated. Note that energy can enter or exit a system by heat, work, and mass. Energy change of a system is the sum of the changes in internal, kinetic, and potential energies. Internal energy is the energy of a unit mass of a stationary fluid while enthalpy is the energy of a unit mass of a flowing fluid. Rate of energy change of a steady-flow system is zero and the rate of total energy input to a steady-flow control volume is equal to the rate of total energy output.

The nature of exergy is opposite to that of entropy in that exergy can be *destroyed*, but it cannot be created. Therefore, the *exergy change* of a system during a process is less than the *exergy transfer* by an amount equal to the *exergy destroyed* during the process within the system boundaries. Then the *decrease of exergy principle* can be expressed as

$$Ex_{\text{in}} - Ex_{\text{out}} - Ex_{\text{destroyed}} = \Delta Ex_{\text{system}} \tag{1.83}$$

This relation can also be written in the rate form, and is referred to as the *exergy balance* and can be stated as *the exergy change of 'a system during a process is equal to the difference between the net exergy transfer through the system boundary and the exergy destroyed within the system boundaries as a result of irreversibilities.* Exergy can be transferred to or from a system by heat, work, and mass.

Irreversibilities such as friction, mixing, chemical reactions, heat transfer through a finite temperature difference, unrestrained expansion, nonquasi-equilibrium compression or expansion always *generate entropy*, and anything that generates entropy always *destroys exergy*. The *exergy destroyed* is proportional to the entropy generated, and is expressed as

$$Ex_{\text{destroyed}} = T_0 S_{\text{gen}} \tag{1.84}$$

Exergy destruction during a process can be determined from an exergy balance on the system (Equation 1.83) or from the entropy generation using Equation 1.84.

A closed system, in general, may possess kinetic and potential energies as the total energy involved. The exergy change of a closed system during a process is simply the exergy difference between the final state 2 and initial state 1 of the system. A closed system involving heat input Q_{in} and boundary work output W_{out} as shown in Figure 1.15, mass, energy, entropy, and exergy balances can be expressed as
Mass balance:

$$m_1 = m_2 = \text{constant} \tag{1.85}$$

Energy balance:

$$Q_{\text{in}} - W_{\text{out}} = m(u_2 - u_1) \tag{1.86}$$

Entropy balance:

$$\frac{Q_{\text{in}}}{T_s} + S_{\text{gen}} = m(s_2 - s_1) \tag{1.87}$$

Exergy balance:

$$Q_{\text{in}}\left(1 - \frac{T_0}{T_s}\right) - [W_{\text{out}} - P_0(V_2 - V_1)] - Ex_{\text{destroyed}} = Ex_2 - Ex_1 \tag{1.88}$$

where u is internal energy, s is entropy, T_s is source temperature, T_0 is the dead state (environment) temperature, S_{gen} is entropy generation, P_0 is the dead-state pressure, and V is volume. For *stationary* closed systems in practice, the kinetic and potential energy terms may drop out. The exergy of a closed system is either *positive* or *zero, and* never becomes negative.

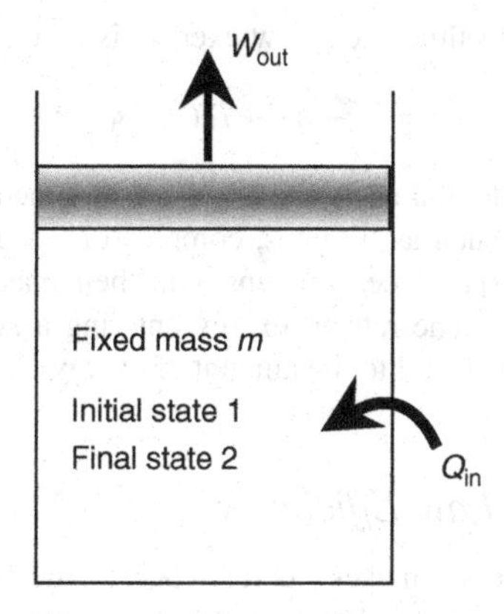

Figure 1.15 A closed system involving heat input Q_{in} and boundary work output W_{out}.

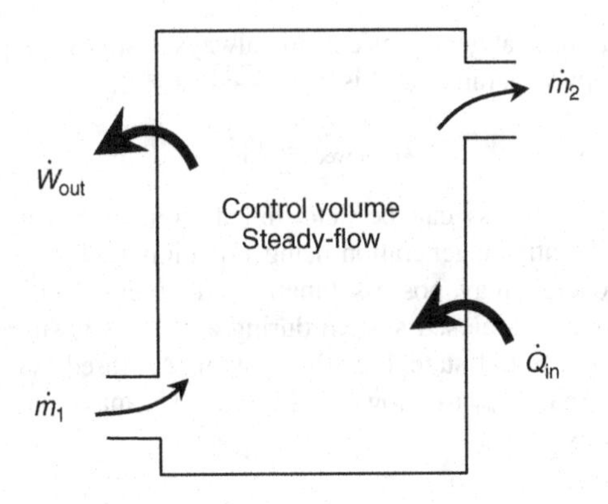

Figure 1.16 A control volume involving heat input and power output.

A control volume involving heat input and power output as shown in Figure 1.16, mass, energy, entropy, and exergy balances can be expressed as
Mass balance:

$$\dot{m}_1 = \dot{m}_2 \tag{1.89}$$

Energy balance:

$$\dot{m}_1 h_1 + \dot{Q}_{in} = \dot{m}_2 h_2 + \dot{W}_{out} \tag{1.90}$$

Entropy balance:

$$\frac{\dot{Q}_{in}}{T_s} + \dot{m}_1 s_1 + \dot{S}_{gen} = \dot{m}_2 s_2 \tag{1.91}$$

Exergy balance:

$$\dot{Q}_{in}\left(1 - \frac{T_0}{T_s}\right) + \dot{m}_1 \psi_1 = \dot{m}\psi_2 + \dot{W}_{out} + \dot{E}x_{destroyed} \tag{1.92}$$

where specific exergy of a flowing fluid (i.e., flow exergy) is given by

$$\psi = h - h_0 - T_0(s - s_0) \tag{1.93}$$

In these equations, kinetic and potential energy changes are assumed to be negligible. Most control volumes encountered in practice such as turbines, compressors, heat exchangers, pipes, and ducts operate steadily, and thus they experience no changes in their mass, energy, entropy, and exergy contents as well as their volumes. The rate of exergy entering a steady-flow system in all forms (heat, work, mass transfer) must be equal to the amount of exergy leaving plus the exergy destroyed.

1.9.5 Exergy or Second Law Efficiency

The first-law (i.e., energy) efficiency makes no reference to the best possible performance, and thus it may be misleading. Consider two heat engines, both having a thermal efficiency of 30%. One of the engines (engine *A*) receives heat from a source at 600 K, and the other one (engine *B*) from a source at 1000 K. After their process, both engines reject heat to a medium at 300 K. At the first glance, both engines seem to be performing equally well. When we take a second look at these engines in light of the SLT, however, we see a totally different picture. These engines, at best, can perform as reversible engines, in which case their efficiencies in terms of the Carnot Cycle become

$$\eta_{th,rev,A} = \left(1 - \frac{T_0}{T_{source}}\right)_A = 1 - \frac{300\ K}{600\ K} = 50\%$$

$$\eta_{th,rev,B} = \left(1 - \frac{T_0}{T_{source}}\right)_B = 1 - \frac{300\ K}{1000\ K} = 70\%$$

Engine A has a 50% useful work potential relative to the heat provided to it, and engine B has 70%. Now it is becoming apparent that engine B has a greater work potential made available to it and thus should do a lot better than engine A. Therefore, we can say that engine B is performing poorly relative to engine A even though both have the same thermal efficiency.

It is obvious from this example that the first-law efficiency alone is not a realistic measure of performance of engineering devices. To overcome this deficiency, we define an *exergy efficiency* (or *second-law efficiency*) for heat engines as the ratio of the actual thermal efficiency to the maximum possible (reversible) thermal efficiency under the same conditions:

$$\eta_{ex} = \frac{\eta_{th}}{\eta_{th,rev}} \tag{1.94}$$

Based on this definition, the energy efficiencies of the two heat engines discussed above become

$$\eta_{ex,A} = \frac{0.30}{0.50} = 60\%$$

$$\eta_{ex,B} = \frac{0.30}{0.70} = 43\%$$

That is, engine A is converting 60% of the available work potential to useful work. This ratio is only 43% for engine B. The second-law efficiency can also be expressed as the ratio of the useful work output and the maximum possible (reversible) work output:

$$\eta_{ex} = \frac{W_{out}}{W_{rev,out}} \tag{1.95}$$

This definition is more general since it can be applied to processes (in turbines, piston–cylinder devices, and so on) and cycles. Note that the exergy efficiency cannot exceed 100%. We can also define an exergy efficiency for work-consuming noncyclic (such as compressors) and cyclic (such as refrigerators) devices as the ratio of the minimum (reversible) work input to the useful work input:

$$\eta_{ex} = \frac{W_{rev,in}}{W_{in}} \tag{1.96}$$

For cyclic devices such as refrigerators and heat pumps, it can also be expressed in terms of the coefficients of performance as

$$\eta_{ex} = \frac{COP}{COP_{rev}} \tag{1.97}$$

In the above relations, the reversible work W_{rev} should be determined by using the same initial and final states as in the actual process.

For general cases where we do not produce or consume work (e.g., thermal energy storage system for a building), a general exergy efficiency can be defined as

$$\eta_{ex} = \frac{\text{Exergy recovered}}{\text{Exergy supplied}} = 1 - \frac{\text{Exergy destroyed}}{\text{Exergy supplied}} \tag{1.98}$$

1.9.6 *Illustrative Examples on Exergy*

Example 1.2

A Geothermal Power Plant

A geothermal power plant uses geothermal liquid water at 160 °C at a rate of 100 kg/s as the heat source, and produces 3500 kW of net power in an environment at 25 °C (Figure 1.17). We will conduct a thermodynamic analysis of this power plant considering both energy and exergy approaches.

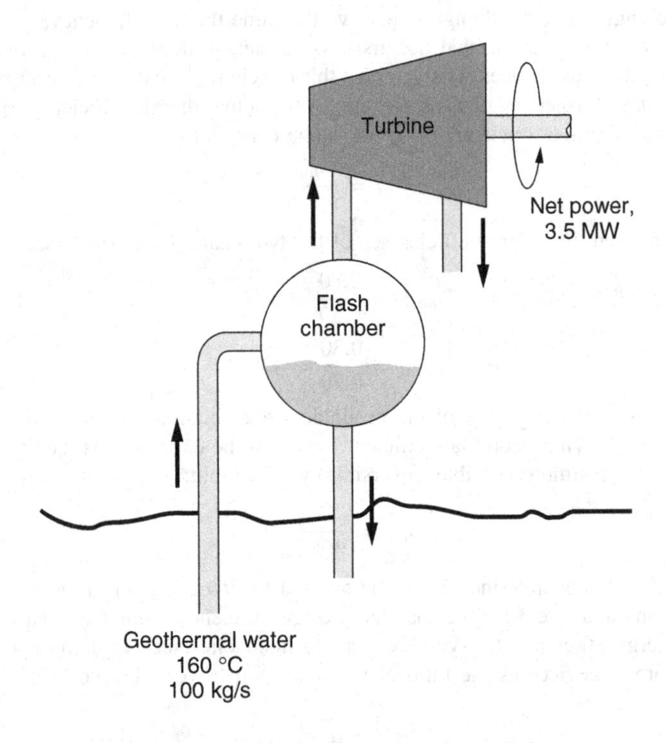

Figure 1.17 A flash-design geothermal power plant.

The properties of geothermal water at the inlet of the plant and at the dead state are obtained from steam tables (not available in the text) to be

$$T_1 = 160\,^\circ\text{C}, \quad \text{liquid} \longrightarrow h_1 = 675.47 \text{ kJ/kg}, \quad s_1 = 1.9426 \text{ kJ/kg} \cdot \text{K}$$

$$T_0 = 25\,^\circ\text{C}, \quad P_0 = 1 \text{ atm} \longrightarrow h_0 = 104.83 \text{ kJ/kg}, \quad s_0 = 0.36723 \text{ kJ/kg} \cdot \text{K}$$

The energy of geothermal water may be taken to be maximum heat that can be extracted from the geothermal water, and this may be expressed as the enthalpy difference between the state of geothermal water and dead state:

$$\dot{E}_{\text{in}} = \dot{m}(h_1 - h_0) = (100 \text{ kg/s}) \left[(675.47 - 104.83) \text{ kJ/kg}\right] = 57{,}060 \text{ kW}$$

The exergy of geothermal water is

$$\dot{E}x_{in} = \dot{m}\,[(h_1 - h_0) - T_0(s_1 - s_0)]$$
$$= (100\ \text{kg/s})\left[(675.47 - 104.83)\ \text{kJ/kg} - (25 + 273\ \text{K})(1.9426 - 0.36723)\text{kJ/kg}\cdot\text{K}\right]$$
$$= 10{,}120\ \text{kW}$$

The thermal efficiency of the power plant is

$$\eta_{th} = \frac{\dot{W}_{net,out}}{\dot{E}_{in}} = \frac{3500\ \text{kW}}{57{,}060\ \text{kW}} = 0.0613 = 6.1\%$$

The exergy efficiency of the plant is the ratio of power produced to the exergy input to the plant:

$$\eta_{ex} = \frac{\dot{W}_{net,out}}{\dot{E}x_{in}} = \frac{3500\ \text{kW}}{10{,}120\ \text{kW}} = 0.346 = 34.6\%$$

The exergy destroyed in this power plant is determined from an exergy balance on the entire power plant to be

$$\dot{E}x_{in} - \dot{W}_{net,out} - \dot{E}x_{dest} = 0$$
$$10{,}120 - 3500 - \dot{E}x_{dest} = 0 \longrightarrow \dot{E}x_{dest} = 6620\ \text{kW}$$

Some of the results of this example are illustrated in Figures 1.18 and 1.19. The exergy of geothermal water (10,120 kW) constitutes only 17.7% of its energy (57,060), owing to its well temperature. The remaining 82.3% is not available for useful work and it cannot be converted to power by even a reversible heat engine. Only 34.6% exergy entering the plant is converted to power and the remaining 65.4% is lost. In geothermal power plants, the used geothermal water typically leaves the power plant at a temperature much greater than the environment temperature and this water is reinjected back to the ground. The total exergy destroyed (6620 kW) includes the exergy of this reinjected brine.

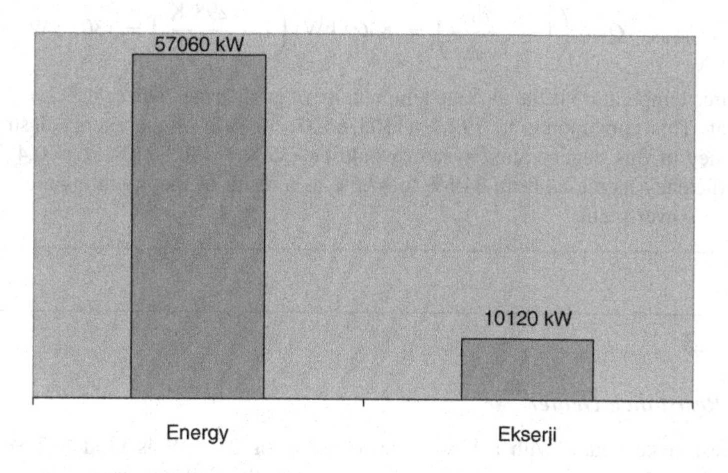

Figure 1.18 Only 18% of the energy of geothermal water is available for converting to power.

In a typical binary-type geothermal power plant, geothermal water would be reinjected back to the ground at about 90 °C. This water can be used in a district heating system. Assuming that

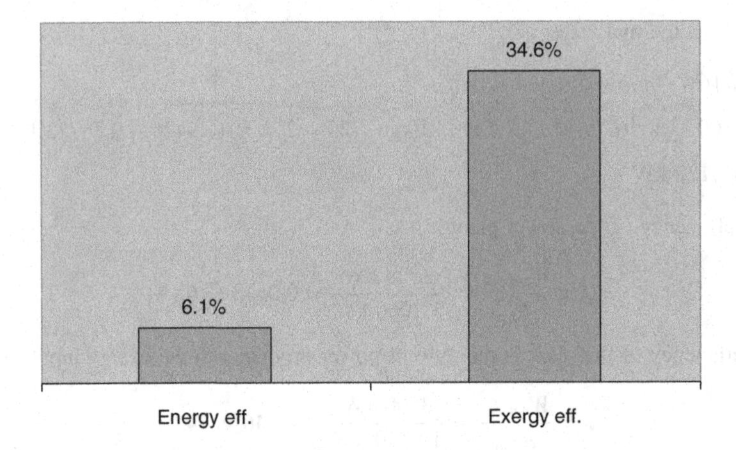

Figure 1.19 Energy and exergy efficiencies of geothermal power plant.

geothermal water leaves the district at 70 °C with a drop of 20 °C during the heat supply, the rate of heat that could be used in the district system would be

$$\dot{Q}_{\text{heat}} = \dot{m}c\Delta T = (100\ \text{kg/s})(4.18\ \text{kJ/kg}\cdot{}^\circ\text{C})(20\,{}^\circ\text{C}) = 8360\ \text{kW}$$

where c is the specific heat of water. This 8360 kW heating is in addition to the 3500-kW power generated. The energy efficiency of this cogeneration system would be $(3500 + 8360)/57{,}060 = 0.208 = 20.8\%$. The energy efficiency increases from 6.1% to 20.8% as a result of incorporating a district heating system into the power plant.

The exergy of heat supplied to the district system is simply the heat supplied times the Carnot efficiency, which is determined as

$$\dot{E}x_{\text{heat}} = \dot{Q}_{\text{heat}}\left(1 - \frac{T_0}{T_{\text{source}}}\right) = (8360\ \text{kW})\left(1 - \frac{298\ \text{K}}{353\ \text{K}}\right) = 1303\ \text{kW}$$

where the source temperature is the average temperature of geothermal water (80 °C = 353 K) when supplying heat. This corresponds to 19.7% $(1303/6620 = 0.197)$ of the exergy destruction. The exergy efficiency of this cogeneration system would be $(3500 + 1303)/10{,}120 = 0.475 = 47.5\%$. The exergy efficiency increases from 34.6% to 47.5% as a result of incorporating a district heating system into the power plant.

Example 1.3

An Electric Resistance Heater

An electric resistance heater with a power consumption of 2.0 kW is used to heat a room at 25 °C when the outdoor temperature is 0 °C (Figure 1.20). We will determine energy and exergy efficiencies and the rate of exergy destroyed for this process.

For each unit of electric work consumed, the heater will supply the house with 1 unit of heat. That is, the heater has a COP of 1. Also, the energy efficiency of the heater is 100% since the energy output (heat supply to the room) and the energy input (electric work consumed by the heater) are

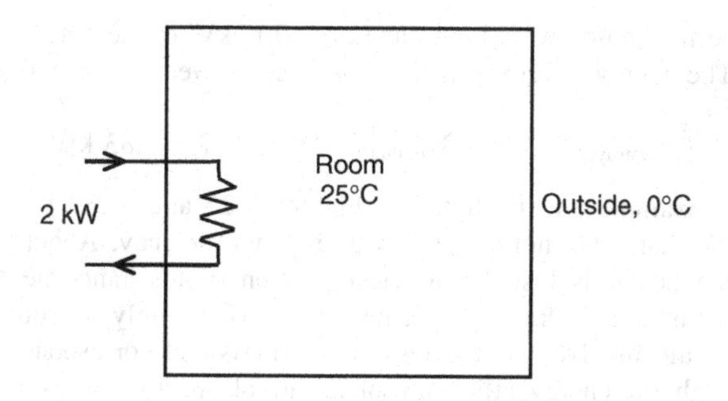

Figure 1.20 An electric resistance heater used to heat a room.

the same. At the specified indoor and outdoor temperatures, a reversible heat pump would have a COP of

$$\mathrm{COP}_{\mathrm{HP,rev}} = \frac{1}{1 - T_L/T_H} = \frac{1}{1 - (273\ \mathrm{K})/(298\ \mathrm{K})} = 11.9$$

That is, it would supply the house with 11.9 units of heat (extracted from the cold outside air) for each unit of electric energy it consumes (Figure 1.21). The exergy efficiency of this resistance heater is

$$\eta_{\mathrm{ex}} = \frac{\mathrm{COP}}{\mathrm{COP}_{\mathrm{HP,rev}}} = \frac{1}{11.9} = 0.084 = 8.4\%$$

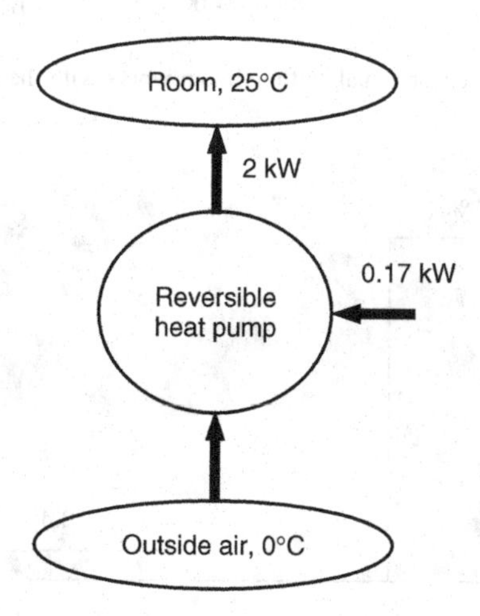

Figure 1.21 A reversible heat pump consuming only 0.17 kW power while supplying 2-kW of heat to a room.

The minimum work requirement to the heater is determined from the COP definition for a heat pump to be

$$\dot{W}_{\mathrm{in,min}} = \frac{\dot{Q}_{\mathrm{supplied}}}{\mathrm{COP}_{\mathrm{HP,rev}}} = \frac{2\ \mathrm{kW}}{11.9} = 0.17\,\mathrm{kW}$$

That is, a reversible heat pump would consume only 0.17 kW of electrical energy to supply the room 2 kW of heat. The exergy destroyed is the difference between the actual and minimum work inputs:

$$\dot{E}x_{destroyed} = \dot{W}_{in} - \dot{W}_{in,min} = 2.0 - 0.17 = 1.83 \text{ kW}$$

The results of this example are illustrated in Figures 1.22 and 1.23. The performance looks perfect with energy efficiency but not so good with exergy efficiency. About 92% of actual work input to the resistance heater is lost during the operation of resistance heater. There must be better methods of heating this room. Using a heat pump (preferably a ground-source one) or a natural gas furnace would involve lower exergy destructions and correspondingly greater exergy efficiencies even though the energy efficiency of a natural gas furnace is lower than that of a resistance heater.

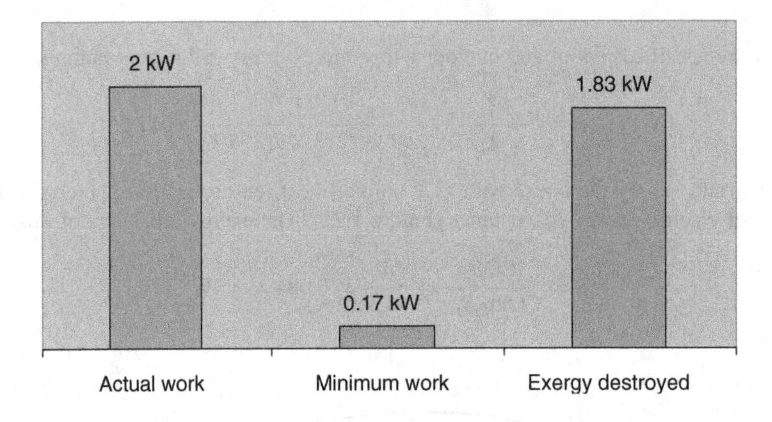

Figure 1.22 Comparison of actual and minimum works with the exergy destroyed.

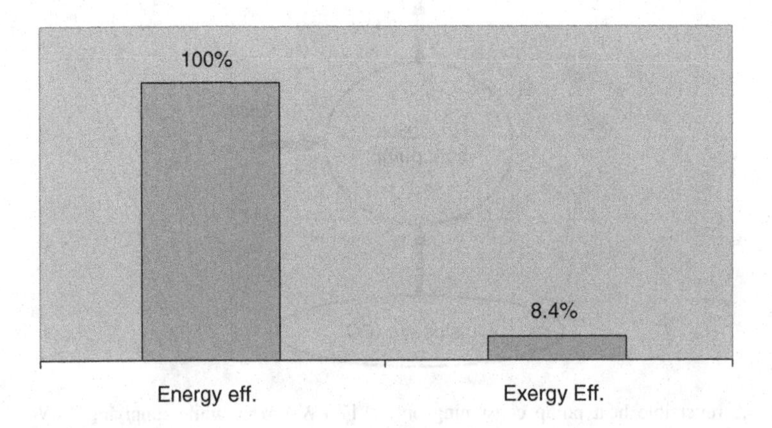

Figure 1.23 Comparison of energy and exergy efficiencies.

Different heating systems may also be compared using primary energy ratio (PER), which is the ratio of useful heat delivered to primary energy input. Obviously, the higher the PER, the more efficient the heating system. The PER for a heat pump is defined as PER $= \eta \times$ COP where η is the thermal efficiency with which the primary energy input is converted into work. For the resistance heater discussed in this example, the thermal efficiency η may be taken to be 0.40 if the electricity is produced from a natural-gas-fueled steam power plant. Since the COP is 1, the PER becomes 0.40. A natural gas furnace with an efficiency of 0.80 (i.e., heat supplied over the heating value of the fuel) would have a PER value of 0.80. Furthermore, for a ground-source heat pump using electricity as the work input, the COP may be taken as 3 and with the same method of electricity production ($\eta = 0.40$), the PER becomes 1.2.

Example 1.4

A Simple Heating Process

In an air-conditioning process, air is heated by a heating coil in which hot water is flowing at an average temperature of 80 °C. Using the values given in Figure 1.24, we will determine the exergy destruction and the exergy efficiency for this process.

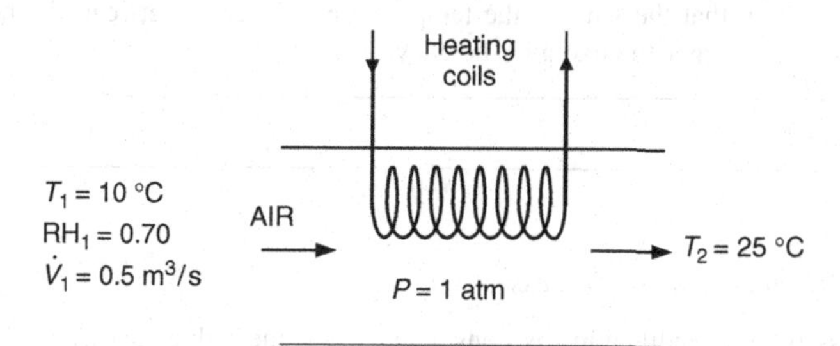

Figure 1.24 Schematic of simple heating process.

The properties of air at various states (including dead state, denoted by the subscript 0) are determined from a software with built-in properties to be

$$v_1 = 0.810 \text{ m}^3/\text{kg}, \quad h_0 = h_1 = 25.41 \text{ kJ/kg}, \quad h_2 = 40.68 \text{ kJ/kg}, \quad s_0 = s_1 = 5.701 \text{ kJ/kg} \cdot \text{K}$$

$$s_2 = 5.754 \text{ kJ/kg} \cdot \text{K}, \quad w_1 = w_2 = 0.00609 \text{ kg water/kg air}, \quad \text{RH}_2 = 0.31$$

The dead-state temperature is taken to be the same as the inlet temperature of air. The mass flow rate of air and the rate of heat input are

$$\dot{m}_a = \frac{\dot{V}_a}{v_1} = 0.617 \text{ kg/s}$$

$$\dot{Q}_{in} = \dot{m}_a(h_2 - h_1) = 9.43 \text{ kW}$$

The exergies of air stream at the inlet and exit are

$$\dot{E}x_1 = 0 \quad \text{and} \quad \dot{E}x_2 = \dot{m}_a\left[(h_2 - h_0) - T_0(s_2 - s_0)\right] = 0.267 \text{ kW}$$

The rates of exergy input and the exergy destroyed are

$$\dot{E}x_{\text{in}} = \dot{Q}_{\text{in}}\left(1 - \frac{T_0}{T_{\text{source}}}\right) = 1.87\ \text{kW}$$

$$\dot{E}x_{\text{destroyed}} = \dot{E}x_{\text{in}} - \dot{E}x_{\text{out}} = 1.87 - 0.267 = 1.60\ \text{kW}$$

where the temperature at which heat is transferred to the air stream is taken as the average temperature of water flowing in the heating coils (80 °C). The exergy efficiency is

$$\eta_{\text{ex}} = \frac{\dot{E}x_{\text{out}}}{\dot{E}x_{\text{in}}} = \frac{0.267\ \text{kW}}{1.87\ \text{kW}} = 0.143 = 14.3\%$$

About 86% of exergy input is destroyed owing to irreversible heat transfer in the heating section. Air-conditioning processes typically involve high rates of exergy destructions as high-temperature (i.e., high quality) heat or high-quality electricity is used to obtain a low-quality product. The irreversibilities can be minimized using lower quality energy sources and less irreversible processes. For example, if heat is supplied at an average temperature of 60 °C instead of 80 °C, the exergy destroyed would decrease from 1.60 to 1.15 kW and the exergy efficiency would increase from 14.3 to 18.8%. The exit temperature of air also affects the exergy efficiency. For example, if air is heated to 20 °C instead of 25 °C, the exergy efficiency would decrease from 14.3 to 10.1%. These two examples also show that the smaller the temperature difference between the heat source and the air being heated, the larger the exergy efficiency.

Example 1.5

A Heating with Humidification Process

A heating process with humidification is considered using the values shown in Figure 1.25. Its psychrometric representation is given in Figure 1.26. Mass, energy, entropy and exergy balances, and exergy efficiency for this process can be expressed as

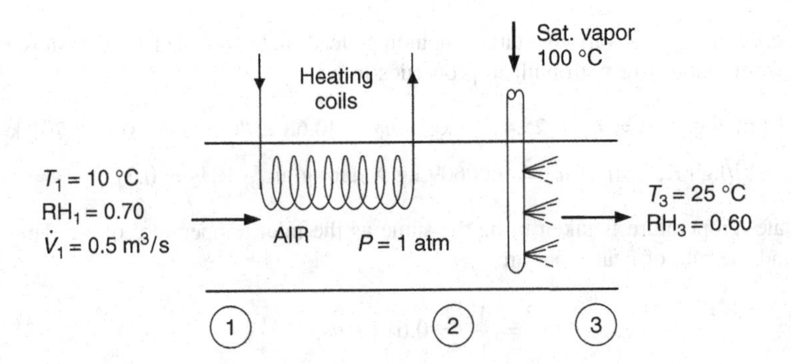

Figure 1.25 A heating with humidification process.

Dry air mass balance:

$$\dot{m}_{a1} = \dot{m}_{a2} = \dot{m}_{a3}$$

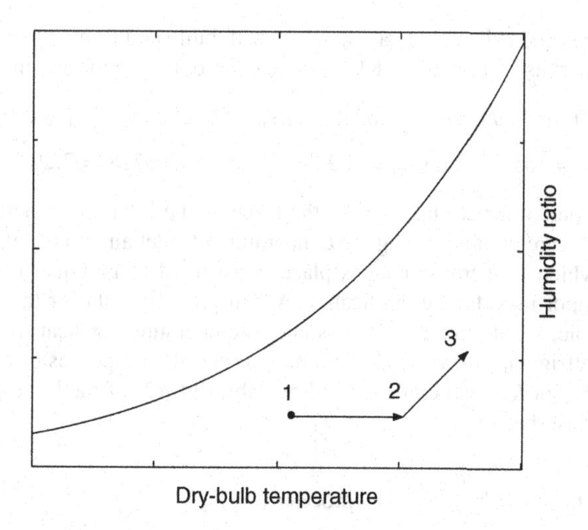

Figure 1.26 Heating with humidification as represented on a psychrometric chart.

Water mass balance:

$$\dot{m}_{w1} = \dot{m}_{w2}$$

$$\dot{m}_{w2} + \dot{m}_{\rm w} = \dot{m}_{w3} \longrightarrow \dot{m}_{a2}\omega_2 + \dot{m}_{\rm w} = \dot{m}_{a3}\omega_3$$

Energy balance:

$$\dot{Q}_{\rm in} + \dot{m}_{a1}h_1 = \dot{m}_{a2}h_2 \text{ (process 1–2)}$$

$$\dot{m}_{a2}h_2 + \dot{m}_{\rm w}h_{\rm w} = \dot{m}_{a3}h_3 \text{ (process 2–3)}$$

$$\dot{Q}_{\rm in} + \dot{m}_{a1}h_1 + \dot{m}_{\rm w}h_{\rm w} = \dot{m}_{a3}h_3 \text{ (process 1–3)}$$

Entropy balance:

$$\dot{m}_{a1}s_1 + \dot{m}_{\rm w}s_{\rm w} + \frac{\dot{Q}_{\rm in}}{T} - \dot{m}_{a3}s_3 + \dot{S}_{\rm gen} = 0 \text{ (process 1–3)}$$

Exergy balance:

$$\dot{Q}_{\rm in}\left(1 - \frac{T_0}{T}\right) + \dot{m}_{a1}\psi_1 - \dot{m}_{a2}\psi_2 - \dot{E}x_{\rm dest} = 0 \text{ (process 1–2)}$$

$$\dot{m}_{a2}\psi_2 + \dot{m}_{\rm w}\psi_{\rm w} - \dot{m}_{a3}\psi_3 - \dot{E}x_{\rm dest} = 0 \text{ (process 2–3)}$$

$$\dot{Q}_{\rm in}\left(1 - \frac{T_0}{T}\right) + \dot{m}_{a1}\psi_1 + \dot{m}_{\rm w}\psi_{\rm w} - \dot{m}_{a3}\psi_3 - \dot{E}x_{\rm dest} = 0 \text{ (process 1–3)}$$

$$\dot{E}x_{\rm dest} = T_0\dot{S}_{\rm gen} = T_0\left(\dot{m}_{a3}s_3 - \dot{m}_{a1}s_1 - \dot{m}_{\rm w}s_{\rm w} - \frac{\dot{Q}_{\rm in}}{T}\right) \text{ (process 1–3)}$$

Exergy efficiency:

$$\eta_{\rm ex} = \frac{\dot{m}_{a3}\psi_3}{\dot{Q}_{\rm in}\left(1 - \frac{T_0}{T}\right) + \dot{m}_{a1}\psi_1 + \dot{m}_{\rm w}\psi_{\rm w}}$$

Based on these balances and using an equation solver with built-in thermodynamic functions including pschrometric properties of moist air (Klein, 2006), we obtain the following results:

$$\dot{m}_{\rm a} = 0.618 \text{ kg/s}, \ \dot{m}_{\rm w} = 0.00406 \text{ kg/s}, \ T_2 = 24.2\,^\circ\text{C}, \ \dot{Q}_{\rm in} = 8.90 \text{ kW}$$

$$\dot{E}x_{\rm in} = 4.565 \text{ kW}, \ \dot{E}x_{\rm dest} = 4.238 \text{ kW}, \ \eta_{\rm ex} = 0.0718 = 7.2\%$$

The dead-state properties of air are taken to be the same as the inlet air properties while the dead-state properties of water are obtained using the temperature of inlet air and the atmospheric pressure. The temperature at which heat transfer takes place is assumed to be equal to the temperature of the saturated water vapor used for humidification. When property data for the fluid flowing in the heating coil is available, we do not have to assume a temperature for heat transfer. For example, let us assume that a refrigerant flows in the heating coil and the properties of the refrigerant at the inlet (denoted by subscript $R1$) and exit (denoted by subscript $R2$) of the heating section are given. The balances in this case become

Energy balance:

$$\dot{Q}_{\rm in} + \dot{m}_{a1}h_1 = \dot{m}_{a2}h_2 \text{ (process 1–2)}$$

$$\dot{m}_{a1}h_1 + \dot{m}_{\rm R}h_{R1} = \dot{m}_{a2}h_2 + \dot{m}_{\rm R}h_{R2} \text{ (process 1–2)}$$

$$\dot{m}_{a1}h_1 + \dot{m}_{\rm R}h_{R1} + \dot{m}_{\rm w}h_{\rm w} = \dot{m}_{a2}h_2 + \dot{m}_{\rm R}h_{R2} \text{ (process 1–3)}$$

Entropy balance:

$$\dot{m}_{a1}s_1 + \dot{m}_{\rm R}s_{R1} + \dot{m}_{\rm w}s_{\rm w} - \dot{m}_{a3}s_3 - \dot{m}_{\rm R}s_{R2} + \dot{S}_{\rm gen} = 0 \text{ (process 1–3)}$$

Exergy balance:

$$\dot{m}_{a1}\psi_1 + \dot{m}_{\rm R}\psi_{R1} + \dot{m}_{\rm w}\psi_{\rm w} - \dot{m}_{a3}\psi_3 - \dot{m}_{\rm R}\psi_{R2} - \dot{E}x_{\rm dest} = 0 \text{ (process 1–3)}$$

$$\dot{E}x_{\rm dest} = T_0\dot{S}_{\rm gen} = T_0\left(\dot{m}_{a3}s_3 + \dot{m}_{\rm R}s_{R2} - \dot{m}_{a1}s_1 - \dot{m}_{\rm R}s_{R1} - \dot{m}_{\rm w}s_{\rm w}\right) \text{ (process 1–3)}$$

Exergy efficiency:

$$\eta_{\rm ex} = \frac{\dot{m}_{a3}\psi_3 + \dot{m}_{\rm R}\psi_{R2}}{\dot{m}_{a1}\psi_1 + \dot{m}_{\rm R}\psi_{R1} + \dot{m}_{\rm w}\psi_{\rm w}}$$

The exergy efficiency of the process is calculated to be 7.2%, which is low. This is typical of air-conditioning processes during which irreversibilities occur mainly because of heat transfer across a relatively high temperature difference and humidification.

1.10 Psychrometrics

Psychrometrics is the science of air and water vapor and deals with the properties of moist air. A thorough understanding of psychrometrics is of great significance, particularly to the HVAC community. It plays a key role, not only in the heating and cooling processes and the resulting comfort of the occupants, but also in building insulation, roofing properties, and the stability, deformation, and fire-resistance of the building materials. That is why understanding of the main concepts and principles involved is essential.

Actually, psychrometry also plays a crucial role in food preservation, especially in cold storage. In order to prevent the spoilage and maintain the quality of perishable products during storage, a proper arrangement of the storage conditions in terms of temperature and relative humidity is extremely important in this regard. Furthermore, the storage conditions are different for each food commodity and should be implemented accordingly.

1.10.1 Common Definitions in Psychrometrics

The following definitions are the most common terms in psychrometrics:

Dry air. Normally, atmospheric air contains a number of constituents, as well as water vapor, along with miscellaneous components (e.g., smoke, pollen, and gaseous pollutants). When we talk about dry air, it no longer contains water vapor and other components.

Moist air. Moist air is the basic medium and is defined as a binary or two-component mixture of dry air and water vapor. The amount of water vapor in moist air varies from nearly zero, referring to dry air, to a maximum of 0.020 kg water vapor/kg dry air under atmospheric conditions depending on the temperature and pressure.

Saturated air. This is known as the saturated mixture (i.e., air and water vapor mixture) where the vapor is given at the saturation temperature and pressure.

Dew point temperature. This is defined as the temperature of moist air saturated at the same pressure and with the same humidity ratio as that of the given sample of moist air (i.e., temperature at state 2 in Figure 1.27). It takes place where the water vapor condenses when it is cooled at constant pressure (i.e., process 1–2).

Relative humidity. This is defined as the ratio of the mole fraction of water vapor in the mixture to the mole fraction of water vapor in a saturated mixture at the same temperature and pressure, based on the mole fraction equation since water vapor is considered to be an ideal gas:

$$\phi = \frac{P_v}{P_s} = \frac{\rho_v}{\rho_s} = \frac{v_s}{v_v} \tag{1.99}$$

where P_v is the partial pressure of vapor, Pa or kPa, and P_s is the saturation pressure of vapor at the same temperature, Pa or kPa, which can be taken directly from saturated water table. The total pressure is $P = P_a + P_v$. According to Figure 1.27, $\phi = P_1/P_3$.

Humidity ratio. The humidity ratio of moist air (so-called *mixing ratio*) is defined as the ratio of the mass of water vapor to the mass of dry air contained in the mixture at the same temperature and pressure:

$$\omega = \frac{m_v}{m_a} = 0.622\frac{P_v}{P_a} \tag{1.100}$$

where $m_v = P_v V/R_v T$ and $m_a = P_a V/R_a T$ since both water vapor and air, as well as their mixtures, are treated as ideal gases.

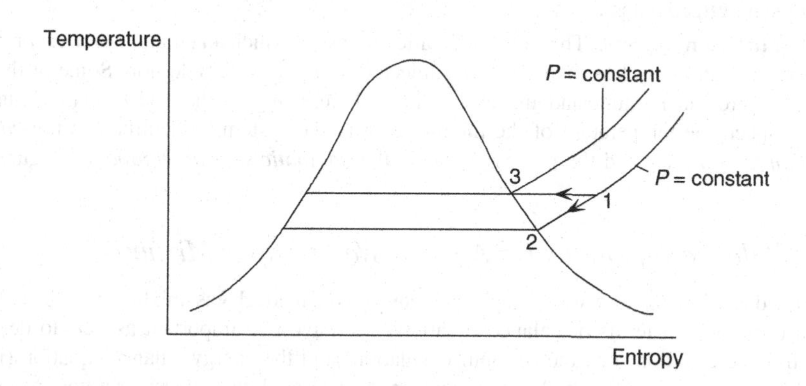

Figure 1.27 Representation of dew point temperature on T–s diagram.

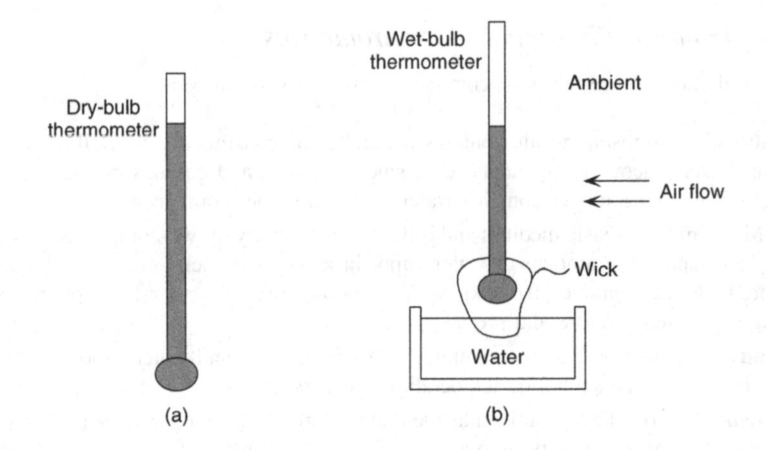

Figure 1.28 Schematic representation of (a) dry-bulb and (b) wet-bulb thermometers.

Since we have the relative humidity and the humidity ratio in terms of the pressure ratio, it is possible to reach the following equation after making the necessary substitutions:

$$\phi = \frac{\omega P_a}{0.622 P_s} \tag{1.101}$$

Degree of saturation. This is defined as the ratio of the actual humidity ratio to the humidity ratio of a saturated mixture at the same temperature and pressure.

Dry-bulb and wet-bulb temperatures. The use of both a dry-bulb thermometer and a wet-bulb thermometer is very old practice to measure the specific humidities of moist air. The dry-bulb temperature is the temperature measured by a dry-bulb thermometer directly. The bulb of the wet-bulb thermometer is covered with a wick which is already saturated with water. When the wick is subjected to an air flow (Figure 1.28), some of the water in the wick gets evaporated into the surrounding air, thereby resulting in a temperature drop in the thermometer. This final temperature is dependent on the moisture content of the air. It is important to mention that in the past there was a convention that the wicks are boiled in distilled water first and allowed to dry before using them in wet-bulb temperature measurements. Nowadays, several new electronic devices and data loggers are preferred to measure the humidity of air due to their simplicity, accuracy, and effectiveness.

Adiabatic saturation process. This is the adiabatic process in which an air and water vapor mixture with a relative humidity less than 100% is subjected to liquid water addition. Some of the water evaporates into the mixture and makes it saturated, referring to the 100% relative humidity. In this respect, the temperature of the mixture exiting the system is identified as the *adiabatic saturation temperature* and the process is called the *adiabatic saturation process* (Figure 1.29).

1.10.2 Balance Equations for Air and Water Vapor Mixtures

As mentioned earlier, air and water vapor is considered an ideal gas mixture which makes the solution a bit easier. In terms of balance equations, we have two important aspects to deal with: the mass balance equation (i.e., the continuity equation) and the energy balance equation (i.e., the FLT). These can be written for both closed and open systems. Let us consider a cooling process, with negligible kinetic and potential energies and no work involved, that has two inputs and one

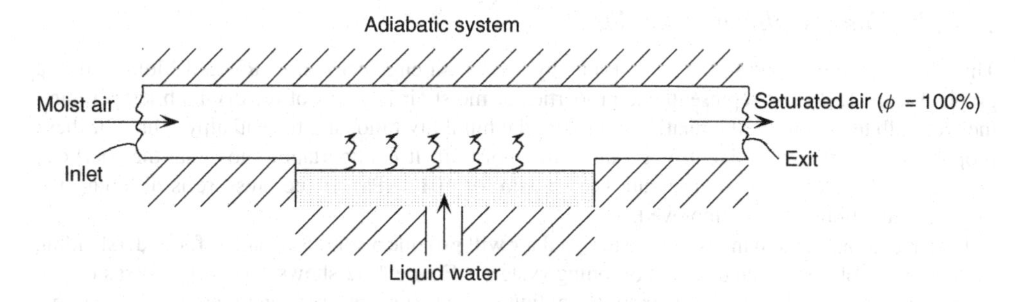

Figure 1.29 Schematic representation of an adiabatic saturation process.

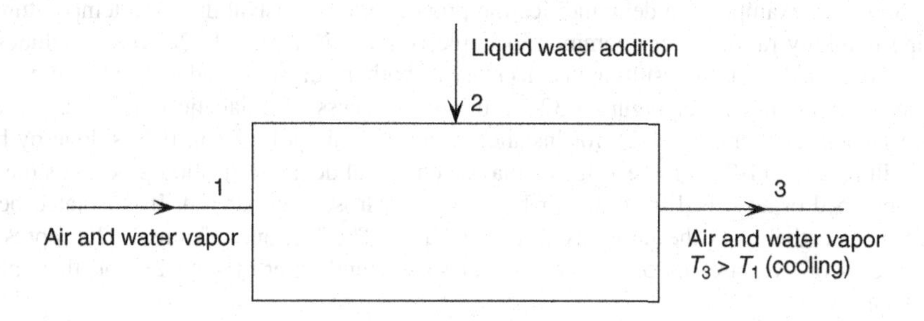

Figure 1.30 Schematic of the system.

output as illustrated in Figure 1.30. Before going into details of analysis of this process, the general mass and energy balance equations may be written as follows:

- The mass balance equations are

$$\Sigma \dot{m}_{a,i} = \Sigma \dot{m}_{a,e} \tag{1.102}$$

$$\Sigma \dot{m}_{v,i} + \Sigma \dot{m}_{l,i} = \Sigma \dot{m}_{a,e} + \Sigma \dot{m}_{l,e} \tag{1.103}$$

- The energy balance equation is

$$\dot{Q}_i + \Sigma \dot{m}_i h_i = \Sigma \dot{m}_e h_e \tag{1.104}$$

Let us now write the respective balance equations for the subject matter system in Figure 1.30 as follows:

$$\dot{m}_{a,1} = \dot{m}_{a,3} = \dot{m}_a \tag{1.105}$$

$$\dot{m}_{v,1} + \dot{m}_{l,2} = \dot{m}_{v,3} \tag{1.106}$$

$$\dot{Q}_i + \dot{m}_a h_{a,1} + \dot{m}_{v,1} h_{v,1} + \dot{m}_{l,2} h_{l,2} = \dot{m}_a h_{a,3} + \dot{m}_{v,3} h_{v,3} \tag{1.107}$$

Equation 1.107 can be arranged in terms of the humidity ratio under $\omega = m_v/m_a$ (Equation 1.100):

$$\frac{\dot{Q}_i}{\dot{m}_a} + h_{a,1} + \omega_1 h_{v,1} + (\omega_1 - \omega_2) h_{l,2} = h_{a,3} + \omega_3 h_{v,3} \tag{1.108}$$

where $\omega_2 = \omega_3$ since there is no more water addition or removal between 2 and 3.

1.10.3 The Psychrometric Chart

This chart was developed in the early 1900s by a German engineer named Richard Mollier. It is a graph (Figure 1.31) that represents the properties of moist air in terms of the dry-bulb temperature, the wet-bulb temperature, the relative humidity, the humidity ratio, and the enthalpy. Three of these properties are sufficient to identify a state of the moist air. It is important to note that the chart can only be used for atmospheric pressure (i.e., 1 atm, or 101.3 kPa). If the pressure is different, the moist air equations can be employed.

Understanding the dynamics of moisture and air will provide a solid foundation for understanding the principles of cooling and air-conditioning systems. Figure 1.32 shows several processes on the psychrometric chart. Figure 1.32a exhibits cooling and heating processes and therefore an example of an increase and decrease in dry-bulb temperature. In these processes, only a change in sensible heat is encountered. There is no latent heat involved due to the constant humidity ratio of the air. Figure 1.32b is an example of a dehumidification process at the constant dry-bulb temperature with decreasing humidity ratio. A very common example is given in Figure 1.32c which includes both cooling and dehumidification, resulting in a decrease of both the dry-bulb and wet-bulb temperatures, as well as the humidity ratio. Figure 1.32d exhibits a process of adiabatic humidification at the constant wet-bulb temperature (1-2), for instance spray type humidification. If it is done by heated water, it will result in (1-2′). Figure 1.32e displays a chemical dehumidification process as the water vapor is absorbed or adsorbed from the air by using a hydroscopic material. It is isolated because of the constant enthalpy as the humidity ratio decreases. The last one (Figure 1.32f) represents a mixing process of two streams of air (i.e., one at state 1 and other at state 2), and their mixture reaches state 3.

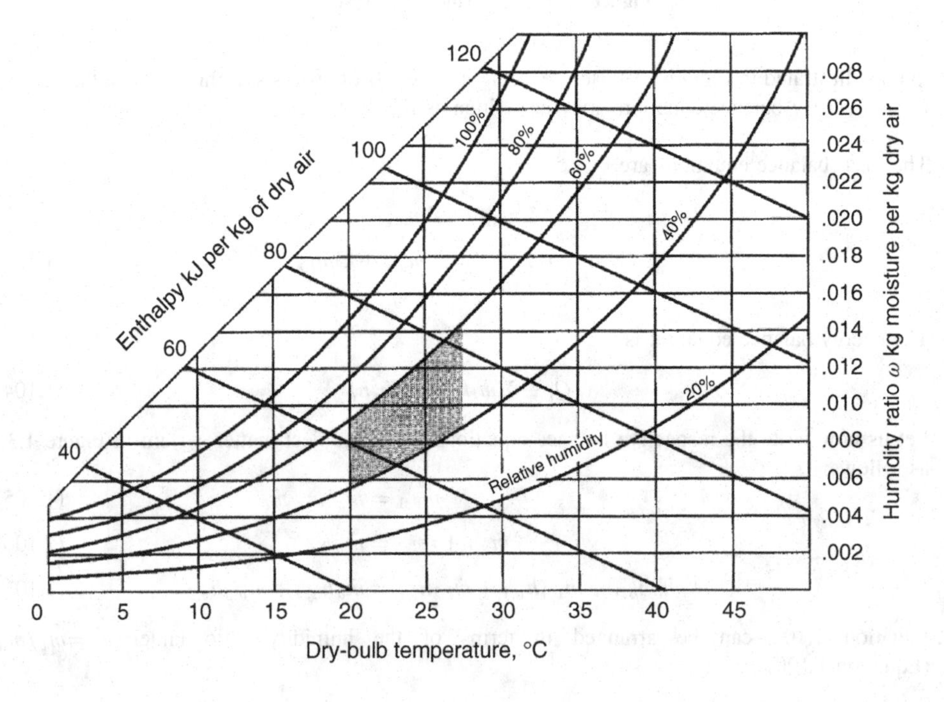

Figure 1.31 Psychrometric chart.

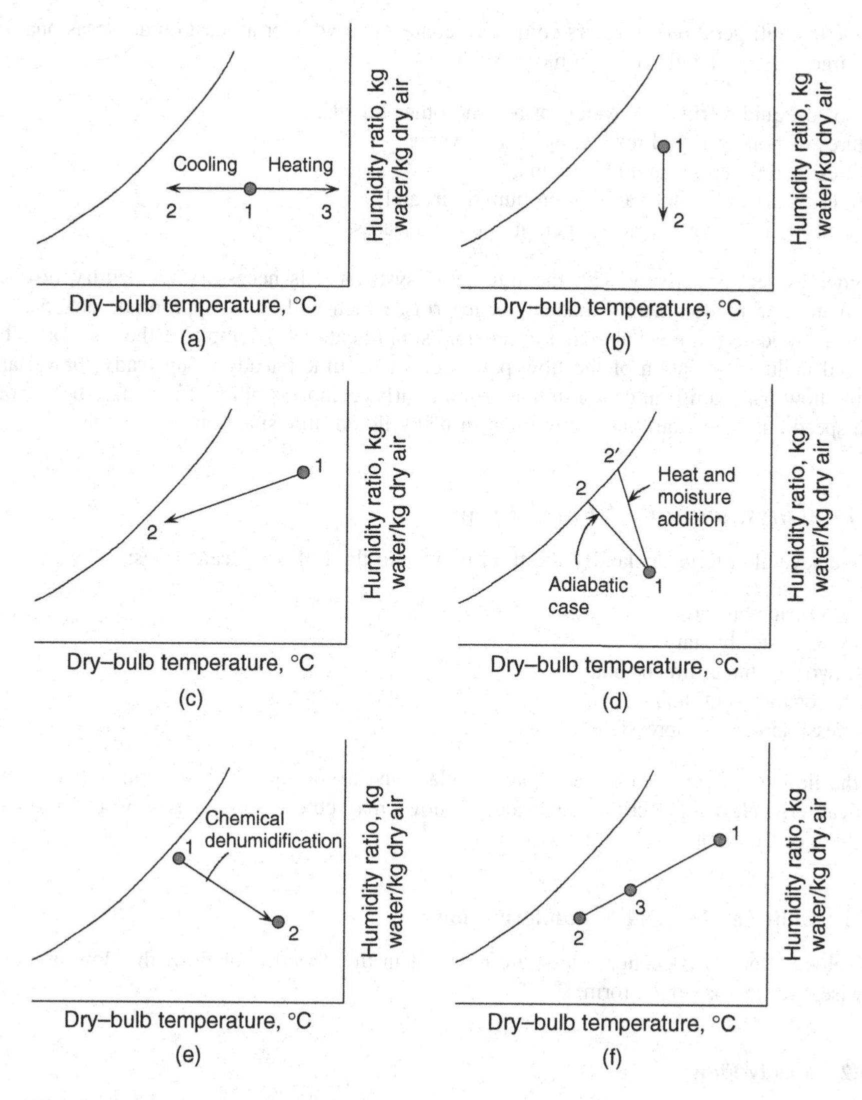

Figure 1.32 Some processes on the psychrometric chart. (a) Cooling and heating. (b) Dehumidification. (c) Cooling and dehumidification. (d) Adiabatic humidification. (e) Chemical dehumidification. (f) Mixture of two moist air flows.

1.11 General Aspects of Fluid Flow

For a good understanding of the operation of refrigeration systems and their components as well as the behavior of fluid flow, an extensive background on fluid mechanics is essential. In addition to learning the principles of fluid flow, the student and/or engineer should develop an understanding of the properties of fluids, which should enable him or her to solve practical refrigeration problems.

In practice, refrigeration engineers come into contact everyday or at least on an occasional basis with a large variety of fluid flow problems such as

- subcooled liquid refrigerant, water, brine, and other liquids,
- mixtures of boiling liquid refrigerant and its vapor,
- mixtures of refrigerants and absorbents,
- mixtures of air and water vapor as in humid air, and
- low- and high-side vaporous refrigerant and other gases.

In order to deal effectively with the fluid flow systems it is necessary to identify flow categories, defined in predominantly mathematical terms, which will allow appropriate analysis to be undertaken by identifying suitable and acceptable simplifications. Example of the categories to be introduced includes variation of the flow parameters with time (steady or unsteady) or variations along the flow path (uniform or non-uniform). Similarly, compressibility effects may be important in high-speed gas flows, but may be ignored in many liquid flow situations.

1.11.1 Classification of Fluid Flows

There are several criteria to classify the fluid flows into the following categories:

- uniform or nonuniform,
- steady or unsteady state,
- one-, two-, or three-dimensional,
- laminar or turbulent, and
- compressible or incompressible.

Also, the liquids flowing in channels may be classified according to their regions, for example, subcritical, critical, or supercritical, and the gas flows may be categorized as subsonic, transsonic, supersonic, or hypersonic.

1.11.1.1 Uniform Flow and Nonuniform Flow

If the velocity and cross-sectional area are constant in the direction of flow, the flow is uniform. Otherwise, the flow is nonuniform.

1.11.1.2 Steady Flow

This is defined as a flow in which the flow conditions do not change with time. However, we may have a steady-flow in which the velocity, pressure, and cross-section of the flow may vary from point to point but do not change with time. Therefore, we need to distinguish this by dividing it into the *steady uniform flow* and the *steady nonuniform flow*. In the steady uniform flow, all conditions (e.g., velocity, pressure, and cross-sectional area) are uniform and do not vary with time or position. For example, uniform flow of water in a duct is considered steady uniform flow. If the conditions (e.g., velocity, cross-sectional area) change from point to point (e.g., from cross-section to cross-section) but not with time, it is called steady nonuniform flow. For example, a liquid flows at a constant rate through a tapering pipe running completely full.

1.11.1.3 Unsteady Flow

If the conditions vary with time, the flow becomes unsteady. At a given time, the velocity at every point in the flow field is the same, but the velocity changes with time, referring to the *unsteady*

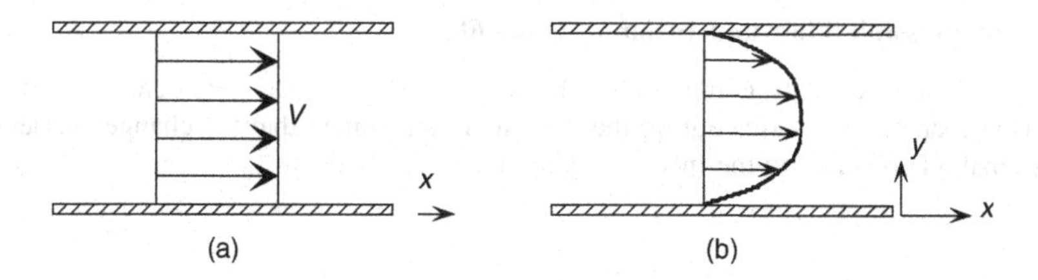

Figure 1.33 Velocity profiles for flows. (a) One-dimensional flow. (b) Two-dimensional flow.

uniform flow, for example, accelerating flow, a fluid through a pipe of uniform bore running full. In the unsteady uniform flow, the conditions in cross-sectional area and velocity vary with time from one point to another, for example, a wave traveling along a channel.

1.11.1.4 One-, Two-, and Three-Dimensional Flow

The flow of real fluids occurs in three dimensions. However, in the analysis the conditions are simplified to either one-dimensional or two-dimensional, depending on the flow problem under consideration. If all fluid and flow parameters (e.g. velocity, pressure, elevation, temperature, density, and viscosity) are considered to be uniform throughout any cross-section and vary only along the direction of flow (Figure 1.33a), the flow becomes one-dimensional. Two-dimensional flow is the flow in which the fluid and flow parameters are assumed to have spatial gradients in two directions, that is, x and y axes (Figure 1.33b). In fact, in a three-dimensional flow the fluid and flow parameters vary in three directions, that is, x, y and z axes, and the gradients of the parameters occur in all three directions.

1.11.1.5 Laminar Flow and Turbulent Flow

This is one of the most important classifications of fluid flow and depends primarily upon the arbitrary disturbances, irregularities, or fluctuations in the flow field, based on the internal characteristics of the flow. In this regard, there are two significant parameters such as velocity and viscosity. If the flow occurs at a relatively low velocity and/or with a highly viscous fluid, resulting in a fluid flow in an orderly manner without fluctuations, the flow is referred to as laminar. As the flow velocity increases and the viscosity of fluid decreases, the fluctuations will take place gradually, referring to a *transition state* which is dependent on the fluid viscosity, the flow velocity, and the geometric details. In this regard, the Reynolds number is introduced to represent the characteristics of the flow conditions relative to the transition state. As the flow conditions deviate more from the transition state, a more chaotic flow field, that is, turbulent flow occurs. It is obvious that increasing Reynolds number increases the chaotic nature of the turbulence. Turbulent flow is therefore defined as a characteristic representative of the irregularities in the flow field.

The differences between laminar flow and turbulent flow can be distinguished by the Reynolds number, which is expressed by

$$\mathrm{Re} = \frac{VD}{\nu} = \frac{\rho VD}{\mu} \tag{1.109}$$

In fact, the Reynolds number indicates the ratio of inertia force to viscous force. One can point out that at high Reynolds numbers the inertia forces predominate, resulting in turbulent flow, while at low Reynolds numbers the viscous forces become dominant, which makes the flow laminar. In a circular duct, the flow is laminar when Re is less than 2100 and turbulent when Re is greater than 4000. In a duct with a rough surface, the flow is turbulent at Re values as low as 2700.

1.11.1.6 Compressible Flow and Incompressible Flow

All actual fluids are normally compressible, leading to the fact that their density changes with pressure. However, in most cases during the analysis it is assumed that the changes in density are negligibly small. This refers to the incompressible flow.

1.11.2 Viscosity

This is known as one of the most significant fluid properties and is defined as a measure of the fluid's resistance to deformation. In gases, the viscosity increases with increasing temperature, resulting in a greater molecular activity and momentum transfer. The viscosity of an ideal gas is a function of molecular dimensions and absolute temperature only, based on the kinetic theory of gases. However, in liquids, molecular cohesion between molecules considerably affects the viscosity, and the viscosity decreases with increasing temperature due to the fact that the cohesive forces are reduced by increasing the temperature of the fluid (causing a decrease in shear stress), resulting in an increase in the rate of molecular interchange; therefore, the net result is apparently a reduction in the viscosity. The coefficient of viscosity of an ideal fluid is zero, meaning that an ideal fluid is inviscid, so that no shear stresses occur in the fluid, despite the fact that shear deformations are finite. Nevertheless, all real fluids are viscous.

There are two types of viscosities, namely, the *dynamic viscosity* which is the ratio of a shear stress to a fluid strain (velocity gradient) and the *kinematic viscosity* which is defined as the ratio of dynamic viscosity to density. The dynamic viscosity is expressed based on Figure 1.34, leading to the fact that the shear stress within a fluid is proportional to the spatial rate of change of fluid strain normal to the flow:

$$\mu = \frac{\tau}{du/dy} \tag{1.110}$$

where the unit of μ is Ns/m^2 or kg/ms in the SI system and lbf $\cdot$ s/ft^2 in the English system.

The kinematic viscosity then becomes

$$\nu = \frac{\mu}{\rho} \tag{1.111}$$

where the units of ν is m^2/s in the SI system and ft^2/s in the English system.

There are some other units (e.g., the cgs system of units) for the dynamic and kinematic viscosities that find applications as follows:

$$1 \text{ pose} = 1 \text{ dyne} \cdot \text{s/cm}^2 = 1 \text{ g/cm} \cdot \text{s} = 0.1 \text{ kg} \cdot \text{m} \cdot \text{s}.$$

$$1 \text{ stoke} = 1 \text{ cm}^2/\text{s} = 10^{-4} \text{ m}^2/\text{s}.$$

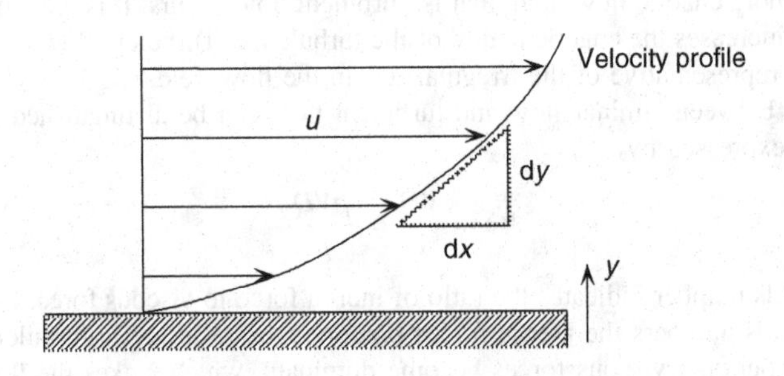

Figure 1.34 Schematic of velocity profile.

From the viscosity point of view, the types of fluids may be classified into Newtonian and non-Newtonian fluids.

1.11.2.1 Newtonian Fluids

These fluids have a dynamic viscosity dependent upon temperature and pressure and are independent of the magnitude of the velocity gradient. For such fluids, Equation 1.110 is applicable. Some examples are water and air.

1.11.2.2 Non-Newtonian Fluids

The fluids which cannot be represented by Equation 1.110 are called non-Newtonian fluids. These fluids are very common in practice and have a more complex viscous behavior due to the deviation from the Newtonian behavior. There are several approximate expressions to represent their viscous behavior. Some examples are slurries, polymer solutions, oil paints, toothpaste, and sludges.

1.11.3 Continuity Equation

This is based on the *conservation of mass* principle. The requirement that mass be conserved at every point in a flowing fluid imposes certain restrictions on the velocity u and density ρ. Therefore, the rate of mass change is zero, referring to that for a steady flow; the mass of fluid in the control volume remains constant and therefore the mass of fluid entering per unit time is equal to the mass of fluid exiting per unit time. Let us apply this to a steady flow in a stream tube (Figure 1.35). The equation of continuity for the flow of a compressible fluid through a stream tube is

$$\rho_1 \delta A_1 u_1 = \rho_2 \delta A_2 u_2 = \text{constant} \quad (1.112)$$

where $\rho_1 \delta A_1 u_1$ is the mass entering per unit time and $\rho_2 \delta A_2 u_2$ is the mass exiting per unit time for the sections 1 and 2.

In practice, for the flow of a real fluid through a pipe or a conduit, the mean velocity is used since the velocity varies from wall to wall. Therefore, Equation 1.112 can be rewritten as

$$\rho_1 A_1 \overline{u}_1 = \rho_2 A_2 \overline{u}_2 = \dot{m} \quad (1.113)$$

where $\overline{u}_1$ and $\overline{u}_2$ are the mean velocities at sections 1 and 2.

For the fluids that are considered as incompressible, Equation 1.113 is simplified to the following, since $\rho_1 = \rho_2$:

$$A_1 \overline{u}_1 = A_2 \overline{u}_2 = \dot{V} \quad (1.114)$$

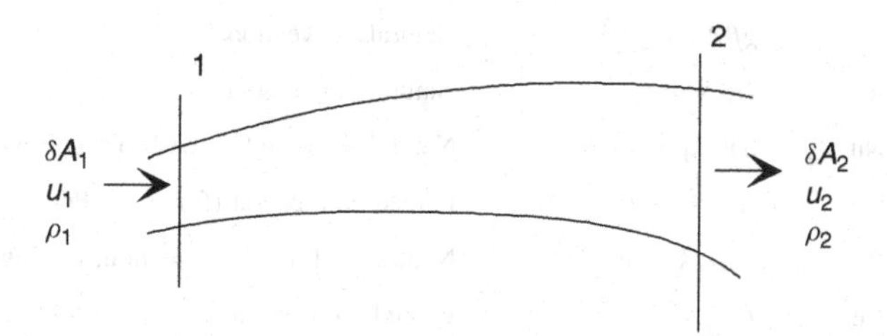

Figure 1.35 Fluid flow in a stream tube.

1.12 General Aspects of Heat Transfer

Thermal processes involving the transfer of heat from one point to another are often encountered in the food industry, as in other industries. The heating and cooling of liquid or solid food products, the evaporation of water vapors, and the removal of heat liberated by a chemical reaction are common examples of processes that involve heat transfer. It is of great importance for food technologists, refrigeration engineers, researchers, and so on, to understand the physical phenomena and practical aspects of heat transfer, along with some knowledge of the basic laws, governing equations, and related boundary conditions.

In order to transfer heat, there must be a driving force, which is the temperature difference between the points where heat is taken and where the heat originates. For example, consider that when a long slab of food product is subjected to heating on the left side, the heat flows from the left-hand side to the right-hand side, which is colder. It is said that heat tends to flow from a point of high temperature to a point of low temperature, with the temperature difference being the driving force.

Many of the generalized relationships used in heat-transfer calculations have been determined by means of dimensional analysis and empirical considerations. It has been found that certain standard dimensionless groups appear repeatedly in the final equations. It is necessary for people working in the food cooling industry to recognize the more important of these groups. Some of the most commonly used dimensionless groups that appear frequently in the heat-transfer literature are given in Table 1.3.

In the utilization of these groups, care must be taken to use equivalent units so that all the dimensions cancel out. Any system of units may be used in a dimensionless group as long as the final result will permit all units to disappear by cancellation.

Basically, heat is transferred in three ways: conduction, convection, and radiation (the so-called modes of heat transfer). In many cases, heat transfer takes place by all three of these methods simultaneously. Figure 1.36 shows the different types of heat-transfer processes as modes. When a temperature gradient exists in a stationary medium, which may be a solid or a fluid, the heat transfer occurring across the medium is by conduction, the heat transfer occurring between a surface and a moving fluid at different temperatures is by convection, and the heat transfer occurring

Table 1.3 Some of the most important heat-transfer dimensionless parameters.

Name	Symbol	Definition	Mode
Biot number	Bi	hY/k	Steady- and unsteady-state conduction
Fourier number	Fo	at/Y^2	Unsteady-state conduction
Graetz number	Gz	GY^2c_p/k	Laminar convection
Grashof number	Gr	$g\beta\Delta TY^3/\nu^2$	Natural convection
Rayleigh number	Ra	$Gr \times Pr$	Natural convection
Nusselt number	Nu	hY/k_f	Natural or forced convection, boiling, or condensation
Peclet number	Pe	$UY/a = Re \times Pr$	Forced convection (for small Pr)
Prandtl number	Pr	$c_p\mu/k = \nu/a$	Natural or forced convection, boiling, or condensation
Reynolds number	Re	UY/ν	Forced convection
Stanton number	St	$h/\rho Uc_p = Nu/RePr$	Forced convection

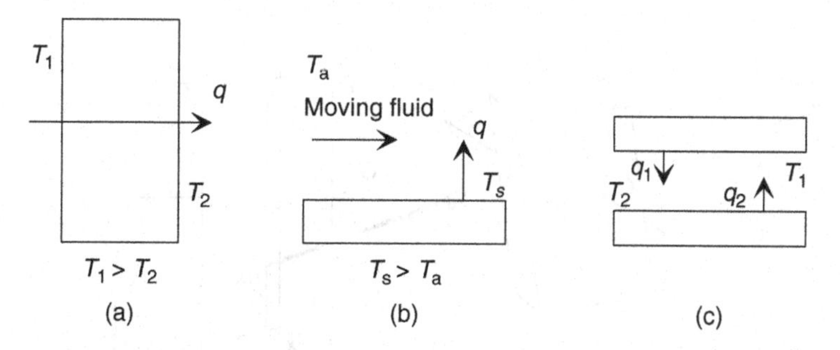

Figure 1.36 Schematic representations of heat-transfer modes. (a) Conduction through a solid. (b) Convection from a surface to a moving fluid. (c) Radiation between two surfaces.

between two surfaces at different temperatures, in the absence of an intervening medium, is by radiation, where all surfaces of finite temperature emit energy in the form of electromagnetic waves.

1.12.1 Conduction Heat Transfer

Conduction is a mode of transfer of heat from one part of a material to another part of the same material, or from one material to another in physical contact with it, without appreciable displacement of the molecules forming the substance. For example, the heat transfer in a food product subject to cooling in a medium is by conduction.

In solid objects, the conduction of heat is partly due to the impact of adjacent molecules vibrating about their mean positions and partly due to internal radiation. When the solid object is a metal, there are also large numbers of mobile electrons which can easily move through the matter, passing from one atom to another, and they contribute to the redistribution of energy in the metal object. Actually, the contribution of the mobile electrons predominates in metals, which explains the relation that is found to exist between the thermal and electrical conductivity of such materials.

1.12.1.1 Fourier's Law of Heat Conduction

Fourier's law states that the instantaneous rate of heat flow through an individual homogeneous solid object is directly proportional to the cross-sectional area A (i.e., the area at right angles to the direction of heat flow) and to the temperature difference driving force across the object with respect to the length of the path of the heat flow, dT/dx. This is an empirical law based on observation.

Figure 1.37 presents an illustration of Fourier's law of heat conduction. Here, a thin slab object of thickness dx and surface area F has one face at a temperature T and the other at a lower temperature $(T - dT)$ where heat flows from the high-temperature side to the low-temperature side, with a temperature change in the direction of the heat flow dT. Therefore, under Fourier's law the heat-transfer equation results in

$$Q = -kA\frac{dT}{dx} \tag{1.115}$$

Here, we have a term *thermal conductivity*, k, of the object that can be defined as the heat flow per unit area per unit time when the temperature decreases by one degree in unit distance. Its units are usually written as W/m · °C or W/m · K.

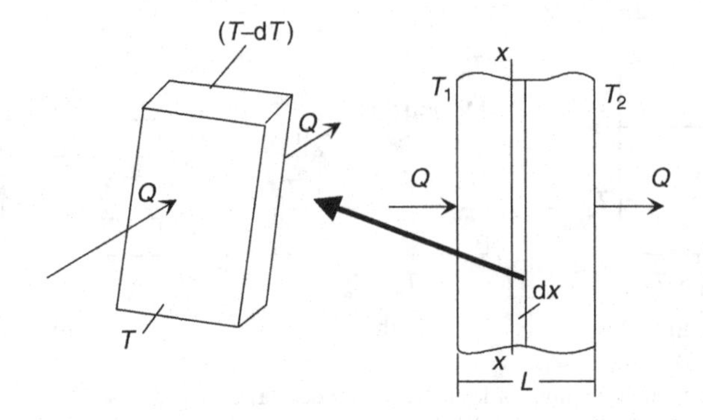

Figure 1.37 Schematic illustration of conduction in a slab object.

Integrating Equation 1.115 from T_1 to T_2 for dT and from 0 to L for dx, the solution becomes

$$Q = -k\frac{A}{L}(T_2 - T_1) = k\frac{A}{L}(T_1 - T_2) \tag{1.116}$$

1.12.2 Convection Heat Transfer

Convection is the heat-transfer mode that takes place within a fluid by mixing one portion of the fluid with another. Convection heat transfer may be classified according to the nature of the flow. When the flow is caused by some mechanical or external means such as a fan, a pump, or atmospheric wind, it is called *forced convection*. On the other hand, for *natural (free) convection* the flow is induced by buoyancy forces in the fluid that arise from density variations caused by temperature variations in the fluid. For example, when a hot food product is exposed to the atmosphere, natural convection occurs, whereas in a cold store forced convection heat transfer takes place between air flow and a food product subject to this flow.

Heat transfer through solid objects is by conduction alone, whereas heat transfer from a solid surface to a liquid or gas takes place partly by conduction and partly by convection. Whenever there is an appreciable movement of the gas or liquid, heat transfer by conduction in the gas or liquid becomes negligibly small compared with the heat transfer by convection. However, there is always a thin boundary layer of liquid on a surface, and through this thin film the heat is transferred by conduction. The convection heat transfer occurring within a fluid is due to the combined effects of conduction and bulk fluid motion. Generally the heat that is transferred is the *sensible*, or internal thermal, heat of the fluid. However, there are convection processes for which there is also *latent* heat exchange, which is generally associated with a phase change between the liquid and vapor states of the fluid.

1.12.2.1 Newton's Law of Cooling

Newton's law of cooling states that the heat transfer from a solid surface to a fluid is proportional to the difference between the surface and fluid temperatures and the surface area. This is a particular characteristic of the convection heat-transfer mode and is defined as

$$Q = hA(T_s - T_f) \tag{1.117}$$

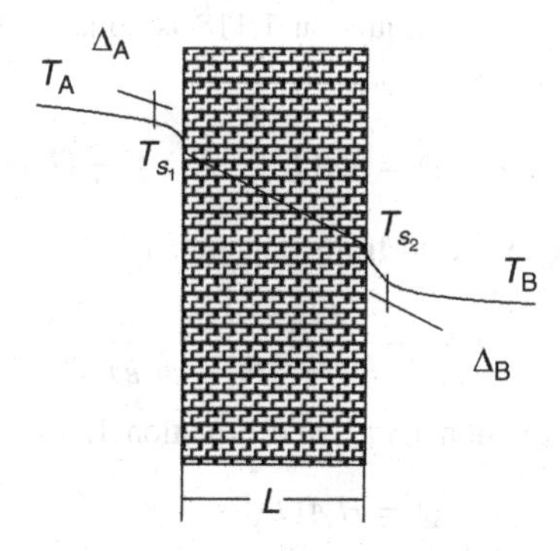

Figure 1.38 A wall subject to convection heat transfer from both sides.

where h is referred to as the *convection heat-transfer coefficient* (the *heat-transfer coefficient*, the *film coefficient*, or the *film conductance*). It encompasses all the effects that influence the convection mode and depends on conditions in the boundary layer, which is affected by factors such as surface geometry, the nature of the fluid motion, and the thermal and physical properties.

In Equation 1.117, a radiation term is not included. The calculation of radiation heat transfer will be discussed later. In many heat-transfer problems, the radiation effect on the total heat transfer is negligible compared with the heat transferred by conduction and convection from the surface to the fluid. When the surface temperature is high, or when the surface loses heat by natural convection, then the heat transfer due to radiation is of a similar magnitude as that lost by convection.

In order to better understand Newton's law of cooling, consider the heat transfer from a high-temperature fluid A to a low-temperature fluid B through a wall of thickness x (Figure 1.38). In fluid A the temperature decreases rapidly from T_A to T_{s1} in the region of the wall, and similarly in fluid B from T_{s2} to T_B. In most cases the fluid temperature is approximately constant throughout its bulk, apart from a thin film (Δ_A or Δ_B) near the solid surface bounding the fluid. The heat transfers per unit surface area from fluid A to the wall and that from the wall to fluid B are

$$q = h_A(T_A - T_{s1}) \tag{1.118}$$

$$q = h_B(T_{s2} - T_B) \tag{1.119}$$

Also, the heat transfer in thin films is by conduction only as follows:

$$q = \frac{k_A}{\Delta_A}(T_A - T_{s1}) \tag{1.120}$$

$$q = \frac{h_B}{\Delta_B}(T_{s2} - T_B) \tag{1.121}$$

Equating Equations 1.118–1.121, the convection heat-transfer coefficients can be found to be $h_A = k_A/\Delta_A$, and $h_B = k_B/\Delta_B$. Thus, the heat transfer in the wall per unit surface area becomes

$$q = \frac{k}{L}(T_{s1} - T_{s2}) \tag{1.122}$$

For a steady-state heat-transfer case, Equation 1.118 is equal to Equation 1.119 and hence to Equation 1.122

$$q = h_A(T_A - T_{s1}) = h_B(T_{s2} - T_B) = \frac{k}{L}(T_{s1} - T_{s2}) \tag{1.123}$$

The following expression can be extracted from Equation 1.123:

$$q = \frac{(T_A - T_B)}{(1/h_A + L/k + 1/h_B)} \tag{1.124}$$

An analogy can be made with Equation 1.117, and Equation 1.124 becomes

$$Q = HA(T_A - T_B) \tag{1.125}$$

where $1/H = [(1/h_A) + (L/k) + (1/h_B)]$. H is the overall heat-transfer coefficient and consists of various heat-transfer coefficients.

1.12.3 Radiation Heat Transfer

An object emits radiant energy in all directions unless its temperature is absolute zero. If this energy strikes a receiver, part of it may be absorbed and part may be reflected. Heat transfer from a hot to a cold object in this manner is known as *radiation heat transfer*. It is clear that the higher the temperature, the greater is the amount of energy radiated. If, therefore, two objects at different temperatures are placed so that the radiation from each object is intercepted by the other, then the body at the lower temperature will receive more energy than it radiates, and thereby its internal energy will increase; in conjunction with this the internal energy of the object at the higher temperature will decrease. Radiation heat transfer frequently occurs between solid surfaces, although radiation from gases also takes place. Certain gases emit and absorb radiation at certain wavelengths only, whereas most solids radiate over a wide range of wavelengths. The radiative properties of some gases and solids may be found in heat-transfer-related books.

Radiation striking an object can be absorbed by the object, reflected from the object, or transmitted through the object. The fractions of the radiation absorbed, reflected, and transmitted are called the absorptivity a, the reflectivity r, and the transmittivity t, respectively. By definition, $a + r + t = 1$. For most solids and liquids in practical applications, the transmitted radiation is negligible and hence $a + r = 1$. A body which absorbs all radiation is called a *blackbody*. For a blackbody $a = 1$ and $r = 0$.

1.12.3.1 The Stefan–Boltzmann Law

This law was found experimentally by Stefan and proved theoretically by Boltzmann. The law states that the emissive power of a blackbody is directly proportional to the fourth power of its absolute temperature. The Stefan–Boltzmann law enables calculation of the amount of radiation emitted in all directions and over all wavelengths simply from knowledge of the temperature of the blackbody. This law is given as follows:

$$E_b = \sigma T_s^4 \tag{1.126}$$

where σ stands for the Stefan–Boltzmann constant, and its value is $5.669 \times 10^{-8}\ W/m^2 \cdot K^4$. T_s stands for the absolute temperature of the surface.

The energy emitted by a non-blackbody becomes

$$E_{nb} = \varepsilon \sigma T_s^4 \tag{1.127}$$

Then the heat transferred from an object's surface to its surroundings per unit area is

$$q = \varepsilon\sigma(T_s^4 - T_a^4) \tag{1.128}$$

It is important to explain that if the emissivity of the object at T_s is much different from the emissivity of the object at T_a, then the gray object approximation may not be sufficiently accurate. In this case, it is a good approximation to take the absorptivity of the object 1 when receiving radiation from a source at T_a as being equal to the emissivity of object 1 when emitting radiation at T_a. This results in

$$q = \varepsilon_{T_s}\sigma T_s^4 - \varepsilon_{T_a}\sigma T_a^4 \tag{1.129}$$

There are numerous applications for which it is convenient to express the net radiation heat transfer (radiation heat exchange) in the following form:

$$Q = h_r A(T_s - T_a) \tag{1.130}$$

After combining Equations 1.120 and 1.121, the radiation heat-transfer coefficient can be found as follows:

$$h_r = \varepsilon\sigma(T_s + T_a)(T_s^2 + T_a^2) \tag{1.131}$$

It is important to note that the radiation heat-transfer coefficient depends strongly on temperature, whereas the temperature dependence of the convection heat-transfer coefficient is generally weak.

The surface within the surroundings may also simultaneously transfer heat by convection to the surroundings. The total rate of heat transfer from the surface is the sum of the convection and radiation modes:

$$Q_t = Q_c + Q_r = h_c A(T_s - T_a) + \varepsilon\sigma A(T_s^4 - T_a^4) \tag{1.132}$$

1.13 Concluding Remarks

In this chapter, some general, but key, aspects of thermodynamics, fluid flow, and heat transfer have been presented. Understanding these topics is important as these will serve as background information for the forthcoming chapters.

Nomenclature

a	acceleration, m/s^2; thermal diffusivity, m^2/s; absorptivity
A	cross-sectional area, m^2; surface area, m^2
c	mass fraction
c_p	constant-pressure specific heat, kJ/kg · K
c_v	constant-volume specific heat, kJ/kg · K
COP	coefficient of performance
E	energy, kJ
$\dot{E}$	rate of energy, kW
Ex	amount of exergy, kJ
$Ex_{\text{destroyed}}$	exergy destruction, kJ
$\dot{E}x$	rate of exergy, kW
F	force; drag force, N
Fo	Fourier number

g	acceleration due to gravity (= 9.81 m/s^2)
Gr	Grashof number
Gz	Graetz number
h	specific enthalpy, kJ/kg; heat-transfer coefficient, W/m$^2\cdot$°C
H	entalpy, kJ; overall heat-transfer coefficient, W/m$^2\cdot$°C; head, m
k	specific heat ratio; thermal conductivity, W/m·°C
KE	kinetic energy, W or kW
L	thickness, m
m	mass, kg
$\dot{m}$	mass flow rate, kg/s
M	molecular weight, kg/kmol
n	mole number, kmol
Nu	Nusselt number
P	pressure, kPa
Pe	Peclet number
PE	potential energy, W or kW
Pr	Prandtl number
$\dot{q}$	heat rate per unit area, W/m^2
Q	amount of heat transfer, kJ
$\dot{Q}$	heat-transfer rate, kW
r	reflectivity; radial coordinate; radial distance, m
R	gas constant, kJ/kg·K; radius, m
$\overline{R}$	universal gas constant, kJ/kg·K
Ra	Rayleigh number
Re	Reynolds number
s	specific entropy, kJ/kg
S	entropy, kJ/K
S_{gen}	entropy generation, kJ/K
St	Stanton number
t	time, s; transmittivity
T	temperature, °C or K
T_s	absolute temperature of the object surface, K
u	specific internal energy, kJ/kg
U	internal energy, kJ; flow velocity, m/s
v	specific volume, m^3/kg
$\overline{v}$	molal specific volume, kmol/kg
V	volume, m^3; velocity, m/s
$\dot{V}$	volumetric flow rate, m^3/s
W	amount of work, kJ
$\dot{W}$	power, W or kW
x	quality, kg/kg
X	length for plate, m
y	mole fraction
Z	compressibility factor

Greek Letters

ΔT	temperature difference, K; overall temperature difference, °C or K
ε	surface emissivity
η	efficiency
η_{th}	thermal efficiency
η_{ex}	exergy (second-law) efficiency

μ	dynamic viscosity, kg/ms; root of the characteristic equation
ψ	flow exergy, kJ/kg
ν	kinematic viscosity, m^2/s
ρ	density, kg/m^3
σ	Stefan–Boltzmann constant, $W/m^2 \cdot K^4$
τ	shear stress, N/m^2
ϕ	relative humidity, %
ω	humidity ratio, kg/kg

Subscripts and Superscripts

a	air; medium; surroundings
av	average
db	dry-bulb
H	high-temperature
in	input
l	liquid
liq	liquid
L	low-temperature
out	output
tot	total
v	vapor
vap	vapor
wb	wet-bulb
0	surroundings; ambient; environment; reference

Study Problems

Introduction, Thermodynamic Properties

1.1 Why are SI units most widely used throughout the world?

1.2 What is the difference between mass and weight?

1.3 What is specific heat? Define two specific heats used. Is specific heat a function of temperature?

1.4 Explain operating principle of thermocouples. What are some typical applications depending on the type of thermocouples? What is the main advantage of thermocouple over other temperature sensors?

1.5 Consider the flow of a refrigerant vapor through a compressor, which is operating at steady-state conditions. Do mass flow rate and volume flow rate of the refrigerant across the compressor remain constant?

1.6 Consider a refrigeration system consisting of a compressor, an evaporator, a condenser, and an expansion valve. Do you evaluate each component as a closed system or as a control volume; as a steady-flow system or unsteady-flow system? Explain.

1.7 What is the difference between an adiabatic system and an isolated system?

1.8 Define intensive and extensive properties. Identify the following properties as intensive or extensive: mass, volume, density, specific volume, energy, specific enthalpy, total entropy, temperature, pressure.

1.9 Define sensible and latent heats, and latent heat of fusion. What are their units.

1.10 What is the weight of a 10-kg substance in N, kN, kg_f, and lbf?

1.11 The vacuum pressure of a tank is given to be 40 kPa. If the atmospheric pressure is 95 kPa, what is the gage pressure and absolute pressure in kPa, kN/m^2, lbf/in^2, psi, and mm Hg.

1.12 Express −40 °C temperature in Fahrenheit (°F), Kelvin (K), and Rankine (R) units.

1.13 The temperature of air changes by 10 °C during a process. Express this temperature change in Kelvin (K), Fahrenheit (°F), and Rankine (R) units.

1.14 The specific heat of water at 25 °C is given to be 4.18 kJ/kg · °C. Express this value in kJ/kg · K, J/g · °C, kcal/kg · °C, and Btu/lbm · °F.

1.15 A 0.2-kg of R134a at 700 kPa pressure initially at 4 °C is heated until 50% of mass is vaporized. Determine the temperature at which the refrigerant is vaporized, and sensible and latent heat transferred to the refrigerant.

1.16 A 0.5-lbm of R134a at 100 psia pressure initially at 40 °F is heated until 50% of mass is vaporized. Determine the temperature at which the refrigerant is vaporized, and sensible and latent heat transferred to the refrigerant.

1.17 A 2-kg ice initially at −18 °C is heated until 75% of mass is melted. Determine sensible and latent heat transferred to the water. The specific heat of ice at 0 °C is 2.11 kJ/kg · °C. The latent heat of fusion of water at 0 °C is 334.9 kJ/kg.

1.18 A 2-kg ice initially at −18 °C is heated until it exists as liquid water at 20 °C. The specific heat of ice at 0 °C is 2.11 kJ/kg · °C. The latent heat of fusion of water at 0 °C is 334.9 kJ/kg. Determine sensible and latent heat transferred to the water.

1.19 Refrigerant-134a enters the evaporator of a refrigeration system at −24 °C with a quality of 25% at a rate of 0.22 kg/s. If the refrigerant leaves evaporator as a saturated vapor, determine the rate of heat transferred to the refrigerant. If the refrigerant is heated by water in the evaporator, which experiences a temperature rise of 16 °C, determine the mass flow rate of water.

Ideal Gases and the First Law of Thermodynamics

1.20 What is compressibility factor?

1.21 What is an isentropic process? Is a constant-entropy process necessarily reversible and adiabatic?

1.22 What is the difference between heat and work.

1.23 An elastic tank contains 0.8 kmol of air at 23 °C and 600 kPa. Determine the volume of the tank. The volume is now doubled at the same pressure. What is the temperature at this state?

1.24 An elastic tank contains 1.4 lb mol of air at 79 °F and 80 psia. Determine the volume of the tank. The volume is now doubled at the same pressure. What is the temperature at this state?

1.25 A 50-liter piston-cylinder device contains oxygen at 52 °C and 170 kPa. Now the oxygen is heated until the temperature reaches 77 °C. What is the amount of heat transfer during this process?

1.26 A 50-liter rigid tank contains oxygen at 52 °C and 170 kPa. Now the oxygen is heated until the temperature reaches 77 °C. What is the amount of heat transfer during this process?

1.27 A 50-liter rigid tank contains oxygen at 52 °C and 170 kPa. Now the oxygen is heated until the temperature reaches 77 °C. What is the entropy change during this process?

1.28 A rigid tank contains 2.5 kg oxygen at 52 °C and 170 kPa. Now the oxygen is heated in an isentropic process until the temperature reaches 77 °C. What is the pressure at final state? What is the work interaction during this process?

1.29 A piston-cylinder device contains 2.5 kg oxygen at 52 °C and 170 kPa. Now the oxygen is heated until the temperature reaches 77 °C. Determine the work done and the amount of heat transfer during this process?

Refrigerators, Heat Pumps, and the Carnot Refrigeration Cycle

1.30 Show that for given values Q_L and Q_H the COPs of a refrigerator and a heat pump are related to each other by $COP_{HP} = COP_R + 1$.

1.31 How can the COP of a Carnot refrigerator be increased?

1.32 A Carnot refrigerator is used to keep a space at 18 °C by rejecting heat to a reservoir at 35 °C. If the heat removal from the cooled space is 12,000 kJ/h, determine the COP of the refrigerator and the power input in kW.

1.33 A Carnot refrigerator is used to keep a space at 65 °F by rejecting heat to a reservoir at 90 °F. If the heat removal from the cooled space is 10,500 Btu/h, determine the COP of the refrigerator and the power input in kW.

1.34 A Carnot refrigerator is used to keep a space at −20 °C. If the COP of the refrigerator is 7.5, what is the temperature of the reservoir to which heat is rejected? For a power input of 3.7 kW, what is the rate of heat rejected to high-temperature reservoir?

Exergy

1.35 What are the two main reasons that cause irreversibility?

1.36 What happens to the parameters mass, energy, entropy, and exergy during an irreversible process: conserved, decreases, or increases?

1.37 How does an exergy analysis help the goal of more efficient energy resource use? What are the advantageous of using an exergy analysis?

Psychrometrics

1.38 What is the difference between humidity ratio and relative humidity?

1.39 Why is heating usually accompanied by humidification and cooling by dehumidification?

1.40 Consider moist air at 24 °C at sea level with a relative humidity of 70%. Using psychrometric chart, determine humidity ratio, wet-bulb temperature, and enthalpy of moist air.

1.41 Consider moist air at 25 °C at sea level with a relative humidity of 40%. What is the partial pressure of water vapor in the air? The saturation pressure of water at 25 °C is 3.17 kPa.

1.42 Consider moist air at 80 °F at sea level with a relative humidity of 40%. What is the partial pressure of water vapor in the air? The saturation pressure of water at 80 °F is 0.507 psia.

1.43 Air at 1 atm and 32 °C with a relative humidity of 20% enters an evaporative cooling section whose effectiveness is 80%. What is the air temperature at the exit of the evaporative cooler?

General Aspects of Fluid Flow

1.44 What is the physical meaning of the Reynolds number? What makes the flow laminar and what makes it turbulent?

1.45 What is viscosity? How does viscosity change with temperature for gases and for liquids?

General Aspects of Heat Transfer

1.46 What are the modes of heat transfer? Explain mechanism of each mode.

1.47 A 20-cm thick wall of a house made of brick ($k = 0.72 \text{W/m} \cdot °\text{C}$) is subjected to inside air at 22 °C with a convection heat-transfer coefficient of 15 W/m^2 · °C. The inner surface temperature of the wall is 18 °C and the outside air temperature is −1 °C. Determine the outer surface temperature of the wall and the heat-transfer coefficient at the outer surface.

1.48 A satellite is subjected to solar energy at a rate of 300 W/m^2. The absorptivity of the surface is 0.75 and its emissivity is 0.60. Determine the equilibrium temperature of the satellite.

1.49 A satellite is subjected to solar energy at a rate of 95 Btu/h · ft^2. The absorptivity of the surface is 0.80 and its emissivity is 0.65. Determine the equilibrium temperature of the satellite.

1.50 An 80-cm-diameter spherical tank made of steel contains liquefied natural gas (LNG) at −160 °C. The tank is insulated with 4-cm-thick insulation ($k = 0.015 \text{W/m} \cdot °\text{C}$). The tank is subjected to ambient air at 18 °C with a convection heat-transfer coefficient of 20 W/m^2 · °C. How long will it take for the temperature of the LNG to drop to −150 °C. Neglect the thermal resistance of the steel tank. The density and the specific heat of LNG are 425 kg/m^3 and 3.475 kJ/kg · °C, respectively.

References

Borgnakke, C. and Sonntag, R. (2008) *Fundamentals of Thermodynamics*, 7th edn, John Wiley & Sons, Ltd., New York.

Cengel, Y.A. and Boles, M.A. (2008) *Thermodynamics: An Engineering Approach*, 6th edn, McGraw Hill, New York..

Dincer, I. (2002) The role of exergy in energy policy making. *Energy Policy*, **30**, 137–149.

Dincer, I. (2003) *Refrigeration Systems and Applications*, 1st edn, John Wiley & Sons, Ltd., New York.

Dincer, I. and Rosen, M.A. (2005) Thermodynamic aspects of renewables and sustainable development. *Renewable and Sustainable Energy Reviews*, **9**, 169–189.

Klein, S.A. (2006) Engineering equation solver (EES), F-Chart Software, www.fChart.com.

Marquand, C. and Croft, D. (1997) *Thermofluids – An Integrated Approach to Thermodynamics and Fluid Mechanics Principles*, John Wiley & Sons, Ltd., New York.

Moran, M.J. and Shapiro, H.N. (2007) *Fundamentals of Engineering Thermodynamics*, 6th edn, John Wiley & Sons, Ltd., New York.

Raznjevic, K. (1995) *Handbook of Thermodynamic Tables*, 2nd edn, Begell House, New York.

2

Refrigerants

2.1 Introduction

The first designers of refrigeration machines, Jacob Perkins in 1834, and others later in the nineteenth century, used ethyl ether (R-610) as the first commercial refrigerant. The reason is easy to understand if one has ever spilt this liquid on the hand and felt the effect. It was not particularly suitable to the purpose however, being dangerous as well as requiring an excessive compressor volume. Other and more appropriate refrigerants, for example, ammonia (R-717), carbon dioxide (R-744), ethyl chloride (R-160), isobutane (R-600a), methyl chloride (R-40), methylene chloride (R-30), and sulfur dioxide (R-764), were soon introduced, including air (R-729). Three of these refrigerants became very popular, that is, ammonia and sulfur dioxide for refrigerators and other small units and carbon dioxide preferably for ships' refrigeration. A large number of substances were tried over the following years, with varying success.

In the early 1930s, the introduction of chlorofluorocarbons (CFCs) was revolutionary as compared with the natural substances. In addition to their use as refrigerants in refrigeration and air-conditioning systems, CFCs were utilized as foam-blowing agents, aerosol propellants, and cleaning solvents since 1950. The main arguments put forward in their favor were complete safety and harmlessness to the environment. Both these claims were proved wrong. Many accidents have occurred because of suffocation in the heavy gas, without warning, in below threshold spaces. It was evident that CFCs and related compounds contribute tremendously to the destruction of the stratospheric ozone layer and to the greenhouse effect (i.e., global climate change), which are considered among the most significant environmental problems. In fact, CFCs are greenhouse gases that give a combined contribution to incremental global warming of the same magnitude. The most abundant greenhouse gas is CO_2 and the others are CH_4, N_2O, CFCs, and so on. The effect of CFCs on global climate change was assumed to vary considerably, roughly contributing in the range 15–20% as compared to 50% for CO_2. The interesting point is that in order to minimize global climate change, making reductions in CFC utilization seemed to be easier than reducing fossil fuel use. Therefore, a full ban on these substances was essential.

Almost a decade ago, CFCs were banned worldwide as a result of their alleged effect on the stratospheric ozone layer and global climate change, despite the fact that CFCs were among the most useful chemical substances ever developed. During the past decade, research activities have been expanded tremendously to conduct ozone level measurements using various types of ground-based or airborne equipment; of course, more recently satellite technology has become a prominent technique providing more accurate findings about the ozone levels in different locations.

It is well known that the stratospheric ozone layer acts as a shield against harmful ultraviolet (UV) solar radiation. More than two decades ago, researchers discovered that chlorine released from

Refrigeration Systems and Applications İbrahim Dinçer and Mehmet Kanoğlu

synthetic CFCs migrates to the stratosphere and destroys ozone molecules and hence the ozone layer, which was recognized as ozone layer depletion, known as one of the biggest environmental problems. In 1987, 24 nations and the European Economic Community signed the Montreal Protocol to regulate the production and trade of ozone-depleting substances. This was considered as a landmark in refrigeration history.

This is by no means a unique experience. Similar predicaments have occurred following release to the environment of many other new chemicals. The extensive use of more new compounds is one of the big problems of our time. In this situation, it does not seem very sensible to replace the CFC/hydrochlorofluorocarbons (HCFCs) with a new family of related halocarbons, equally foreign to nature, to be used in quantities of hundreds of thousands of tons every year.

2.1.1 Refrigerants

In general, refrigerants are well known as the fluids absorbing heat during evaporation. These refrigerants, which provide a cooling effect during the phase change from liquid to vapor, are commonly used in refrigeration, air conditioning, and heat pump systems, as well as process systems.

2.2 Classification of Refrigerants

This section is focused only on the primary refrigerants, which can be classified into the following five main groups (Dincer, 2003):

- halocarbons,
- hydrocarbons (HCs),
- inorganic compounds,
- azeotropic mixtures, and
- nonazeotropic mixtures.

2.2.1 Halocarbons

The halocarbons contain one or more of the three halogens – chlorine, fluorine, or bromine – and are widely used in refrigeration and air-conditioning systems as refrigerants. These are more commonly known by their trade names, such as Freon, Arcton, Genetron, Isotron, and Uron. Numerical indication is preferable in practice.

In this group, the halocarbons, consisting of chlorine, fluorine, and carbon, were the most commonly used refrigerants (so-called *chlorofluorocarbons, CFCs*). CFCs were commonly used as refrigerants, solvents, and foam-blowing agents. The most common CFCs have been CFC-11 or R-11, CFC-12 or R-12, CFC-113 or R-113, CFC-114 or R-114, and CFC-115 or R-115.

Although CFCs such as R-11, R-12, R-22, R-113, and R-114 were very common refrigerants in refrigeration and air-conditioning equipment, they were used in several industries as aerosols, foams, solvents, etc. Their use rapidly decreased, because of their environmental impact. In the past decade CFC phaseout in refrigeration became a primary political issue as well as, technically speaking, a more and more difficult problem. In addition to ozone layer depletion, the refrigeration and air-conditioning industry faces another problem – the increase in the greenhouse effect, which will be explained later.

It is well known that CFCs are odorless, nontoxic, and heavier than air, as well as dangerous if not handled properly. Inhalation of high concentrations is not detectable by human senses and can prove fatal because of oxygen exclusion caused by CFC leakages in an enclosed area. The combustion products of CFCs include phosgene, hydrogen fluoride, and hydrogen chloride, which are all highly poisonous if inhaled. Although these CFCs are not identical in performance and composition, they are part of the same basic family of chemicals.

In this family, there are some other components such as halons, carbon tetrachlorides, and perfluorocarbons (PFCs). Halons are the compounds consisting of bromine, fluorine, and carbon. The halons (i.e., halon 1301 and halon 1211) are used as fire extinguishing agents, both in built-in systems and in handheld portable fire extinguishers. Halon production was banned in many countries; for example, in the United States it ended on December 31, 1993 because of the contribution of halons to ozone depletion. They cause ozone depletion because they contain bromine. Bromine is many times more effective at destroying ozone than chlorine. Carbon tetrachloride (CCl_4) is a compound consisting of one carbon atom and four chlorine atoms. Carbon tetrachloride was widely used as a raw material in many industrial applications, including the production of CFCs, and as a solvent. Solvent use ended when it was discovered to be carcinogenic. It is also used as a catalyst to deliver chlorine ions to certain processes. PFC is a compound consisting of carbon and fluorine. PFCs have an extremely high effect on global climate change and very long lifetimes. However, they do not deplete stratospheric ozone; but the concern is about their impact on global warming.

2.2.2 *Hydrocarbons*

HCs are the compounds that mainly consist of carbon and hydrogen. HCs include methane, ethane, propane, cyclopropane, butane, and cyclopentane. Although HCs are highly flammable, they may offer advantages as alternative refrigerants because they are inexpensive to produce and have zero ozone depletion potential (ODP), very low global warming potential (GWP), and low toxicity. There are several types of HC families such as the following:

- Hydrobromofluorocarbons (HBFCs) are the compounds that consist of hydrogen, bromine, fluorine, and carbon.
- HCFCs are the compounds that consist of hydrogen, chlorine, fluorine, and carbon. The HCFCs are one class of chemicals being used to replace the CFCs. They contain chlorine and thus deplete stratospheric ozone, but to a much lesser extent than CFCs. HCFCs have ODPs ranging from 0.01 to 0.1. Production of HCFCs with the highest ODPs will be phased out first, followed by other HCFCs.
- Hydrofluorocarbons (HFCs) are the compounds that consist of hydrogen, fluorine, and carbon. These are considered a class of replacements for CFCs, because of the fact that they do not contain chlorine or bromine and do not deplete the ozone layer. All HFCs have an ODP of 0. Some HFCs have high GWPs. HFCs are numbered according to a standard scheme.
- Methyl bromide (CH_3Br) is a compound consisting of carbon, hydrogen, and bromine. It is an effective pesticide and is used to fumigate soil and many agricultural products. Because it contains bromine, it depletes stratospheric ozone and has an ODP of 0.6. Its production is banned in several countries, for example, in the United States since the end of December 2000.
- Methyl chloroform (CH_3CCl_3) is a compound consisting of carbon, hydrogen, and chlorine. It is used as an industrial solvent. Its ODP is 0.11.

For refrigeration applications, a number of HCs such as methane (R-50), ethane (R-170), propane (R-290), *n*-butane (R-600), and isobutane (R-600a) that are suitable as refrigerants can be used.

2.2.3 *Inorganic Compounds*

In spite of the early invention of many inorganic compounds, today they are still used in many refrigeration, air conditioning, and heat pump applications as refrigerants. Some examples are ammonia (NH_3), water (H_2O), air ($0.21O_2 + 0.78N_2 + 0.01Ar$), carbon dioxide ($CO_2$), and sulfur dioxide ($SO_2$). Among these compounds, ammonia has received the greatest attention for practical applications and, even today, is of interest. Below, we will briefly focus on three compounds of this family – ammonia, carbon dioxide, and air.

2.2.3.1 Ammonia (R-717)

Ammonia is a colorless gas with a strong pungent odor which may be detected at low levels (e.g., 0.05 ppm). Liquid ammonia boils at atmospheric pressure at −33 °C. The gas is lighter than air and very soluble in water. Despite its high thermal capability to provide cooling, it may cause several technical and health problems including the following (Dincer, 1997):

- Gaseous ammonia is irritating to the eyes, throat, nasal passages, and skin. Although workers apparently develop a tolerance to ammonia, exposure to levels in the range of 5 to 30 ppm may cause eye irritation.
- Exposure to levels of 2500 ppm causes permanent eye damage, breathing difficulties, and asthmatic spasm and chest pain.
- Potentially fatal accumulation of fluid in the lung may develop some hours after the exposure. Nonfatal poisoning may lead to the development of bronchitis, pneumonia, and an impaired lung function.
- Skin exposure to gaseous ammonia at very high levels causes skin irritation, skin burns, and the formation of fluid-filled blisters.
- Eye contact with liquid ammonia may lead to blindness and skin contact may lead to potentially fatal chemical burns.
- Ammonia is a flammable gas and forms potentially explosive mixtures in the range of 16 to 25% with air. Ammonia which is dissolved in water is not flammable.
- Ammonia reacts or produces explosive products with fluorine, chlorine, bromine, iodine, and some other related chemical compounds.
- Ammonia reacts with acids and produces some heat.
- Ammonia vapors react with the vapors of acid (e.g., HCl) to produce an irritating white smoke.
- Ammonia and ammonia-contaminated oil must be disposed of in a proper way approved by local regulatory agencies.

Despite its disadvantages, Lorentzen (1988) considered that ammonia is an excellent refrigerant and indicated that these possible disadvantages can be eliminated with proper design and control of the refrigeration system.

2.2.3.2 Carbon Dioxide (R-744)

Carbon dioxide is one of the oldest inorganic refrigerants. It is a colorless, odorless, nontoxic, nonflammable, and nonexplosive refrigerant and can be used in cascade refrigeration systems and in dry-ice production, as well as in food freezing applications.

2.2.3.3 Air (R-729)

Air is generally used in aircraft air conditioning and refrigeration systems. Its coefficient of performance (COP) is low because of the light weight of the air system. In some refrigeration plants, it may be used in the quick freezing of food products.

2.2.4 Azeotropic Mixtures

An azeotropic refrigerant mixture consists of two substances having different properties but behaving as a single substance. The two substances cannot be separated by distillation. The most common azeotropic refrigerant is R-502, which contains 48.8% R-22 and 51.2% R-115. Its COP is higher than that of R-22 and its lesser toxicity provides an opportunity to use this refrigerant in household refrigeration systems and the food refrigeration industry. Some other examples of azeotropic mixtures are R-500 (73.8% R-12 + 26.2% R-152a), R-503 (59.9% R-13 + 40.1% R-23), and R-504 (48.2% R-32 + 51.8% R-115).

2.2.5 Nonazeotropic Mixtures

Nonazeotropic mixture is a fluid consisting of multiple components of different volatiles that, when used in refrigeration cycles, change composition during evaporation (boiling) or condensation. Recently, nonazeotropic mixtures have been called *zeotropic mixtures* or *blends*. The application of nonazeotropic mixtures as refrigerants in refrigeration systems has been proposed since the beginning of the twentieth century. A great deal of research on these systems with nonazeotropic mixtures and on their thermophysical properties has been done since that time. Great interest has been shown in nonazeotropic mixtures, especially for heat pumps, because their adaptable composition offers a new dimension in the layout and design of vapor-compression systems. Much work has been done since the first proposal to use these fluids in heat pumps. Through the energy crises in the 1970s, nonazeotropic mixtures became more attractive in research and development on advanced vapor-compression heat pump systems. They offered the following advantages:

- energy improvement and saving,
- capacity control, and
- adaptation of hardware components regarding capacity and applications limits.

In the past, studies showed that widely used refrigerants such as R-11, R-12, R-22, and R-114 became most popular for the pure components of the nonazeotropic mixtures. Although many nonazeotropic mixtures (e.g., R-11 + R-12, R-12 + R-22, R-12 + R-114, R-13B1 + R-152a, R-22 + R-114, and R-114 + R-152a, etc.) were well known, a decade ago research and development mainly focused on three mixtures, R-12 + R-114, R-22 + R-114, and R-13B1 + R-152a. It is clear that the heat-transfer phenomena during the phase change of nonazeotropic mixtures are more complicated than with single-component refrigerants.

2.3 Prefixes and Decoding of Refrigerants

Various refrigerants (e.g., CFCs, HCs, HCFCs, HBFCs, HFCs, PFCs, and halons) are numbered according to a system devised several decades ago and now used worldwide. Although it may seem confusing, in fact it provides very complex information about molecular structure and also easily distinguishes among various classes of chemicals. In practice, it is of great importance to first understand the prefixes of refrigerants and their meanings, as well as decoding for them. In this section, we have three subsections as prefixes, decoding the number, and isomers. Further information on these aspects is found in EPA (2009).

2.3.1 Prefixes

Some of the most common refrigerants' prefixes are CFC, HCFC, HFC, PFC, and Halon, respectively. In CFCs and HCFCs, the first "C" is for chlorine (Cl), and in all of them, "F" is for

Table 2.1 The prefixes and atoms in refrigerants.

Name	Prefix	Atoms Contained
Chlorofluorocarbon	CFC	Cl, F, C
Hydrochlorofluorocarbon	HCFC	H, Cl, F, C
Hydrobromofluorocarbon	HBFC	H, Br, F, C
Hydrofluorocarbon	HFC	H, F, C
Hydrocarbon	HC	H, C
Perfluorocarbon	PFC	F, C
Halon	Halon	Br, Cl (in some), F, H (in some), C

fluorine (F), "H" is for hydrogen (H), and the final "C" is for carbon (C). PFC is a special prefix meaning *perfluorocarbon*. "Per" means "all," so PFCs have all bonds occupied by fluorine atoms. Consequently, halons are a general term for compounds that contain C, F, Cl, H, and bromine (atomic symbol: Br). Halon numbers are different from the others and will be discussed later. For example, an HFC contains no chlorine, so your results should not show any Cl atoms. Table 2.1 summarizes the prefixes and atoms contained in each refrigeration commodity.

Compounds used as refrigerants may be described using either the appropriate prefix as given in the table or with the prefixes "R-" or "Refrigerant." For example, CFC-12 may also be written as R-12 or Refrigerant 12.

Blends of refrigerants are assigned numbers serially, with the first zeotropic blend numbered R-400 and the first azeotropic blend numbered R-500. Blends that contain the same components but in differing percentages are distinguished by capital letters. For example, R-401A contains 53% HCFC-22, 13% HFC-152a, and 34% HCFC-124, but R-401B contains 61% HCFC-22, 11% HFC-152a, and 28% HCFC-124.

2.3.2 Decoding the Number

The prefix describes the kinds of atoms in a particular molecule, and the next step is to calculate the number of each type of atom. The key to the code is to add 90 to the number; the result shows the number of C, H, and F atoms. For HCFC-141b:

$141 + 90 = 2$ (#C) 3 (#H) 1(#F)

Additional information is needed to decipher the number of Cl atoms. All these chemicals are saturated, so that they contain only single bonds. The number of bonds available in a carbon-based molecule is $2C + 2$. Thus, for HCFC-141b, which has two-C atoms, there are six bonds. Cl atoms occupy bonds remaining after the F and H atoms. So HCFC-141b has two C, three H, one F, and two Cl atoms.

HCFC-141b = $C_2H_3FCl_2$

where the HCFC designation is a good double-check on the decoding, containing H, Cl, F, and C. The "b" at the end describes how these atoms are arranged; different "isomers" contain the same atoms, but they are arranged differently.

Let us see this time HFC-134a as an example.

134 + 90 = 2 (#C)   2 (#H)   4(#F)

where there are six bonds. But in this case, there are no bonds leftover after F and H, so there are no Cl atoms. Thus,

HFC-134a = $C_2H_2F_4$

where the prefix is accurate: this is an HFC, so it contains only H, F, and C, but no Cl.

Note that any molecule with only one C (e.g., CFC-12) will have a two-digit number, while those with two C or three C will have a three-digit number.

Halon numbers directly show the number of C, F, Cl, and Br atoms. The numbering scheme above does not give a direct number for the number of Cl atoms, but that can be calculated. Similarly, Halon numbers do not specify the number of H atoms directly and there is no need to add anything to decode the number.

Halon = 1 (#C)   2 (#F)   1(#Cl)   1(#Br)

For this molecule, there are $2 \times 1 + 2 = 4$ bonds, all of which are taken by Cl, F, and Br, leaving no room for any H atoms. Thus,

Halon 1211 = CF_2ClBr

2.3.3 Isomers

Isomers of a given compound contain the same atoms but they are arranged differently. Isomers usually have different properties; only one isomer may be useful. Since all of the compounds under discussion are based on carbon chains (1–3 carbon atoms attached in a line of single bonds: e.g., C—C—C), the naming system is based on how H, F, Cl, and Br atoms are attached to that chain. A single C atom can only bond with four other atoms in one way, so there are no isomers of those compounds. For two-C molecules, a single lowercase letter following the number designates the isomer. For three-C molecules, a lowercase two-letter code serves this purpose.

Consider two-C molecules, for example, HCFC-141, HCFC-141a, and HCFC-141b in which all have the same atoms (two carbon, three hydrogen, one flourine, and two chlorine) but are organized differently. To determine the letter, total the atomic weights of the atoms bonded to each of the carbon atoms. The arrangement that most evenly distributes atomic weights has no letter. The next most even distribution is the “a” isomer, the next is “b,” and so on, until no more isomers are possible. A common way of writing isomers’ structure is to group atoms according to the carbon atom with which they bond. Thus, the isomers of HCFC-141 are as follows:

- HCFC-141: $CHFCl$—CH_2Cl (atomic weights on the 2C = 37.5 and 55.5)
- HCFC-141a: $CHCl_2$—CH_2F (atomic weights on the 2C = 21 and 72)
- HCFC-141b: $CFCl_2$—CH_3 (atomic weights on the 2C = 3 and 90)

For HFC-134, the isomers are as follows:

- HFC-134: CHF_2—CHF_2
- HFC-134a: CF_3—CH_2F

In order to specify the chemical structures for each of the Cl, F, and Br atoms, we use the ordinal number of the C to which they are bonded and numerical prefixes (i.e., 2 = di, 3 = tri,

4 = tetra, etc.) to specify the total number of each kind of atom. The suffix for the molecular name is dependent on the number of carbons. Molecules with one C end in "methane" (since there are no isomers of methane-derived molecules, they have no letter designation), two-C end in "ethane," and three-C end in "propane." It is assumed that any bonds not occupied by Cl, F, or Br are occupied by H, so H atoms are not specified. So, the isomers of HCFC-141 can be written in the following way:

- HCFC-141: $CHFCl—CH_2Cl$: 1,2-dichloro-1-fluoroethane
- HCFC-141a: $CHCl_2—CH_2F$: 1,1-dichloro-2-fluoroethane
- HCFC-141b: $CFCl_2—CH_3$: 1,1-dichloro-1-fluoroethane

For HFC-134, the isomers are as follows:

- HFC-134: $CHF_2—CHF_2$: 1,1,2,2-tetrafluoroethane
- HFC-134a: $CF_3—CH_2F$: 1,1,1,2-tetrafluoroethane

CFC-12 does not have any isomers, since it contains only one C. In addition, there is no need to number the carbons, even in a case such as difluorodichloromethane.

Molecules with three-C atoms are more complicated to name. The first letter designates the atoms attached to the middle C atom, and the second letter designates decreasing symmetry in atomic weights of atoms attached to the outside C atoms. Unlike 2C chains, however, the most symmetric distribution is the "a" isomer, instead of omitting the letter entirely. The code letters for the atoms on middle C are a for Cl_2, b for Cl and F, c for F_2, d for Cl and H, e for H and F, and f for H_2; for example,

HCFC-225ca: $C_3HF_5Cl_2$ (3C = 8 bonds), $CF_3—CF_2—CHCl_2$, and 1, 1,1,2,2-pentafluoro-3,3-dichloropropane

When no isomers are possible, no letters are used. For example, there is only one way to arrange three C and eight F, so it is written as PFC-218 and not PFC-218ca.

2.4 Secondary Refrigerants

Secondary refrigerants play a role in carrying heat from an object or a space being cooled to the primary refrigerant or the evaporator of a refrigeration system. During this process, the secondary refrigerant has no phase change. In the past, the most common secondary refrigerants were brines, which are water–salt (e.g., sodium chloride and calcium chloride) solutions, and even today they are still used in spite of their corrosive effects. Also, the antifreezes, which are solutions of water and ethylene glycol, propylene glycol, or calcium chloride, are widely used as secondary refrigerants. Of these fluids, propylene glycol has the unique feature of being safe when in contact with food products. A decade ago dichloromethane (CH_2Cl_2), trichloroethylene (C_2HCl_3), alcohol solutions, and acetone were also used in some special applications. The following features are considered as main criteria in the selection of a proper secondary refrigerant (Dincer, 2003).

- satisfactory thermal and physical properties,
- stability,
- noncorrosiveness,
- nontoxicity,
- low cost, and
- usability.

2.5 Refrigerant–Absorbent Combinations

The refrigerant–absorbent combinations (so-called *working fluids*) are basically used in absorption refrigeration and heat pump systems. Inorganic and organic groups are major sources of the refrigerants and absorbents. Some organic groups for refrigerants are amines, alcohols, halogens, and HCs, and for absorbents, alcohols, ethers, alcohol-ethers, amides, amines, amine-alcohols, esters, ketones, acids, or aldehydes can be used.

Two well-known examples are ammonia–water and water–lithium bromide. In some literature, the absorbent is also called the *solvent*. The absorbent should have a greater chemical affinity for the refrigerant than that indicated by the ordinary law of solubility. Very little heat is released when the freons, nitrogens, or certain other gases are dissolved in water. However, water has a high chemical affinity for ammonia, and considerable heat is evolved during absorption. For example, at 15 °C one unit of water can absorb approximately 800 units of ammonia. Thus the quantity of heat released in absorption is a crude measure of the chemical affinity.

In practical absorption–refrigeration applications, besides ammonia–water and water–lithium bromide combinations, various refrigerant–absorbent combinations have been considered such as:

- ammonia/calcium chloride,
- ammonia/strontium chloride,
- ammonia/heptanoyl,
- ammonia/triethanolamine,
- ammonia/glycerin,
- ammonia/silicon oil,
- ammonia/lithium nitrate,
- ammonia/lithium bromide,
- ammonia/zinc bromide,
- ammonia/dimethyl ether tetraethylene glycol (DMETEG),
- ammonia/dimethyl formamide (DMF),
- methyl amine/water,
- methyl chloride/tetraethylene glycol,
- R12/dimethyl acid amide,
- R12/cyclohexanone,
- R21/dimethyl ester,
- R22/DMETEG,
- R22/DMF, and
- R22/dimethyl acid amide.

The interest in finding new refrigerants and working fluids brought R134a as an alternative refrigerant with DMETEG and DMF forefront for absorption refrigeration and heat pump systems.

A desirable refrigerant–absorbent combination should have the property of high solubility at conditions in the absorber but should have low solubility at conditions in the generator. In absorption refrigeration and heat pump systems, the following summarizes the desirable properties and influences of the absorbents (Dincer, 1997):

- negligible vapor pressure at the generator, compared to the vapor pressure of the refrigerant at 37.5 °C, influencing rectifier losses and operating cost;
- good temperature, pressure, and concentration relations (absorbent should remain liquid throughout the cycle, and the relations must be in conformity with practical condenser, absorber, and generator temperatures and pressures);

- high stability, influencing the ability to withstand the heating operation at the maximum temperatures encountered in the generator;
- low specific heat, influencing heat-transfer requirements;
- low surface tension, influencing heat transfer and absorption; and
- low viscosity, influencing heat transfer and power for pumping.

Also, refrigerant–absorbent combinations are expected to meet the following requirements:

- solubility (high solubility of the refrigerant at the temperatures of the cooling medium, e.g., air or water, and at a pressure corresponding to the vapor pressure of the refrigerant at 5 °C, plus low solubility of the refrigerant in the absorbent at generator temperatures and at a pressure corresponding to the vapor pressure of the refrigerant at the temperature of the cooling medium);
- stability (refrigerant and absorbent must be incapable of any nonreversible chemical action with each other within a practical temperature range, for example, from −5 to 120 °C); and
- superheating and supercooling (influencing operation).

The properties of the combinations relate to the liquid and/or vapor state, as encountered in normal operation of absorption systems using such combinations, and to the crystallization boundary of the liquid phase, where applicable. The property data can be classified as follows:

- vapor–liquid equilibria,
- crystallization temperature,
- corrosion characteristics,
- heat of mixing,
- liquid-phase densities,
- vapor–liquid-phase densities,
- specific heat,
- thermal conductivity,
- viscosity,
- stability,
- heat mass transfer rates,
- entropy,
- refractive index,
- surface tension,
- toxicity, and
- flammability.

2.6 Stratospheric Ozone Layer

Here, we explain some very important definitions and names before going into details.

UV radiation is a portion of the electromagnetic spectrum with wavelengths shorter than visible light. The sun produces UV, which is commonly split into three bands known as UVA, UVB, and UVC.

- **UVA.** This is a band of UV radiation with wavelengths from 320–400 nm produced by the sun and is not absorbed by ozone. This band of radiation has wavelengths just shorter than visible violet light.
- **UVB.** This is a band of UV radiation with wavelengths from 280–320 nm produced by the sun. UVB is a kind of UV light from the sun (and sun lamps) that has several harmful effects, particularly effective at damaging DNA. It is a cause of melanoma and other types of skin cancer. It has also been linked to damage to some materials, crops, and marine organisms. The ozone layer protects the earth against most UVB coming from the sun. It is always important to protect

oneself against UVB, even in the absence of ozone depletion, by wearing hats, sunglasses, and sunscreen. However, these precautions will become more important as ozone depletion worsens.

- **UVC.** This is a band of UV radiation with wavelengths shorter than 280 nm. Despite being extremely dangerous, it is completely absorbed by ozone and normal oxygen (O_2).

Stratosphere is a region of the atmosphere above the troposphere and extends from about 15 to 50 km in altitude. As a matter of fact, in the stratosphere, temperature increases with altitude because of the absorption of UV light by oxygen and ozone. This creates a global inversion layer which impedes vertical motion into and within the stratosphere – since warmer air lies above colder air in the upper stratosphere, convection is inhibited. The word *stratosphere* is related to the word *stratification* or *layering*.

Troposphere is a region of the atmosphere closest to the earth and extends from the surface up to about 10 km in altitude, although this height varies with latitude. Almost all weather takes place in the troposphere. Mt Everest, the highest mountain on earth, is only 8.8 km high. Temperatures decrease with altitude in the troposphere. As warm air rises up, it gets cooled, falling back to the earth. This process, known as *convection*, means that there are huge air movements that mix the troposphere very efficiently.

Ozone is a gas composed of three atoms of oxygen, known as a *bluish gas*, that is harmful to breathe. Nearly 90% of the earth's ozone is situated in the stratosphere and is referred to as the *ozone layer*. Ozone absorbs a band of UVB that is particularly harmful to living organisms. The ozone layer prevents most UVB from reaching the ground.

Ozone layer is a region of the stratosphere containing the bulk of atmospheric ozone. The ozone layer lies approximately 15–40 km above the earth's surface, in the stratosphere. The ozone layer is between 2 and 5 mm thick in the stratosphere under normal temperature and pressure conditions and its concentration varies depending on the season, hour of the day, and location. The concentration is greatest at an altitude of about 25 km near the equator and at about an altitude of 16 km near the poles. The ozone comes mostly from the photodisassociation of oxygen by UV radiation of very short wavelength (i.e., 200 μm).

Column ozone is the ozone between the earth's surface and outer space. Ozone levels can be described in several ways. One of the most common measures is the amount of ozone in a vertical column of air. The Dobson unit (DU) is a measure of column ozone, which is described in the next paragraph. Other measures include partial pressure, number density, and concentration of ozone, and can represent either column ozone or the amount of ozone at a particular altitude.

Dobson unit (DU) is a measure of column ozone levels. If 100 DU of ozone were brought to the earth's surface, it would form a layer of 1 mm thickness. In the tropics, ozone levels are typically between 250 and 300 DU year-round. In temperate regions, seasonal variations can produce large swings in ozone levels. For instance, measurements in St Petersburg have recorded ozone levels as high as 475 DU and as low as 300 DU (Wayne, 1991). These variations occur even in the absence of ozone depletion. Ozone depletion refers to reductions in ozone below normal levels after accounting for seasonal cycles and other natural effects. A DU is convenient for measuring the total amount of ozone occupying a column overhead. If the ozone layer over the United States were compressed to 0 °C and 1 atm pressure, it would be about 3 mm thick. So, 0.01 mm thickness at 0 °C and 1 atm is defined to be 1 DU; this makes the ozone layer over the United States measures ~300 DU. In absolute terms, 1 DU is about 2.7×10^{16} mol/cm^2 (Wayne, 1991). In total, there are about 3 billion metric tons, or 3×10^{15} g, of ozone in the earth's atmosphere; about 90% of this is in the stratosphere. In absolute terms, ozone is distributed at about 10^{12} mol/cm^3 at 15 km, rising to nearly 10^{13} at 25 km, then falling to 10^{11} at 45 km; and in relative terms, ozone is distributed at about 0.5 parts per million by volume (ppmv) at 15 km, rising to ~8 ppmv at ~35 km and falling to ~3 ppmv at 45 km (Wayne, 1991).

In the past, ozone measurements were made from the ground utilizing an accurately calibrated Dobson instrument and, more recently, other types. The fluctuation in the concentration was extremely large depending on the season and the seasonal activities of weather due to varying solar activity, some of which was of an apparently stochastic nature with the reason yet unexplained.

2.6.1 Stratospheric Ozone Layer Depletion

The stratospheric ozone layer depletion is a chemical destruction beyond natural reactions and is known as one of the global environmental problems (Figure 2.1). It has been shown that this issue is mainly caused by the ozone depletion substances (ODSs). Stratospheric ozone is constantly being created and destroyed through natural cycles. Various ODSs, however, accelerate the destruction processes, resulting in less than normal ozone levels. Depletion of this layer by ODS will lead to higher UVB levels, which in turn will cause increased skin cancers and cataracts and potential damage to some marine organisms, plants, and plastics.

The *ozone-depleting substances* are the compounds that contribute to stratospheric ozone depletion. ODSs include CFCs, HCFCs, halons, methyl bromide, carbon tetrachloride, and methyl chloroform. ODSs are generally very stable in the troposphere and only degrade under intense UV light in the stratosphere. When they break down, they release chlorine or bromine atoms, which then deplete ozone.

Three decades ago, Rowland and Molina first launched a theory that CFCs and some other anthropogenic trace gases in the atmosphere may act to deplete the stratospheric ozone layer by catalytic action of free chlorine. They predicted very rapid reduction of ozone concentration despite having ozone measurements almost steady for nearly 50 years, without any serious consideration. This theory brought in a new phase in the modeling of stratospheric chemistry and gave rise to renewed activities in the field. In fact, the most significant point that makes the conditions quite complicated is the natural air movement in all directions, air having nearly 40 different compounds giving several hundred possible reactions. That is why the models were extremely complex. The reduction in mean ozone level was estimated in the range between 0 and 10%, depending on the assumptions.

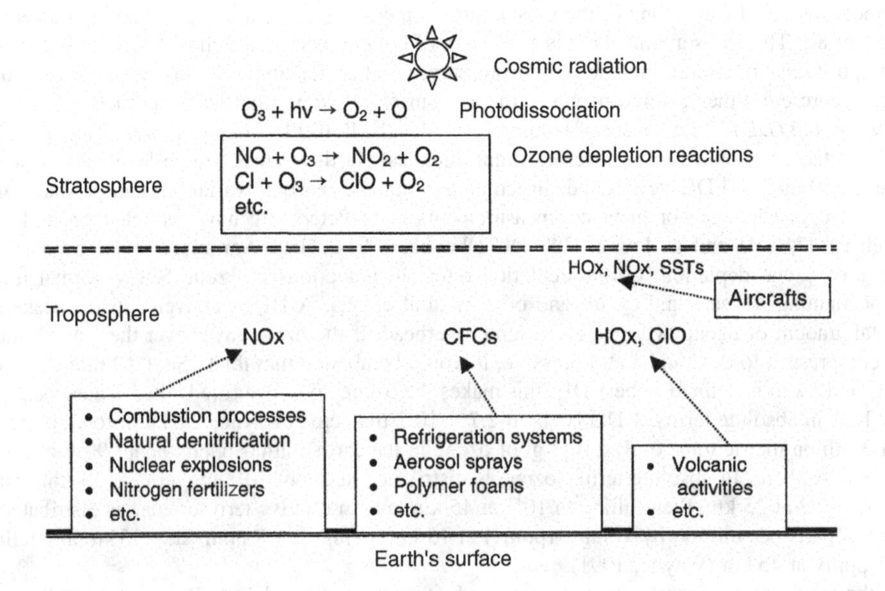

Figure 2.1 A schematic representation of stratospheric ozone depletion.

In the 1930s, Chapman described the following reactions: ozone is created in the upper stratosphere by short-wavelength UV radiation (less than ~240 nm) when it is absorbed by oxygen molecules (O_2), which dissociate to give oxygen (O) atoms. These atoms combine with other oxygen molecules and make ozone as follows (Rowland, 1991; Dincer, 1997):

$O_2 + UV \rightarrow 2O$ and $O + O_2 \rightarrow O_3$

Sunlight with wavelengths ranging between 240 and 320 nm is absorbed by ozone, which then falls apart to give an O atom and an O_2 molecule. Ozone is transformed back into oxygen if an O atom comes together with an O_3 as follows:

$O_3 + \text{sunlight} \rightarrow O + O_2$ and $O + O_3 \rightarrow 2O_2$

This cycle seems to combine with many actions, particularly catalytic destructive actions. An example of ozone depletion is as follows:

$R + O_3 \rightarrow RO + O_2$ and $RO + O \rightarrow R + O_2$

where R may be nitrogen or hydroxide or chlorine radicals.

CFCs are compounds with at least one chlorine, one fluorine, and one carbon atom in their molecule. Chlorine from the CFCs has been understood to lead to the depletion of ozone in the stratosphere. It is the chlorine that makes a substance ozone depleting; CFCs and HCFCs are a threat to the ozone layer but HFCs are not.

If the ozone depletion continues, it is likely to have effects on the following:

- human skin, with the development of skin tumors and more rapid aging of the skin,
- human eyes, with an increase in cataracts,
- human immunological system, and
- land and sea biomass, with a reduction in crop yields and in the quantity of phytoplankton.

2.6.2 *Ozone Depletion Potential*

The ODP is a number that refers to the amount of stratospheric ozone depletion caused by a substance. The ODP is the ratio of the impact on ozone of a chemical compared to the impact of a similar mass of R-11. Thus, the ODP of R-11 is defined to be 1.0. Other CFCs and HCFCs have ODPs that range from 0.01 to 1.0. The halons have ODPs ranging up to 10. Carbon tetrachloride has an ODP of 1.2 and methyl chloroform's ODP is 0.11. HFCs have zero ODP because they do not contain chlorine. The ODP data of all ozone-depleting substances are tabulated in Table 2.2.

As an example, a compound with an ODP of 0.2 is roughly about one-fifth as harmful as R-11. The ODP of any refrigerant (i.e., R-X) is defined as the ratio of the total amount of ozone destroyed by a fixed amount of R-X to the amount of ozone destroyed by the same mass of R-11, as follows:

ODP (R-X) = (Ozone loss because of R-X)/(Ozone loss because of R-11)

CFCs are considered to be fully halogenated. This means that there are no hydrogen atoms, only halogens (chlorine, fluorine, bromine, etc.). As mentioned earlier, the refrigerants with hydrogen atoms are known as *HCFCs* (e.g., R-22, R-123, R-124, R-141b, and R-142b); they are not fully halogenated and are less stable than CFCs. The computed ODP values for HCFC refrigerants are very low (on the order of 0.01 to 0.08) compared to the values estimated for CFCs (on the order of 0.7 to 1, for R-11, R-12, R-113, and R-114 and about 0.4 for R-115). It is for this reason that the Montreal Protocol had a main goal of phasing out CFC-type refrigerants. There is a family of refrigerants with an estimated ODP value of zero and without any chlorine, called *HFCs*. Some

Table 2.2 ODPs, GWPs, and CAS numbers of Class I and II ODSs.

Chemical Name	ODP	GWP	CAS Number
Class I			
Group I			
CFC-11 Trichlorofluoromethane	1.0	4,000	75-69-4
CFC-12 Dichlorodifluoromethane	1.0	8,500	75-71-8
CFC-113 1,1,2-Trichlorotrifluoroethane	0.8	5,000	76-13-1
CFC-114 Dichlorotetrafluoroethane	1.0	9,300	76-14-2
CFC-115 Monochloropentafluoroethane	0.6	9,300	76-15-3
Group II			
Halon 1211 Bromochlorodifluoromethane	3.0	1,300	353-59-3
Halon 1301 Bromotrifluoromethane	10.0	5,600	75-63-8
Halon 2402 Dibromotetrafluoroethane	6.0	–	124-73-2
Group III			
CFC-13 Chlorotrifluoromethane	1.0	11,700	75-72-9
CFC-111 Pentachlorofluoroethane	1.0	–	354-56-3
CFC-112 Tetrachlorodifluoroethane	1.0	–	76-12-0
CFC-211 Heptachlorofluoropropane	1.0	–	422-78-6
CFC-212 Hexachlorodifluoropropane	1.0	–	3182-26-1
CFC-213 Pentachlorotrifluoropropane	1.0	–	2354-06-5
CFC-214 Tetrachlorotetrafluoropropane	1.0	–	29255-31-0
CFC-215 Trichloropentafluoropropane	1.0	–	1599-41-3
CFC-216 Dichlorohexafluoropropane	1.0	–	661-97-2
CFC-217 Chloroheptafluoropropane	1.0	–	422-86-6
Group IV			
CCl_4 Carbon tetrachloride	1.1	1,400	56-23-5
Group V			
Methyl chloroform 1,1,1-trichloroethane	0.1	110	71-55-6
Group VI			
CH_3Br Methyl bromide	0.7	5	7-55-6
Group VII		–	–
$CHFBr_2$	1.0	–	–
CHF_2Br (HBFC-12B1)	0.74	–	–

Table 2.2 (*continued*)

Chemical Name	ODP	GWP	CAS Number
CH_2FBr	0.73	–	–
C_2HFBr_4	0.3–0.8	–	–
$C_2HF_2Br_3$	0.5–1.8	–	–
$C_2HF_3Br_2$	0.4–1.6	–	–
C_2HF_4Br	0.7–1.2	–	–
$C_2H_2FBr_3$	0.1–1.1	–	–
$C_2H_2F_2Br_2$	0.2–1.5	–	–
$C_2H_2F_3Br$	0.7–1.6	–	–
$C_2H_3FBr_2$	0.1–1.7	–	–
$C_2H_3F_2Br$	0.2–1.1	–	–
C2H_4FBr	0.07–0.1	–	–
C_3HFBr_6	0.3–1.5	–	–
$C_3HF_2Br_5$	0.2–1.9	–	–
$C_3HF_3Br_4$	0.3–1.8	–	–
$C_3HF_4Br_3$	0.5–2.2	–	–
$C_3HF_5Br_2$	0.9–2.0	–	–
C_3HF_6Br	0.7–3.3	–	–
$C_3H_2FBr_5$	0.1–1.9	–	–
$C_3H_2F_3Br_4$	0.2–2.1	–	–
$C_3H_2F_3Br_3$	0.2–5.6	–	–
$C_3H_2F_4Br_2$	0.3–7.5	–	–
$C_3H_2F_5Br$	0.9–1.4	–	–
$C_3H_3FBr_4$	0.08–1.9	–	–
$C_3H_3F_2Br_3$	0.1–3.1	–	–
$C_3H_3F_3Br_2$	0.1–2.5	–	–
$C_3H_3F_4Br$	0.3–4.4	–	–
$C_3H_4FBr_3$	0.03–0.3	–	–
$C_3H_4F_2Br_2$	0.1–1.0	–	–
$C_3H_4F_3Br$	0.07–0.8	–	–
$C_3H_5FBr_2$	0.04–0.4	–	–
$C_3H_5F_2Br$	0.07–0.8	–	–
C_3H_6FBr	0.02–0.7	–	–

(*continued overleaf*)

Table 2.2 (*continued*)

Chemical Name	ODP	GWP	CAS Number
Class II			
HCFC-21 dichlorofluoromethane	0.04	210	75-43-4
HCFC-22 monochlorodifluoromethane	0.055	1,700	75-45-6
HCFC-31 monochlorofluoromethane	0.02	–	593-70-4
HCFC-121 tetrachlorofluoroethane	0.01–0.04	–	354-14-3
HCFC-122 trichlorodifluoroethane	0.02–0.08	–	354-21-2
HCFC-123 dichlorotrifluoroethane	0.02	93	306-83-2
HCFC-124 monochlorotetrafluoroethane	0.022	480	2837-89-0
HCFC-131 trichlorofluoroethane	0.01–0.05	–	359-28-4
HCFC-132b dichlorodifluoroethane	0.01–0.05	–	1649-08-7
HCFC-133a monochlorotrifluoroethane	0.02–0.06	–	75-88-7
HCFC-141b dichlorofluoroethane	0.11	630	1717-00-6
HCFC-142b monochlorodifluoroethane	0.065	2,000	75-68-3
HCFC-221 hexachlorofluoropropane	0.01–0.07	–	422-26-4
HCFC-222 pentachlorodifluoropropane	0.01–0.09	–	422-49-1
HCFC-223 tetrachlorotrifluoropropane	0.01–0.08	–	422-52-6
HCFC-224 trichlorotetrafluoropropane	0.01–0.09	–	422-54-8
HCFC-225ca dichloropentafluoropropane	0.025	180	422-56-0
HCFC-225cb dichloropentafluoropropane	0.033	620	507-55-1
HCFC-226 monochlorohexafluoropropane	0.02–0.1	–	431-87-8
HCFC-231 pentachlorofluoropropane	0.05–0.09	–	421-94-3
HCFC-232 tetrachlorodifluoropropane	0.008–0.1	–	460-89-9
HCFC-233 trichlorotrifluoropropane	0.007–0.2	–	7125-84-0
HCFC-234 dichlorotetrafluoropropane	0.01–0.28	–	425-94-5
HCFC-235 monochloropentafluoropropane	0.03–0.52	–	460-92-4
HCFC-241 tetrachlorofluoropropane	0.004–0.09	–	666-27-3
HCFC-242 trichlorodifluoropropane	0.005–0.13	–	460-63-9
HCFC-243 dichlorotrifluoropropane	0.007–0.12	–	460-69-5
HCFC-244 monochlorotetrafluoropropane	0.009–0.14	–	–
HCFC-251 trichlorofluoropropane	0.001–0.01	–	421-41-0
HCFC-252 dichlorodifluoropropane	0.005–0.04	–	819-00-1
HCFC-253 monochlorotrifluoropropane	0.003–0.03	–	460-35-5
HCFC-261 dichlorofluoropropane	0.002–0.02	–	420-97-3
HCFC-262 monochlorodifluoropropane	0.002–0.02	–	421-02-03
HCFC-271 monochlorofluoropropane	0.001–0.03	–	430-55-7

Source: U.S. Environmental Protection Agency, EPA (2009).

examples of HFCs mentioned above are R-125, R-134a, R-143a, and R-152a. Research and development activities have focused on the use of these ozone- and environment-friendly refrigerants.

2.6.3 Montreal Protocol

The world's leading ecologists and their counterparts in industry and commerce all agree that CFCs are the primary cause of ozone-layer depletion in the atmosphere. Ozone layer depletion and the greenhouse effect (direct or indirect) are the first environmental problems that have arisen from the use of CFCs. In 1974, Molina and Rowland observed a hole in the ozone layer over Antarctica, which they thought to be abnormal. There seemed to be a direct connection with CFCs. In 1977, 3 years after Molina and Rowland presented their hypothesis of ozone destruction by CFCs, the United Nations Environment Program organized a crucial conference to initiate action. Since then, this situation has been discussed at several meetings and symposia. On September 19, 1987, 24 countries meeting in Montreal signed the *Protocol on Substances Depleting the Ozone Layer*. The Montreal Protocol was the international treaty governing the protection of stratospheric ozone. The Montreal Protocol and its amendments control the phaseout of ODS production and use. Under the Protocol, several international organizations report on the science of ozone depletion, implement projects to help move away from ODS, and provide a forum for policy discussions. In addition, the Multilateral Fund provides resources to developing nations to promote the transition to ozone-safe technologies.

This protocol provided for a reduction in consumption (of 20% of the 1986 consumption by July 1, 1993 and of 50% by July 1, 1998), with later deadlines for developing countries. In addition, many countries (over 70) signed the Protocol and accepted the regulations in the following Helsinki Conference (May 1989) and London Conference (June 1990) and so on. Later, many countries have adopted regulations stricter than those of the Montreal Protocol.

After the Montreal Protocol, there was a tremendous effort within the refrigeration and air-conditioning industry to find proper replacements for CFCs now under phaseout. In this respect, the thermodynamic aspects of replacement refrigerants, in particular, the consequences for system operating efficiencies and the desired operating temperatures and pressures for conventional refrigeration equipment, are being investigated. Recently, there has been increasing interest in research and development in many areas, for example, ecological phenomena, fluid toxicology, thermodynamic and technological properties of the alternative refrigerants and equipment, and use of the new cycles and systems.

2.7 Greenhouse Effect (Global Warming)

Although, the term *greenhouse effect* has generally been used for the role of the whole atmosphere (mainly water vapor and clouds) in keeping the surface of the earth warm, it has been increasingly associated with the contribution of CO_2 (currently, it is estimated that CO_2 contributes about 50% to the anthropogenic greenhouse effect). However, several other gases such as CH_4, CFCs, halons, N_2O, ozone, and peroxyacetylnitrate (so-called *greenhouse gases*) produced by the industrial and domestic activities can also contribute to this effect, resulting in a rise in the earth's temperature. A schematic representation of this global problem is illustrated in Figure 2.2.

In the greenhouse effect phenomenon, the rays of the sun reach the earth and maintain an average temperature level of around +15 °C. A large part of the infrared rays reflected off the earth are caught by CO_2, H_2O, and other substances (including CFCs) present in the atmosphere and kept from going back into space. The increase in the greenhouse effect would result in a sudden rise in temperature and it is very likely linked with human activity, in particular, the emissions from fossil fuel consumption.

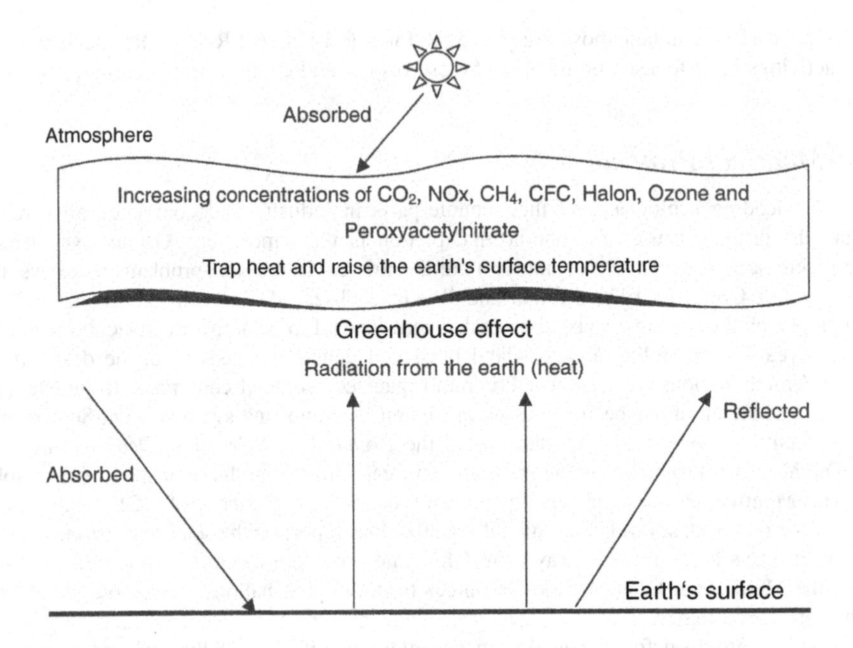

Figure 2.2 A schematic representation of the greenhouse effect.

Also, increasing the greenhouse effect may result in the following (Dincer, 2003):

- an intermediate warming of the atmosphere (estimated as 3 to 5 °C by 2050),
- a rise in the level of the oceans (estimated as 20 cm by 2050), and
- climatic effects (increases in drought, rain, snow, warming, and cooling).

Increasing atmospheric concentrations of CFCs have accounted for about 24% of the direct increase in the radiative heating from greenhouse gases over the last decade. However, an observed decrease in stratospheric ozone, thought to be connected to increasing stratospheric chlorine from CFCs, suggests a negative radiative heating or cooling tendency over the last decade.

The release of CFCs into the atmosphere affects climate in two different ways (Badr *et al.*, 1990):

- CFCs are highly harmful greenhouse gases (relative to CO_2) because of their stronger IR band intensities, stronger absorption features, and longer atmospheric lifetimes.
- CFCs deplete the stratospheric ozone layer that affects earth's surface temperature in two ways: more solar radiation reaching the surface–lower troposphere system, resulting in a warmer climate and leading to lower stratospheric temperatures and, therefore, less IR radiation being passed to the earth's surface–lower troposphere system, resulting in lower ground-level temperatures.

Therefore, the net effect is dependent on the altitudes where the ozone change takes place.

2.7.1 Global Warming Potential

GWP is a number that refers to the amount of global warming caused by a substance. GWP is the ratio of the warming caused by a substance to the warming caused by a similar mass of

CO_2. Thus, the GWP of CO_2 is defined to be 1.0. CFC-12 has a GWP of 8500, while CFC-11 has a GWP of 5000. Various HCFCs and HFCs have GWPs ranging from 93 to 12,100. Water, a substitute in numerous end uses, has a GWP of 0. GWP represents how much a given mass of a chemical contributes to global warming, over a given time period, compared to the same mass of CO_2. CO_2's GWP is defined as 1.0. These values are calculated over a 100-year time horizon. HFCs are numbered according to the ASHRAE Standard 34 scheme. Table 2.2 lists many ozone-depleting substances showing their ODPs, GWPs, and Chemical Abstract Service (CAS) numbers.

As can be seen in Table 2.2, PFCs do not deplete stratospheric ozone, but the US Environmental Protection Agency is concerned about their impact on global warming. Recent scientific studies, however, indicate that the ODPs for halon 1301 and halon 1211 are at least 13 and 4, respectively. Although HBFCs were not originally regulated under the "Clean Air Act (CAA)," subsequent regulation added HBFCs to the list of class I substances. Although HCs are highly flammable, they may offer advantages as ODS substitutes because they are inexpensive to produce and have zero ODP, very low GWP, and low toxicity. HCFCs have ODPs ranging from 0.01 to 0.1. Production of HCFCs with the highest ODPs will be phased out first, followed by other HCFCs. All HFCs have an ODP of 0. Some HFCs have high GWPs.

2.8 Clean Air Act (CAA)

Scientists worldwide have concluded that R-12 and other CFCs deplete the ozone layer. As a result, over 150 countries have signed a treaty called the *Montreal Protocol* to protect the earth's ozone layer as mentioned earlier. For example, the Protocol was implemented in the United States by the CAA as the law amended by Congress in 1990 and regulations issued under the act ended the production of R-12 for air conditioning and refrigeration uses on December 31, 1995. The CAA directs the U.S. EPA to protect the ozone layer through several regulatory and voluntary programs and covers the production of ODSs, the recycling and handling of ODSs, the evaluation of substitutes, and efforts to educate the public. A detailed list of class I and class II substances with their ODPs, GWPs, and CAS numbers has been given in Table 2.2. A class I substance is any chemical with an ODP of 0.2 or greater. Class II substances include all of the HCFCs. These compounds are numbered according to the ASHRAE Standard 34 scheme. In fact, CFCs are numbered according to a standard scheme.

2.8.1 Significant New Alternatives Policy (SNAP)

In 1994, EPA established the Significant New Alternatives Policy (SNAP) Program to review alternatives to ODSs like CFC-12. Under the authority of the 1990 CAA, EPA examines new substitutes for their ozone depleting, global warming, flammability, and toxicity characteristics. EPA has determined that several refrigerants are acceptable for use as CFC-12 replacements, subject to certain use conditions. This section lists the use conditions in detail and provides information about the current crop of refrigerants.

It is important to understand the meaning of "acceptable subject to use conditions." EPA believes that such refrigerants, when used in accordance with the conditions, are safer for human health and the environment than CFC-12. This designation does not mean that the refrigerant will work in any specific system nor does it mean that the refrigerant is perfectly safe regardless of how it is used. Finally, note that EPA does not approve or endorse any one refrigerant that is acceptable, subject to use conditions, over others also in that category.

Also, note that EPA does not test refrigerants. Rather, it reviews information submitted to it by manufacturers and various independent testing laboratories. Therefore, it is important to discuss any new refrigerant before deciding to use it, and, in particular, to determine what the effects of using a new refrigerant will be. Before choosing a new refrigerant, one should also consider whether it is readily and widely available, and one should consider the cost of buying recovery equipment for blends or recovery/recycling equipment for HFC-134a.

Many companies use the term *drop-in* to mean that a substitute refrigerant will perform identically to CFC-12, that no modifications need to be made to the system, and that the alternative can be used alone or mixed with CFC-12. However, EPA believes that the term confuses and obscures several important regulatory and technical points. First, charging one refrigerant into a system before extracting the old refrigerant is a violation of the SNAP use conditions and is, therefore, illegal. Second, certain components may be required by law, such as hoses and compressor shut-off switches. If these components are not present, they must be installed. Third, it is impossible to test one refrigerant among the thousands of refrigeration systems in existence to demonstrate identical performance. In addition, system performance is strongly affected by outside temperature, humidity, driving conditions, etc., and it is impossible to ensure equal performance under all of these conditions. Finally, it is very difficult to demonstrate that system components will last as long as they would have if CFC-12 were used. For all of these reasons, EPA does not use the term "drop-in" to describe any alternative refrigerant.

Under the SNAP rule, each new refrigerant must be used in accordance with the conditions listed below. If an alternative is chosen, one should make sure that the service shop meets these requirements and that it has dedicated recovery equipment for blends or recovery/recycling equipment for HFC-134a.

The following (e.g., Table 2.3 and 2.4) is the SNAP glossary of the EPA (2009):

- **Acceptable.** This designation means that a substitute may be used, without restriction, to replace the relevant ODS within the end use specified. For example, HCFC-22 is an acceptable substitute for R-502 in industrial process refrigeration.
- **Acceptable subject to use conditions.** This designation means that a substitute would be unacceptable unless it is used under certain conditions. An example is the set of use conditions placed on the refrigerants, requiring the use of unique fittings and labels and requiring that the original refrigerant be removed before charging with an alternative. Use of the substitute in the end use is legal, provided the conditions are fully met.
- **Acceptable subject to narrowed use limits.** This designation means that a substitute would be unacceptable unless its use was restricted to specific applications within an end use. This designation is generally used when the specific characteristics of different applications within an end use result in differences in risk. Use of the substitute in the end use is legal only in those applications included within the narrowed use limit.
- **Application.** This refers to the most specific category of equipment. This description is generally used in sectors where the end uses are fairly broad. In order of increasing specificity, a particular system is part of an industrial use sector, an end use, and an application.
- **End use.** Processes or classes of specific applications within major industrial sectors where a substitute is used to replace an ODS. The specific definition varies by sector, but examples are refrigeration, air conditioning, electronics cleaning, flooding fire extinguishing systems, and polyurethane integral skin foam. Substitutes are listed by end use in the SNAP lists. In order of increasing specificity, a particular system is part of an industrial use sector, an end use, and an application.
- **Industrial use sector.** This refers to a user community that uses an ODS in similar ways. SNAP reviews substitutes in nine sectors: (i) refrigerants, (ii) foam blowing, (iii) solvent cleaning, (iv) fire and explosion protection, (v) aerosols, (vi) sterilants, (vii) tobacco expansion, (viii) adhesives, coatings, and inks, and (ix) pesticides. In order of increasing specificity, a particular system is part of an industrial use sector, an end use, and an application.

Table 2.3 Acceptable substitutes for commercial refrigeration under SNAP program.

Substitutes	Trade Name	ODS Replaced	CSW	RT	RFR	IM	VM	WC	LTR
HCFC-22	22	12; 502	R, N	R, N	R, N	N	R, N	N	–
HFC-23	23	12; 13; 13B1; 503	–	–	–	–	–	–	R, N
HFC-134a	134a	12	R, N	R, N	R, N	N	R, N	R, N	–
HFC-227ea	–	12	N	–	N	–	–	–	–
R-401A; R-401B	MP39; MP66	12	R, N	R, N	R, N	R, N	R, N	R, N	–
R-402A; R-402B	HP80; HP81	502	R, N	R, N	R, N	R, N	–	–	–
R-404A	HP62; 404A	502	R, N	R, N	R, N	R, N	R, N	–	–
R-406A	GHG	12; 500	R	R	R	R	R	R	–
R-407A; R-407B	Klea407A; 407B	502	R, N	R, N	R, N	R, N	–	–	–
R-408A	408A	502	R	R	R	R	–	–	–
R-409A	409A	12	–	R	R	R	R	R	–
R-411A; R-411B	411A; 411B	12; 500; 502	R, N	R, N	R, N	R, N	R, N	R, N	–
R-507	AZ-50	502	R, N	R, N	R, N	R, N	R, N	–	–
R-508A	KLE 5R3	13; 13BB1; 503	–	–	–	–	–	–	R, N
R-508B	SUVA95	13; 13BB1; 503	–	–	–	–	–	–	R, N
FRIGC	FRIGC FR-12	12; 500	R, N	R, N	R, N	R, N	R, N	R, N	–
Free Zone	RB-276	12	R, N	R, N	R, N	R, N	R, N	R, N	–
Hot Shot	Hot Shot	12; 500	R, N	R, N	R, N	R, N	R, N	R, N	–
GHG-X4	GHG-X4	12; 500	R, N	R, N	R, N	R, N	R, N	R, N	–
GHG-X5	GHG-X5	12; 500	R, N	R, N	R, N	R, N	R, N	R, N	–
GHG-HP	GHG-HP	12	R, N	R, N	R, N	R, N	R, N	R, N	–
FREEZE 12	FREEZE 12	12	R, N	R, N	R, N	R, N	R, N	R, N	–
G2018C	411C	12; 500; 502	R, N	R, N	R, N	R, N	R, N	R, N	–
HCFC-22/ HCFC-142b	–	12	R, N	R, N	R, N	R, N	R, N	R, N	–
Ammonia Vapor Compression	–	ALL	N	–	N	N	–	–	–
Evaporative/ Desiccant Cooling	–	ALL	N	–	–	–	–	–	–
Stirling Cycle	–	ALL	–	N	–	–	–	–	–
Direct Nitrogen Expansion	–	ALL	–	N	–	–	–	–	–
Pressure Step-down	–	ALL	N	–	–	–	–	–	–

(*continued overleaf*)

Table 2.3 (*continued*)

Substitutes	Trade Name	ODS Replaced	CSW	RT	RFR	IM	VM	WC	LTR
CO_2	–	11; 12; 13; 113; 114; 115; 13B1; 502; 503	R, N	R, N	R, N	R, N	R, N	R, N	R, N
NARM-502	–	13; 13B1; 503	–	–	–	–	–	–	R, N
THR-04	THR-04	502	R, N	R, N	R, N	R, N	R, N	R, N	–

CSW, cold storage warehouses; RT, refrigerated transport; RFR, retail food refrigeration; IM, ice machines; VM, vending machines; WC, water coolers; LTR, low-temperature refrigeration (cryogenics); R, retrofit uses; N, new uses.
Source: US Environmental Protection Agency, EPA (2009).

- **Unacceptable.** This designation means that it is illegal to use a product as a substitute for an ODS in a specific end use. For example, HCFC-141b is an unacceptable substitute for CFC-11 in building chillers.
- **Use restriction.** A general term that includes both use conditions and narrowed use limits.

2.8.2 Classification of Substances

There are two significant classes of substance as follows:

- Class I substance is one of several groups of chemicals with an ODP of 0.2 or higher. Class I substances listed in the CAA include CFCs, halons, carbon tetrachloride, and methyl chloroform. EPA later added HBFCs and methyl bromide to the list by regulation. Table 2.3 shows the ODPs, GWPs, and CAS numbers of class I substances.
- Class II substance is a chemical with an ODP of less than 0.2. Currently, all of the HCFCs are class II substances. Lists of class II substances with their ODPs are given in Table 2.2.

In addition to these lists, the Administrator may add, by rule, in accordance with the criteria set forth, as the case may be, any substance to the list of class I or class II substances. Whenever a substance is added to the list of class I substances, the Administrator, to the extent consistent with the Montreal Protocol, assigns such substance to existing group I, II, III, IV, or V or places such substance in a new group.

Regarding the regulations for production and consumption of class I substances, the Administrator promulgated regulations within 10 months after the enactment of the Clean Air Act Amendments of 1990 phasing out the production of class I substances in accordance with this section and other applicable provisions of this title. The Administrator also promulgated regulations to ensure that the consumption of class I substances in the United States is phased out and terminated in accordance with the same schedule (subject to the same exceptions and other provisions) as is applicable to the phaseout and termination of production of class I substances.

Regarding the regulations for production and consumption, as well as restriction of class I substances, from January 1, 2015 it shall be unlawful for any person to introduce into interstate commerce or use any class II substance unless such substance (i) has been used, recovered, and recycled; (ii) is used and entirely consumed (except for trace quantities) in the production of other chemicals; or (iii) is used as a refrigerant in appliances manufactured prior to January 1, 2020.

Table 2.4 Acceptable substitutes for noncommercial refrigeration under SNAP program.

Substitutes	Trade Name	ODS Replaced	Industrial Process Refrigeration	Ice Skating Rinks	Household Refrigerators	Household Freezers
HCFC-123	123	11	R, N	–	–	–
HFC-22	22	12; 502	R, N	R, N	R, N	R, N
HFC-23	134a	13; 13B1, 503	R, N	–	–	–
HFC-134a	134a	12	R, N	–	R, N	R, N
HFC-152a	–	12	–	–	N	N
HFC-227ea	–	12	N	–	–	–
HFC-236fa	–	114	R,N	–	–	–
R-401A; R-401B	MP39; MP66	12	R, N	R,	R, N	R, N
R-402A; R-402B	HP80, HP81	502	R, N	–	–	R, N
R-403B	Isceon 69-L	13; 13B1; 503	R, N[a]	–	–	–
R-404A	HP62; 404A	502	R, N	–	–	R, N
R-406A	GHG	12; 500	R	–	R	R
R-407A; R-407B	Klea407A; 407B	502	R, N	R, N	–	–
R-408A	408A	502	R	–	–	–
R-409A	409A	12	–	–	R	R
R-411A; R-411B	411A; 411B	12; 500; 502	R, N	–	–	–
R-507	AZ-50	502	R, N	–	–	–
R-508A	KLE 5R3	13; 13BB1; 503	R, N	–	–	–
R-508B	SUVA95	13; 13BB1; 503	R, N	–	–	–
FRIGC	FRIGC FR-12	12; 500	R, N	–	R, N	R, N
Free Zone	RB-276	12	R, N	R, N	R, N	R, N
Hot Shot	Hot Shot	12; 500	R, N	R, N	R, N	R, N
GHG-X4	GHG-X4	12; 500	R, N	R, N	R, N	R, N
GHG-X5	GHG-X5	12; 500	R, N	–	R, N	R, N
GHG-HP	GHG-HP	12	R, N	–	R, N	R, N
FREEZE 12	FREEZE 12	12	R, N	R, N	R, N	R, N
G2018C	411C	12; 500; 502	R, N	R, N	–	–
NARM-502	NARM-502	13; 503	R, N	–	–	–
THR-01	THR-01	12	–	–	N	N
THR-04	THR-04	502	R, N	R, N	R, N	–
HCFC-22/ HCFC-142B	–	12	R, N	–	R, N	R, N

(*continued overleaf*)

Table 2.4 (*continued*)

Substitutes	Trade Name	ODS Replaced	Industrial Process Refrigeration	Ice Skating Rinks	Household Refrigerators	Household Freezers
CO_2	–	12; 13; 13B1; 502; 503	R, N	–	R, N	–
Ammonia Refrigeration	–	12; 502	R, N	R, N	–	–
Ammonia Absorption	–	12	–	–	N	N
Propane; Propylene; Butane; HC Blend A, B	HC-12a; OZ-12	ALL	R, N[a]	–	–	–
Chlorine	–	ALL	R, N	–	–	–
Evaporative/ Desiccant Cooling	–	ALL	N	–	–	–

R, retrofit uses; N, new uses.
[a]Prohibited for other end uses.
Source: US Environmental Protection Agency, EPA (2009).

As used, the term *refrigerant* means any class II substance used for heat transfer in a refrigerating system. From January 1, 2015, it shall be unlawful for any person to produce any class II substance in an annual quantity greater than the quantity of such substance produced by such person during the baseline year. From January 1, 2030, it shall be unlawful for any person to produce any class II substance, EPA (2009). Before December 31, 2002, the Administrator will promulgate regulations phasing out the production, and restricting the use, of class II substances in accordance with this section, subject to any acceleration of the phaseout of production. The Administrator also promulgates regulations to insure that the consumption of class II substances in the United States is phased out and terminated in accordance with the same schedule (subject to the same exceptions and other provisions) as is applicable to the phaseout and termination of production of class II substances under this title.

2.9 Alternative Refrigerants

New, alternative substances are required to replace the fully halogenated refrigerants that are believed to contribute to atmospheric ozone depletion. In the past decade, many research and development studies on the synthesis and characterization of alternative refrigerants were undertaken. The replacement of restricted ODSs by any alternative may involve substantial changes in the design of various components such as insulation, lubricants, heat exchangers, and motors. Tests must be done to optimize the system performance and to ensure the reliability and safety of the system. Several alternative refrigerants are already available on the market. Several people (e.g., Lorentzen, 1993) have suggested natural refrigerants, that is, ammonia, propane, and CO_2 to replace ODSs. Here, we present R-134a as the most common alternate as well as some other potential replacements, including the natural substances, for ODSs, although there are so many alternates available ranging from application to application.

2.9.1 *R-134a*

R-134a is an HFC refrigerant which has a boiling temperature of −26.2 °C (−29.8 °C for R-12) and a latent heat of 205 kJ/kg (159 kJ/kg for R-12) (Pearson, 1991). This is a nonflammable and nontoxic substitute for R-12. R-134a has been widely used in household refrigerators and in automotive air conditioning, but there appears to be little benefit in using it in conventional air conditioning or refrigeration where reasonable condensing temperatures can be specified. R-134a is suggested as a potential replacement for R-22 in packaged systems. However, the volumetric displacement of a compressor for R-134a must be about 50% larger than the displacement of an R-22 compressor of the same cooling capacity. Pressure drops in refrigerant tubing can have a significant effect on the COP of an R-134a system, so larger tubing may be needed than in R-22 systems. R-134a packaged systems will tend to be physically larger and more costly than their R-22 counterparts (Hickman, 1994). Currently, R-134a is not considered for wide use in reciprocating chillers; however, it is offered by several manufacturers as an alternative to R-22 for screw and centrifugal chillers with flooded evaporators.

R-134a is also acceptable as a substitute for R-400 ((60/40)% by weight) and R-114 in new industrial process air conditioning. EPA recommends that R-134a only be used where ambient temperatures are lower than 70 °C because of very high system pressures. R-134a does not contribute to ozone depletion. R-134a's GWP and atmospheric lifetime are close to those of other alternatives which are acceptable in this end use. While R-134a is compatible with most existing refrigeration and air-conditioning equipment parts, it is not compatible with the mineral oils currently used in such systems. An appropriate ester-based, polyalkylene glycol-based, or other type of lubricant should be used.

2.9.1.1 Guidance on Retrofitting to R-134a

Some systems have a device that automatically releases refrigerant to the atmosphere to prevent extremely high pressures. When retrofitting any system with such a device to use a new refrigerant, a high-pressure shut-off switch must be installed. This switch will prevent the compressor from increasing the pressure to the point where the refrigerant is vented.

The term *retrofit* describes special procedures required to convert an R-12 system to use an alternative refrigerant. This section describes some facts about aftermarket options and procedures for retrofitting air conditioning and refrigeration systems to R-134a. As known, R-134a is chosen worldwide to be the long-term replacement for R-12 in air conditioning and refrigeration systems.

Manufacturers (also known as original equipment manufacturers) have developed retrofit kits or guidelines for some of their models. These procedures were designed to provide the best level of performance with the new R-134a system. Although using these kits and guidelines will provide the greatest assurance that comparable system performance will be achieved, the costs of these procedures will in many instances be relatively high. Procedures required for a least-cost retrofit are simple and do not require major component changes. Generally, the process calls for removal of the old refrigerant, installation of new fittings and a new label, and the addition of either a polyalkylene glycol (PAG) or polyester (POE or ester) lubricant as well as the R-134a refrigerant.

According to EPA regulations (EPA, 2009), the use of any alternative refrigerant to replace R-12 requires at a minimum that

- unique service fittings be used in order to minimize the risk of cross-contamination of either the refrigeration system or the service facility's recycling equipment;
- the new refrigerant be identified by a uniquely colored label in order to identify the refrigerant in the system;
- all R-12 be properly removed from the system before filling the system with an alternative refrigerant;

- in order to prevent release of refrigerant to the atmosphere, a high-pressure compressor shut-off switch be installed on any system equipped with a pressure relief device; and
- separate equipment be used to recover the R-12 from the system.

Also, alternative refrigerant blends that contain R-22 must be used with barrier hoses.

2.9.1.2 Technical Aspects on Retrofitting to R-134a

The following technical aspects are of practical importance in retrofitting applications and should be taken into consideration wherever applicable.

- **Toxicity, flammability, corrosion.** R-134a is regarded as one of the safest refrigerants yet introduced, based on current toxicity data. The chemical industry's program for Alternative Fluorocarbon Toxicity Testing (PAFT) tested R-134a in a full battery of laboratory animal toxicity studies. The results indicate that R-134a does not pose cancer or birth defects hazard. In addition, R-134a is being used in metered dose inhalers in Europe. The flammability and corrosivity of each potential R-12 substitute has been examined by the chemical manufacturers and various institutes. Like R-12, R-134a is not flammable at ambient temperatures and atmospheric pressures. Some mixtures of air and R-134a have been shown to be combustible at elevated pressures. These mixtures may be potentially dangerous, causing injury or property damage. R-134a is not corrosive on standard steel, aluminum, and copper samples.
- **Handling.** When handling R-134a, as with any other chemical, service staff should be sure to work in a well-ventilated area. It is never a good idea to inhale any vapor to such an extent that it replaces the oxygen in the lungs.
- **Charging into the system.** The amount of R-134a charged into the system should normally be 80–90% of the amount of R-12 in the system. Most system manufacturers provide guidelines regarding the amount of R-134a to be used.
- **Lubricants (PAGs versus esters).** The mineral oil used with R-12 cannot be sufficiently transported throughout the system by R-134a. Manufacturers tested both PAGs and esters for refrigerant/lubricant miscibility, lubricity, chemical stability, and materials compatibility. In the process of developing recommendations, they also considered the additives and conditioners present in the oils. Most chose to use PAG lubricants in new systems equipped with R-134a, and also recommended PAG lubricants for retrofits. Some compressor manufacturers are shipping new compressors with PAGs, some with esters, and some are shipping them empty. PAGs are hygroscopic, which means that they will draw water from the atmosphere when exposed. Many aftermarket specialists are choosing to use ester lubricants because they believe that the hygroscopic characteristics of PAGs may limit their lubricating ability and may introduce corrosion into the system. Esters are also hygroscopic (although less so than PAGs), and care must still be taken to ensure that excess moisture does not go into the system. It is good practice to use PVC-coated gloves (or, if that is impractical, barrier creams) and safety goggles when handling these lubricants, since prolonged skin contact and/or even brief eye contact can cause irritations such as stinging and burning sensations. One should also avoid breathing vapors produced by the lubricants, and make sure to use them in well-ventilated areas and keep both PAGs and esters in tightly sealed containers, both so that humidity does not contaminate the oil and so that vapors do not escape.
- **Flushing.** The amount of mineral oil that can safely remain in a system after retrofitting, without affecting performance, is still being debated. It was originally thought that any mineral oil left in the system might cause system failure. As long as the technical staff has removed as much of the old mineral oil as possible, any residual R-12 left in the system should not have a significant effect on the performance of the system. Removing the mineral oil may require draining certain components.

- **Hoses and O-rings.** When R-134a was first introduced, it was thought that all nonbarrier/nitrile hoses would have to be replaced during a retrofit. Early laboratory tests showed that the small R-134a molecules leaked through the walls of nonbarrier hoses more readily than the larger R-12 molecules did. In the lab, this caused unacceptably high leakage rates. More recent testing, however, has shown that oil used in automotive air-conditioning systems is absorbed into the hose to create a natural barrier to R-134a permeation. In most cases, the R-12 system hoses will perform well, provided they are in good condition. Cracked or damaged hoses should always be replaced with barrier hoses. Unless a fitting has been disturbed during the retrofit process, replacement should not be necessary. Most retrofit instructions call for lubricating replaced O-rings with mineral oil to provide this protection.
- **Compressors.** Industry experts once thought that a retrofit would require compressor replacement. This belief helped create some of the horror stories about the expense of retrofitting. Now it is routinely accepted that most compressors that are functioning well in R-12 systems will continue to function after the systems have been retrofitted. When a compressor is first run with R-12, a thin film of metal chloride forms on the bearing surfaces and acts as an excellent antiwear agent. This film continues to protect after the system has been converted to R-134a. This helps explain why a new R-12 compressor may fail more quickly if it is installed in an R-134a system without the benefit of a break-in period on R-12. A few older compressors use seals that are not compatible with either R-134a or the new lubricants. The compressor manufacturer can identify which compressors need special attention.
- **Condensers and pressure cutout switches.** When retrofits were first studied several years ago, it was thought that the condenser and perhaps the evaporator would have to be replaced to maintain an acceptable level of cooling performance on a retrofitted system. Now, it is generally accepted that if an R-12 system is operating within the manufacturer's specifications, there may be no need to replace either part. It is true, however, that the higher vapor pressures associated with R-134a may result in lost condenser capacity. When retrofitting, one should consider how the airflow and condenser design on the particular vehicle will affect the success of the retrofit. It should be noted that bent, misshapen, or improperly positioned airflow dams and directors may affect performance. In addition, systems that are not equipped with a high-pressure cutout switch should have one installed to prevent damage to system parts and to prevent refrigerant emissions. It is recommended that the installation of a high-pressure cutout switch will shut off the compressor when high pressures are encountered, reducing the possibility of venting the refrigerant and overheating the engine cooling system.
- **Control devices.** Refrigerant controls (whether they are orifice tubes or expansion valves that meter refrigerant flow, or pressure cycling switches or other pressure controls designed to protect against freezing) may have to be changed during the course of a retrofit.

2.9.2 R-123

R-123 is an HCFC which has a boiling point of 27.1 °C (23.8 °C for R-11), a latent heat of 175 kJ/kg at 15 °C (194 kJ/kg for R-11), and a molecular weight of 153. It is suitable for use in chiller systems with centrifugal compressors, air conditioning, refrigeration, and heat pump systems. R-123's environmental suitability as a replacement for R-11 is not in doubt. It has an ODP and a GWP less than 0.02, compared with 1.0 for R-11, and does not present any flammability problem. About its toxicology, R-123 was assigned an allowable exposure limit of 100 ppm (Anon, 1991). This means that a worker should not be continuously exposed to more than 100 ppm of R-123 during any working day of 8–12 hours. Regarding the safety issue, the requirements for handling both R-123 and R-11 refrigerants in charging and servicing equipment are the same. It was suggested that the paraffinic oils which are used with R-11 could also be used with R-123.

Despite the concern that R-123 is not safe to use in centrifugal chillers and causes cancer in rats, EPA lists it as acceptable. Although all refrigerants pose a certain amount of toxicity, R-123 actually poses less acute risk than R-11. In other words, in the event of a major leak, it is safer with R-123 than with R-11. As for chronic toxicity, it is true that R-123 caused tumors in several organs in rats, including the testes. However, the critical facts are that in all cases (i) the tumors were benign, (ii) they only appeared after long exposures to very high concentrations, (iii) the tumors were never life-threatening, and (iv) the exposed rats actually lived longer at these higher concentrations. The acceptable exposure limit set by the R-123 manufacturers is 30 ppm. This represents the concentration to which a worker could be exposed for 8 hours/day for a working lifetime without effects. EPA conducted a study to determine the typical exposure level found in actual equipment rooms. The study concluded that if appropriate measures are taken (for instance, complying with ASHRAE 15), the concentration of R-123 can be kept below 1 ppm. EPA believes that R-123 is a necessary transition refrigerant as the world phases out the CFCs and SNAP lists it as acceptable for use in chillers. It is safe to be used in the long term, and is actually safer in emergencies than R-11.

2.9.3 Nonazeotropic (Zeotropic) Mixtures

Table 2.5 shows refrigerant mixtures in the 400 series (zeotropic mixtures). Pederson 2001 calculated their GWP values on the basis of the values in the table for single substances, weighting on the basis of the mix ratio between the individual substances.

Table 2.5 Some common zeotropic mixtures in the 400 series.

R-number	Substances	GWP	Concentration (weight %)
R-401A	HCFC-22/HFC-152a/HCFC-124	1082	53/13/34
R-402A	HCFC-22/HFC-125/HC-290	2326	38/60/2
R-403A	HCFC-22/PFC-218/HC-290	2675	75/20/5
R-403B	HCFC-22/PFC-218/HC-290	3682	56/39/5
R-404A	HFC-143a/HFC-125/HFC-134a	3260	52/44/4
R-406A	HCFC-22/HC-600a/HCFC-142b	1755	55/4/41
R-407C	HFC-32/HFC-125/HFC-134a	1526	23/25/52
R-408A	HCFC-22/HFC-143a/HFC-125	2743	47/46/7
R-409A	HCFC-22/HCFC-142b/HCFC-124	1440	60/15/25
R-410A	HFC-32/HFC-125	1725	50/50
R-412A	HCFC-22/HCFC-142b/PFC-218	2040	70/25/5
R-413A	HFC-134a/PFC-218/HC-600a	1774	88/9/3
R-414A	HCFC-22/HCFC-124/HCFC-142b/HC-600a	1329	51/28.5/16.5/4
R-415A	HCFC-22/HFC-23/HFC-152a	1966	80/5/15

Source: Pederson (2001).

In the 400 series, R-401A and R-401B are acceptable as substitutes for R-11, R-12, R-500, and R-502 in the following end uses:

- new and retrofitted reciprocating chillers,
- new industrial process refrigeration,
- new cold storage warehouses,
- new refrigerated transport,
- new retail food refrigeration,
- new commercial ice machines,
- new vending machines,
- new water coolers,
- new air coolers,
- new household refrigerators,
- new household freezers, and
- new residential dehumidifiers.

R-401A and R-401B appear to be acceptable as a substitute for R-400 ((60/40)% by weight) and R-114 in retrofitted industrial process air conditioning. Note that different temperature regimes may affect the applicability of the above substitutes within these end uses.

R-404A is acceptable as a substitute for R-12 in new household refrigerators. None of this blend's constituents contains chlorine, and thus this blend poses no threat to stratospheric ozone. However, R-125 and R-143a have very high GWPs, and the GWP of HFC-134a is somewhat high. In addition, Pearson (1991) suggested R-410A to replace R-22 in high-pressure unitary air-conditioning applications.

2.9.4 *Azeotropic Mixtures*

During the past several years, a number of new azeotropic refrigerant mixtures have been proposed as substitutes to replace harmful CFCs, for example, R-22 in air-conditioning systems and R-12, R-22, and R-502 in household and industrial refrigeration systems. Table 2.6 lists some common azeotropic refrigeration mixtures in the 500 series. Among these, R-507 is acceptable as a substitute for R-12 in new household refrigerators and for R-502 in cold store plants. It is an azeotropic blend by weight of 53% R-125 and 47% R-134a at atmospheric pressure. At lower temperatures, the azeotropic range of the blend may vary from 40 to 60% R-134a by weight (Konig, 1996). Note

Table 2.6 Some common azeotropic mixtures in the 500 series.

R-number	Substances	GWP	Concentration (weight %)
R-502	CFC-115/HCFC-22	5,576	51/49
R-507	HFC-143a/HFC-125	3,300	50/50
R-508A	HFC-23/PFC-116	10,175	39/61
R-508B	HFC-23/PFC-116	10,350	46/54
R-509A	HCFC-22/PFC-218	4,668	44/56

Source: Pederson (2001).

that neither R-125 nor R-134a contains any chlorine, and ester oils can be used for lubrication. In terms of material compatibility (i.e., corrosiveness) with seals and metals including copper and aluminum, R-507 is comparable to R-502 and retrofitting R-502 refrigeration plants for R-507 can easily be carried out in accordance with the already familiar procedures for conversion from R-12 to R-134a. None of this blend's constituents contains chlorine, and thus this blend poses no threat to stratospheric ozone. However, R-125 and R-143a have very high GWPs. In many countries, recycling and reclamation of this blend is strongly recommended to reduce its direct global warming impact. Although R-143a is flammable, the blend is not. Leak testing has demonstrated that its composition never becomes flammable.

It is important to mention that substituting HFC refrigerants for CFC (or HCFC) refrigerants in household refrigerator and freezers, auto air conditioners, and residential air-conditioning systems with compressors located below the evaporator and no liquid receivers has a high probability of success with the original mineral or alkyl benzene oil (Kramer, 1999). The application of HFC refrigerants to receiver-based refrigeration systems and systems with suction risers has the greatest probability of success where reduced viscosity alkyl benzene or mineral oils are used along with effective oil separators, with the suction riser design implementing appropriate refrigerant vapor velocity.

2.9.5 Ammonia (R-717)

This has been the most widely used one among the classic alternative refrigerants. Two characteristics of R-717, the saturation pressure–temperature relationship and the volume flow rate per unit refrigeration capacity, are quite similar to those of R-22 and R-502. On the other hand, R-717 has some advantages over R-22 and R-502, such as lower cost, better cycle efficiency, higher heat-transfer coefficients, higher critical temperature, greater detectability in the event of leaks, lower liquid pumping costs for liquid recirculation systems, more tolerance of water contamination, more favorable behavior with oil, zero ODP and GWP, and smaller refrigerant piping.

After 120 years of extensive usage, a tremendous amount of practical experience exists with this refrigerant. There is no doubt about its excellent thermodynamic and transfer properties, much superior to those of the halocarbons, and its important practical advantages such as tolerance to normal lubricating oils and limited pollution with water, easy leak detection, and low price. All these factors contribute to its sustained popularity and wide application, in spite of the often expressed doubts about its safety.

It is true that ammonia is poisonous and can burn with air, although it is quite difficult to ignite and will hardly sustain a flame by itself. The risk is strongly counteracted by the fact that it has an extremely strong odor and that it is much lighter than air. A leak is easily detected by smell at a concentration far below a dangerous level, and a massive escape of ammonia rapidly disappears upward in the atmosphere. Accidents are therefore extremely rare.

It has not been possible to find any reliable statistics on refrigeration accidents. Serious cases are reported from time to time, and it is the impression that they are fairly evenly divided between ammonia and halocarbon plant (Lorentzen, 1993). A recent effort to survey the situation in Norway has brought to attention a considerable number of incidents, some of them quite serious. Over a 20-year period, altogether four people were killed by refrigerants, two each by ammonia and R-22. One further died in a nitrogen-cooled truck. Great caution and sound professionalism are clearly needed in design, erection, and operation of large refrigeration systems, regardless of the refrigerant used.

Whenever there is a need to avoid any risk of ammonia smell, this can easily be arranged by enclosing the plant in a reasonably gastight room or casing, absorbing any fumes in a water spray or venting them over the roof in a safe place. A secondary refrigerant (brine) has to be used for distribution outside the enclosure. In this way it is possible to use this excellent medium safely in nearly all practical applications.

Ammonia has been used successfully for generations and has demonstrated its reliability and superior thermodynamic and transport properties. It can be applied to advantage over practically the whole field of medium and large refrigeration systems, when the necessary safety requirements are taken care of by simple and obvious means. The total consumption for this purpose will only be a tiny fraction of the ammonia production for other uses.

2.9.6 Propane (R-290)

Propane has been used as a working medium in large refrigeration plants for many years, notably in the petrochemical processing industry. It has excellent thermodynamic properties approaching those of ammonia, but the explosion and fire hazard are much more severe. This will certainly limit its application in the normal refrigeration field, although the risk should not be overestimated. Combustible gases are commonplace in many technical applications and do not cause many problems when simple precautions are observed. Propane has a special advantage for use in turbo compressors because of its near ideal molar mass.

Another area where propane or mixtures of HCs may have a future is as a substitute for R-12 in household refrigerators and freezers, and also perhaps in small air-conditioning units. The charge in a normal refrigerator need not be more than about 50 g, propane being a much lighter fluid than R-12, and this is less than what may be used as the drive gas in a hair spray. Recent extensive studies have testified that the risk involved is negligible (Lorentzen, 1993).

The few hundred grams needed in a small air conditioner would also seem to be perfectly acceptable with proper design. A propane–butane mixture may be advantageous in order to achieve a temperature glide to match the limited air volume flow of the evaporator and condenser.

In a study undertaken by James and Missenden (1992), the implications of using propane in domestic refrigerators were examined in relation to energy consumption, compressor lubrication, costs, availability, environmental factors, and safety, and compared with the results obtained for R-12, R-22, and R-134a. From the experiments they found that propane can substitute for R-12 with a similar performance at a lower charge and concluded that propane is an attractive and environmentally friendly alternative to ODSs. Although propane is very stable and meets the refrigeration requirements (e.g., COP, pressure ratio, comparative discharge, etc.) easily, the only concern is its reactive characteristics such as combustion and halogenation.

2.9.7 CO_2 (R-744)

Carbon dioxide was a commonly used refrigerant from the late nineteenth and well into the twentieth century. Owing to its complete harmlessness it was the generally preferred choice for usage onboard ships, while ammonia was more common in stationary applications. With the advent of the "Freons" and R-12 in the first place, CO_2 was rapidly abandoned, and it has nearly been forgotten in the course of the last 40–50 years. The main reasons for this development were certainly the rapid loss of capacity at high cooling water temperatures in the tropics, and, not the least, the failure of the manufacturers to follow modern trends in compressor design toward more compact and price-effective high-speed types. Time is now ripe for a reassessment of this refrigerant for application with present-day technology.

CO_2 is naturally present everywhere in our environment. Air contains about 0.35 parts per thousand of it, in total nearly 300 billion tons for the whole atmosphere, and several hundred billion tons per year circulate in the living biosphere. No complicated and time-consuming research is needed to ascertain its complete harmlessness.

One may possibly object that CO_2 is also a greenhouse gas, and this is of course correct, although its GWP is defined as 1, and the GWPs of other refrigerants are indexed to it. But in reality, gas will be used, which is already available as a waste product in unlimited quantity from other activities. What we do is just to postpone its release. This is in principle good for the environment, like planting a tree to bind carbon for a period of time.

With regard to personal safety, CO_2 is at least as good as the best of halocarbons. It is nontoxic and incombustible, of course. On release from the liquid form about half will evaporate while the remainder becomes solid in the form of snow and can be removed with broom and dustpan, or just left to sublimate. Most people are already familiar with the handling of "dry ice." In the case of accidental loss of a large quantity, a good ventilation system is required to eliminate any risk of suffocation, in particular, in spaces below ground level. In this respect, the situation is the same as for any large halocarbon plant.

It is sometimes asserted that the high pressure of CO_2 could constitute a special danger in the case of accidental rupture. Actually this is not so since the volume is so small. The explosion energy, similar to the product $P \times V$, is approximately the same for all systems with the same capacity, regardless of the refrigerant used.

CO_2 also has a number of further advantages such as the following:

- nonflammable, nonexplosive, nontoxic,
- low cost and good availability,
- having zero ODP and 1 as GWP,
- thermal stability,
- pressure close to the economically optimal level,
- greatly reduced compression ratio compared to conventional refrigerants,
- complete compatibility to normal lubricants and common machine construction materials,
- easy availability everywhere, independent of any supply monopoly, and
- simple operation and service, no "recycling" required, very low price.

Its only technical disadvantage is the high triple-point temperature and the low critical temperature. Therefore, CO_2 as a pure substance cannot be an alternative refrigerant.

CO_2 is a refrigerant with a great potential for development of energy- and cost-effective systems. Examples have demonstrated this for some applications and appropriate technology for other fields will certainly be found. This substance comes very close to the ideal refrigerant and a rapid revival of this popularity for usage over a wide field can be expected. CO_2 now appears to be a substitute for the following:

- R-13, R-13B1, and R-503 in very low-temperature refrigeration (retrofit and new),
- R-13, R-13B1, and R-503 in industrial process refrigeration (retrofit and new), and
- R-11, R-12, R-113, R-114, and R-115 in nonmechanical systems (retrofit and new).

2.10 Selection of Refrigerants

In the selection of an appropriate refrigerant for use in a refrigeration or heat pump system, there are many criteria to be considered. Briefly, the refrigerants are expected to meet the following conditions (Dincer, 1997):

- ozone- and environment friendly,
- low boiling temperature,
- low volume of flow rate per unit capacity,
- vaporization pressure lower than atmospheric pressure,
- high heat of vaporization,
- nonflammable and nonexplosive,

- noncorrosive and nontoxic,
- nonreactive and nondepletive with the lubricating oils of the compressor,
- nonacidic in case of a mixture with water or air,
- chemically stable,
- suitable thermal and physical properties (e.g., thermal conductivity, viscosity),
- commercially available,
- easily detectable in case of leakage, and
- low cost.

When selecting the refrigerant, the saturation properties of the refrigerant should be taken into account. To have heat transfer at a reasonable rate, a temperature difference of 5 to 10 °C should be maintained between the refrigerant and the medium with which it is exchanging heat. If a refrigerated space is to be maintained at 0 °C, for example, the temperature of the refrigerant should remain at about −10 °C while it absorbs heat in the evaporator. The lowest pressure in a refrigeration cycle occurs in the evaporator, and this pressure should be above atmospheric pressure to prevent any air leakage into the refrigeration system. Therefore, a refrigerant should have a saturation pressure of 1 atm or higher at −10 °C in this particular case. Ammonia and R-134a are two such substances. Also, the temperature (and thus the pressure) of the refrigerant on the condenser side depends on the medium to which heat is rejected. Lower temperatures in the condenser (thus higher COPs) can be maintained if the refrigerant is cooled by a lower temperature medium such as liquid water (Cengel and Boles, 2008).

Example 2.1

A refrigerator using R-134a is used to maintain a space at −10 °C while rejecting heat to a reservoir at 20 °C. If a temperature difference of 10 °C is desired, which evaporating and condensing pressures should be used?

Solution

The evaporating temperature should be (−10) − (10) = −20 °C and the condensing temperature should be 20 + 10 = 30 °C. Then from Table B.3,

$$P_{sat\,@\,-20\,^\circ C} = \mathbf{132.8\ kPa}\ \text{(evaporator pressure)}$$

$$P_{sat\,@\,30\,^\circ C} = \mathbf{770.6\ kPa}\ \text{(condenser pressure)}$$

Tables 2.7 through 2.10 tabulate the percentage compositions of refrigerant blends (for R-12, R-502, R-22, R-113, R-13B1, and R-503) found acceptable subject to narrowed use limits or acceptable subject to use conditions by the Clean Air Act of the United States, along with their trade names and ASHRAE numbers.

2.11 Thermophysical Properties of Refrigerants

It is essential to have sufficient knowledge and information on the thermophysical properties of the refrigerants and their mixtures in order to provide optimum system design and optimum operating conditions. As is obvious, an effective and efficient use of ecologically and environmentally friendly refrigerants can only be achieved if, among others, their thermophysical properties are known to a sufficiently high degree of accuracy.

Table 2.7 Percentage compositions of substitutes for R-12.

Trade Name	ASHRAE Number	HCFCs			HFCs			HCs	
		22	124	142b	134a	152a	227ea	Butane (R-600)	Isobutane (R-600a)
MP-39	401A	53%	34%	–	–	13%	–	–	–
MP-66	401B	61%	28%	–	–	11%	–	–	–
MP-52	401C	33%	52%	–	–	15%	–	–	–
GHG	406A	55%	–	41%	–	–	–	–	4%
FX-56	409A	60%	25%	15%	–	–	–	–	–
FRIGC FR-12	–	–	39%	–	59%	–	–	2%	–
GHG-HP	–	65%	–	31%	–	–	–	–	4%
Hot Shot	414B	50%	39%	9.5%	–	–	–	–	1.5%
GHG-X4	414A	51%	28.5%	16.5%	–	–	–	–	4%
Freeze 12	–	–	–	20%	80%	–	–	–	–
GHG-X5	–	41%	–	15%	–	–	40%	–	4%

Source: US Environmental Protection Agency, EPA (2009).

Table 2.8 Percentage compositions of substitutes for R-502.

Trade Name	ASHRAE Number	HCFC-22	HFCs					HCs	
			32	125	134a	143a	152a	Propane	Propylene
HP-80	402A	38%	–	60%	–	–	–	2%	–
HP-81	402B	60%	–	38%	–	–	–	2%	–
HP-62, FX-70	404A	–	–	44%	4%	52%	–	–	–
KLEA 407A	407A	–	20%	40%	40%	–	–	–	–
KLEA 407B	407B	–	10%	70%	20%	–	–	–	–
FX-10	408A	46%	–	7%	–	47%	–	–	–
R-411A	411A	87.5%	–	–	–	–	11%	–	1.5%
R-411B	411B	94%	–	–	–	–	3%	–	3%
G2018C	–	95.5%	–	–	–	–	1.5%	–	4%
AZ-50	507	–	–	50%	–	50%	–	–	

Source: US Environmental Protection Agency, EPA (2009).

Table 2.9 Percentage compositions of substitutes for R-22.

Trade Name	ASHRAE Number	HFC-32	HFC-125	HCFC-134a
KLEA 407C, AC9000	407C	23%	25%	52%
AZ-20, Puron, Suva 9100	410A	50%	50%	–
AC9100	410B	45%	55%	–

Source: US Environmental Protection Agency, EPA (2009).

Table 2.10 Percentage compositions of substitutes for CFC-113, R-13B1, and R-503.

Trade Name	ASHRAE Number	HCFC-22	HFC-23	HFC-152a	Propane	PFC-116 (Perfluoro-ethane)	PFC-218 (Perfluoro-propane)
R-403B	403B	56%	–	–	5%	–	39%
KLEA 5R3	508A	–	39%	–	–	61%	–
Suva97	508B	–	46%	–	–	54%	–
NARM-502	–	90%	5%	5%	–	–	–

Source: US Environmental Protection Agency, EPA (2009).

In general, thermophysical properties are conventionally classified into three categories such as equilibrium or thermodynamic properties, nonequilibrium or transport properties, and other miscellaneous properties including radiation, optical, and electrical properties. However, Watanabe and Sato (1990) provide an excellent categorized and prioritized grouping of thermophysical property data needed to permit an assessment of the environmentally acceptable refrigerants as follows:

- **Group Zero.** Normal boiling point, molecular structure;
- **Group I.** Vapor pressure, critical parameters, saturated liquid density, vapor-phase pressure–volume–temperature (PVT) properties, ideal gas heat capacity;
- **Group II.** Viscosity and thermal conductivity at saturated states, liquid-phase PVT properties, surface tension, dielectric strength;
- **Group III.** Liquid-phase heat capacity including saturated liquid, more extensive PVT properties in single-phase region, miscellaneous molecular properties such as dipole moment;
- **Group IV.** More extensive viscosity and thermal conductivity in single-phase region, velocity of sound, second virial coefficient, vapor-phase heat capacity, additional transport properties such as thermal diffusivity.

In recent years, increasing attention has been paid to research activities related to dynamic simulation of refrigeration systems which require a substantial amount of thermodynamic property evaluations. Cleland (1986) stated that many refrigeration design and simulation works mainly need the following:

- boiling (saturation) temperature from vapor pressure,
- vapor pressure from boiling (saturation) temperature,
- liquid refrigerant enthalpy from saturation temperature and liquid subcooling,

- vaporized refrigerant enthalpy from saturation temperature and vapor superheating,
- specific volume of vapor from saturation temperature and vapor superheating, and
- enthalpy change due to isentropic compression from vapor superheating prior to compression and suction and discharge pressures (or other equivalent saturation temperature).

In an experimental work undertaken by Heide and Lippold (1990), thermophysical properties of the refrigerants R-134a and R-152a in terms of dynamic viscosity, thermal conductivity, and surface tension were determined experimentally depending on the saturation temperature and pressure, and the following correlations were developed with high accuracy (i.e., average relative derivation less than 1.5%):

- For dynamics viscosity:

$$\ln \eta = 2.7847 - 0.1576T + 3.5324 \times 10^{-6}T^2 + 2.5718 \times 10^{-9}T^3 \text{ for R-134a and}$$

$$\ln \eta = 12.119 - 0.1215T + 3.8074 \times 10^{-4}T^2 - 4.3428 \times 10^{-7}T^3 \text{ for R-152a}$$

- For thermal conductivity:

$$\ln \lambda = 194.6 - 0.3626T \text{ for R-134a and}$$

$$\ln \lambda = 241.0 - 0.4523T \text{ for R-152a}$$

- For surface tension:

$$\sigma = 58.96(1 - T/374.26)^{1.276} \text{ for R-134a and}$$

$$\sigma = 59.23(1 - T/386.65)^{1.235} \text{ for R-152a}$$

where η is dynamic viscosity, mPa·s; λ is thermal conductivity, mW/mK; σ is surface tension, mN/m; and T is temperature, K.

2.12 Lubricating Oils and Their Effects

It is known that the lubricating oil contained in the crankcase of the compressor is generally in contact with the refrigerant. When oil dissolves in the refrigerant, it affects the thermodynamic properties of the refrigerant. The main effect is the reduction of the vapor pressure by the amount depending on the nature of the oil and the refrigerant and on how much oil dissolves. It is important to state that the refrigerants are expected to be chemically and physically stable in the presence of oil, so that neither the refrigerant nor the oil is adversely affected by the relationship.

For example, in ammonia systems the amount of oil in the solution with liquid ammonia is extremely small to cause any effect. However, with HC refrigerants the amount of oil in the solution is much larger and some HC refrigerants therefore react with the oils to some extent. The magnitude of the effect is dependent upon the operating conditions – at normal operating conditions with high-quality oil in dry and clean system the reaction becomes minor to cause any effect. However, if contaminants such as air and moisture are present in the system with low-quality oil, various problems may appear including decomposition of the oil and formation of corrosive acids and sludges. The other aspect is that high discharge temperatures accelerate such causes tremendously.

As far as the lubricating oil and refrigerant relationship are concerned, one of the differentiating characteristics for various refrigerants is the oil miscibility, which is defined as the ability of the refrigerant to be dissolved into the oil and vice versa. With reference to oil miscibility, refrigerants may be divided into three groups as follows (Dossat, 1997):

- those that are miscible with oil in all proportions under the conditions found in the refrigerating system,
- those that are miscible under the conditions normally found in the condensing section, but separate from the oil under the conditions normally found in the evaporator section, and
- those that are not miscible with oil at all (or only very slightly) under the conditions found in the system.

The viscosity of the lubricating oil is also one of the significant thermophysical aspects and should be maintained within certain limits in order to form a protective film between the various rubbing surfaces and to keep them separated. For example, if the viscosity is too low, the oil cannot do so; if it is too high, the oil cannot have enough fluidity to make the necessary penetration. In both cases, the lubrication of the compressor is not adequate. It is important to mention that in order to minimize the circulation of oil in the refrigerant, an oil separator or trap is sometimes installed in the discharge line of the compressor.

One of the more recent works (Kramer, 1999) addresses the fact that HFC refrigerants and blends are not miscible with mineral oil. It also explores factors that favor using or retaining mineral oil when retrofitting existing systems by discussing the potential problems arising from such use or retention. In practice, one of the big questions is why do we retain mineral oil? The following factors favor retaining or using mineral oil in the systems with HFC refrigerants:

- lower lubricant cost,
- direct refrigerant replacement,
- lower refrigerant solubility,
- improved working viscosity,
- reduced refrigerant charge,
- faster refrigeration on start,
- reduced slugging and oil carry-over on start (less need for pump down cycles or heaters),
- less distortion of the composition of refrigerant blends,
- reduced oil separator flooding,
- reduced hygroscopicity,
- reduced chemical reactivity,
- reduced electrical resistivity, and
- reduced dirt transfer.

The utilization of mineral oils with HFCs is preferred over the polyol ester lubricants because of several advantages including (Kramer, 1999)

- nonsticking suction need,
- better visual detection,
- water solubility and less environmental impact, and
- improved foaming characteristics (to promote bearing lubrication and reduce compressor noise).

2.13 Concluding Remarks

In this chapter, a large number of aspects related to refrigerants, alternative refrigerants, and their environmental impact have been studied for refrigeration systems and applications. Various technical criteria for selecting and evaluating alternative refrigerants are also presented.

Study Problems

Introduction

2.1 Are CFCs greenhouse gases? What is the effect of CFCs on greenhouse effect in comparison with carbon dioxide?

2.2 What is a refrigerant? What are the application areas of refrigerants?

Classification of Refrigerants

2.3 What are the main classifications of refrigerants?

2.4 What is a halocarbon? What are the most commonly used halocarbons in refrigeration applications?

2.5 What are the most commonly used CFCs?

2.6 What are the health effects of CFCs?

2.7 What are the advantages of HCs as alternative refrigerants? Which HCs are suitable as refrigerants?

2.8 What are the three most commonly used inorganic compounds as refrigerants? Which refrigerant has the highest COP and which one has the lowest COP? Which inorganic compound is used as refrigerant in aircraft air conditioning?

2.9 What is an azeotropic refrigerant? What is the most common azeotropic refrigerant?

2.10 What is a nonazeotropic refrigerant? Is there any other name for it?

Prefixes and Decoding of Refrigerants

2.11 What are the advantages of using prefixes and decoding of refrigerants?

2.12 What is the meaning of each letter in CFC and HCFC?

2.13 Are CFC-12 and R-12 the same refrigerant? What does "R" stand for?

2.14 Determine the number of atoms for each substance in HCFC-124 and HFC-152a.

2.15 Determine the number of atoms for each substance in CFC-12 and Halon 1301.

Secondary Refrigerants

2.16 What is a secondary refrigerant? What are the commonly used secondary refrigerants?

Refrigerant–Absorbent Combinations

2.17 Which refrigerant–absorber combinations are used in absorption refrigeration systems?

Stratospheric Ozone Layer

2.18 What is UVB? What are the harmful effects of UVB?

2.19 What is ozone? What is ozone layer? What is the effect of ozone layer on UVB?

2.20 What is column ozone? What is DU used for? What does a 300 DU mean?

2.21 What is the stratospheric ozone layer depletion? What are the ODS? What are the consequences of the stratospheric ozone layer depletion?

2.22 What are the common ozone-depleting substances? How do they deplete ozone?

2.23 Which substance in CFCs is responsible for ozone depletion? Do HCFCs and HFCs deplete ozone?

2.24 What is ODP? What is the ODP of R-11? What are the typical ODP ranges of CFCs, HCFCs, halons, and HFCs?

2.25 R134a is commonly used as refrigerant in household refrigerators. What is the ODP of R-134a?

2.26 What is the Montreal Protocol? What were the outcomes of this protocol?

Greenhouse Effect (Global Climate Change)

2.27 What are the greenhouse effect and the global warming? Which substances cause greenhouse effect?

2.28 What is GWP? What are the GWP of CO_2, R-11, R-12, and water?

2.29 What are the GWP and ODP characteristics of HCs and HFCs?

Clean Air Act

2.30 What is CAA? What were the outcomes of this act?

Alternative Refrigerants

2.31 Why are alternative refrigerants required?

2.32 Why was R-134a developed? What are the applications of R-134a?

2.33 What replacements are done during retrofitting to R-134a?

2.34 When a system using R-12 is retrofitted to R-134a, do compressor, condenser, and evaporator need to be replaced? What is the amount of R-134a that needs to be charged in comparison with the amount of R-12?

2.35 What are the suitable applications of R-123? It is used for the replacement of which refrigerant? What are the ODP and GWP of R-123?

2.36 What are the advantages of using ammonia as refrigerant?

2.37 Despite its superior characteristics as a refrigerant, why is ammonia not used in household refrigerators?

2.38 What are the dangers associated with using ammonia as a refrigerant? Assess the level of risks involved.

2.39 What are the suitable refrigeration applications of ammonia?

2.40 Compare propane to ammonia as a refrigerant in terms of thermodynamic properties and risks associated with its use. What have been the applications of propane?

2.41 It is known that carbon dioxide (CO_2) emission is responsible for at least 50% of greenhouse emissions. Do we need to be concerned about GWP of CO_2 when using it as a refrigerant? What is the ODP of CO_2?

Selection of Refrigerants

2.42 What are some criteria that need to be considered in the selection of a refrigerant?

2.43 A refrigerator using R-134a is used to maintain a space at −6 °C. Would you recommend an evaporator pressure of 140, 200, or 240 kPa? Why?

2.44 A refrigerator using R-134a is used to maintain a space at −6 °C while rejecting heat to a reservoir at 30 °C. Would you recommend a condenser pressure of 700, 850, or 1000 kPa? Why?

2.45 A refrigerator using R-134a is used to maintain a space at 0 °C while rejecting heat to a reservoir at 14 °C. If a temperature difference of 10 °C is desired, which evaporating and condensing pressures should be used?

2.46 A heat pump using R-134a is used to maintain a space at 25 °C while absorbing heat from a medium at 5 °C. If a temperature difference of 5 °C is desired, which evaporating and condensing pressures should be used?

2.47 The evaporator and condenser pressures of an R-134a refrigerator are 200 kPa and 600 kPa, respectively. Heat is rejected to lake water running through the condenser. If water enters the condenser at 12 °C, what is the maximum temperature rise of water in the condenser?

2.48 It is known that lower temperatures in the condenser (thus higher COPs) can be maintained if the refrigerant is cooled by a lower temperature medium such as liquid water. Based on this, would you recommend designing a household refrigerator with water cooling for the condenser? Explain.

Thermophysical Properties of Refrigerants

2.49 Determine surface tension of R-134a at −20, 0, and 20 °C.

Lubricating Oils and Their Effects

2.50 What is oil miscibility? How may refrigerants be grouped in terms of oil miscibility?

2.51 Compare the amount of oil in the refrigerant for ammonia and HCs.

References

Anon. (1991) R123 – A promising future. Refrigeration and Air Conditioning, July, 32–33.

Badr, O., Probert, S.D. and O'Callaghan, P.W. (1990) Chlorofluorocarbons and the environment: scientific, economic, social and political issues. Applied Energy, 37, 247–327.

Cengel, Y.A. and Boles, M.A. (2008) Thermodynamics: An Engineering Approach, McGraw-Hill, New York.

Cleland, A.C. (1986) Computer subroutines for rapid evaluation of refrigerant thermodynamic properties. International Journal of Refrigeration, 9, 346–351.

Dincer, I. (1992) Chlorofluorocarbons and environment: I (in Turkish). Bulten, 1 (1), 7–8.

Dincer, I. (1997) Heat Transfer in Food Cooling Applications, Taylor & Francis, Washington, DC.

Dincer, I. (2003) Refrigeration Systems and Applications, 1st edn John Wiley & Sons, Ltd., New York.

Dossat, R.J. (1997) Principles of Refrigeration, Prentice Hall, Englewood Cliffs, New Jersey.

EPA (2009) Significant New Alternatives Policy, under section 612 of the Clean Air Act Amendments, United States Environmental Protection Agency.

Heide, R. and Lippold, H. (1990) Thermophysical Properties of the Refrigerants R134a and R152a, Proceedings of the Meeting of I.I.R. Commissions B2, C2, D1, D2/3, September 24–28, Dresden, Germany, pp. 237–240.

Hickman, K.E. (1994) Redesigning equipment for R-22 and R-502 alternatives. ASHRAE Journal, 36, 42–47.

James, R.W. and Missenden, J.F. (1992) The use of propane in domestic refrigerators. International Journal of Refrigeration, 15 (2), 95–100.

Konig, H. (1996) Performance Comparison of R-507 and R-404A in a Cold Store Refrigeration Installation, Solvay Fluor und Derivate GmbH, Product Bulletin No: C/04.96/01/E.

Kramer, D.E. (1999) CFC to HFC conversion issues. Why not mineral oil? ASHRAE Journal, 41, 19–28.

Lorentzen, G. (1988) Ammonia, an excellent alternative. International Journal of Refrigeration, 11 (4), 248–252.

Lorentzen, G. (1993) Application of Natural Refrigerants, Proceedings of the Meeting of I.I.R. Commission B1/2, May 12–14, Ghent, Belgium, pp. 55–64.

Pearson, S.F. (1991) Which refrigerant? Refrigeration and Air Conditioning, July, 21–23.

Pederson, P.H. (2001) Ways of Reducing Consumption and Emission of Potent Greenhouse Gases (HFCs, PFCs and SF6), Project for the Nordic Council of Ministers, DTI Energy, Denmark.

Rowland, F.S. (1991) Stratospheric ozone depletion. Annual Review of Physical Chemistry, 42, 731–734.

Watanabe, K. and Sato, H. (1990) Thermophysical Properties Research on Environmentally Acceptable Refrigerants, Proceedings of the Meeting of I.I.R. Commission B1, March 5–7, Herzlia, Israel, pp. 29–36.

Wayne, R.P. (1991) Chemistry of Atmospheres, 2nd edn, Oxford University Press, Oxford.

References

3

Refrigeration System Components

3.1 Introduction

Refrigeration is the process of removing heat from matter which may be a solid, a liquid, or a gas. Removing heat from the matter cools it, or lowers its temperature. There are a number of ways of lowering temperatures, some of which are of historical interest only. In some older methods, lowering of temperature may be accomplished by the rapid expansion of gases under reduced pressures. Thus, cooling may be brought about by compressing air, removing the excess heat produced in compressing it, and then permitting it to expand.

A lowering of temperatures is also produced by adding certain salts, such as sodium nitrate, sodium thiosulfate (hypo), and sodium sulfite to water. The same effect is produced, but to a lesser extent, by dissolving common salt or calcium chloride in water.

As known, two common methods of refrigeration are natural and mechanical. In the natural refrigeration, ice has been used in refrigeration since ancient times and it is still widely used. In this natural technique, the forced circulation of air passes around blocks of ice. Some of the heat of the circulating air is transferred to the ice, thus cooling the air, particularly for air-conditioning applications. In the mechanical refrigeration, the refrigerant is a substance capable of transferring heat that it absorbs at low temperatures and pressures to a condensing medium; in the region of transfer, the refrigerant is at higher temperatures and pressures. By means of expansion, compression, and a cooling medium, such as air or water, the refrigerant removes heat from a substance and transfers it to the cooling medium.

In this chapter, we provide information on refrigeration system components (e.g., compressors, condensers, evaporators, throttling devices) and discuss various technical and operational aspects. Auxiliary refrigeration system components are also covered.

3.2 History of Refrigeration

For centuries, people have known that the evaporation of water produces a cooling effect. At first, they did not attempt to recognize and understand the phenomenon, but they knew that any portion of the body that became wet felt cold as it dried in the air. At least as early as the second century, evaporation was used in Egypt to chill jars of water, and it was employed in ancient India to make ice (Neuberger, 1930).

The first attempts to produce refrigeration mechanically depended on the cooling effects of the evaporation of water. In 1755, William Cullen, a Scottish physician, obtained sufficiently low temperatures for ice making. He accomplished this by reducing the pressure on water in

Refrigeration Systems and Applications İbrahim Dinçer and Mehmet Kanoğlu

a closed container with an air pump. At a very low pressure the liquid evaporated or boiled at a low temperature. The heat required for a portion of water to change phase from liquid to vapor was taken from the rest of the water, and at least part of the water remaining turned to ice. Since Cullen, many engineers and scientists have created a number of inventions for clarifying the main principles of mechanical refrigeration (Goosman, 1924). In 1834, Jacob Perkins, an American residing in England, constructed and patented a vapor-compression machine with a compressor, a condenser, an evaporator, and a cock between the condenser and the evaporator (Critchell and Raymond, 1912). He made it by evaporating under reduced pressure a volatile fluid obtained by the destructive distillation of India rubber. It was used to produce a small quantity of ice, but not commercially. Growing demand over the 30 years after 1850 brought great inventive accomplishments and progress. New substances, for example, ammonia and carbon dioxide, which were more suitable than water and ether, were made available by Faraday, Thilorier, and others, and they demonstrated that these substances could be liquefied. The theoretical background required for mechanical refrigeration was provided by Rumford and Davy, who had explained the nature of heat, and by Kelvin, Joule, and Rankine, who were continuing the work begun by Sadi Carnot in formulating the science of thermodynamics (Travers, 1946). Refrigerating machines appeared between 1850 and 1880, and these could be classified according to the substance (*refrigerant*). Machines using air as a refrigerant were called compressed-air or cold-air machines and played a significant role in refrigeration history. Dr John Gorrie, an American, developed a real commercial cold-air machine and patented it in England in 1950 and in America in 1951 (DOI, 1952).

Refrigerating machines using cold air as a refrigerant were divided into two types, closed cycle and open cycle. In the closed cycle, air confined to the machine at a pressure higher than the atmospheric pressure was utilized repeatedly during the operation. In the open cycle, air was drawn into the machine at atmospheric pressure and, when cooled, was discharged directly into the space to be refrigerated. In Europe, Dr Alexander C. Kirk commercially developed a closed-cycle refrigerating machine in 1862, and Franz Windhausen invented a closed-cycle machine and patented it in America in 1870. The open-cycle refrigerating machines theoretically outlined by Kelvin and Rankine in the early 1850s were invented by a Frenchman, Paul Giffard, in 1873 and by Joseph J. Coleman and James Bell in Britain in 1877 (Roelker, 1906).

In 1860, a French engineer, Ferdinand P. Edmond Carre, invented an intermittent crude ammonia absorption apparatus based on the chemical affinity of ammonia for water, which produced ice on a limited scale. Despite its limitations, it represented significant progress. His apparatus had a hand pump and could freeze a small amount of water in about 5 minutes (Goosman, 1924). It was widely used in Paris for a while, but it suffered from a serious disadvantage in that the sulfuric acid quickly became diluted with water and lost its affinity. The real inventor of a small, hand-operated absorption machine was H.A. Fleuss, who designed an effective pump for this machine. A comparatively large-scale ice-making absorption unit was constructed in 1878 by F. Windhausen. It operated continuously by drawing water from sulfuric acid with additional heat to increase the affinity (Goosman, 1924).

One of the earliest of the vapor-compression machines was invented and patented by an American professor, Alexander C. Twining, in 1853. He established an ice production plant using this system in Cleveland, Ohio, and could produce close to a ton per day. After that, a number of other inventors experimented with vapor-compression machines which used ether or its compounds (Woolrich, 1947). In France, F.P.E. Carre developed and installed an ether-compression machine and Charles Tellier (who was a versatile pioneer of mechanical refrigeration) constructed a plant using methyl ether as a refrigerant. In Germany, Carl Linde, financed by brewers, established a methyl ether unit in 1874. Just before this, Linde had paved the way for great improvements in refrigerating machinery by demonstrating how its thermodynamic efficiency could be calculated and increased (Goosman, 1924). Inventors of compression machines also experimented with ammonia, which became the most popular refrigerant and was used widely for many years. In the 1860s, Tellier developed an ammonia-compression machine. In 1872, David Boyle made satisfactory equipment for ice making and patented it in 1872 in America. Nevertheless, the most important figure in the development of ammonia-compression machines was Linde, who obtained a patent in 1876 for the one which was installed in Trieste brewery the following year. Later, Linde's model became very popular and was considered excellent in its mechanical details (Awberry, 1942). The use of ammonia in the compression refrigerating machines was a significant step forward. In addition to its thermodynamic advantage, the pressures it required were easy to produce, and machines which used it could be small in size. In the late 1860s, P.H. Van der Weyde of Philadelphia got a patent for a compression unit which featured a refrigerant composed of petroleum products (Goosman, 1924). In 1875, R.P. Pictet at the University of Geneva introduced a compression machine that used sulfuric acid. In 1866, T.S.C. Lowe, an American, developed refrigerating equipment that used carbon dioxide. Carbon dioxide compression machines became important, because of the gas' harmlessness, in installations where safety was the primary concern, although they were not used extensively until the 1890s (Awberry, 1942). Between 1880 and 1890, ammonia-compression installations became more common. By 1890, mechanical refrigeration had proved to be both practical and economical for the food refrigeration industry. Europeans provided most of the theoretical background for the development of mechanical refrigeration, but Americans participated vigorously in the widespread inventive activity between 1850 and 1880 (Dincer, 1997; 2003).

Steady technical progress in the field of mechanical refrigeration marked the years after 1890. Revolutionary changes were not the rule, but many improvements were made, in several countries, in the design and construction of refrigerating units, as well as in their basic components, compressors, condensers, and evaporators.

3.3 Main Refrigeration Systems

The main goal of a refrigeration system which performs the reverse effect of a heat engine is to remove the heat from a low-level temperature medium (*heat source*) and to transfer this heat to a higher level temperature medium (*heat sink*). Figure 3.1 shows a thermodynamic system acting as

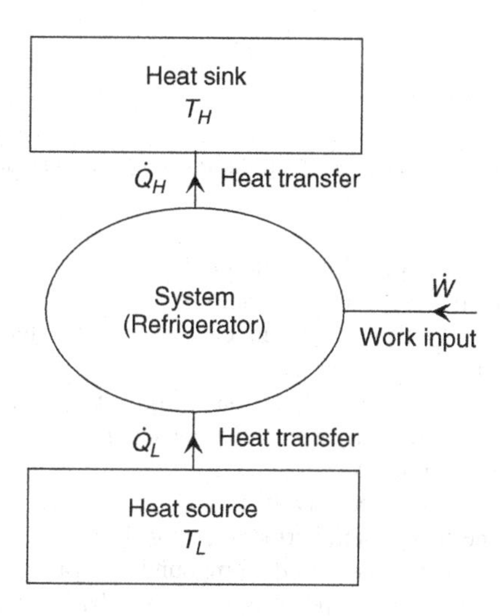

Figure 3.1 A thermodynamic system acting as a refrigerator.

a refrigeration machine. The absolute temperature of the source is T_L and the heat transferred from the source is the refrigeration effect (refrigeration load) Q_L. On the other side, the heat rejection to the sink at the temperature T_H is Q_H. Both effects are accomplished by the work input W. For continuous operation, the first law of thermodynamics is applied to the system.

Refrigeration is one of the most important thermal processes in various practical applications, ranging from space conditioning to food cooling. In these systems, the refrigerant is used to transfer the heat. Initially, the refrigerant absorbs heat because its temperature is lower than the heat source's temperature and the temperature of the refrigerant is increased during the process to a temperature higher than the heat sink's temperature. Therefore, the refrigerant delivers the heat.

In this chapter, the main refrigeration systems and cycles that we deal with are

- vapor-compression refrigeration systems,
- absorption refrigeration systems,
- air-standard refrigeration systems,
- jet ejector refrigeration systems,
- thermoelectric refrigeration, and
- thermoacoustic refrigeration.

Before commencing on these refrigeration systems, we first introduce the refrigeration system components and discuss their technical and operational aspects.

3.4 Refrigeration System Components

There are several mechanical components required in a refrigeration system. In this part, we discuss the four major components of a system and some auxiliary equipment associated with these major components. These components include condensers, evaporators, compressors, refrigerant lines and piping, refrigerant capacity controls, receivers, and accumulators.

Major components of a vapor-compression refrigeration system are as follows:

- compressor,
- condenser,
- evaporator, and
- throttling device.

In the selection of any component for a refrigeration system, there are a number of factors that need to be considered carefully, including

- maintaining total refrigeration availability while the load varies from 0 to 100%;
- frost control for continuous performance applications;
- variations in the affinity of oil for refrigerant caused by large temperature changes, and oil migration outside the compressor crankcase;
- selection of cooling medium: (i) direct expansion refrigerant, (ii) gravity or pump recirculated or flooded refrigerant, or (iii) secondary coolant (brines, e.g., salt and glycol);
- system efficiency and maintainability;
- type of condenser: air, water, or evaporatively cooled;
- compressor design (open, hermetic, semihermetic motor drive, reciprocating, screw, or rotary);
- system type (single stage, single economized, compound or cascade arrangement); and
- selection of refrigerant (note that the type of refrigerant is basically chosen based on operating temperature and pressures).

3.5 Compressors

In a refrigeration cycle, the compressor has two main functions within the refrigeration cycle. One function is to pump the refrigerant vapor from the evaporator so that the desired temperature and pressure can be maintained in the evaporator. The second function is to increase the pressure of the refrigerant vapor through the process of compression, and simultaneously increase the temperature of the refrigerant vapor. By this change in pressure the superheated refrigerant flows through the system.

Refrigerant compressors, which are known as the heart of the vapor-compression refrigeration systems, can be divided into two main categories:

- displacement compressors and
- dynamic compressors.

Note that both displacement and dynamic compressors can be hermetic, semihermetic, or open types.

The compressor both pumps refrigerant round the circuit and produces the required substantial increase in the pressure of the refrigerant. The refrigerant chosen and the operating temperature range needed for heat pumping generally lead to a need for a compressor to provide a high pressure difference for moderate flow rates, and this is most often met by a positive displacement compressor using a reciprocating piston. Other types of positive displacement compressor use rotating vanes or cylinders or intermeshing screws to move the refrigerant. In some larger applications, centrifugal or turbine compressors are used, which are not positive displacement machines but accelerate the refrigerant vapor as it passes through the compressor housing. These various compressor types are illustrated in Figure 3.2.

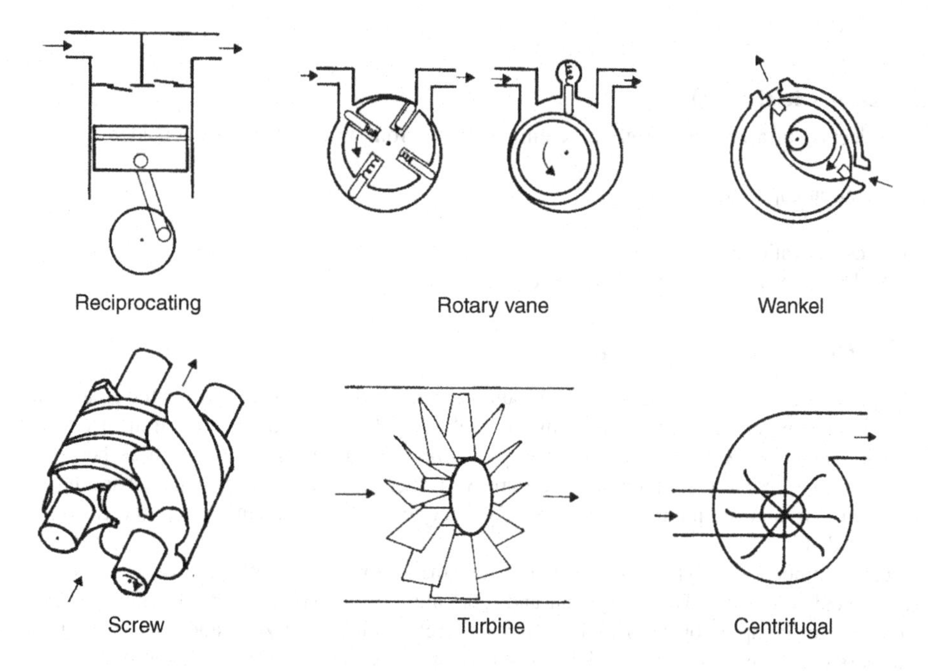

Figure 3.2 Compressor types (Heap, 1979).

In the market, there are many different types of compressors available, in terms of both enclosure type and compression system. Here are some options for evaluating the most common types (DETR, 1999):

- Reciprocating compressors are positive displacement machines, available for every application. The efficiency of the valve systems has been improved significantly on many larger models. Capacity control is usually by cylinder unloading (a method which reduces the power consumption almost in line with the capacity).
- Scroll compressors are rotary positive displacement machines with a constant volume ratio. They have good efficiencies for air conditioning and high-temperature refrigeration applications. They are only available for commercial applications and do not usually have inbuilt capacity control.
- Screw compressors are available in large commercial and industrial sizes and are generally fixed volume ratio machines. Selection of a compressor with the incorrect volume ratio can result in a significant reduction in efficiency. Part-load operation is achieved by a slide valve or lift valve unloading. Both types give a greater reduction in efficiency on part load than the reciprocating capacity control systems.

Expectations from the compressors

The refrigerant compressors are expected to meet the following requirements:

- high reliability,
- long service life,
- easy maintenance,
- easy capacity control,
- quiet operation,
- compactness, and
- cost effectiveness.

Compressor selection criteria

In the selection of a proper refrigerant compressor, the following criteria are considered:

- refrigeration capacity,
- volumetric flow rate,
- compression ratio, and
- thermal and physical properties of the refrigerant.

3.5.1 Hermetic Compressors

Compressors are preferable on reliability grounds to units primarily designed for the smaller range of temperatures required in air conditioning or cooling applications. In small equipment where cost is a major factor and on-site installation is preferably kept to a minimum, such as hermetically sealed motor/compressor combinations (Figure 3.3), there are no rotating seals separating motor and compressor, and the internal components are not accessible for maintenance, the casing being factory welded.

In these compressors, which arc available for small capacities, motor and drive are sealed in compact welded housing. The refrigerant and lubricating oil are contained in this housing. Almost all small motor-compressor pairs used in domestic refrigerators, freezers, and air conditioners are of the hermetic type. An internal view of a hermetic-type refrigeration compressor is shown in Figure 3.3. The capacities of these compressors are identified with their motor capacities. For

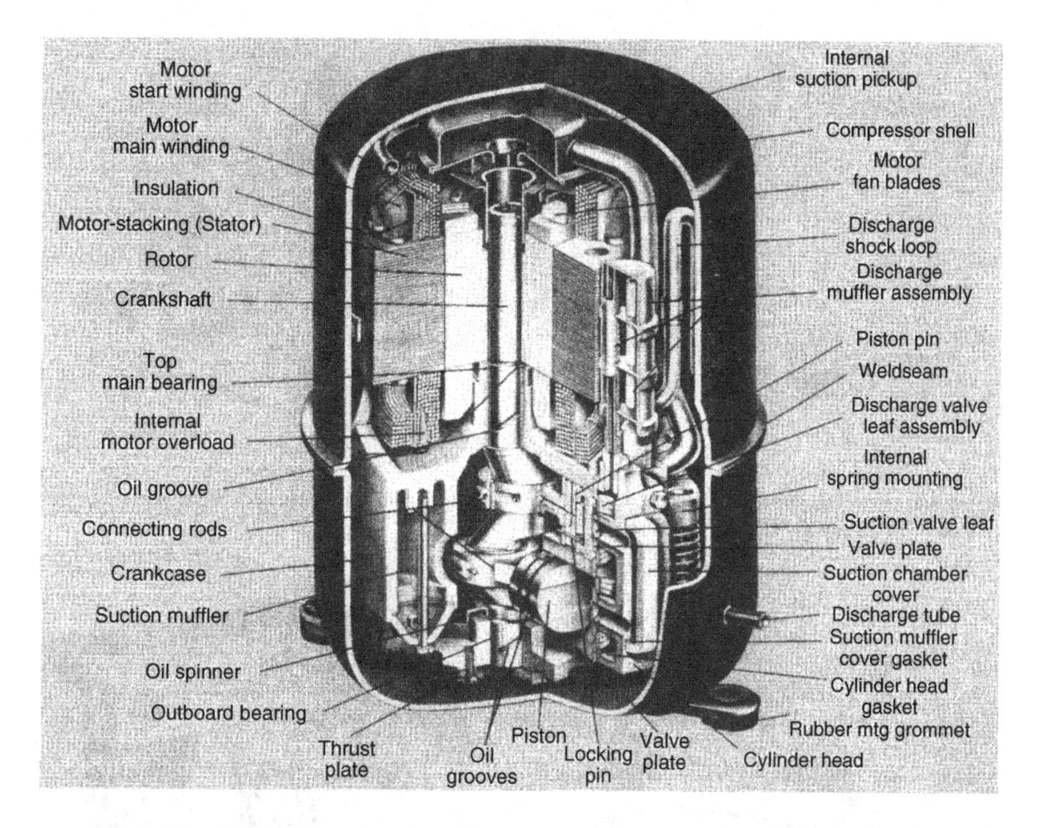

Figure 3.3 A typical hermetic reciprocating compressor (*Courtesy of Tecumseh Products Co*).

example, the compressor capacity ranges from 1/12 HP to 30 BG in household refrigerators. Their revolutions per minute are either 1450 or 2800 rpm. Hermetic compressors can work for a long time in small-capacity refrigeration systems without any maintenance requirement and without any gas leakage, but they are sensitive to electric voltage fluctuations, which may make the copper coils of the motor burn. The cost of these compressors is very low. Also, Figure 3.4 shows two air-cooled condensing units using a hermetic-type refrigeration compressor.

3.5.2 Semihermetic Compressors

In larger sizes, refrigeration compressors are often semihermetic, that is, although motor and compressor are within one casing, this casing may be unbolted, and the refrigerant does not flow over the motor windings. Access for maintenance is straightforward, but the need for external motor cooling which aids efficiency in cooling applications is no advantage in refrigeration operations, and the cost is substantially higher than for hermetic units. As large motors are more efficient than small ones, overall efficiencies of up to 70% or more are theoretically possible, and in multicylinder compressors, capacity may be controlled by making one or more cylinders ineffective (e.g., by holding the inlet valve open). Cylinder unloading at start-up is also a convenient way of reducing starting torque.

These compressors (single or double acting) were developed to avoid the disadvantages of the hermetic compressors. Semihermetic compressors are identical to the hermetic types, but the motor

Figure 3.4 New, high-efficient compact coil air-cooled condensing units using hermetic compressors (*Courtesy of Tecumseh Products Co*).

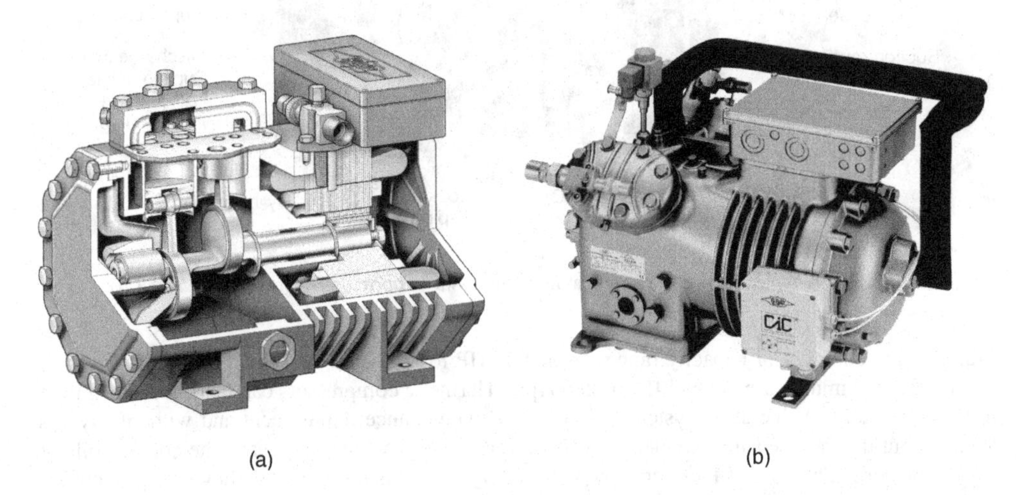

Figure 3.5 Semihermetic reciprocating compressors. (a) Single stage. (b) Two stage (*Courtesy of Bitzer Kühlmaschinenbau GmbH*).

and compressor are constructed in a fabricated enclosure with bolted sections or access panels to facilitate servicing. These compressors are manufactured in small and medium capacities and their motor capacities can reach 300 kW. For this reason they are cheap and another advantage is that they are compact. Also, they do not have a leakage problem. Figure 3.5 shows new-type semihermetic reciprocating compressors for medium- and low-temperature commercial refrigeration applications. These compressors are available for alternative refrigerants (e.g., R-134a, R-404A, and R-507). Figure 3.5a shows the cutaway view of a single-stage, octagon series semihermetic reciprocating compressor with nominal motor powers of 60 and 70 hp. With integrated pulsation mufflers and capacity control (100-50%), smooth running, efficient, and compact reciprocating semihermetics are also available now for this category of capacity. They can be operated with the refrigerants R-134a, R-407C, R-404A, R-507A, and R-22. Figure 3.5b shows a two-stage semihermetic reciprocating compressor for extremely low-temperature applications and its main feature is the two-stage compression in one housing. In two-stage compression, the compression ratio is divided, thus avoiding extreme operating temperatures and achieving very reliable operation. Particularly for commercial refrigeration applications with high load variations, an energy-efficient operation at full and part load (up to four capacity stages) with all common refrigerants is possible at reasonable cost. In addition to that, there are the recognized features of the octagon compressors which even pay off double in tandem configuration.

3.5.3 Open Compressors

Open reciprocating compressors with a shaft seal and an external drive motor suitable for a range of prime movers are also available up to about 2 MW duty (e.g., compressor in the condensing unit given in Figure 3.6b). In these compressors, the crankshafts, which are externally coupled with electric motors, extend through the compressor housings. Appropriate seals must be used where the shafts come through the compressor housings to prevent refrigerant gas from leaking out or air from leaking in (when the crankcase pressure is lower than atmospheric pressure). In order to prevent leakage at the seal, the motor and compressor are rarely enclosed in the same housing.

Figure 3.6a shows an open-type reciprocating compressor which is suitable for all kinds of refrigerants, including NH_3, and Figure 3.6b shows a compact air-cooled condensing unit with an open reciprocating compressor.

3.5.4 Displacement Compressors

These compressors use the shaft work to increase the refrigerant pressure by reducing the compression volume in the chamber. The compressors of this group are reciprocating, vane (rotary), and screw (helical rotary) compressors.

3.5.4.1 Reciprocating Compressors

A great majority of reciprocating compressors which compress the refrigerant gas only on the forward stroke of a piston are built to be single acting in a large-capacity range, up to hundreds of

Figure 3.6 (a) Open type reciprocating compressor and (b) air-cooled condensing unit with an open type reciprocating compressor (*Courtesy of Bitzer Kühlmaschinenbau GmbH*).

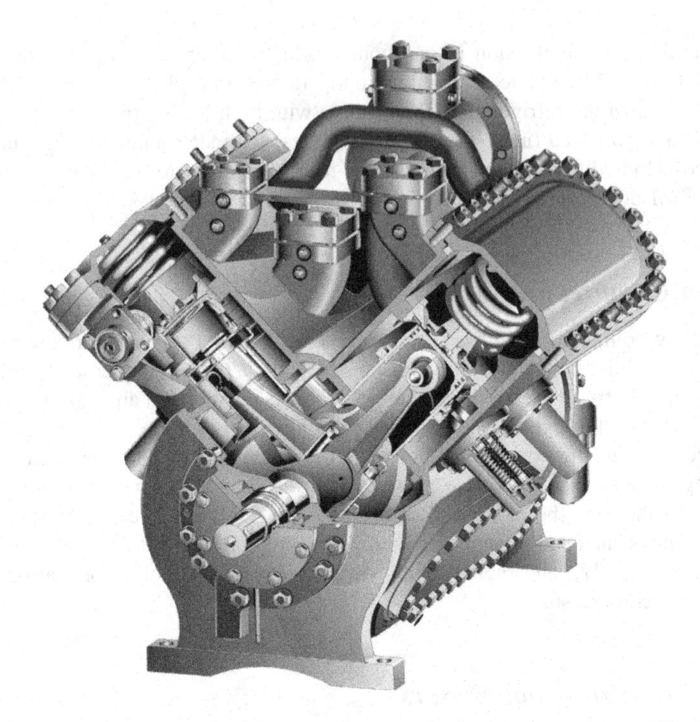

Figure 3.7 An internal view of V-type six-cylinder reciprocating compressor (*Courtesy of Grasso Products b.v.*).

kilowatts. Models of these compressors may be single-cylinder or multicylinder in V (Figure 3.7), W, radial, or line form. The power required for the compressor can be provided either directly by a motor or indirectly by a belt or a gear drive. In these compressors, cylinder clearance volume, compression ratio, amount of suction superheat, valve pressure drops, and the refrigerant-oil characteristics are the main parameters which affect their efficiencies. The selection of cooling method is dependent on the discharge temperature. For example, when the discharge temperature is low, as in R-134a compressors, air cooling is usually chosen. Water cooling is used where high discharge temperatures occur.

Note that Danfoss Maneurop hermetic reciprocating compressors are specially designed for applications with a wide range of operating conditions. The concept has proven its reliability and durability in low-, medium-, and high-temperature applications. Suction gas enters the compressor and cools the electrical motor. The circular valve design and profiled piston provide for an efficient compression process. Discharge gas passes through an internal muffler to eliminate gas pulsation which reduces sound level and vibration. The internal discharge line runs through the oil sump taking care of an oil temperature high enough to evaporate eventual liquid refrigerant entering the compressor.

3.5.4.2 Rotary Compressors

In reality, rotary compressors are of four general design configurations: (i) rolling piston, (ii) rotating vane, (iii) screw, and (iv) scroll. Therefore, rotary compressors have a rotary or circular motion instead of a reciprocating motion. They operate on rotors which rotate on an eccentric shaft. Gas enters through a space between the rotor and the cylinder through a suction port. The gas is compressed as the rotor revolves because of the eccentrical assembly of the rotor and the cylinder. A discharge port on the opposite releases the compressed air. The two more commonly used rotary compressors include the rolling piston-type and the rotating-vane-type. Both are very similar in size, performance, and applications. Rotary compressors are popular in domestic refrigeration and suited for applications where large volumes of vapor are circulated and where a low-compression ratio is desired. In fact, these work as positive displacement pumps.

3.5.4.3 Vane Compressors

There are two major types of vane compressors, single-vane (rotary) and multivane. A rotary compressor simply consists of a bladed, eccentric rotor in a cavity. As the rotor turns, the blades extend and retract, sealing off the cavity into segments of varying size. The gas enters the intake port where the segments are large, is compressed as the cavities are reduced, and is discharged where the segments are small. These compressors are commonly used in domestic refrigerators, freezers, and air conditioners. The possible maximum compression ratios achieved are on the order of 7:1. Small systems and some ammonia systems also employ compressors of this type. In multistage systems in which each stage has a low-compression ratio, vane compressors can be used as boosters. Figure 3.8 shows the cutaway view of a rotary vane compressor. These compressors have some basic advantages which are as follows:

- **Simple, compact design.** Sturdy construction with few moving parts, easy to access and maintain, easy to replace parts, very reliable, and durable.
- **Single-stage compression.** The nature of the design produces sufficient compression in a single stage, resulting in a very high-compression ratio during cycle, as well as better energy efficiency, reduced risk of fault, and reduced maintenance requirements.
- **Direct axial coupling to the motor.** Direct coupling is possible because the high-compression ratio permits low-rotation speeds, eliminating the need for transmission or gears. Fewer parts mean lower energy dissipation and simplified maintenance.
- **Low-rotation speeds.** Lower speeds reduce vibration, thus diminishing noise and wear, lowering temperature, and eliminating the need for foundations.
- **Low cycle temperature.** Lower temperatures reduce wear, oil consumption, and leakage caused by distension of parts. Less energy is needed for cooling and the purity of delivered air is enhanced.
- **Low need for maintenance.** With fewer parts suffering little wear, single-stage rotary vane units offer cleaner and more reliable operation, significantly reducing maintenance needs.

3.5.4.4 Screw Compressors

Surprisingly enough, the screw compressor was invented in 1878. However, commercial application developed slowly because of its inability to match tight tolerances with existing manufacturing equipment of the time. Over the past 10 years, several manufacturers have introduced chillers with screw compressors and have moved away from older reciprocating technology. Screw compressor technology offers many benefits over reciprocating types, including higher reliability and improved performance. In addition to these benefits, some noteworthy characteristics make the screw compressor the compressor of choice for future chiller developments and designs (Duncan, 1999).

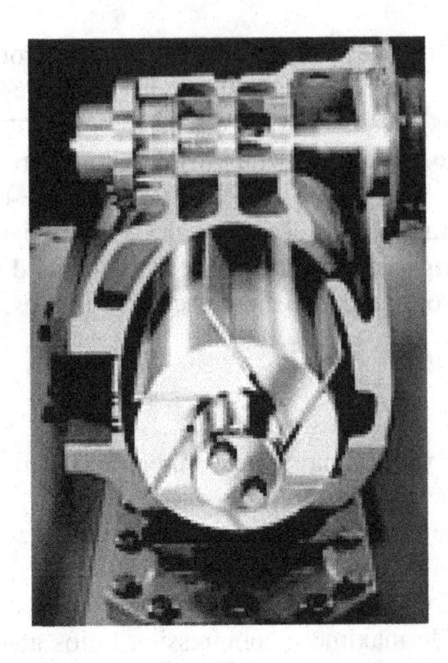

Figure 3.8 Cutaway view of a rotary vane compressor (*Courtesy of Pneumofore SpA*).

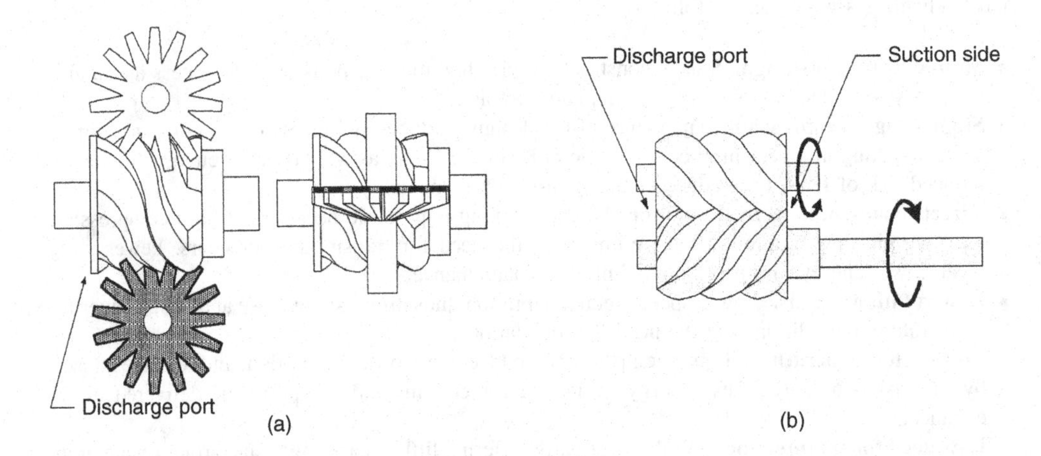

Figure 3.9 Screw compressor. (a) Dual rotor. (b) Mono rotor (Duncan, 1999) (*Courtesy of ASHRAE*).

Screw compressors are also positive displacement refrigeration system components. Both single-screw and twin-screw compressors are widely used in refrigeration applications. A single-screw compressor consists of a single helical rotor (shaft) and a pair of gate rotors that then mesh together, and with the casing form a sealed volume wherein compression takes place. There are two different rotary screw compressor designs. One is a twin rotary screw design, in which there is a male and a female rotor that mesh together (see Figure 3.9a). The other is a single rotary screw design, in which two gate rotors are placed on both sides of the main compressor rotor (Figure 3.9b).

Reciprocating compressors have, until now, carried the workload in applications requiring temperatures below −35 °C. This was the technology of choice, mainly because cascading refrigeration systems was the only choice. The screw compressors were developed specifically for use in applications of −40 °C and below (down to −50 °C). Originally designed for larger applications, this technology is now available in chambers requiring only a single 15 hp or larger compressor. The development of this advanced screw-style refrigeration system offers the following benefits:

- better performance per hp,
- improved reliability,
- reduced costs,
- fewer moving parts,
- less vibration, and
- less refrigerant loss.

By design and function, the screw compressor has far fewer moving parts than the reciprocating style. Engineered with no valves and rolling element bearings, the total number of parts is drastically reduced as well. This reduction of parts is important because it dramatically improves the compressor's reliability rate and increases its expected life span.

Note that screw compressor technology greatly reduces the risk of refrigerant loss because of the decrease in vibration within the entire system. Any structural breakdown within a refrigeration unit may cause loss of its valuable refrigerant. With the accelerating costs of R-22, R-134a, and R-507/404A, product loss becomes a crucial operating factor.

A twin-screw compressor consists of two helically grooved rotors (containing a pair of intermeshing screws) and operates like a gear pump (Figure 3.10). The male screw is directly coupled to the electric motor and this drives the compressor. With the absence of the suction or discharge valves, the gas is drawn into the compression chambers between the gear teeth and the cylinder wall and the helical movement of the gears forces the gas to travel parallel to the rotor shaft. Single-screw compressors are also available and these consist of a helical gear on the main rotor shaft and a pair of planet wheels, one on either side to separate the high and low pressures. Oil flooding

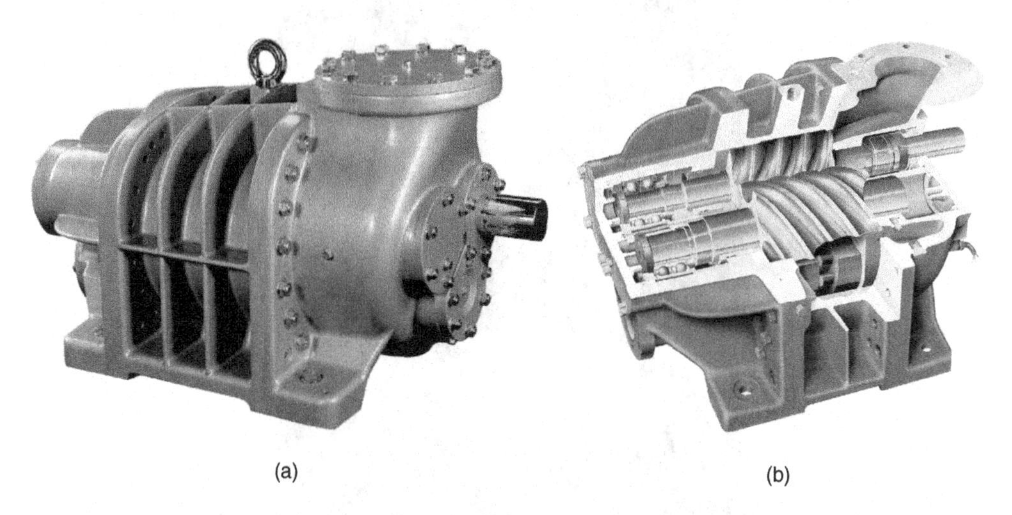

Figure 3.10 A large-capacity double-screw compressor. (a) Complete view. (b) Internal view (*Courtesy of Grasso Products b.v.*).

provides lubrication and restricts leakage of refrigerant gas. These compressors are used mainly in heat reclaim and heat pumping applications.

A screw compressor (Figure 3.10) has a male rotor (with four lobes) which drives a female rotor (with gullies) in a stationary casing with the inlet port at one end and the outlet port at the other. The rotating elements open a void to the suction inlet of the vane compressor, take in a volume of gas, and then seal the port. More rotation decreases the volume between the rotors and compresses the refrigerant gas. The gas is discharged at the low-volume, high-pressure end of the compressor through the outlet port. In general, the male rotor is directly driven. On the other hand, the female rotor rotates along with the male rotor, either through a gear drive or through direct rotor contact.

For industrial refrigeration applications, such as process chillers, the high-temperature compact screw compressors provide an ideal solution. The integrated oil separator and oil reservoir significantly reduce the installation time, complexity, cost, and space required. Such compressors are available in sizes ranging from 50–140 hp and are equipped with the dual capacity control system and auto-economizer and can be used with the common refrigerants R-134a, R-407C, and R-22 (R-404A, R-507A in special applications). The operation with or without economizing is possible.

Figure 3.11 shows the cutaway view of a hermetic rotary type screw compressor which is commonly used in small-scale refrigeration applications, particularly in household and commercial units.

Figure 3.11 Internal view of a hermetic rotary screw compressor (*Courtesy of Hartford Compressors*).

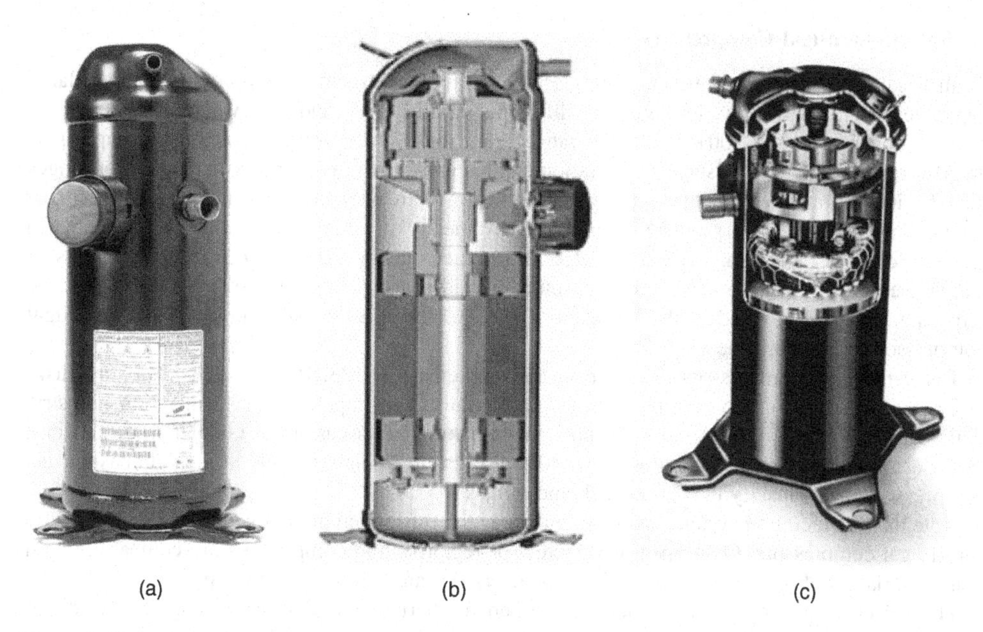

(a) (b) (c)

Figure 3.12 A hermetic scroll compressor. (a) Complete view. (b) Cutaway view. (c) Internal view (*Courtesy of Carlyle Compressor Company*).

3.5.4.5 Scroll Compressors

The scroll compressor (Figure 3.12) uses one stationary (fixed) and one orbiting scroll to compress refrigerant gas vapors from the evaporator to the condenser of the refrigerant path. The upper scroll is stationary and contains the refrigerant gas discharge port. The lower scroll is driven by an electric motor shaft assembly imparting an eccentric or orbiting motion to the driven scroll. That is, the rotation of the motor shaft causes the scroll to orbit (not rotating) about the shaft center.

This orbiting motion gathers refrigerant vapors at the perimeter, pockets the refrigerant gas, and compresses it as the orbiting proceeds. The trapped pocket works progressively toward the center of the stationary scroll and leaves through the discharge port. Maximum compression is achieved when a pocket reaches the center where the discharge port is located. This happens after three complete orbits. The compression is a continuous process. When gas is being compressed in the second orbit, another quantity of gas enters the scrolls and a quantity of gas is being discharged at the same time. This ensures a smooth compression process with low noise and low vibration compared to other compression technologies. Studying this time lapse series carefully gives a true picture on how the trapped gases are progressively compressed as they proceed toward the discharge port.

Scroll compressors are a relatively recent compressor development and are expected to eventually replace reciprocating compressors in many cooling system applications, where they often achieve higher efficiency and better part-load performance and operating characteristics.

3.5.5 *Dynamic Compressors*

These compressors increase the refrigerant pressure through a continuous exchange of angular momentum between a rotating mechanical element and the fluid subject to compression. The main types are centrifugal and turbo compressors.

3.5.5.1 Centrifugal Compressors

Centrifugal compressors are often used in place of positive displacement compressors for very large capacities, or for high-flow, low-pressure difference applications, and are available, designed for refrigeration use, in the 300 kW–20 MW range (e.g., 400–10,000 tons). Centrifugal compressors are also appropriate to multistage refrigeration applications, where two or more compression stages may be incorporated within the same turbine housing with interstage gas injection between the rotors. These compressors produce compression by means of a high-speed impeller connected to an electric motor or gas engine. Figure 3.13a shows the cutaway view of a centrifugal compressor which uses hybrid bearings. The incorporation of hybrid bearings in compressor designs allows the refrigerant itself to be used as the lubricant. Figure 3.13b shows a chiller unit with a centrifugal compressor using hybrid bearings.

The centrifugal compressors available in the market use R-123, R-22, and R-134a. This usually calls for semihermetic designs, with single or multistage impellers. In refrigeration industry, multistage centrifugal compressors are now manufactured with cast iron, nodular iron, and cast steel casings for discharge pressures up to 40 bar. With up to eight wheels in a single casing, the compressor has a capacity of 42,000 m^3/h and 9000 kW.

Note that refrigeration systems using ammonia as the refrigerant are not generally available with centrifugal compressors. Only open-drive screw or reciprocating compressors are compatible with ammonia, largely because of its corrosive characteristics and reactions with copper.

The selection of single stage, multistage, open, or hermetic designs is largely a function of individual manufacturer preference and the application. For example, centrifugal compressors are limited in their compression ratio per impeller. Therefore, applications calling for high-temperature lifts (such as with ice thermal storage) may require multistage designs.

The operating principle of a centrifugal compressor is the same as that of a centrifugal pump, but the refrigerant gas is pumped instead of a liquid. A rotating impeller imparts velocity to the gas, flinging it outward. The housing slows the gas flow, converting a portion of the kinetic energy (the velocity pressure) into a static pressure. These compressors are commonly used for large-capacity

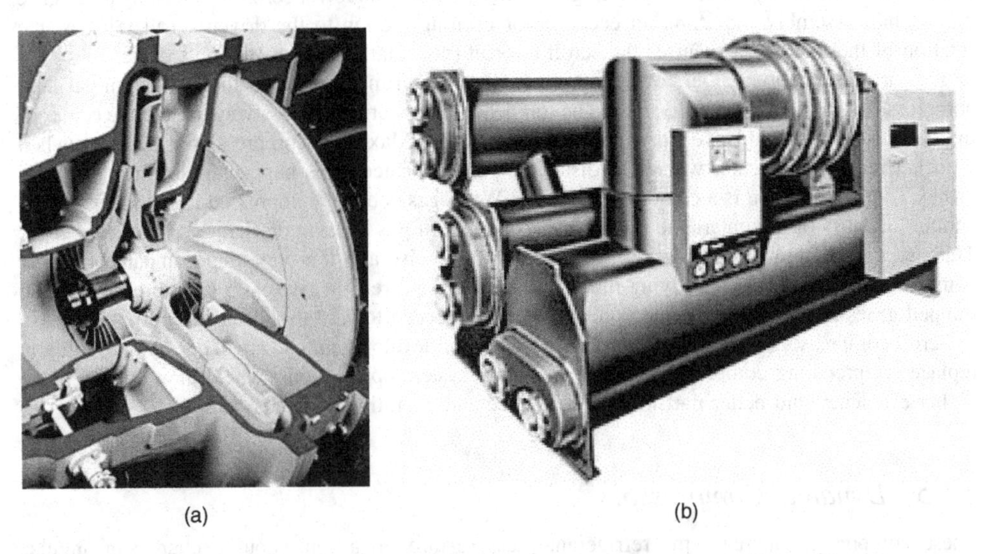

Figure 3.13 (a) Cutaway view of a centrifugal compressor. (b) A chiller unit with centrifugal compressor (*Courtesy of Trane Company*).

Figure 3.14 A centrifugal compressor (*Courtesy of York International*).

refrigeration systems with low-pressure ratios and operate with adiabatic compression efficiencies of up to 80%. Evaporator temperatures may reach −100 °C.

Packaged water-cooled centrifugal compressors are available in sizes ranging from 85 to over 5000 tons. Larger sizes, typically 1200 to 1500 tons and larger, are shipped in subassemblies. Smaller sizes are shipped as a factory-assembled package (Figure 3.14).

Centrifugal compressors use one or more rotating impellers to increase the refrigerant vapor pressure from the evaporator enough to make it condense in the condenser. Unlike the positive displacement, reciprocating, scroll or screw compressors, the centrifugal compressor uses the combination of rotational speed (rpm) and tip speed to produce this pressure difference. The refrigerant vapors from the chiller evaporator are commonly prerotated using variable inlet guide vanes. The consequent swirling action provides extended part-load capacity and improved efficiency. The vapors then enter the centrifugal compressor along the axis of rotation.

The vapor passageways in the centrifugal compressor are bounded by vanes extending from the compressor hub, which may be shrouded for flow-path efficiency. The combination of rotational speed and wheel diameter combines to create the tip speed necessary to accelerate the refrigerant vapor to the high-pressure discharge where they move on to the condenser. Due to their very high vapor-flow capacity characteristics, centrifugal compressors dominate the 200-ton level, where they are the least costly and most efficient cooling compressor design. Centrifugal forces are most commonly driven by electric motors, but can also be driven by steam turbines and gas engines. Depending on the manufacturer's design, centrifugal compressors used in the packages may be one, two, or three stages and use a semihermetic or an open motor with shaft seal.

Figure 3.15 shows a new type of centrifugal compressor which has recently been developed by York International, providing completely oil-free compression using magnetic bearings, particularly for large-tonnage refrigeration and gas compression applications. Tested extensively using S2M magnetic bearing experience and technology, the magnetic bearing option eliminates all the negatives of purchasing and maintaining a lubrication system. In fact, the magnetic bearings enhance centrifugal compressor efficiency and operation. These compressors are available in all sizes and options.

3.5.5.2 Turbo Compressors

In refrigeration technology, turbo compressors usually denote centrifugal compressors, but their efficiencies are low. In this type of compressor, the discharge pressure is limited by the maximum

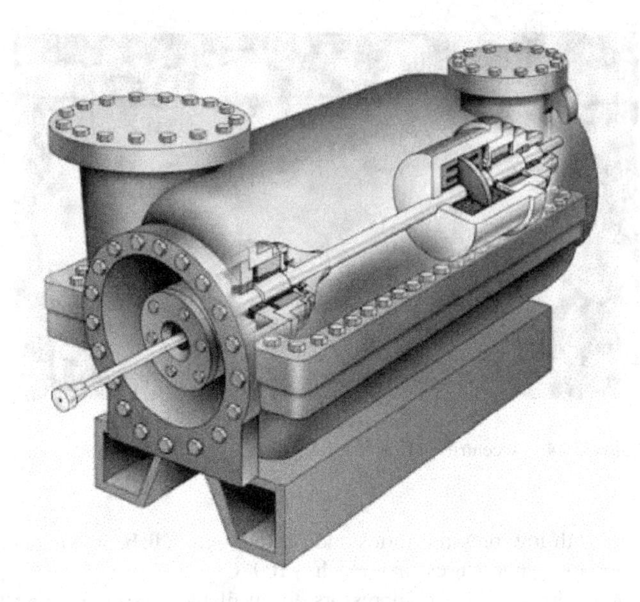

Figure 3.15 Cutaway view of a centrifugal compressor using magnetic bearings (*Courtesy of York International*).

permitted tip speed. A set of impellers is arranged for high-compression pressures. These compressors have found applications in air conditioning and water chilling systems where high suction volumes at high suction pressures are required.

3.5.6 Energy and Exergy Analyses of Compressors

Compressors are used to increase the pressure of a fluid. In the case of a refrigeration cycle, it is used to compress the refrigerant. Compressors operate continuously and the compression process can be modeled as a steady-flow process. Then, referring to Figure 3.16, conservation of mass principle requires that

$$\dot{m}_1 = \dot{m}_2 \longrightarrow \rho_1 A_1 V_1 = \rho_2 A_2 V_2 \longrightarrow \frac{1}{v_1} A_1 V_1 = \frac{1}{v_2} A_2 V_2 \longrightarrow \frac{\dot{V}_1}{v_1} = \frac{\dot{V}_2}{v_2} \tag{3.1}$$

where $\dot{m}$ is the mass flow rate (kg/s), ρ is density (kg/m^3), A is cross-sectional area (m^2), V is velocity (m/s), v is specific volume (m^3/kg), and $\dot{V}$ is the volume flow rate (m^3/s).

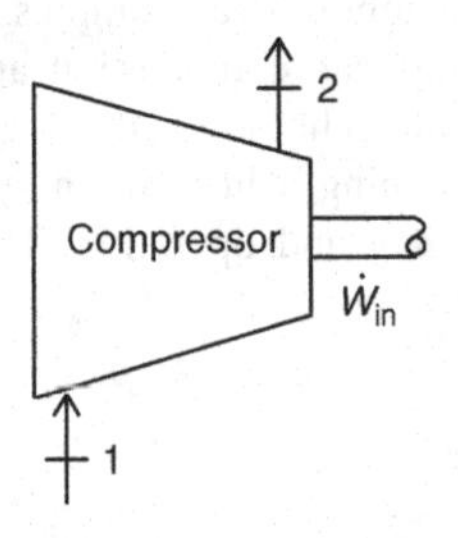

Figure 3.16 The schematic of a compressor considered for mass and energy analysis.

A compressor involves power input $\dot{W}_{in}$ and energy entering and leaving by the fluid stream. The steady-flow energy balance can be written as (with negligible kinetic and potential energies)

$$\begin{aligned} \dot{E}_{in} &= \dot{E}_{out} \\ \dot{W}_{in} + \dot{m}h_1 &= \dot{m}h_2 \\ \dot{W}_{in} &= \dot{m}(h_2 - h_1) \end{aligned} \tag{3.2}$$

where h is enthalpy (kJ/kg). Compressors are normally not insulated and there can be heat transfer between the fluid being compressed and the surrounding air. Depending on the temperature of the refrigerant across the compression process and the temperature of the surrounding air, the net heat transfer could be from the compressor or to the compressor. However, the magnitude of this heat transfer is small and it is usually neglected. Assuming that there is a net heat transfer from the compressor, the energy balance equation becomes

$$\begin{aligned} \dot{W}_{in} + \dot{m}h_1 &= \dot{Q}_{out} + \dot{m}h_2 \\ \dot{W}_{in} - \dot{Q}_{out} &= \dot{m}(h_2 - h_1) \end{aligned} \tag{3.3}$$

Considering again an adiabatic compressor with a steady-flow compression process, an entropy balance may be written as

$$\begin{aligned} \dot{S}_{in} - \dot{S}_{out} + \dot{S}_{gen} &= \Delta\dot{S}_{sys} = 0 \\ \dot{S}_{gen} &= \dot{S}_{out} - \dot{S}_{in} \\ \dot{S}_{gen} &= \dot{m}s_2 - \dot{m}s_1 = \dot{m}(s_2 - s_1) \end{aligned} \tag{3.4}$$

Then the exergy destruction during the compression process becomes

$$\dot{E}x_{dest} = T_0\dot{S}_{gen} = \dot{m}T_0(s_2 - s_1) \tag{3.5}$$

The exergy destruction can also be determined by writing an exergy balance on the compressor:

$$\begin{aligned} \dot{E}x_{in} - \dot{E}x_{out} - \dot{E}x_{dest} &= 0 \\ \dot{E}x_{dest} &= \dot{E}x_{in} - \dot{E}x_{out} \\ \dot{E}x_{dest} &= \dot{W}_{in} + \dot{E}x_1 - \dot{E}x_2 \\ &= \dot{W}_{in} - \Delta\dot{E}x_{12} \\ &= \dot{W}_{in} - \dot{m}\,[h_2 - h_1 - T_0(s_2 - s_1)] \\ &= \dot{W}_{in} - \dot{W}_{rev} \\ &= \dot{m}(h_2 - h_1) - \dot{m}\,[h_2 - h_1 - T_0(s_2 - s_1)] \\ &= \dot{m}T_0(s_2 - s_1) \end{aligned} \tag{3.6}$$

where the reversible work input to the compressor is

$$\dot{W}_{rev} = \dot{E}x_2 - \dot{E}x_1 = \dot{m}\,[h_2 - h_1 - T_0(s_2 - s_1)] \tag{3.7}$$

The exergy efficiency of the compressor may be expressed as the ratio of the reversible work to the actual work.

$$\eta_{Comp,ex} = \frac{\dot{W}_{rev}}{\dot{W}_{in}} = 1 - \frac{\dot{E}x_{dest}}{\dot{W}_{in}} \tag{3.8}$$

3.5.7 Compressor Capacity and Performance

All compressors are rated in terms of how much flow they produce at a given ratio of outlet to inlet pressure (compression ratio). This flow is obviously a function of compressor size (e.g., the number of cylinders and volume displacement for reciprocating compressors) and operating speed (rpm).

Compression ratio is defined by the discharge pressure divided by the suction pressure (both in absolute pressure, Pa or kPa).

The limits of clearance volumes and valve pressure differentials force some of the compressor's flow volume capability to be lost as useful compression. This is referred to as volumetric efficiency. For example, at a compression ratio of 3 to 1, 82% of the volume of the compressor is useful. Thus, if the refrigeration effect required 10 cfm of vapor flow from the evaporator, the compressor would have to produce 10/0.82 or 12.2 cfm of flow.

3.5.7.1 Compression Ratio

The compression ratio is defined as the ratio of discharge pressure to suction pressure at saturated conditions, expressed in absolute terms, for example, Pa or kPa.

$$CR = \frac{P_d}{P_s} \tag{3.9}$$

where CR is compression ratio; P_d is saturated discharge pressure, kPa; and P_s is saturated suction pressure, kPa.

The performance of a compressor is influenced by numerous parameters including the following:

- compressor speed,
- suction pressure and temperature,
- discharge pressure and temperature, and
- type of refrigerant and its flow rate.

3.5.7.2 Compressor Efficiency

In practice, ARI Standard 500-2000 defines the compressor efficiency as the ratio of isentropic work to the actual measured input power. This is also called isentropic efficiency or adiabatic efficiency. Referring to Figure 3.16, the compressor isentropic efficiency becomes

$$\eta_{Comp,isen} = \frac{\dot{W}_{isen}}{\dot{W}_{act}} = \frac{\dot{m}(h_{2s} - h_1)}{\dot{m}(h_2 - h_1)} = \frac{h_{2s} - h_1}{h_2 - h_1} \tag{3.10}$$

where $\dot{m}$ is the mass flow rate of refrigerant, kg/s; h_{2s} is specific enthalpy of refrigerant vapor at discharge pressure at constant entropy ($s_1 = s_{2s}$), kJ/kg; h_1 and h_2 are specific enthalpies of refrigerant at the inlet and exit of the compressor, kJ/kg; $\dot{W}_{isen}$ and $\dot{W}_{act}$ are isentropic and actual powers, kW.

Note that the other compressor efficiency, the volumetric efficiency, can be approximately represented in terms of the ratio of the clearance volume to the displacement volume (R), and the refrigerant-specific volumes at the compressor inlet (suction) and exit (discharge) (v_1 and v_2), as given below:

$$\eta_{Comp,vol} = 1 - R\left(\frac{v_1}{v_2} - 1\right) \tag{3.11}$$

Also, note that the refrigeration capacity can be defined in terms of the compressor volumetric displacement rate ($\dot{V}$, m^3/s), compressor volumetric efficiency ($\eta_{comp,vol}$), density of the refrigerant at the compressor inlet (ρ_1, kg/m^3), and specific enthalpies of the refrigerant at the inlet (state 4) and exit (state 1) of the evaporator. It is then written as

$$\dot{Q}_R = \dot{V}\eta_{\text{Comp,vol}}\rho_1(h_1 - h_4) \tag{3.12}$$

Further details on the practical performance evaluations and ratings of compressors, and definitions of compressor related items, are given extensively in ARI (2000).

Although there are a number of issues that affect the compressor efficiency, the most significant one is the temperature lift (or compression ratio). To a lesser extent, the suction temperature, lubrication, and cooling also play an important role. Therefore, the following solutions to increase the efficiency of the compressor become crucial (DETR, 1999):

- **Minimization of temperature lift.** The compressor is most efficient when the condensing pressure is low and the evaporating pressure is high, leading to the minimum temperature lift and compression ratio. The effect of operating conditions is illustrated by the compressor data example in Figure 3.17. In conjunction with this, a good system design should ensure that the condensing pressure is as low as possible and the evaporating temperature is as high as possible. Designing

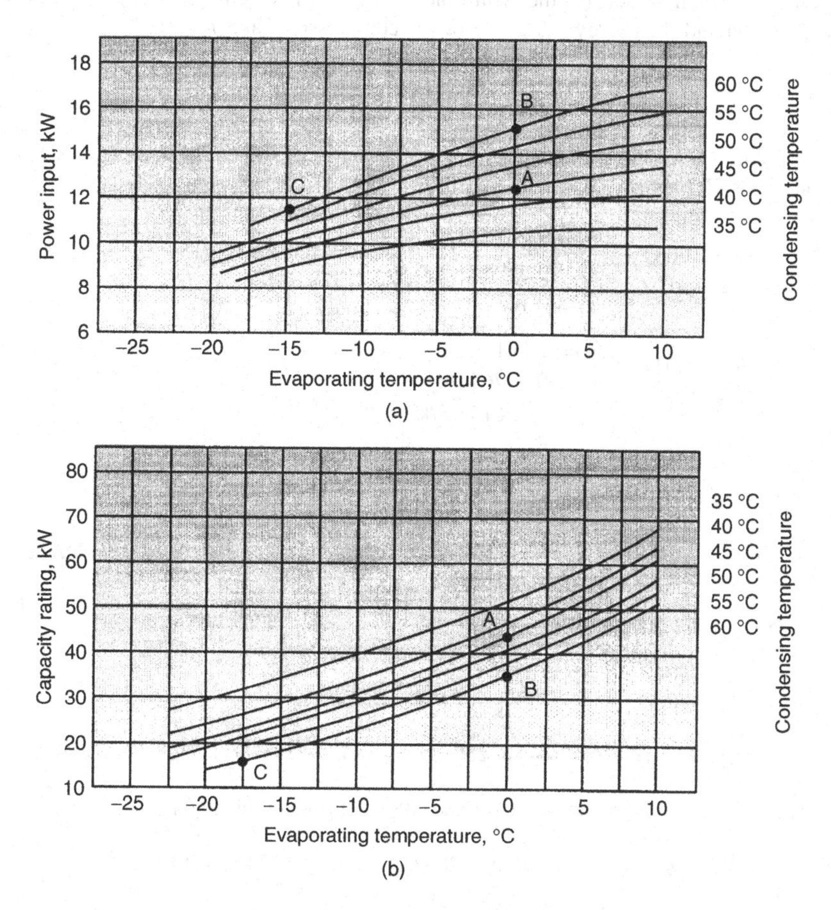

Figure 3.17 Compressor performance profiles at different evaporator and condenser temperatures (DETR, 1999).

a system with a small condenser and evaporator to save capital cost is always false economy. Using a larger evaporator and condenser often means that a smaller compressor can be used and that it always reduces running costs. The additional benefit is that the compressor will be more reliable because it does not have to work as hard and operates with lower discharge temperatures.

- **Lowering suction temperature.** The lower the suction gas temperature the higher the capacity with no effect on power input. The discharge temperature will also be lower, thus increasing reliability. Suction line insulation is essential.
- **Effective lubrication and cooling.** The compressor must be lubricated and efficiently cooled. Insufficient lubrication increases bearing friction and temperature and reduces compressor efficiency, often resulting in failure.

Example 3.1

Refrigerant-134a enters the compressor of a refrigeration cycle at 160 kPa and −10 °C with a flow rate of 0.25 m^3/min and leaves at 900 kPa and 60 °C. The ratio of the clearance volume to the displacement volume is 0.05. Determine (a) the volumetric efficiency of the compressor, (b) the power input to the compressor, (c) the isentropic efficiency of the compressor, and (d) the rate of exergy destruction and the exergy efficiency of the compressor. Take $T_0 = 25\,°\text{C}$.

Solution

(a) The properties of refrigerant at the inlet and exit states of the compressor are obtained from R-134a tables (Tables B.3, B.4, and B.5):

$$\left.\begin{array}{l} P_1 = 160\,\text{kPa} \\ T_1 = -10\,°\text{C} \end{array}\right\} \begin{array}{l} h_1 = 245.77\,\text{kJ/kg} \\ s_1 = 0.9598\,\text{kJ/kg}\cdot\text{K} \\ v_1 = 0.1269\,\text{m}^3/\text{kg} \end{array}$$

$$\left.\begin{array}{l} P_2 = 900\,\text{kPa} \\ T_2 = 60\,°\text{C} \end{array}\right\} \begin{array}{l} h_2 = 295.13\,\text{kJ/kg} \\ s_2 = 0.9976\,\text{kJ/kg}\cdot\text{K} \\ v_2 = 0.02615\,\text{m}^3/\text{kg} \end{array}$$

$$\left.\begin{array}{l} P_2 = 900\,\text{kPa} \\ s_2 = s_1 = 0.9598\,\text{kJ/kg}\cdot\text{K} \end{array}\right\} h_{2s} = 282.76\,\text{kJ/kg}$$

$$\eta_{\text{Comp,vol}} = 1 - R\left(\frac{v_1}{v_2} - 1\right) = 1 - 0.05\left(\frac{0.1269}{0.02615} - 1\right) = 0.807 = \mathbf{80.7\%}$$

(b) The mass flow rate of the refrigerant and the actual power input are

$$\dot{m} = \frac{\dot{V}_1}{v_1} = \frac{(0.25/60)\,\text{m}^3/\text{s}}{0.1269\,\text{m}^3/\text{kg}} = 0.03285\,\text{kg/s}$$

$$\dot{W}_{\text{act}} = \dot{m}(h_2 - h_1) = (0.03285\,\text{kg/s})(295.13 - 245.77)\,\text{kJ/kg} = \mathbf{1.621\,kW}$$

(c) The power input for the isentropic case and the isentropic efficiency are

$$\dot{W}_{\text{isen}} = \dot{m}(h_{2s} - h_1) = (0.03285\,\text{kg/s})(282.76 - 245.77)\,\text{kJ/kg} = 1.215\,\text{kW}$$

$$\eta_{\text{Comp,isen}} = \frac{\dot{W}_{\text{isen}}}{\dot{W}_{\text{act}}} = \frac{1.215\,\text{kW}}{1.621\,\text{kW}} = 0.749 = \mathbf{74.9\%}$$

(d) The exergy destruction is

$$\dot{E}x_{\text{dest}} = \dot{m}T_0(s_2 - s_1) = (0.03285\,\text{kg/s})(298\ \text{K})(0.9976 - 0.9598)\,\text{kJ/kg}\cdot\text{K} = \mathbf{0.370\ kW}$$

The reversible power and the exergy efficiency are

$$\dot{W}_{\text{rev}} = \dot{W}_{\text{act}} - \dot{E}x_{\text{dest}} = 1.621 - 0.370 = 1.251\,\text{kW}$$

$$\eta_{\text{Comp,ex}} = \frac{\dot{W}_{\text{rev}}}{\dot{W}_{\text{act}}} = \frac{1.251\,\text{kW}}{1.621\,\text{kW}} = \mathbf{0.772 = 77.2\%}$$

3.5.7.3 Compressor Capacity Control

A capacity controlled refrigeration unit is a unit in which the compression ability of the compressor can be controlled to reduce or increase refrigerant mass flow rate. The concept of compressor flow modulation achieves improved performance in two ways. First, by using efficient compressor capacity reduction to prevent the increase in mass flow rate of refrigerant at high ambient temperatures, the coefficient of performance (COP) at higher ambient temperatures can be significantly increased. Reliability will also be increased because of reduced load on the compressor. The second improvement in performance is realized by a change in system sizing strategy. Conventional heat pumps are sized for the cooling load so that comfortable air conditioning is obtained. With compressor capacity control, the heat pump can be sized for a greater heating capacity, thereby having a lower balance point and eliminating some of the auxiliary heating. Then, via the capacity control which is inherent in the concept, the capacity of the unit during cooling can be controlled to achieve proper comfort control.

One method of system capacity control frequently in use today is hot gas bypass. Hot gas bypass, where discharge gas from the compressor is vented back to the suction side of the compressor, is an easy retrofit to most systems, but is disastrous from an energy savings viewpoint because capacity is reduced without reducing compressor work, and is probably best avoided. Other possible capacity control methods fall into essentially three categories:

- **Speed control.** Speed control can be done either continuously or stepwise. Continuously variable speed control is one of the most efficient methods of capacity control, and it offers good control down to about 50% of rated speed of normal compressors. More than 50% speed reduction is unacceptable because of lubrication requirements of the compressors. Continuously variable speed control is also an expensive process, though not necessarily prohibitively expensive, for it might be possible to replace some of the conventional starting controls with the motor controls and hence reduce the cost increment. Stepwise speed control, as achieved, for example, by using multipoled electric motors and switching the number of active poles, is another viable alternative. It might be possible to achieve satisfactory improvements in performance by using a finite number of stepped changes to vary compressor capacity. Step control is less costly than continuously variable speed control, but is also limited to 50% of rated compressor speed because of lubrication requirements. Also, step changes in load on the compressor could put high stresses on compressor components.
- **Clearance volume control.** This requires substantial amounts of additional clearance volume to achieve the amount of flow reduction desirable. For example, to reduce the mass flow rate by 50%, the clearance volume must be equal to about half of the displacement volume, adding substantially to the bulk of the compressor. Moreover, the large amount of residual mass causes unacceptably high discharge temperatures with large amounts of flow reduction. For this reason, clearance volume control is considerably less attractive than some other types of control.

- **Valve control.** Suction valve unloading, a compressor capacity control method often used in large air conditioning and refrigeration systems to reduce cooling capacity when load decreases, can achieve some energy savings but has a number of drawbacks. In unloading, the suction valve of one or more cylinders is held open so that gas is pumped into and out of the cylinder through the valve without being compressed. Substantial losses can occur because of this repeated throttling through the suction valve. In addition, stepwise cylinder unloading causes uneven stresses on the crankshaft and provides inadequate, if not totally unacceptable, control in smaller compressors. The method is, however, relatively inexpensive. Two newer methods of compressor flow regulation via valve control are late suction-valve closing and early suction-valve closing. Late suction-valve closing again incurs the throttling loss by pumping gas back out of the suction valve for part of the stroke. Late valve closing, however, gives more acceptable, smoother control than complete valve unloading. At present, however, the method is limited to a maximum of 50% capacity reduction and to large low-speed compressors. Early suction-valve closing eliminates losses due to throttling gas back out of the suction valves. Instead, the suction valve, or a secondary valve just upstream of the suction valve, is closed prematurely on the intake stroke, limiting the amount of gas taken in. The gas inside the cylinder is expanded and then recompressed, resulting in much lower losses. Continuously variable capacity control over a wide range is possible with the early valve closing approach. The early suction-valve closing approach requires the most development of the capacity control methods discussed above, but it also holds promise for being one of the most efficient and inexpensive approaches.

3.5.7.4 Capacity Control for Varying Loads to Provide Better Efficiency

There are several ways to meet varying loads, each with different efficiency, as summarized below (DETR, 1999):

- **Case 1.** Single large compressor. This cannot meet variable load and results in wasted capacity and lower efficiency when at part load.
- **Case 2.** Single large compressor with inbuilt capacity control. This is a good option to meet variable load as long as load stays above 50%.
- **Case 3.** Three small compressors (two with same capacity and one with capacity control). This allows fairly close matching to demand.
- **Case 4.** Three small compressors with different capacities. This is a good option to meet variable load. The aim is to mix and match to varying load with sequence control.
- **Case 5.** Three compressors with parallel control. This is often used, but is not always recommended due to nonlinear input power with capacity turn-down. For example, at 180% capacity (i.e., 3 at 60%), it requires ~240% power due to inefficiencies, which brings an additional input of about 60%.
- **Case 6.** Three compressors (two are on and one is off). In this case, one compressor is used at 100%, and one is used to trim to exact demand (e.g., 80% in the above case), giving 180% capacity with 188% power (22% saving over the above case).

In the selection of one of the above cases, two main criteria are power demand and budget. Note that the load profile must be available to select the best compressor option. Different options should be compared at the most common operating conditions as well as throughout the load range. The efficiency of the different options varies enormously and there is no hard and fast rule to selecting the best solution. Switching a compressor off to reduce the system capacity is the most efficient method of meeting a reduced load. The efficiency of a compressor operating on inbuilt capacity control is always lower than when it operates at full load. The efficiency of the different methods of capacity control varies. In general, any method which recirculates compressed gas back into the suction of a compressor is very poor. When considering compressors with capacity control, we should compare the options accurately. Compressors are often oversized for an application because so many safety factors are used when calculating the load. This should be avoided as oversized compressors often operate with a lower power factor.

Regardless of the configuration option selected to meet a load, the control of the compressors is important. The control strategy should be designed to

- select the most efficient mix of compressors to meet the load,
- avoid operation on inbuilt capacity control when possible, and
- avoid operation at low suction pressures when possible.

Selecting compressors of different sizes and designing a good control strategy to cycle them to accurately match the most common loads is often the most efficient option.

3.6 Condensers

There are several condensers to be considered when making a selection for installation. They are air-cooled, water-cooled, shell and tube, shell and coil, tube within a tube, and evaporative condensers. Each type of condenser has its own unique application. Some determining factors include the size and the weight of the unit, weather conditions, location (city or rural), availability of electricity, and availability of water.

A wide variety of condenser configurations are employed in the process industry. Selection of condenser type is not easy and depends on the following criteria:

- condenser heat capacity,
- condensing temperature and pressure,
- the flow rates of refrigerant and coolant,
- design temperature for water and/or air,
- operation period, and
- climatic conditions.

Condensers utilized in the refrigeration industry are commonly of three types, as follows:

- water-cooled condensers,
- air-cooled condensers, and
- evaporative condensers.

Common types of water- and air-cooled refrigerant condensers for commercial refrigeration use are

- shell and tube, blow-through, horizontal airflow,
- shell and coil, draw-through, vertical airflow, and
- tube in tube, static, or forced airflow.

The type of condenser selected depends largely on the following considerations:

- size of the cooling load,
- refrigeration used,
- quality and temperature of available cooling water (if any), and
- amount of water that can be circulated, if water use is acceptable.

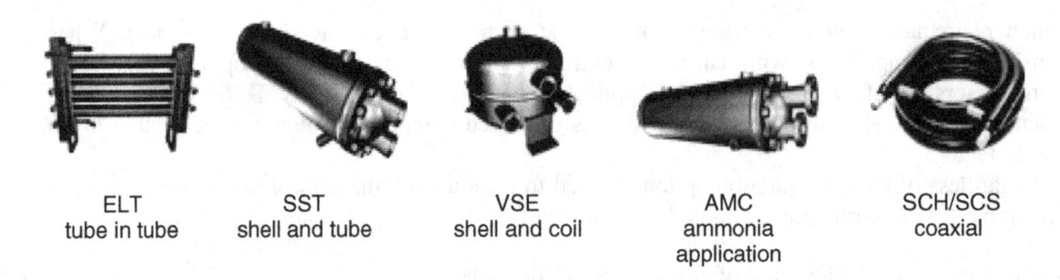

Figure 3.18 Various water-cooled condensers (*Courtesy of Standard Refrigeration Company*).

3.6.1 Water-Cooled Condensers

Water-cooled condensers are of many different types as shown in Figure 3.18. The most common condensers are generally shell and tube type heat exchangers with refrigerant flow through the shell and water (as coolant) flow through the tubes (SST type in Figure 3.18). The lower portion of the shell acts as a liquid receiver. These condensers are widely used in large heat capacity refrigerating and chilling applications. If a water-cooled condenser is used, the following criteria must be examined:

- requirement of cooling water for heat rejection,
- utilization of a cooling tower if inexpensive cooling water is available,
- requirement of auxiliary pumps and piping for recirculating cooling water,
- requirement of water treatment in water recirculation systems,
- space requirements,
- maintenance and service situations, and
- provision of freeze protection substances and tools for winter operation.

In general, water-cooled condensers are used with cooling towers or groundwater (well, lake, river, etc.).

3.6.2 Air-Cooled Condensers

The air-cooled condensers find applications in domestic, commercial, and industrial refrigerating, chilling, freezing, and air-conditioning systems with a common capacity of 20–120 tons (Figure 3.19). The centrifugal fan air-cooled condensers (with a capacity of 3–100 tons) are particularly used for heat recovery and auxiliary ventilation applications. In fact, they employ outside air as the cooling medium. Fans draw air past the refrigerant coil and the latent heat of the refrigerant is removed as sensible heat by the air stream. The advantages of air-cooled condensers are the following:

- no water requirement,
- standard outdoor installation,
- elimination of freezing, scaling, and corrosion problems,
- elimination of water piping, circulation pumps, and water treatment,
- low installation cost, and
- low maintenance and service requirement.

Figure 3.19 A typical air-cooled condenser (*Courtesy of Trane Company*).

On the other hand, they have some disadvantages as given below:

- high condensing temperatures,
- high refrigerant cost because of long piping runs,
- high power requirements per kW of cooling,
- high noise intensity, and
- multiple units required for large-capacity systems.

3.6.3 Evaporative Condensers

Evaporative condensers are apparently water-cooled designs and work on the principle of cooling by evaporating water into a moving air stream. The effectiveness of this evaporative cooling process depends upon the wet-bulb temperature of the air entering the unit, the volume of airflow, and the efficiency of the air/water interface.

Evaporative condensers use water sprays and airflow to condense refrigerant vapors inside the tubes. The condensed refrigerant drains into a tank called a liquid receiver. Refrigerant subcooling can be accomplished by piping the liquid from the receiver back through the water sump where additional cooling reduces the liquid temperature even further.

In an evaporative condenser (Figure 3.20a), the fluid to be cooled is circulated inside the tubes of the unit's heat exchanger. Heat flows from the process fluid through the coil tubes to the water outside, which is cascading downward over the tubes. Air is forced upward through the coil evaporating a small percentage of the water, absorbing the latent heat of vaporization, and discharging the heat to the atmosphere. The remaining water falls to the sump to be recirculated by the pump, while water entrained in the air stream is reclaimed and returned to the sump by the mist eliminators at the unit discharge. The only water consumed is the amount evaporated plus a small amount which is intentionally bled off to limit the concentration of impurities in the pan. With the optional extended surface coil, the recirculating water pump can be shut-off and the unit operated dry during periods of below-design ambient temperatures. Air is still

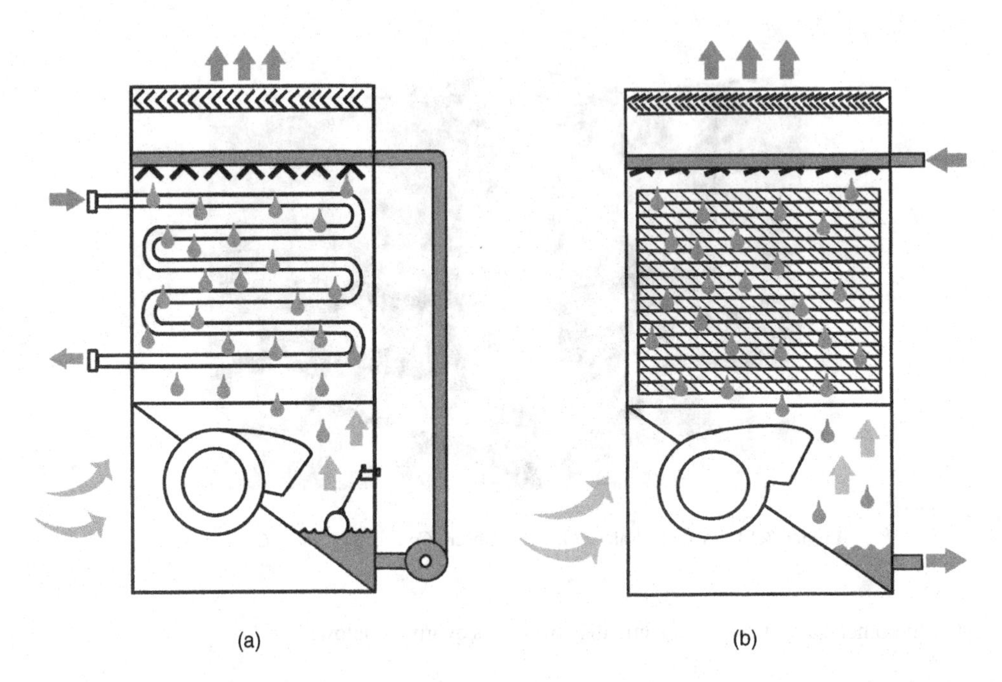

Figure 3.20 (a) An evaporative condenser. (b) A counter-flow cooling tower (*Courtesy of Baltimore Aircoil International*).

forced upward through the coil, but the heat is now dissipated to the atmosphere by sensible cooling alone.

The following are some characteristics of these condensers:

- reduced circulating water for a given capacity,
- water treatment is necessary,
- reduced space requirement,
- small piping sizes and short overall lengths,
- small system pumps, and
- availability of large-capacity units and indoor configurations.

The volume of water used by evaporative condensers is significant. Not only does water evaporate just to reject the heat, but water must be added to avoid the buildup of dissolved solids in the basins of the evaporative condensers. If these solids build up to the point that they foul the condenser surfaces, the performance of the unit can be greatly reduced.

3.6.4 Cooling Towers

Cooling towers (Figure 3.20b) are like evaporative condensers, working on the principle of cooling by evaporating water into a moving air stream. The effectiveness of this evaporative cooling process depends upon the wet-bulb temperature of the air entering the unit, the volume of airflow, and the efficiency of the air or water interface.

As mentioned above, cooling towers are essentially large evaporative coolers where the cooled water is circulated to a remote shell and tube refrigerant condenser. Note that the cooling water is circulating through the tubes while refrigerant vapor condenses and gathers in the lower region of the heat exchanger. Notice also that this area subcools the refrigerant below the temperature of condensation by bringing the coldest cooling tower water into this area of the condenser. The warmed cooling water is sprayed over a fill material in the tower. Some of it evaporates in the moving air stream. The evaporative process cools the remaining water.

The volume of water used by cooling towers is significant. Not only does water evaporate just to reject the heat but it must also be added to avoid the buildup of dissolved solids in the basins of the cooling towers. If these solids build up to the point that they foul the condenser surfaces, the performance of the unit can be greatly reduced.

3.6.5 Energy and Exergy Analyses of Condensers

Condensers are used to reject heat from a refrigeration system. In a vapor-compression refrigeration cycle, the refrigerant is cooled and condensed as it flows in the condenser coils as shown in Figure 3.21a. The conservation of mass principle requires that

$$\dot{m}_1 = \dot{m}_2 \tag{3.13}$$

Referring to Figure 3.21a, energy is entering and leaving by the refrigerant stream and heat is rejected from the condenser, ($\dot{Q}_{\text{out}}$ or $\dot{Q}_H$). The steady-flow energy balance can be written as (with negligible kinetic and potential energies)

$$\begin{aligned} \dot{m}h_1 &= \dot{m}h_2 + \dot{Q}_{\text{out}} \\ \dot{Q}_{\text{out}} &= \dot{m}(h_1 - h_2) \end{aligned} \tag{3.14}$$

In the condenser of a household refrigerator, heat is rejected from the refrigerant to the kitchen air as it flows in the condenser coils. Figure 3.21a represents operation of such a compressor. In some refrigeration applications (especially large ones), heat is rejected from the refrigerant into water (water-cooled condenser) in a refrigerant-to-water heat exchanger as shown in Figure 3.21b. Assuming that heat exchanger is insulated, the energy balance in this case becomes

$$\begin{aligned} \dot{m}_R h_1 + \dot{m}_w h_3 &= \dot{m}_R h_2 + \dot{m}_w h_4 \\ \dot{m}_R(h_1 - h_2) &= \dot{m}_w(h_4 - h_3) \end{aligned} \tag{3.15}$$

where $\dot{m}_R$ and $\dot{m}_w$ are the mass flow rates of the refrigerant and the water. The rate of heat rejected to water is

$$\dot{Q}_{\text{out}} = \dot{m}_R(h_1 - h_2) = \dot{m}_w(h_4 - h_3) \tag{3.16}$$

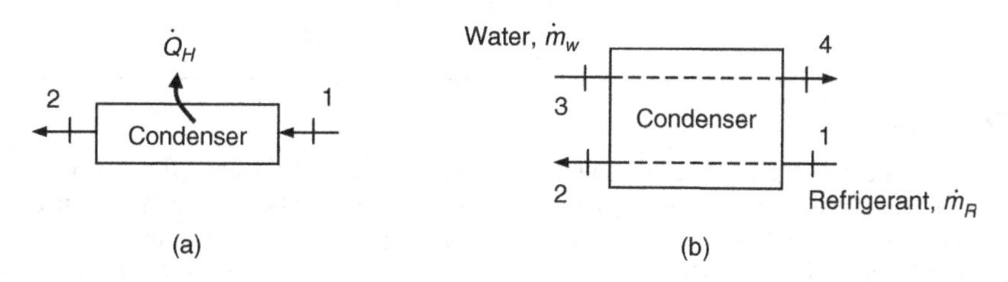

Figure 3.21 The schematic of condensers considered for mass and energy analysis. (a) Air-cooled condenser. (b) Water-cooled condenser.

Referring to Figure 3.21a, an entropy balance on the condenser may be written as

$$\dot{S}_{in} - \dot{S}_{out} + \dot{S}_{gen} = \Delta\dot{S}_{sys} = 0$$
$$\dot{S}_{gen} = \dot{S}_{out} - \dot{S}_{in}$$
$$\dot{S}_{gen} = \frac{\dot{Q}_H}{T_H} + \dot{m}s_2 - \dot{m}s_1 \tag{3.17}$$
$$= \dot{m}\left(s_2 - s_1 + \frac{q_H}{T_H}\right)$$

Then the exergy destruction in the condenser becomes

$$\dot{E}x_{dest} = T_0\dot{S}_{gen} = \dot{m}T_0\left(s_2 - s_1 + \frac{q_H}{T_H}\right) \tag{3.18}$$

The exergy destruction can also be determined by writing an exergy balance on the condenser:

$$\dot{E}x_{in} - \dot{E}x_{out} - \dot{E}x_{dest} = 0$$
$$\dot{E}x_{dest} = \dot{E}x_{in} - \dot{E}x_{out}$$
$$\dot{E}x_{dest} = (\dot{E}x_1 - \dot{E}x_2) - \dot{E}x_{\dot{Q}_H} \tag{3.19}$$
$$= \dot{m}\left[h_1 - h_2 - T_0(s_1 - s_2)\right] - \dot{Q}_H\left(1 - \frac{T_0}{T_H}\right)$$

The exergy efficiency of the condenser may be expressed as the ratio of the exergy of the heat transferred to the high-temperature medium to the exergy decrease of the refrigerant across the condenser:

$$\eta_{ex,Cond} = \frac{\dot{E}x_{\dot{Q}_H}}{\dot{E}x_1 - \dot{E}x_2} = \frac{\dot{Q}_H\left(1 - \frac{T_0}{T_H}\right)}{\dot{m}\left[h_1 - h_2 - T_0(s_1 - s_2)\right]} = 1 - \frac{\dot{E}x_{dest}}{\dot{E}x_1 - \dot{E}x_2} \tag{3.20}$$

If we consider Figure 3.21b for the operation of an evaporator, an entropy balance may be written as

$$\dot{S}_{gen} = \dot{S}_{out} - \dot{S}_{in}$$
$$\dot{S}_{gen} = (\dot{m}_R s_2 + \dot{m}_w s_4) - (\dot{m}_R s_1 + \dot{m}_w s_3) \tag{3.21}$$
$$= \dot{m}_R(s_2 - s_1) - \dot{m}_w(s_3 - s_4)$$

Then the exergy destruction in the evaporator becomes

$$\dot{E}x_{dest} = T_0\dot{S}_{gen} = T_0\left[\dot{m}_R(s_2 - s_1) - \dot{m}_w(s_3 - s_4)\right] \tag{3.22}$$

Example 3.2

Refrigerant-134a enters the condenser of a refrigeration cycle at 800 kPa and 60 °C with a flow rate of 0.095 kg/s and leaves at the same pressure subcooled by 3.3 °C. The refrigerant is condensed by rejecting its heat to water which experiences a temperature rise of 11 °C. Determine (a) the rate of heat rejected in the condenser, (b) the mass flow rate of water, (c) the COP of this refrigeration cycle if the cooling load at these conditions is 12 kW, and (d) the rate of exergy destruction in the condenser. Take $T_0 = 25\,°C$.

Solution

(a) We refer to Figure 3.21b for the schematic of the condenser. The properties of refrigerant at the inlet and exit states of the condenser are (from Tables B.3, B.4, and B.5)

$$\left.\begin{array}{l} P_1 = 800\,\text{kPa} \\ T_1 = 60\,°\text{C} \end{array}\right\} \begin{array}{l} h_1 = 296.81\,\text{kJ/kg} \\ s_1 = 1.011\,\text{kJ/kg}\cdot\text{K} \end{array}$$

$$\left.\begin{array}{l} P_2 = 800\,\text{kPa} \\ T_2 = T_{\text{sat @ 800 kPa}} - \Delta T_{\text{subcool}} = 31.3 - 3.3 = 28\,°\text{C} \end{array}\right\} \begin{array}{l} h_2 \cong h_{f@28\,°\text{C}} = 90.69\,\text{kJ/kg} \\ s_2 \cong s_{f@28\,°\text{C}} = 0.3383\,\text{kJ/kg}\cdot\text{K} \end{array}$$

The rate of heat rejected in the condenser is

$$\dot{Q}_H = \dot{m}_R(h_1 - h_2) = (0.095\,\text{kg/s})(296.81 - 90.69)\,\text{kJ/kg} = \mathbf{19.58\,kW}$$

(b) The mass flow rate of water can be determined from an energy balance on the condenser:

$$\dot{Q}_H = \dot{m}_R(h_1 - h_2) = \dot{m}_w c_p \Delta T_w$$

$$19.58\,\text{kW} = \dot{m}_w(4.18\,\text{kJ/kg}\cdot°\text{C})(11\,°\text{C})$$

$$\dot{m}_w = \mathbf{0.426\,kg/s}$$

The specific heat of water is taken to be 4.18 kJ/kg · °C.

(c) From the definition of COP for a refrigerator,

$$\text{COP} = \frac{\dot{Q}_L}{\dot{W}_{\text{in}}} = \frac{\dot{Q}_L}{\dot{Q}_H - \dot{Q}_L} = \frac{12\,\text{kW}}{(19.58 - 12)\,\text{kW}} = \mathbf{1.58}$$

(d) The entropy generation and the exergy destruction in the condenser are

$$\dot{S}_{\text{gen}} = \dot{m}_R(s_2 - s_1) + \frac{\dot{Q}_H}{T_H}$$

$$= (0.095\,\text{kg/s})(0.3383 - 1.011)\,\text{kJ/kg}\cdot\text{K} + \frac{19.58\,\text{kW}}{298\,\text{K}} = 0.001794\,\text{kW/K}$$

$$\dot{E}x_{\text{dest}} = T_0\dot{S}_{\text{gen}} = (298\,\text{K})(0.001794\,\text{kJ/kg}\cdot\text{K}) = \mathbf{0.5345\,kW}$$

3.7 Evaporators

Evaporator can be considered as the point of heat capture in a refrigeration system and provides the cooling effect required for any particular application. There are almost as many different types of evaporators as there are applications of heat exchangers. However, evaporators are divided into two categories such as (i) direct cooler evaporators that cool air that, in turn, cools the product and (ii) indirect cooler evaporators that cool a liquid such as brine solution that, in turn, cools the product. Normally, the proper evaporator comes with the system. However, there may be an occasion when designing a system, that one will need to determine the requirements and select the proper evaporator from a manufacturer's catalog or manual.

In practice, the following evaporators are commonly used for cooling, refrigerating, freezing, and air-conditioning applications:

- liquid coolers,
- air coolers, and/or gas coolers.

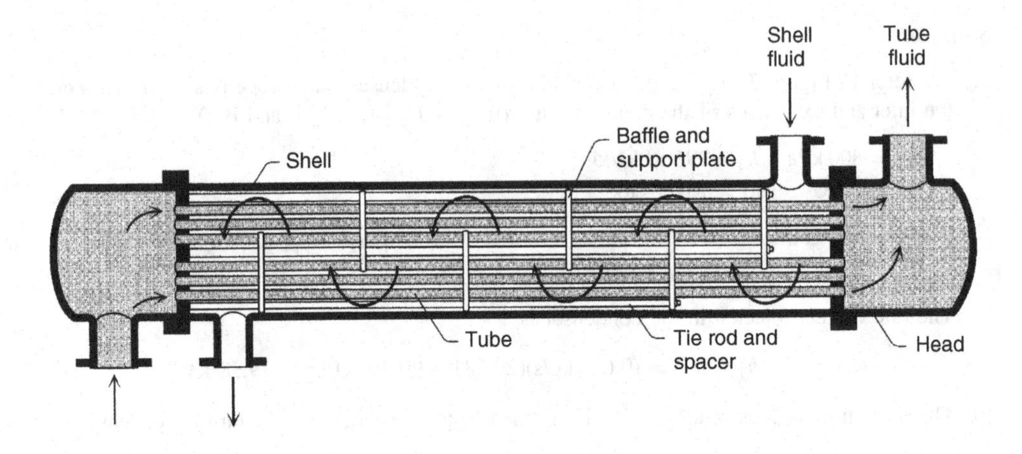

Figure 3.22 A shell-tube type evaporator (Bejan, 2004).

3.7.1 Liquid Coolers

Shell and tube type heat exchangers (Figure 3.22) are the more common form of evaporation units for water cooling and chilling applications. These are utilized to cool liquids, which can be used as the secondary refrigerant or to cool the final products directly. In practice, these types of heat exchangers are known as liquid coolers or chillers.

Some example applications in food and refrigeration industry are

- chilling of drinkable water,
- chilling of water for air-conditioning coils,
- chilling of milk after pasteurization, and
- process cooling operations.

Chilled water systems can use either a flooded evaporator or a direct expansion evaporator which are typically shell and tube type heat exchangers. In a flooded evaporator, refrigerant floods the shell side of the heat exchanger and is controlled by a level valve. Water being chilled passes through the tubes. Conversely, in a direct expansion evaporator, water is carried in the shell and refrigerant is boiled inside the tubes. The rate of refrigerant flow is throttled to insure that only refrigerant gas exits the evaporator. Copper tubes mounted within a carbon steel shell is the most common construction used for chilled water evaporators.

It is important to note that if the refrigerant vaporizes on the outside surface of the tubes the evaporator is a *flooded cooler*; if it vaporizes inside the tubes the evaporator is a *dry cooler* (note that in this more common type, the mixture of liquid and vapor is evaporated completely), usually with some degree of superheating (Hewitt, Shires and Bott, 1994). In a flooded cooler the water or brine is circulated through the tubes, which are usually finned to provide an increment in the heat-transfer rate and a decrease in the evaporator size. In a dry cooler the liquid refrigerant is contained within the tubes, and water or brine is circulated through the shell of the cooler, which serves as an evaporator. Flooded coolers are often specified for applications where shell-side vaporization of refrigerant of other liquids is desirable. Due to rapid boiling in the shell, in order to obtain high purity vapors, a vapor disengagement vessel is often welded to the main shell. Flooded coolers are particularly employed in multiple compressor systems.

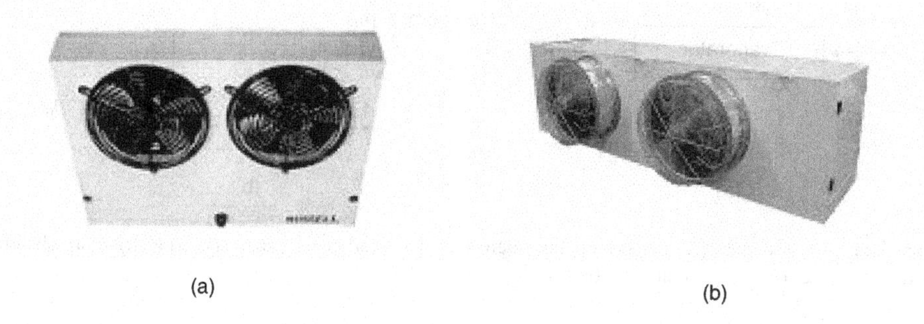

(a) (b)

Figure 3.23 Air coolers. (a) Room type. (b) Large-scale industrial type (*Courtesy of Super Radiator Coils*).

3.7.2 Air and Gas Coolers

These coolers are generally called *direct expansion coils* and consist of a series of tubes through which refrigerant flows (Figure 3.23). The tubes, which are finned to increase the heat-transfer rate from the medium to be cooled (e.g., air) to the boiling point, are normally arranged into a number of parallel circuits fed from a single throttling valve. The hot refrigerant vapor is accumulated in the outlet (suction) gas header. These direct expansion coils are used only in the positive displacement compressor systems, owing to quite low-pressure ratios. Like liquid coolers, these coolers are also classified as *flooded* and *dry* types. In a flooded coil, a float valve is used to maintain the preset level in the coil, meaning that evaporator coil is kept close to full of the liquid refrigerant. This full contact of the liquid with the tube walls provides a high heat-transfer rate. In practical applications, flooded-type evaporators are not preferable, because they require large amounts of refrigerant. A dry coil requires only a small amount of refrigerant and this reduces the cost of the refrigerant charge. Sometimes a metering device (*thermal expansion valve*) regulates the amount of the liquid entering the coil to maintain a predetermined amount of superheat in the refrigerant at the coil outlet. The dry expansion coil contains mostly liquid at the inlet and only superheated vapor at the outlet, after absorbing heat from the medium to be cooled. In the air coolers, when the surface temperatures fall below 0 °C, frosting occurs. Thick layers of frost act as insulation and reduce the airflow rate (in the forced convection coils) and the available inner space.

Several methods are used for defrosting, for example, hot-gas defrost and water defrost. But recently, frost-free refrigeration systems have become popular because of the problems mentioned above.

3.7.3 Energy and Exergy Analyses of Evaporators

Evaporators are used to absorb heat from the refrigerated space. In a vapor-compression refrigeration cycle, the refrigerant is evaporated as it flows in the evaporator coils as shown in Figure 3.24a. The conservation of mass principle requires that

$$\dot{m}_1 = \dot{m}_2 \tag{3.23}$$

Referring to Figure 3.24a, energy is entering and leaving by the refrigerant stream and heat is absorbed from the cooled space, ($\dot{Q}_{in}$ or $\dot{Q}_L$). The steady-flow energy balance can be written as (with negligible kinetic and potential energies)

$$\begin{aligned} \dot{m}h_1 + \dot{Q}_{in} &= \dot{m}h_2 \\ \dot{Q}_{in} &= \dot{m}(h_2 - h_1) \end{aligned} \tag{3.24}$$

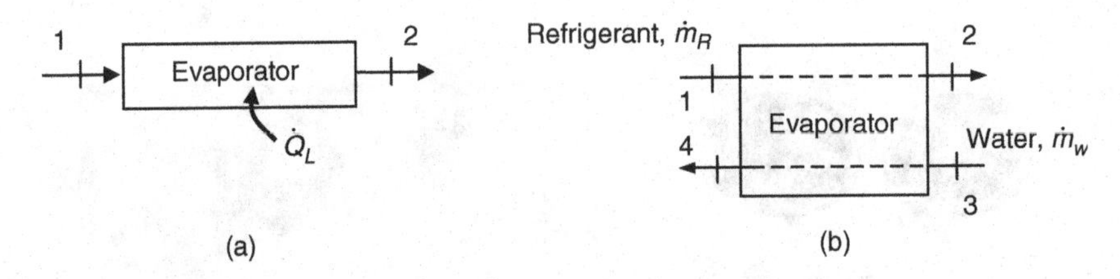

Figure 3.24 The schematic of evaporators considered for mass and energy analysis. (a) Refrigerant absorbing heat from a space. (b) Refrigerant absorbing heat from water.

In the evaporator of a household refrigerator, heat is absorbed from the freezer section as it flows in the evaporator coils. Figure 3.24a represents operation of such a compressor. Some refrigeration systems are used to cool a fluid stream in the evaporator. Figure 3.24b shows a heat exchanger operating as an evaporator in which water is cooled as the refrigerant is evaporated. Assuming that heat exchanger is insulated, the energy balance in this case becomes

$$\begin{aligned} \dot{m}_R h_1 + \dot{m}_w h_3 &= \dot{m}_R h_2 + \dot{m}_w h_4 \\ \dot{m}_R (h_2 - h_1) &= \dot{m}_w (h_3 - h_4) \end{aligned} \tag{3.25}$$

where $\dot{m}_R$ and $\dot{m}_w$ are the mass flow rates of the refrigerant and the water. The rate of heat absorbed by the refrigerant (and rejected from the water) is

$$\dot{Q}_{\text{in}} = \dot{m}_R (h_2 - h_1) = \dot{m}_w (h_3 - h_4) \tag{3.26}$$

Referring to Figure 3.24a, an entropy balance on the evaporator may be written as

$$\begin{aligned} \dot{S}_{\text{in}} - \dot{S}_{\text{out}} + \dot{S}_{\text{gen}} &= \Delta \dot{S}_{\text{sys}} = 0 \\ \dot{S}_{\text{gen}} &= \dot{S}_{\text{out}} - \dot{S}_{\text{in}} \\ \dot{S}_{\text{gen}} &= \dot{m} s_2 - \dot{m} s_1 - \frac{\dot{Q}_H}{T_H} \\ &= \dot{m} \left(s_2 - s_1 - \frac{q_L}{T_L} \right) \end{aligned} \tag{3.27}$$

Then the exergy destruction in the evaporator becomes

$$\dot{E}x_{\text{dest}} = T_0 \dot{S}_{\text{gen}} = \dot{m} T_0 \left(s_2 - s_1 - \frac{q_L}{T_L} \right) \tag{3.28}$$

The exergy destruction can also be determined by writing an exergy balance on the evaporator:

$$\begin{aligned} \dot{E}x_{\text{in}} - \dot{E}x_{\text{out}} - \dot{E}x_{\text{dest}} &= 0 \\ \dot{E}x_{\text{dest}} &= \dot{E}x_{\text{in}} - \dot{E}x_{\text{out}} \\ \dot{E}x_{\text{dest}} &= -\dot{E}x_{\dot{Q}_L} + \dot{E}x_1 - \dot{E}x_2 \\ \dot{E}x_{\text{dest}} &= (\dot{E}x_1 - \dot{E}x_2) - \dot{E}x_{\dot{Q}_L} \\ &= \dot{m} \left[h_1 - h_2 - T_0 (s_1 - s_2) \right] - \left[-\dot{Q}_L \left(1 - \frac{T_0}{T_L} \right) \right] \end{aligned} \tag{3.29}$$

The exergy efficiency of the evaporator may be expressed as the ratio of the exergy increase of the cold space as a result of losing heat to the exergy decrease of the refrigerant due to receiving heat from the cold reservoir.

$$\eta_{ex,Evap} = \frac{\dot{E}x_{\dot{Q}_L}}{\dot{E}x_1 - \dot{E}x_2} = \frac{-\dot{Q}_L\left(1 - \frac{T_0}{T_L}\right)}{\dot{m}\left[h_1 - h_2 - T_0(s_1 - s_2)\right]} = 1 - \frac{\dot{E}x_{dest}}{\dot{E}x_1 - \dot{E}x_2} \tag{3.30}$$

If we consider Figure 3.24b for the operation of an evaporator, an entropy balance may be written as

$$\dot{S}_{gen} = \dot{S}_{out} - \dot{S}_{in}$$

$$\dot{S}_{gen} = (\dot{m}_R s_2 + \dot{m}_w s_4) - (\dot{m}_R s_1 + \dot{m}_w s_3) \tag{3.31}$$

$$= \dot{m}_w(s_4 - s_3) - \dot{m}_R(s_1 - s_2)$$

Then the exergy destruction in the evaporator becomes

$$\dot{E}x_{dest} = T_0\dot{S}_{gen} = T_0\left[\dot{m}_w(s_4 - s_3) - \dot{m}_R(s_1 - s_2)\right] \tag{3.32}$$

Example 3.3

Heat is absorbed from a cooled space at 32 °F at a rate of 320 Btu/min by refrigerant-22 that enters the evaporator at −12 °F with a quality of 0.3 and leaves as saturated vapor at the same pressure. Determine (a) the volume flow rates of R-22 at the evaporator inlet and outlet and (b) the rate of exergy destruction in the evaporator and the exergy efficiency of the evaporator. Take $T_0 = 77\,°F$. The properties of R-22 at the inlet and exit of the evaporator are as follows:

$$h_1 = 102.67\,\text{Btu/lbm},\ s_1 = 0.2776\,\text{Btu/lbm}\cdot\text{R},\ v_1 = 0.5332\,\text{ft}^3/\text{lbm}$$

$$h_2 = 169.82\,\text{Btu/lbm},\ s_2 = 0.4276\,\text{Btu/lbm}\cdot\text{R},\ v_2 = 1.750\,\text{ft}^3/\text{lbm}$$

Solution

(a) The mass flow rate of R-22 may be determined from an energy balance on the evaporator to be (see Figure 3.24a)

$$\dot{Q}_L = \dot{m}(h_2 - h_1) \longrightarrow 320/60\,\text{Btu/s} = \dot{m}(169.82 - 102.67)\,\text{Btu/lbm} \longrightarrow \dot{m} = 0.0794\,\text{lbm/s}$$

The volume flow rate at the evaporator inlet and outlet are

$$\dot{V}_1 = \dot{m}v_1 = (0.0794\,\text{lbm/s})(0.5332\,\text{ft}^3/\text{lbm}) = 0.04235\,\text{ft}^3/\text{s} = \mathbf{2.54\,ft^3/min}$$

$$\dot{V}_2 = \dot{m}v_2 = (0.0794\,\text{lbm/s})(1.750\,\text{ft}^3/\text{lbm}) = 0.139\,\text{ft}^3/\text{s} = \mathbf{8.34\,ft^3/min}$$

(b) The entropy generation and the exergy destruction are

$$\dot{S}_{gen} = \dot{m}(s_2 - s_1) - \frac{\dot{Q}_L}{T_L}$$

$$= (0.0794\,\text{lbm/s})(0.4276 - 0.2776)\,\text{Btu/lbm}\cdot\text{R} - \frac{5.33\,\text{Btu/s}}{492\,\text{R}} = 0.001073\,\text{Btu/s}\cdot\text{R}$$

$$\dot{E}x_{dest} = T_0\dot{S}_{gen} = (537\,\text{R})(0.001073\,\text{Btu/s}\cdot\text{R}) = \mathbf{0.576\,Btu/s}$$

The exergy decrease of the refrigerant as it flows in the evaporator is

$$\dot{E}x_1 - \dot{E}x_2 = \dot{m}(h_1 - h_2) - \dot{m}T_0(s_1 - s_2)$$
$$= 5.33 - (0.0794\,\text{lbm/s})(537\,\text{R})(0.2776 - 0.4276)\,\text{Btu/lbm}\cdot\text{R}$$
$$= 1.06\,\text{Btu/s}$$

The exergy efficiency is then

$$\eta_{ex,Evap} = 1 - \frac{\dot{E}x_{dest}}{\dot{E}x_1 - \dot{E}x_2} = 1 - \frac{0.576}{1.06} = 0.458 = \mathbf{45.8\%}$$

3.8 Throttling Devices

In practice, throttling devices, called either *expansion valves* or *throttling valves*, are used to reduce the refrigerant condensing pressure (high pressure) to the evaporating pressure (low pressure) by a throttling operation and regulate the liquid-refrigerant flow to the evaporator to match the equipment and load characteristics. These devices are designed to proportion the rate at which the refrigerant enters the cooling coil to the rate of evaporation of the liquid refrigerant in the coil; the amount depends, of course, on the amount of heat being removed from the refrigerated space. The most common throttling devices are

- thermostatic expansion valves,
- constant-pressure expansion valves,
- float valves, and
- capillary tubes.

Note that a practical refrigeration system may consist of a large range of mechanical and electronic expansion valves and other flow-control devices for small- and large-scale refrigeration systems, comprising thermostatic expansion valves, solenoid valves, thermostats and pressostats, modulating pressure regulators, filter driers, liquid indicators, nonreturn valves and water valves, and furthermore, decentralized electronic systems for full regulation and control.

3.8.1 *Thermostatic Expansion Valves*

The thermostatic expansion valves are essentially reducing valves between the high-pressure side and the low-pressure side of the system. These valves, which are the most widely used devices, automatically control the liquid-refrigerant flow to the evaporator at a rate that matches the system capacity to the actual load. They operate by sensing the temperature of the superheated refrigerant vapor leaving the evaporator. For a given valve type and refrigerant, the associated orifice assembly is suitable for all versions of the valve body and in all evaporating temperature ranges.

When the thermostatic expansion valve is operating properly, the temperature at the outlet side of the valve is much lower than that at the inlet side. If this temperature difference does not exist when the system is in operation, the valve seat is probably dirty and clogged with foreign matter. Once a valve is properly adjusted, further adjustment should not be necessary. The major problem can usually be traced to moisture or dirt collecting at the valve seat and orifice. Figure 3.25 shows a common type of electrically driven expansion valve.

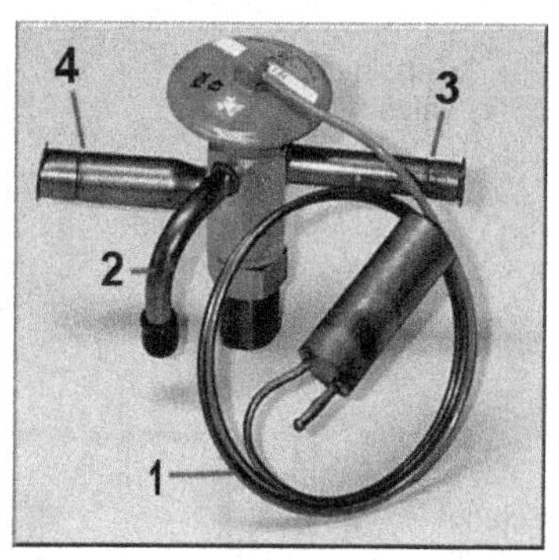

Figure 3.25 An electronic expansion valve (*Courtesy of Danfoss A/S*).

3.8.2 Constant-Pressure Expansion Valves

The constant-pressure valve is the forerunner of the thermostatic expansion valve. It is called an automatic expansion valve because of the fact that it opens and closes automatically without the aid of any external mechanical device. These expansion valves are basically pressure regulating devices. These valves maintain a constant pressure at outlet. They sense and keep the evaporated pressure at a constant value by controlling the liquid-refrigerant flow into the evaporator, based on the suction pressure. The refrigerant flows at a rate that exactly matches compressor capacity. Their applications are limited because of the constant cooling load.

3.8.3 Float Valves

These valves are divided into high-side float valves and low-side float valves. They are employed to control the refrigerant flow to a flooded-type liquid cooler. A high-side float valve is located on the high-pressure side of the throttling device. It is used in a refrigeration system with a single evaporator, compressor, and condenser. A low-side float valve is particularly located on the low-pressure side of the throttling device and may be used in refrigeration systems with multiple evaporators. In some cases, a float valve operates an electrical switch controlling a solenoid valve which periodically admits the liquid refrigerant to the evaporator, allowing the liquid level to fluctuate within preset limits.

3.8.4 Capillary Tubes

The capillary tube is the simplest type of refrigerant-flow-control device and may be used in place of an expansion valve. The capillary tubes are small-diameter tubes through which the refrigerant flows into the evaporator. These devices, which are widely used in small hermetic-type refrigeration systems (up to 30 kW capacity), reduce the condensing pressure to the evaporating pressure in a copper tube of small internal diameter (0.4–3 mm diameter and 1.5–5 m long), maintaining a constant evaporating pressure independently of the refrigeration load change. These tubes are used to transmit pressure from the sensing bulb of some temperature control device to the operating element. A capillary tube may also be constructed as a part of a heat exchanger, particularly in household refrigerators.

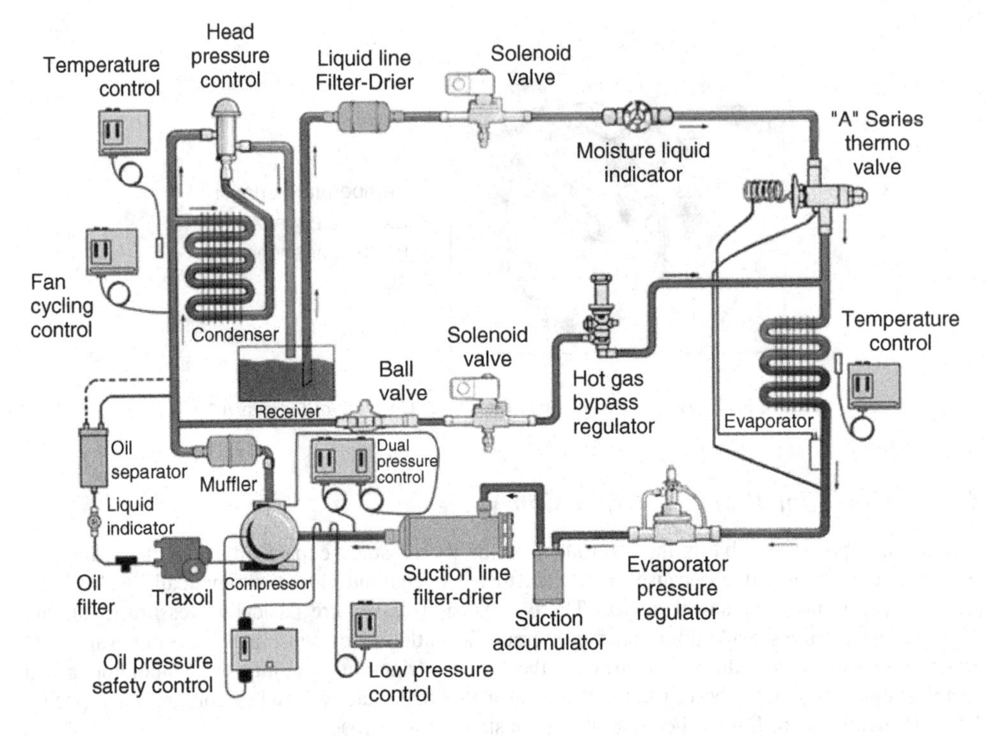

Figure 3.26 A practical vapor-compression refrigeration system with all control devices (*Courtesy of ALCO Controls*).

With capillary tubes, the length of the tube is adjusted to match the compressor capacity. Other considerations in determining capillary tube size include condenser efficiency and evaporator size. Capillary tubes are most effective when used in small-capacity systems.

Figure 3.26 shows an excellent diagram of a practical vapor-compression refrigeration system with all control devices.

3.8.5 *Energy and Exergy Analyses of Throttling Devices*

Throttling devices are used to decrease pressure of a fluid. In a vapor-compression refrigeration cycle, the refrigerant enters the throttling valve as a liquid and leaves as a saturated liquid–vapor mixture (Figure 3.27). The conservation of mass principle requires that

$$\dot{m}_1 = \dot{m}_2 \tag{3.33}$$

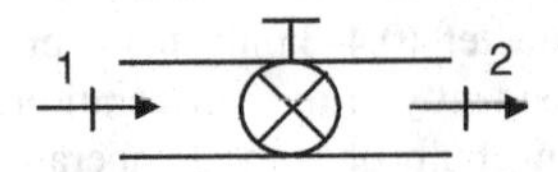

Figure 3.27 The schematic of a throttling valve considered for mass and energy analysis.

Referring to Figure 3.27, energy is entering and leaving by the refrigerant stream. Heat transfer with the surroundings is negligible and there is no work interaction. Then the steady-flow energy balance can be written as (with negligible kinetic and potential energies)

$$\dot{m}h_1 = \dot{m}h_2 \rightarrow h_1 = h_2 \tag{3.34}$$

That is, a throttling valve is essentially an isenthalpic (i.e., constant enthalpy) device. Noting that enthalpy is defined as the sum of internal energy u and flow energy Pv, we have

$$h_1 = h_2 \rightarrow u_1 + P_1v_1 = u_2 + P_2v_2 \tag{3.35}$$

In a throttling valve, pressure decreases and specific volume increases. The internal energy must decrease in order to achieve a drop in temperature. This requires that the increase in specific volume is greater than the decrease in pressure. The large increase in specific volume is made possible by turning some of the liquid into vapor.

Referring to Figure 3.27, an entropy balance on the throttling valve may be written as

$$\begin{aligned}\dot{S}_{in} - \dot{S}_{out} + \dot{S}_{gen} &= \Delta\dot{S}_{sys} = 0\\ \dot{S}_{gen} &= \dot{S}_{out} - \dot{S}_{in}\\ &= \dot{m}s_2 - \dot{m}s_1\\ &= \dot{m}(s_2 - s_1)\end{aligned} \tag{3.36}$$

Then the exergy destruction in the throttling valve becomes

$$\dot{E}x_{dest} = T_0\dot{S}_{gen} = \dot{m}T_0(s_2 - s_1) \tag{3.37}$$

The exergy destruction can also be determined by writing an exergy balance on the condenser:

$$\begin{aligned}\dot{E}x_{in} - \dot{E}x_{out} - \dot{E}x_{dest} &= 0\\ \dot{E}x_{dest} &= \dot{E}x_{in} - \dot{E}x_{out}\\ \dot{E}x_{dest} &= \dot{E}x_1 - \dot{E}x_2\\ &= \dot{m}\,[h_1 - h_2 - T_0(s_1 - s_2)]\end{aligned} \tag{3.38}$$

The exergy efficiency of the throttling valve may be expressed as the ratio of the exergy recovered to the exergy expended.

$$\eta_{ex,ExpValve} = 1 - \frac{\dot{E}x_{dest}}{\dot{E}x_1 - \dot{E}x_2} = 1 - \frac{\dot{E}x_1 - \dot{E}x_2}{\dot{E}x_1 - \dot{E}x_2} \tag{3.39}$$

Note that there is no exergy recovered in an expansion valve, and thus the exergy efficiency is zero.

Example 3.4

Refrigerant-134a enters the throttling valve of a heat pump system at 800 kPa as a saturated liquid and leaves at 140 kPa. Determine (a) the temperature of R-134a at the outlet of the throttling valve and (b) the entropy generation and the exergy destruction during this process. Take $T_0 = 25\,°C$.

Solution

(a) The properties of refrigerant at the inlet and exit states of the throttling valve are (from Table B.4)

$$\left.\begin{array}{l} P_1 = 800\,\text{kPa} \\ x_1 = 0 \end{array}\right\} \begin{array}{l} h_1 = 95.47\,\text{kJ/kg} \\ s_1 = 0.35404\,\text{kJ/kg}\cdot\text{K} \end{array}$$

$$\left.\begin{array}{l} P_2 = 140\,\text{kPa} \\ h_2 = h_1 = 95.447\,\text{kJ/kg} \end{array}\right\} \begin{array}{l} T_2 = \mathbf{-18.8\,°C} \\ s_2 = 0.3797\,\text{kJ/kg}\cdot\text{K} \end{array}$$

(b) Noting that the throttling valve is adiabatic, the entropy generation is determined from

$$s_{gen} = s_2 - s_1 = (0.3797 - 0.35404)\,\text{kJ/kg}\cdot\text{K} = \mathbf{0.0257\ kJ/kg\cdot K}$$

Then the irreversibility (i.e., exergy destruction) of the process becomes

$$ex_{dest} = T_0 s_{gen} = (298\ \text{K})(0.0257\ \text{kJ/kg}\cdot\text{K}) = \mathbf{7.65\ kJ/kg}$$

3.9 Auxiliary Devices

3.9.1 Accumulators

It is well known that compressors are designed to compress vapors, not liquids. Many refrigeration systems are subject to the return of excessive quantities of liquid refrigerant to the compressor. Liquid refrigerant returning to the compressor dilutes the oil, washes out the bearings, and in some cases causes complete loss of oil in the compressor crankcase. This condition is known as oil pumping or slugging and results in broken valve reeds, pistons, rods, crankshafts, and the like. The purpose of the accumulator is to act as a reservoir to temporarily hold the excess oil-refrigerant mixture and to return it at a rate that the compressor can safely handle. Some accumulators include a heat-exchanger coil to aid in boiling of the liquid refrigerant while subcooling the refrigerant in the liquid line (Figure 3.28), thus helping the system to operate more efficiently. Note that proper installation of a suction accumulator in the suction line just after the reversing valve and before the compressor helps eliminate the possible damage.

In large holdover plate refrigerator and freezer systems, refrigerant can accumulate in the plates and suction line when the compressor is not running. On start-up, this liquid refrigerant can be suddenly dumped into the compressor, creating a situation of liquid slugging of refrigerant and oil. This can cause damage to the compressor. When installed in the suction line of the compressor, a suction accumulator protects the compressor from this liquid slugging by gradually feeding liquid refrigerant into the compressor.

Note that accumulators should be selected according to the tonnage, evaporator temperature, and holding capacity.

3.9.2 Receivers

Some of the refrigeration units have enough space within the condenser to accommodate the entire refrigerant charge of the system. If the condenser does not have sufficient space, a receiver tank should be provided. The amount of refrigerant required for proper operation of the system determines whether or not a receiver is required. In practice, when proper unit operation requires approximately 3.6 kg or more of refrigerant, the use of a receiver is essential (Langley, 1982).

Figure 3.28 An accumulator (*Courtesy of Standard Refrigeration Company*).

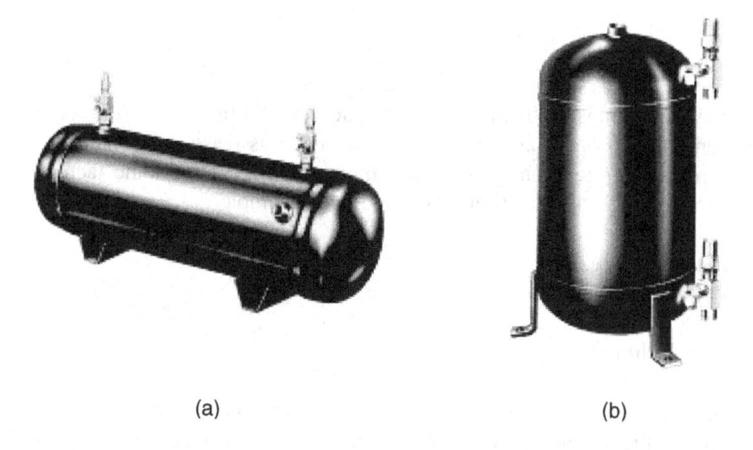

(a) (b)

Figure 3.29 Receivers. (a) Horizontal design. (b) Vertical design (*Courtesy of Standard Refrigeration Company*).

Receivers (Figure 3.29) are required on refrigeration systems that use an expansion valve for refrigerant control. The receiver provides a place to store the excess refrigerant in the system when the expansion valve restricts the flow to the evaporator. Receivers are not required, however, when using a capillary metering system. In addition to accommodating fluctuations in the refrigerant charge, the receiver aims to maintain the condenser drained of liquid, thereby preventing the liquid level from building up in the condenser and reducing the amount of effective condenser surface area.

Figure 3.30 A coalescing oil separator (*Courtesy of Standard Refrigeration Company*).

3.9.3 Oil Separators

Oil separators (Figure 3.30) provide oil separation and limit oil carry-over to approximately 0.0003–0.001% of the total amount of refrigerant, depending on various system characteristics, for example, operating conditions, refrigerant, start/stop, load/unload frequency, and so on. These separators are normally used for a large variety of refrigerants, for example, ammonia, R-134a and propane. Note that all the separators require the mounting of an external float assembly to control return from the separator to the compressor.

3.9.4 Strainers

Strainers remove foreign matter such as dirt and metal chips from the refrigerant lines. If left in the system, unwanted matter could clog the small orifices of the flow-control devices and check valves and also enter the compressor. Various types are available such as straight-through sealed type, cleanable angle type, and the cleanable Y type.

3.9.5 Driers

In refrigeration systems, moisture is the single most detrimental factor in a refrigeration system. A unit can stand only a very small amount of moisture. For this reason, the majority of both field- and factory-assembled refrigeration systems are equipped with driers. Some factors influence the selection of the correct size of drier (Langley, 1983), for example,

- type and amount of refrigerant,
- refrigeration system tonnage,
- line size, and
- allowable pressure drop.

When the refrigerant type, line size, and equipment application are known, the drier is generally selected on the basis of recommended capacities, which take into account both drying and refrigerant flow capacity.

3.9.6 Check Valves

Check valves are used for two essential goals: (i) to cause the refrigerant to flow through the flow-control device and (ii) to allow the refrigerant to bypass the flow-control device. These valves are installed in a loop that bypasses the flow-control device and only open when pressure is exerted in the right direction; therefore, they should be installed with the arrow pointing in the proper direction of refrigerant flow at the point of installation. In operation, the refrigerant pushes either against the valve seat to close it tighter or against its face to cause it to open and allow refrigerant to pass through. These valves are usually spring loaded and will open when the pressure difference on the seat reaches about 100 to 135 kPa.

3.9.7 Solenoid Valves

Solenoid valves are extensively used in all types of refrigeration applications. These valves are employed as electrically operated line stop valves and perform in the same manner as hand shut-off valves. These valves are convenient for remote applications because of the fact that these are electrically operated and controlled easily.

3.9.8 Defrost Controllers

A defrost controller with timer (Figure 3.31) operates various control valves and fan relays to quickly and efficiently remove frost and ice accumulation from evaporator surfaces. There are four easy-to-set defrost steps:

- pump out,
- hot gas,
- equalize, and
- fan delay.

This controller uses reliable, solid-state electronics with a precision quartz time clock and time interval adjusting slide knobs to sequentially operate through the four steps for smooth defrosting.

Figure 3.31 A defrost controller with timer (*Courtesy of Hansen Technologies Corporation*).

Each step is clearly indicated by a bright light emitting diode (LED) during operation. Terminals for optional sensor defrost initiation and termination are provided. A 24-hour quartz time clock facilitates simple setting in 15-minute increments of defrost start times. A 7-day quartz time clock for weekly scheduling is also available. All time clocks have 72-hour battery backup in case of short-term power failure. Because of its time-adjustable four-step defrost operation, this controller is suitable for almost every defrost application including top and bottom feed unit coolers, blast freezer evaporators, ice makers, etc.

3.10 Concluding Remarks

This chapter has dealt with a large number of theoretical and practical topics in refrigeration systems, covering the history of refrigeration, refrigeration system components, and auxiliary equipment and their technical and operational aspects, and their mass and energy analyses along with the representative examples.

Nomenclature

A area, m^2
COP coefficient of performance
c_p constant-pressure specific heat, kJ/kgK
CR compression ratio
ex specific exergy, kJ/kg
$\dot{E}$ energy rate, kW
h enthalpy, kJ/kg
$\dot{m}$ mass flow rate, kg/s
P pressure, kPa
$\dot{Q}$ rate of heat transfer, kW
R ratio of clearance volume to displacement volume
s entropy, kJ/kg
s_{gen} entropy generation rate, kJ/kg · K
T temperature, °C or K
u internal energy, kJ/kg
v specific volume, m^3/kg
V volume, m^3
$\dot{V}$ volumetric flow rate, m^3/s
$\dot{W}$ work input to compressor, kW
x quality

Greek Letters

η efficiency
ρ density, kg/m^3

Study Problems

Refrigeration System Components

3.1 What are the major components of a vapor-compression refrigeration system?

Compressors

3.2 What are the two main functions of a compressor in a refrigeration cycle?

3.3 What are the two main categories of refrigerant compressors?

3.4 What are the desirable characteristics of a compressor?

3.5 What criteria are considered in the selection of a proper refrigerant compressor?

3.6 What are the main characteristics of hermetic compressors?

3.7 What are the main applications of semihermetic compressors?

3.8 What is the difference between hermetic and semihermetic compressors?

3.9 What are the three types of positive displacement compressors?

3.10 What are the main parameters affecting the efficiencies of reciprocating compressors?

3.11 What are the general design configurations of rotary compressors?

3.12 Describe operating principle of rotary compressors.

3.13 What are the suitable applications of rotary compressors?

3.14 What are the basic advantages of vane compressors?

3.15 Is screw compressor a positive displacement compressor? What is the temperature range for screw compressors?

3.16 Describe the operating principle of a screw compressor?

3.17 What are the basic advantages of screw compressors?

3.18 What is the basic operating principle of dynamic compressors? What are the main types?

3.19 What are the suitable applications of centrifugal compressors in place of positive displacement compressors? What is the suitable load range for centrifugal compressors?

3.20 What is the basic operating principle of a centrifugal compressor?

3.21 How are compressors rated? Define compression ratio for a compressor.

3.22 What are the factors influencing the performance of a compressor?

3.23 Does lowering suction temperature decrease the power input to a refrigerant compressor? Explain.

3.24 It is known that the higher the compression ratio of a compressor the lower the efficiency. Explain how the higher compression ratios be avoided.

3.25 List the methods of compressor capacity control.

3.26 Refrigerant-134a enters the compressor of a refrigeration cycle at 120 kPa gage pressure. The condenser is maintained at an absolute pressure of 800 kPa. If the atmospheric pressure is 95 kPa, determine the compression ratio of the compressor.

3.27 Refrigerant-134a enters the compressor of a refrigeration cycle at 100 kPa and −20 °C with a flow rate of 1.8 m^3/min and leaves at 700 kPa and 50 °C. Determine (a) the power input, (b) the isentropic efficiency, and (c) the exergy destruction and the exergy efficiency of the compressor. Take $T_0 = 25\,°C$.

3.28 Refrigerant-134a enters the compressor of a refrigeration cycle at 160 kPa as a saturated vapor with a flow rate of 6.5 m^3/min and leaves at 900 kPa. The compressor isentropic efficiency is 75%. Determine (a) the temperature of R-134a at the exit of the compressor and (b) the exergy destruction and the exergy efficiency of the compressor. Take $T_0 = 25\,°C$.

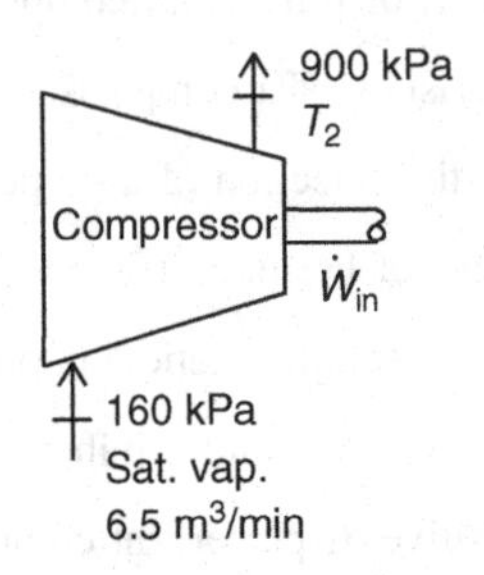

3.29 Refrigerant-134a enters the compressor of a refrigeration cycle at 100 kPa and −20 °C with a flow rate of 0.18 m^3/min and leaves at 700 kPa and 50 °C. The ratio of the clearance volume to the displacement volume is 0.05. Determine the volumetric efficiency of the compressor.

3.30 Refrigerant-134a enters the compressor of a refrigeration cycle at 160 kPa as a saturated vapor and leaves at 900 kPa. The compressor volumetric efficiency is 85% and the ratio of the clearance volume to the displacement volume is 0.04. Determine the temperature of R-134a at the exit of the compressor.

3.31 Refrigerant-134a enters the evaporator of a refrigeration cycle at 200 kPa with a vapor mass fraction of 0.15 and leaves at 1200 kPa as a saturated vapor with a flow rate of 0.045 m^3/min. The volumetric efficiency of the compressor is 92%. Determine the refrigeration capacity of the system.

Condensers

3.32 What criteria are used in the selection of condensers?

3.33 What are the main types of condensers?

3.34 What are the advantages and disadvantages of air-cooled condensers?

3.35 Describe the operating principle of a cooling tower.

3.36 What is the effect of climatic conditions on the effectiveness of evaporative condensers?

3.37 Refrigerant-134a enters the condenser of a refrigeration cycle at 1000 kPa and 80 °C with a flow rate of 0.038 kg/s and leaves at the same pressure subcooled by 4.4 °C. The refrigerant is condensed by rejecting its heat to water, which experiences a temperature rise of 9 °C. Determine (a) the rate of heat rejected in the condenser, (b) the mass flow rate of water, and (c) the rate of cooling if the COP of this refrigeration cycle at these conditions is 1.4.

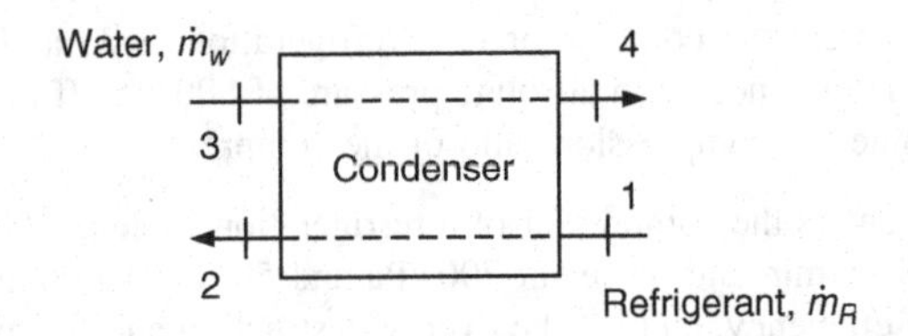

3.38 Heat is rejected from the condenser of a heat pump cycle by refrigerant-134a entering at 700 kPa and 50 °C at a rate of 105 kg/h and leaves as a saturated liquid. Determine (a) the temperature of R-134a at the condenser exit, (b) the volume flow rate at the exit of the condenser in L/min, (c) the COP of the heat pump if the rate of heat absorbed in the evaporator is 12,000 Btu/h, and (d) the rate of exergy destruction. Take $T_0 = 77\,°F$.

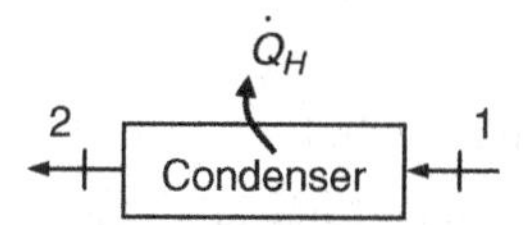

3.39 A vapor-compression refrigeration cycle uses ammonia as the working fluid. Heat is rejected from ammonia to air in the condenser. The air enters at 70 °F at a rate of 45 lbm/min and leaves at 85 °F. Ammonia experiences an enthalpy change of 86 Btu/lbm as it flows through the condenser. Determine (a) the rate of heat rejected in the condenser in Btu/h and (b) the ratio of mass flow rates of air and ammonia. Take the specific heat of air to be 0.240 Btu/lbm·°F.

Evaporators

3.40 How can evaporators be classified?

3.41 List some applications of liquid coolers in refrigeration.

3.42 What is the difference between the operation of a flooded evaporator and a direct expansion evaporator (also called dry cooler)? Which one is more preferable?

3.43 Heat is absorbed from a cooled space at a rate of 320 kJ/min by refrigerant-22 that enters the evaporator at −10 °C with a quality of 0.3 and leaves as saturated vapor at the same pressure. Determine the volume flow rates of R-22 at the compressor inlet and outlet. The properties of R-22 at the inlet and exit of the evaporator are as follows: $h_1 = 252.16$ kJ/kg, $v_1 = 0.02010$ m^3/kg, $h_2 = 401.10$ kJ/kg, $v_2 = 0.06523$ m^3/kg

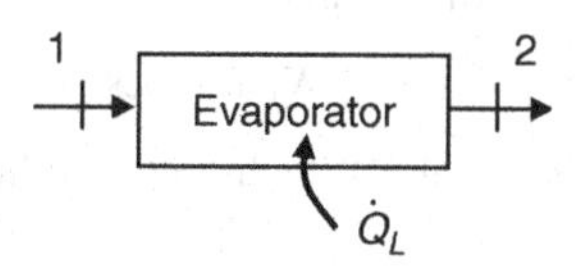

3.44 Refrigerant-134a enters the expansion valve of a refrigeration cycle at 900 kPa as a saturated liquid with a flow rate of 150 L/h. R-134a leaves the evaporator at 100 kPa superheated by 6.4 °C. The refrigerant is evaporated by absorbing heat from air which is cooled from 15 to 2 °C. Determine (a) the rate of heat absorbed in the evaporator, (b) the mass flow rate of air, (c) the COP of the cycle if the compressor work input is 72.5 kJ/kg, and (d) the rate of entropy generation and exergy destruction in the evaporator. Take $T_0 = 25\,°C$.

3.45 A heat pump operates on a vapor-compression refrigeration cycle with R-134a as the refrigerant. R-134a enters the evaporator at −12.7 °C with a vapor mass fraction of 27% and leaves at the same pressure as a saturated vapor. The refrigerant is evaporated by absorbing heat from ambient air at 0 °C. Determine (a) the amount of heat absorbed from the ambient air and (b) the exergy destruction in the evaporator, both per unit mass flow rate of the refrigerant.

Throttling Devices

3.46 List the most common throttling devices.

3.47 Can thermostatic expansion valves control the rate of liquid-refrigerant flow to the evaporator? If so, how is this done?

3.48 If there is no temperature drop across a thermostatic expansion valve, what could be the reason? Explain.

3.49 Explain characteristics of capillary tubes.

3.50 Refrigerant-134a enters the throttling valve of a heat pump system at 200 psia as a saturated liquid and leaves at 20 psia. Determine (a) the temperature drop across the throttling valve and (b) the entropy generation and the exergy destruction during this process. Take $T_0 =$ 77 °F.

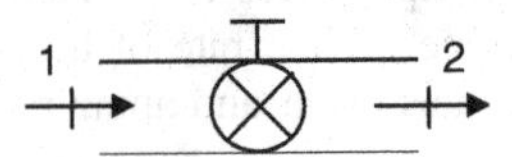

3.51 Refrigerant-502 (a blend of R-115 and R-22) enters the throttling valve of a heat pump system at 45 °C as a saturated liquid and leaves at −22 °C as a mixture of saturated liquid and vapor. Determine (a) the pressures at the inlet and exit of the valve and the vapor mass fraction at the exit and (b) the entropy generation during this process. R-502 properties are not available in the book. Use other sources to solve this problem.

Auxiliary Devices

3.52 List auxiliary devices used in refrigeration systems.

3.53 What is the purpose of using an accumulator?

3.54 What is the purpose of using a receiver?

3.55 What is the purpose of using an oil separator?

3.56 What is the purpose of using a strainer? What types are available?

3.57 What is the purpose of using a drier? Which factors influence the selection of the correct size of a drier?

3.58 What is the purpose of using a check valve?

3.59 Describe the operation of a defrost controller with timer.

References

ARI (2000) *Variable Capacity Positive Displacement Refrigerant Compressors and Compressor Units for Air Conditioning and Heat Pump Applications*, Standard 500-2000, Air Conditioning and Refrigeration Institute, Arlington, VA.

Bejan, A. (2004) *Convection Heat Transfer*, 3rd edn, John Wiley & Sons, Ltd., London.

Awberry, J.H. (1942) Carl von Linde: a pioneer of deep refrigeration. *Nature*, **149**, 630.

Critchell, J.T. and Raymond, J. (1912) *A History of Frozen Meat Trade*, 2nd edn, London, pp. 4–5.

DETR (1999) *The Engine of the Refrigeration System: Selecting and Running Compressors for Maximum Efficiency*, vol. 52, The Department of the Environment, Transport and Regions' Energy Efficiency Best Practice Programme, London General Information Leaflet, p. 8.

Dincer, I. (1997) *Heat Transfer in Food Cooling Applications*, Taylor & Francis, Washington, DC.

Dincer, I. (2003) *Refrigeration Systems and Applications*, 1st edn, John Wiley & Sons, Ltd., New York.

DOI (1952) *Report of the Commissioner of Patents for the Year 1951*, U.S. Department of Interior, Patent Office, Washington, DC, p. 76.

Duncan, T. (1999) The rotary screw compressor. *ASHRAE Journal*, **41**, 34–36.

Goosman, J.C. (1924) History of refrigeration. *Ice and Refrigeration*, **67**, 329.

Heap, R.D. (1979) American heat pumps in British houses . *Elektrowärme International 35*, **A2**, A77–A81.

Hewitt, G.F., Shires, G.L. and Bott, T.R. (1994) *Process Heat Transfer*, CRC Press, Boca Raton, FL.

Langley, B.C. (1982) *Basic Refrigeration*, Reston Publishing Company, Reston, VA.

Langley, B.C. (1983) *Heat Pump Technology*, Reston Publishing Company, Reston, VA.

Neuberger, A. (1930) *The Technical Arts and Sciences of the Ancients*, (ed. H.L. Brose), tr. New York, p. 123.

Roelker, H.B. (1906) The Allen dense air refrigerating machine . *Transactions American Soc. Refrig. Engineers*, **2**, 52–54.

Travers, M.W. (1946) *Liquefaction of Gases*, vol. 14, Encyclopaedia Britannica, Chicago, pp. 172–173.

Woolrich, W.R. (1947) Mechanical refrigeration – its American birthright. *Refrigerating Engineering*, **53**, 250.

4

Refrigeration Cycles and Systems

4.1 Introduction

Refrigeration is used in industry for cooling and freezing of products, condensing vapors, maintaining environmental conditions, and for cold storage. The number of different applications is huge and they are a major consumer of electricity. In some sectors, particularly food, drink, and chemicals it represents a significant proportion of overall site energy costs (up to 90% in the case of some cold storage facilities) (Dincer, 2003).

Presently, the refrigeration industry urgently needs (i) technical information on the refrigeration systems, system components, and technical and operational aspects of such systems and components; (ii) procedures for energy and exergy analyses of refrigeration systems for system design and optimization; (iii) application of optimum refrigeration techniques; (iv) techniques for the measurement and evaluation of the components' performance; and (v) methodology for the use of the cooling data to design an efficient and effective refrigeration system and/or to improve the existing refrigeration systems.

The primary objective of this chapter is to discuss refrigeration cycles and their energy and exergy analyses, some new refrigeration techniques for more efficient and effective refrigeration, and to provide some illustrative and practical examples to highlight the importance of the topic and show how to conduct energy and exergy analyses for the refrigeration systems.

4.2 Vapor-Compression Refrigeration Systems

In practical applications, vapor-compression refrigeration systems are the most commonly used refrigeration systems, and each system employs a compressor. In a basic vapor-compression refrigeration cycle as shown in Figure 4.1, four major thermal processes take place as follows:

- evaporation,
- compression,
- condensation, and
- expansion.

4.2.1 Evaporation

Unlike freezing and melting, evaporation and condensation occur at almost any temperature and pressure combination. Evaporation is the gaseous escape of molecules from the surface of a liquid and is accomplished by the absorption of a considerable quantity of heat without any change in temperature. Liquids (e.g., refrigerants) evaporate at all temperatures with increased rates of

Refrigeration Systems and Applications İbrahim Dinçer and Mehmet Kanoğlu

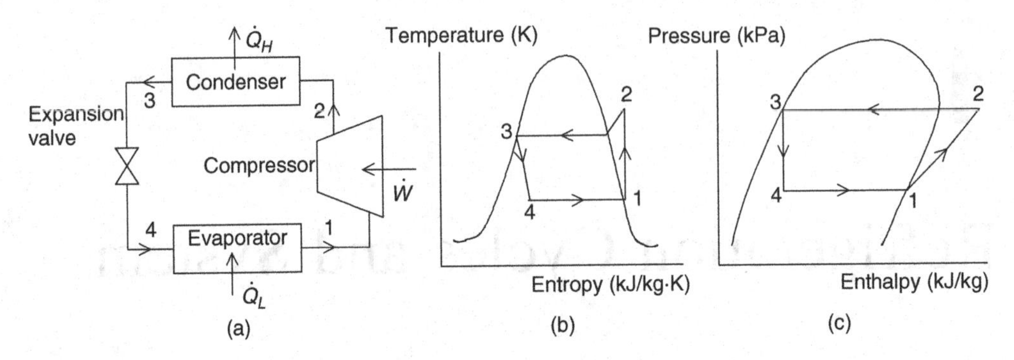

Figure 4.1 (a) A basic vapor-compression refrigeration system, (b) its T-s diagram, and (c) its log P-h diagram.

evaporation occurring at higher temperatures. The evaporated gases exert a pressure called the vapor pressure. As the temperature of the liquid rises, there is a greater loss of the liquid from the surface, which increases the vapor pressure. In the evaporator of a refrigeration system, a low-pressure cool refrigerant vapor is brought into contact with the medium or matter to be cooled (i.e., heat sink), absorbs heat, and hence boils, producing a low-pressure saturated vapor.

4.2.2 Compression

Using shaft work of a compressor raises the pressure of the refrigerant vapor obtained from the evaporator. The addition of heat may play a role in raising the pressure. Increasing the gas pressure raises the boiling and condensing temperature of the refrigerant. When the gaseous refrigerant is sufficiently compressed, its boiling point temperature is higher than the heat sink's temperature.

4.2.3 Condensation

This is a process of changing a vapor into a liquid by extracting heat. The high-pressure gaseous refrigerant, which carries the heat energy absorbed in the evaporator and the work energy from the compressor, is brought into the condenser. The condensing temperature of the refrigerant is higher than that of the heat sink and therefore heat transfer condenses the high-pressure refrigerant vapor to the high-pressure saturated liquid. The heat source has been cooled by pumping heat to the heat sink. Instead of using a condenser to reject heat, the refrigerant vapor can be discharged to the atmosphere, but this technique is impractical. Condensing the refrigerant gas allows reuse at the beginning of the next cycle. In some practical applications, it is desired that the condenser cools the refrigerant further, below the condensation temperature. This is called *subcooling*, which is usually observed in the condenser to reduce flashing when the refrigerant pressure is reduced in the throttling device. This method provides a reduction in the amount of gas entering the evaporator and hence an improvement in the system performance (Dincer, 1997).

4.2.4 Expansion

The condensed refrigerant liquid is returned to the beginning of the next cycle. A throttling device such as a valve, orifice plate, or capillary tube for the expansion process is used to reduce the pressure of the refrigerant liquid to the low-pressure level and the boiling temperature of the refrigerant to below the temperature of the heat source. Energy losses through this pressure reduction must be offset by additional energy input at the pressurization stage.

Figure 4.1a shows a schematic diagram of a basic vapor-compression refrigeration system. For better understanding, this refrigeration cycle is shown by temperature–entropy (T–s) and pressure–enthalpy (log P–h) diagrams as given in Figure 4.1b and c. Along the lines of the steps given above, the operation of this system is as follows:

- **(1–2) Reversible adiabatic compression.** From the evaporator, low-pressure saturated refrigerant vapor comes to the compressor and is compressed into the condenser by volume reduction and increased pressure and temperature.
- **(2–3) Reversible heat rejection at constant pressure.** From the compressor, high-pressure refrigerant vapor enters the condenser and is liquefied by employing water or air.
- **(3–4) Irreversible expansion at constant enthalpy.** From the condenser, high-pressure saturated refrigerant liquid passes through an expansion valve and its pressure and temperature are reduced.
- **(4–1) Reversible heat addition at constant pressure.** From the expansion valve, low-pressure refrigerant liquid arrives in the evaporator. It boils here and in the process absorbs heat from the surrounding medium, thereby providing a cooling effect.

As shown in Figure 4.1, the essential components of a simple vapor-compression refrigeration system, as explained earlier, are as follows:

- **Evaporator.** This is the device where there is heat exchange for providing refrigeration, and therefore it boils the liquid refrigerant at a low temperature, which causes the refrigerant to absorb heat.
- **Suction line.** This is the tube between the evaporator and the compressor. After the liquid has absorbed the heat, the suction line carries the refrigerant to the compressor. In this line, the refrigerant is a superheated gas.
- **Compressor.** This device separates the low-pressure side of the system from the high-pressure side and has two main goals: (i) to remove vapor from the evaporator to keep the evaporator's boiling point low and (ii) to compress the low-temperature refrigerant vapor into a small volume, creating a high-temperature, high-pressure superheated vapor.
- **Hot gas discharge line.** This tube connects the compressor with the condenser. After the compressor has discharged the high-pressure, high-temperature superheated refrigerant vapor, the hot gas discharge line carries it to the condenser.
- **Condenser.** This device is used for heat exchange, similar to the evaporator, except that its job is to expel heat, not absorb it. The condenser changes the state of the superheated refrigerant vapor back into a liquid. This is done by creating a high pressure that raises the boiling point of the refrigerant and removes enough heat to cause the refrigerant to condense back into a liquid.
- **Liquid line.** This line connects the condenser with the refrigerant control device, including the expansion valve. Only liquid refrigerant should be in this line. Also, the line will be somewhat warm because the refrigerant is still under high pressure.
- **Refrigerant control.** This last control works as a metering device. It monitors the liquid refrigerant that enters the evaporator and makes sure that all the liquid is boiled off before the refrigerant goes to the suction line. If liquid refrigerant enters the suction line, it will enter the compressor and cause it to fail.

In addition to the above listed components, there are some additional features, for example, liquid receiver, service valves, suction service valve, discharge service valve, and liquid receiver service valve, which can enhance the refrigeration system's operation.

4.3 Energy Analysis of Vapor-Compression Refrigeration Cycle

A vapor-compression refrigeration cycle consists of a number of flow processes as mentioned above and can be analyzed by applying steady-state flow according to the first law of thermodynamics, as applied to each of the four components individually (Figure 4.2a), since energy must be conserved by each component and also by the whole system. Therefore, the energy balance equation for each component of the system is as follows (with the assumption that the changes in kinetic and potential energies are negligible)

For compressor:

$$\dot{E}_{\text{in}} = \dot{E}_{\text{out}}$$

$$\dot{m}h_1 + \dot{W} = \dot{m}h_2 \tag{4.1}$$

$$\dot{W} = \dot{m}(h_2 - h_1)$$

where $\dot{m}$ is mass flow rate of refrigerant, kg/s; h is enthalpy, kJ/kg; and $\dot{W}$ is compressor power input, kW.

For condenser:

$$\dot{m}h_2 = \dot{m}h_3 + \dot{Q}_H$$

$$\dot{Q}_H = \dot{m}(h_2 - h_3) \tag{4.2}$$

where $\dot{Q}_H$ is the heat rejection from the condenser to the high-temperature environment.

For expansion valve:

$$\dot{m}h_3 = \dot{m}h_4$$

$$h_3 = h_4 \tag{4.3}$$

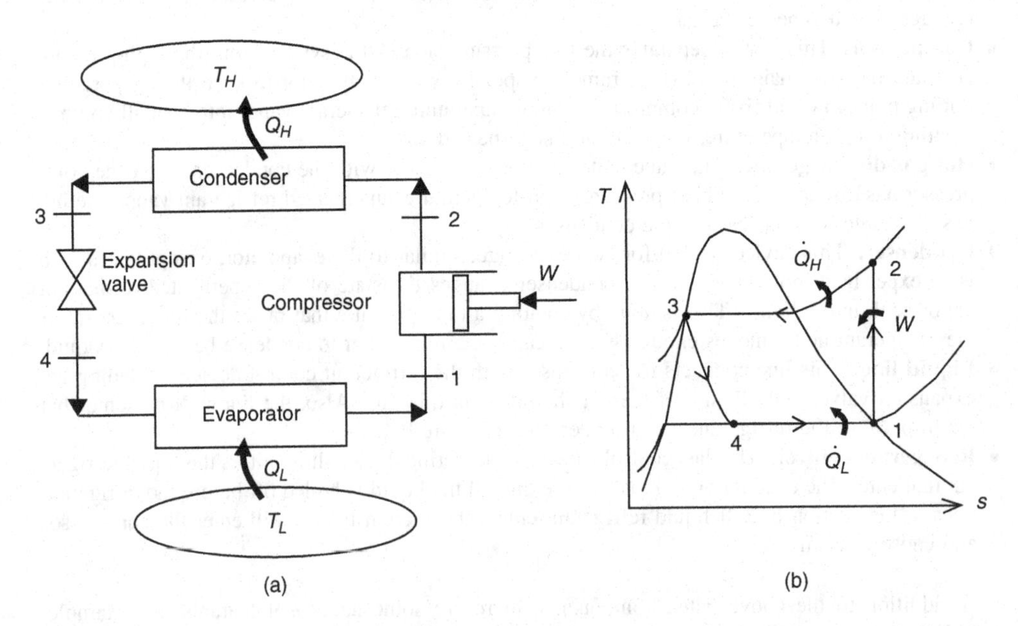

Figure 4.2 An ideal vapor-compression refrigeration system for analysis and its temperature–entropy diagram.

For evaporator:

$$\dot{m}h_4 + \dot{Q}_L = \dot{m}h_1$$

$$\dot{Q}_L = \dot{m}(h_1 - h_4) \tag{4.4}$$

where $\dot{Q}_L$ is the heat taken from the low-temperature environment to the evaporator.

For the entire refrigeration system, the energy balance can be written as

$$\dot{W} + \dot{Q}_L = \dot{Q}_H \tag{4.5}$$

The coefficient of performance (COP) of the refrigeration system becomes

$$\text{COP} = \frac{\dot{Q}_L}{\dot{W}} \tag{4.6}$$

The isentropic efficiency of an adiabatic compressor is defined as

$$\eta_{\text{Comp}} = \frac{\dot{W}_{\text{isen}}}{\dot{W}} = \frac{h_{2s} - h_1}{h_2 - h_1} \tag{4.7}$$

where h_{2s} is the enthalpy of the refrigerant at the turbine exit, if the compression process is isentropic (i.e., reversible and adiabatic).

The temperature–entropy diagram of an ideal vapor-compression refrigeration cycle is given in Figure 4.2b. In this cycle, the refrigerant enters the compressor as a saturated vapor. It is compressed isentropically in a compressor; it is cooled and condensed at constant pressure by rejecting heat to a high-temperature medium until it exists as a saturated vapor at the exit of the condenser. The refrigerant is expanded in an expansion valve, during which the enthalpy remains constant; it is evaporated in the evaporator at constant pressure by absorbing heat from the refrigerated space and it leaves the evaporator as a saturated vapor.

Note that in the energy analysis of this kind of vapor-compression system, it is required to obtain the enthalpy values. Three practical methods are available:

- using log *P–h* (pressure–enthalpy) diagrams, which provide the thermodynamic properties of the refrigerants,
- using the tabulated numerical values of the thermodynamic properties of the refrigerants, and
- using known values of the latent heats and specific heats of the refrigerants and making use of the fact that areas on the *T–s* diagrams represent heat quantities.

Thermodynamic property tables for Refrigerant-134a is given in the Appendix for both SI (Tables B.3–B.5) and English (Tables B.6–B.8) unit systems.

Example 4.1

Refrigerant-134a enters the compressor of a vapor-compression refrigeration cycle at 120 kPa as a saturated vapor and leaves at 900 kPa and 75 °C (Figure 4.2a) . The refrigerant leaves the condenser as a saturated liquid. The rate of cooling provided by the system is 18,000 Btu/h. Determine (a) the mass flow rate of R-134a and (b) the COP of the cycle. (c) Also, determine the COP of the cycle if the expansion valve is replaced by an isentropic turbine. Do you recommend such a replacement for refrigeration systems? (d) Determine the COP if the evaporator pressure is 160 kPa and other values remain the same. (e) Determine the COP if the condenser pressure is 800 kPa and other values remain the same.

Solution

Temperature–entropy diagram of the cycle is given in Figure 4.3.

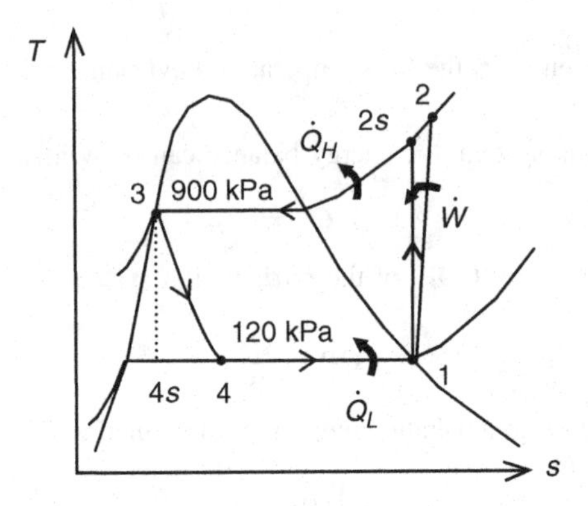

Figure 4.3 Temperature–entropy diagram of vapor-compression refrigeration cycle considered in Example 4.1.

(a) The properties of R-134a are (from Tables B.3–B.5)

$$\left.\begin{array}{l} P_1 = 120 \text{ kPa} \\ x_1 = 1 \end{array}\right\} h_1 = 236.97 \text{ kJ/kg}$$

$$\left.\begin{array}{l} P_2 = 900 \text{ kPa} \\ T_2 = 75\ ^\circ\text{C} \end{array}\right\} h_2 = 310.51 \text{ kJ/kg}$$

$$\left.\begin{array}{l} P_3 = 900 \text{ kPa} \\ x_3 = 0 \end{array}\right\} h_3 = 101.61 \text{ kJ/kg}$$

$$h_4 = h_3 = 101.61 \text{ kJ/kg}$$

The work input and heat removal per unit mass of the refrigerant are

$$w = h_2 - h_1 = 310.51 - 236.97 = 73.54 \text{ kJ/kg}$$

$$q_L = h_1 - h_4 = 236.97 - 101.61 = 135.4 \text{ kJ/kg}$$

The mass flow rate of R-134a is

$$\dot{m}_R = \frac{\dot{Q}_L}{q_L} = \frac{(18,000 \text{ Btu/h})\left(\frac{1 \text{ kW}}{3412.14 \text{ Btu/h}}\right)}{135.4 \text{ kJ/kg}} = \frac{5.275 \text{ kW}}{135.4 \text{ kJ/kg}} = \mathbf{0.0390\,kg/s}$$

(b) The COP of the refrigerator is

$$\text{COP} = \frac{q_L}{w} = \frac{135.4 \text{ kJ/kg}}{73.54 \text{ kJ/kg}} = \mathbf{1.84}$$

(c) If the expansion valve is replaced by an isentropic turbine

$$\left.\begin{array}{l} P_3 = 900 \text{ kPa} \\ x_3 = 0 \end{array}\right\} s_3 = 0.3738 \text{ kJ/kg} \cdot \text{K}$$

$$\left.\begin{array}{l} P_4 = 120 \text{ kPa} \\ s_4 = s_3 = 0.3738 \text{ kJ/kg} \cdot \text{K} \end{array}\right\} h_{4s} = 92.98 \text{ kJ/kg}$$

$$w_{\text{Turb,out}} = h_3 - h_{4s} = 101.61 - 92.98 = 8.63 \text{ kJ/kg}$$

$$w_{\text{net,in}} = w_{\text{Comp,in}} - w_{\text{Turb,out}} = 73.54 - 8.63 = 64.91 \text{ k/kg}$$

$$q_L = h_1 - h_{4s} = 236.97 - 92.98 = 144.0 \text{ kJ/kg}$$

$$\text{COP} = \frac{q_L}{w_{\text{net,in}}} = \frac{144.0 \text{ kJ/kg}}{64.91 \text{ kJ/kg}} = \mathbf{2.21}$$

The COP increases by 20.1% by replacing the expansion valve by a turbine. This replacement makes thermodynamic sense since it decreases the work requirement and thus increases COP. However, this is not practical for household refrigerators and most other refrigeration systems. In natural gas liquefaction plants, the liquefied natural gas is expanded by cryogenic turbines, which is proven to be feasible.

(d) If the evaporator pressure is 160 kPa,

$$\text{COP} = \frac{q_L}{w} = \frac{h_1 - h_4}{h_2 - h_1} = \frac{241.12 - 101.61}{310.51 - 241.12} = \mathbf{2.01}$$

Increasing evaporator pressure from 120 to 160 kPa (increasing evaporating temperature from −22.3 to −15.6 °C) increases the COP from 1.84 to 2.01, an increase of 9.2%.

(e) If the condenser pressure is 800 kPa,

$$\text{COP} = \frac{q_L}{w} = \frac{h_1 - h_4}{h_2 - h_1} = \frac{236.97 - 95.47}{311.92 - 236.97} = \mathbf{1.89}$$

Decreasing condenser pressure from 900 to 800 kPa (decreasing condensing temperature from 35.5 to 31.3 °C) increases the COP from 1.84 to 1.89, an increase of 2.7%.

4.4 Exergy Analysis of Vapor-Compression Refrigeration Cycle

Figure 4.2 is a schematic of a vapor-compression refrigeration cycle operating between a low-temperature medium (T_L) and a high-temperature medium (T_H). The maximum COP of a refrigeration cycle operating between temperature limits of T_L and T_H based on the Carnot refrigeration cycle was given in Chapter 1 as

$$\text{COP}_{\text{Carnot}} = \frac{T_L}{T_H - T_L} = \frac{1}{T_H/T_L - 1} \tag{4.8}$$

Practical refrigeration systems are not as efficient as ideal models like the Carnot cycle, because of the lower COP due to irreversibilities in the system. As a result of Equation 4.8, a smaller temperature difference between the heat sink and the heat source ($T_H - T_L$) provides greater refrigeration system efficiency (i.e., COP). The Carnot cycle has certain limitations, because it represents the cycle of the maximum theoretical performance.

The aim in an exergy analysis is usually to determine the exergy destructions in each component of the system and to determine exergy efficiencies. The components with greater exergy destructions are also those with more potential for improvements. Exergy destruction in a component can be determined from an exergy balance on the component. It can also be determined by first calculating the entropy generation and using

$$\dot{E}x_{\text{dest}} = T_0 \dot{S}_{\text{gen}} \tag{4.9}$$

where T_0 is the dead-state temperature or environment temperature. In a refrigerator, T_0 is usually equal to the temperature of the high-temperature medium T_H. Exergy destructions and exergy efficiencies for major components of the cycle are as follows (state numbers refer to Figure 4.2):

Compressor:

$$\dot{E}x_{in} - \dot{E}x_{out} - \dot{E}x_{dest,1-2} = 0$$

$$\dot{E}x_{dest,1-2} = \dot{E}x_{in} - \dot{E}x_{out} \tag{4.10}$$

$$\dot{E}x_{dest,1-2} = \dot{W} + \dot{E}x_1 - \dot{E}x_2$$

$$= \dot{W} - \Delta\dot{E}x_{12} = \dot{W} - \dot{m}\,[h_2 - h_1 - T_0(s_2 - s_1)] = \dot{W} - \dot{W}_{rev}$$

or

$$\dot{E}x_{dest,1-2} = T_0\dot{S}_{gen,1-2} = \dot{m}T_0(s_2 - s_1) \tag{4.11}$$

$$\eta_{ex,Comp} = \frac{\dot{W}_{rev}}{\dot{W}} = 1 - \frac{\dot{E}x_{dest,1-2}}{\dot{W}} \tag{4.12}$$

Condenser:

$$\dot{E}x_{dest,2-3} = \dot{E}x_{in} - \dot{E}x_{out}$$

$$\dot{E}x_{dest,2-3} = (\dot{E}x_2 - \dot{E}x_3) - \dot{E}x_{\dot{Q}_H} \tag{4.13}$$

$$= \dot{m}\,[h_2 - h_3 - T_0(s_2 - s_3)] - \dot{Q}_H\left(1 - \frac{T_0}{T_H}\right)$$

or

$$\dot{E}x_{dest,2-3} = T_0\dot{S}_{gen,2-3} = \dot{m}T_0\left(s_3 - s_2 + \frac{q_H}{T_H}\right) \tag{4.14}$$

$$\eta_{ex,Cond} = \frac{\dot{E}x_{\dot{Q}_H}}{\dot{E}x_2 - \dot{E}x_3} = \frac{\dot{Q}_H\left(1 - \frac{T_0}{T_H}\right)}{\dot{m}\,[h_2 - h_3 - T_0(s_2 - s_3)]} = 1 - \frac{\dot{E}x_{dest,2-3}}{\dot{E}x_2 - \dot{E}x_3} \tag{4.15}$$

Expansion valve:

$$\dot{E}x_{dest,3-4} = \dot{E}x_{in} - \dot{E}x_{out}$$

$$\dot{E}x_{dest,3-4} = \dot{E}x_3 - \dot{E}x_4 = \dot{m}\,[h_3 - h_4 - T_0(s_3 - s_{43})] \tag{4.16}$$

or

$$\dot{E}x_{dest,3-4} = T_0\dot{S}_{gen,3-4} = \dot{m}T_0(s_4 - s_3) \tag{4.17}$$

$$\eta_{ex,ExpValve} = 1 - \frac{\dot{E}x_{dest,3-4}}{\dot{E}x_3 - \dot{E}x_4} = 1 - \frac{\dot{E}x_3 - \dot{E}x_4}{\dot{E}x_3 - \dot{E}x_4} \tag{4.18}$$

Evaporator:

$$\dot{E}x_{dest,4-1} = \dot{E}x_{in} - \dot{E}x_{out}$$

$$\dot{E}x_{dest,4-1} = -\dot{E}x_{\dot{Q}_L} + \dot{E}x_4 - \dot{E}x_1$$

$$\dot{E}x_{dest,4-1} = (\dot{E}x_4 - \dot{E}x_1) - \dot{E}x_{\dot{Q}_L} \tag{4.19}$$

$$= \dot{m}\,[h_4 - h_1 - T_0(s_4 - s_1)] - \left[-\dot{Q}_L\left(1 - \frac{T_0}{T_L}\right)\right]$$

or

$$\dot{E}x_{\text{dest},4-1} = T_0 \dot{S}_{\text{gen},4-1} = \dot{m} T_0 \left(s_1 - s_4 - \frac{q_L}{T_L} \right) \tag{4.20}$$

$$\eta_{\text{ex,Evap}} = \frac{\dot{E}x_{\dot{Q}_L}}{\dot{E}x_1 - \dot{E}x_4} = \frac{-\dot{Q}_L \left(1 - \frac{T_0}{T_L}\right)}{\dot{m}\,[h_1 - h_4 - T_0(s_1 - s_4)]} = 1 - \frac{\dot{E}x_{\text{dest},4-1}}{\dot{E}x_1 - \dot{E}x_4} \tag{4.21}$$

The total exergy destruction in the cycle can be determined by adding exergy destructions in each component:

$$\dot{E}x_{\text{dest,total}} = \dot{E}x_{\text{dest},1-2} + \dot{E}x_{\text{dest},2-3} + \dot{E}x_{\text{dest},3-4} + \dot{E}x_{\text{dest},4-1} \tag{4.22}$$

It can be shown that the total exergy destruction in the cycle can also be expressed as the difference between the exergy supplied (power input) and the exergy recovered (the exergy of the heat transferred from the low-temperature medium):

$$\dot{E}x_{\text{dest,total}} = \dot{W} - \dot{E}x_{\dot{Q}_L} \tag{4.23}$$

where the exergy of the heat transferred from the low-temperature medium is given by

$$\dot{E}x_{\dot{Q}_L} = -\dot{Q}_L \left(1 - \frac{T_0}{T_L}\right) \tag{4.24}$$

The minus sign is needed to make the result positive. Note that the exergy of the heat transferred from the low-temperature medium is in fact the minimum power input to accomplish the required refrigeration load $\dot{Q}_L$:

$$\dot{W}_{\min} = \dot{E}x_{\dot{Q}_L} \tag{4.25}$$

The second-law efficiency (or exergy efficiency) of the cycle is defined as

$$\eta_{\text{II}} = \frac{\dot{E}x_{\dot{Q}_L}}{\dot{W}} = \frac{\dot{W}_{\min}}{\dot{W}} = 1 - \frac{\dot{E}x_{\text{dest,total}}}{\dot{W}} \tag{4.26}$$

Substituting $\dot{W} = \dfrac{\dot{Q}_L}{\text{COP}}$ and $\dot{E}x_{\dot{Q}_L} = -\dot{Q}_L \left(1 - \dfrac{T_0}{T_L}\right)$ into the second-law efficiency relation (Equation 4.26)

$$\eta_{\text{II}} = \frac{\dot{E}x_{\dot{Q}_L}}{\dot{W}} = \frac{-\dot{Q}_L \left(1 - \frac{T_0}{T_L}\right)}{\frac{\dot{Q}_L}{\text{COP}}} = -\dot{Q}_L \left(1 - \frac{T_0}{T_L}\right) \frac{\text{COP}}{\dot{Q}_L} = \frac{\text{COP}}{\frac{T_L}{T_H - T_L}} = \frac{\text{COP}}{\text{COP}_{\text{Carnot}}} \tag{4.27}$$

since $T_0 = T_H$. Thus, the second-law efficiency is also equal to the ratio of actual and maximum COPs for the cycle. This second-law efficiency definition accounts for irreversibilities within the refrigerator since heat transfers with the high- and low-temperature reservoirs are assumed to be reversible.

Example 4.2

A refrigerator using R-134a as the refrigerant is used to keep a space at −10 °C by rejecting heat to ambient air at 22 °C. R-134a enters the compressor at 140 kPa at a flow rate of 375 L/min as a saturated vapor. The isentropic efficiency of the compressor is 80%. The refrigerant leaves the condenser at 46.3 °C as a saturated liquid. Determine (a) the rate of cooling provided by the system, (b) the COP, (c) the exergy destruction in each component of the cycle, (d) the second-law efficiency of the cycle, and (e) the total exergy destruction in the cycle.

Solution

Temperature–entropy diagram of the cycle is given in Figure 4.4.

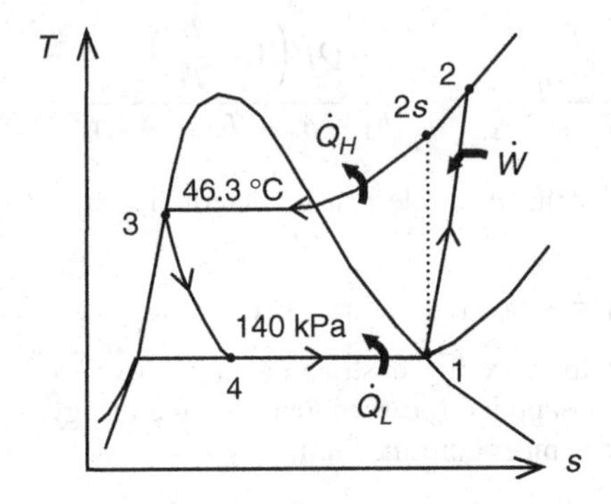

Figure 4.4 Temperature–entropy diagram of vapor-compression refrigeration cycle considered in Example 4.2.

(a) The properties of R-134a are (from Tables B.4 and B.5)

$$\left.\begin{array}{l} P_1 = 140 \text{ kPa} \\ x_1 = 1 \end{array}\right\} \begin{array}{l} h_1 = 239.17 \text{ kJ/kg} \\ s_1 = 0.9446 \text{ kJ/kg}\cdot\text{K} \\ v_1 = 0.1402 \text{ m}^3\text{/kg} \end{array}$$

$$P_3 = P_{\text{sat@46.3 °C}} = 1200 \text{ kPa}$$

$$\left.\begin{array}{l} P_2 = 1200 \text{ kPa} \\ s_2 = s_1 = 0.9446 \text{ kJ/kg}\cdot\text{K} \end{array}\right\} h_{2s} = 284.09 \text{ kJ/kg}$$

$$\left.\begin{array}{l} P_3 = 1200 \text{ kPa} \\ x_3 = 0 \end{array}\right\} \begin{array}{l} h_3 = 117.77 \text{ kJ/kg} \\ s_3 = 0.4244 \text{ kJ/kg}\cdot\text{K} \end{array}$$

$$h_4 = h_3 = 117.77 \text{ kJ/kg}$$

$$\left.\begin{array}{l} P_4 = 140 \text{ kPa} \\ h_4 = 117.77 \text{ kJ/kg} \end{array}\right\} s_4 = 0.4674 \text{ kJ/kg}\cdot\text{K}$$

$$\eta_C = \frac{h_{2s} - h_1}{h_2 - h_1}$$

$$0.80 = \frac{284.09 - 239.17}{h_2 - 239.17} \longrightarrow h_2 = 295.32 \text{ kJ/kg}$$

$$\left.\begin{array}{l} P_2 = 1200 \text{ kPa} \\ h_2 = 295.32 \text{ kJ/kg} \end{array}\right\} s_2 = 0.9783 \text{ kJ/kg}\cdot\text{K}$$

The mass flow rate of the refrigerant is

$$\dot{m} = \frac{\dot{V}_1}{v_1} = \frac{(0.375/60) \text{ m}^3\text{/s}}{0.1402 \text{ m}^3\text{/kg}} = 0.04458 \text{ kg/s}$$

The refrigeration load, the rate of heat rejected, and the power input are

$$\dot{Q}_L = \dot{m}(h_1 - h_4) = (0.04458\text{ kg/s})(239.17 - 117.77)\text{ kJ/kg} = \mathbf{5.41\,kW}$$

$$\dot{Q}_H = \dot{m}(h_2 - h_3) = (0.04458\text{ kg/s})(295.32 - 117.77)\text{ kJ/kg} = 7.92\text{ kW}$$

$$\dot{W} = \dot{m}(h_2 - h_1) = (0.04458\text{ kg/s})(295.32 - 239.17)\text{ kJ/kg} = 2.50\text{ kW}$$

(b) The COP of the cycle is

$$\text{COP} = \frac{\dot{Q}_L}{\dot{W}_{\text{in}}} = \frac{5.41\text{ kW}}{2.50\text{ kW}} = \mathbf{2.16}$$

(c) Noting that the dead-state temperature is $T_0 = T_H = 295$ K, the exergy destruction in each component of the cycle is determined as follows:
Compressor:

$$\dot{S}_{\text{gen},1-2} = \dot{m}(s_2 - s_1) = (0.04458\text{ kg/s})(0.9783 - 0.9446)\text{ kJ/kg}\cdot\text{K} = 0.001502\text{ kW/K}$$

$$\dot{E}x_{\text{dest},1-2} = T_0\dot{S}_{\text{gen},1-2} = (295\text{ K})(0.001502\text{ kW/K}) = \mathbf{0.4432\,kW}$$

Condenser:

$$\dot{S}_{\text{gen},2-3} = \dot{m}(s_3 - s_2) + \frac{\dot{Q}_H}{T_H}$$

$$= (0.04458\text{ kg/s})(0.4244 - 0.9783)\text{ kJ/kg}\cdot\text{K} + \frac{7.92\text{ kW}}{295\text{ K}} = 0.002138\text{ kW/K}$$

$$\dot{E}x_{\text{dest},2-3} = T_0\dot{S}_{\text{gen},2-3} = (295\text{ K})(0.002138\text{ kJ/kg}\cdot\text{K}) = \mathbf{0.6308\,kW}$$

Expansion valve:

$$\dot{S}_{\text{gen},3-4} = \dot{m}(s_4 - s_3) = (0.04458\text{ kg/s})(0.4674 - 0.4244)\text{ kJ/kg}\cdot\text{K} = 0.001916\text{ kW/K}$$

$$\dot{E}x_{\text{dest},3-4} = T_0\dot{S}_{\text{gen},3-4} = (295\text{ K})(0.001916\text{ kJ/kg}\cdot\text{K}) = \mathbf{0.5651\,kW}$$

Evaporator:

$$\dot{S}_{\text{gen},4-1} = \dot{m}(s_1 - s_4) - \frac{\dot{Q}_L}{T_L}$$

$$= (0.04458\text{ kg/s})(0.9446 - 0.4674)\text{ kJ/kg}\cdot\text{K} - \frac{5.41\text{ kW}}{263\text{ K}} = 0.0006964\text{ kW/K}$$

$$\dot{E}x_{\text{dest},4-1} = T_0\dot{S}_{\text{gen},4-1} = (295\text{ K})(0.0006964\text{ kW/K}) = \mathbf{0.2054\,kW}$$

(d) The exergy of the heat transferred from the low-temperature medium is

$$\dot{E}x_{\dot{Q}_L} = -\dot{Q}_L\left(1 - \frac{T_0}{T_L}\right) = -(5.41\text{ kW})\left(1 - \frac{295}{263}\right) = 0.6585\text{ kW}$$

This is also the minimum power input for the cycle. The second-law efficiency of the cycle is

$$\eta_{\text{II}} = \frac{\dot{E}x_{\dot{Q}_L}}{\dot{W}} = \frac{0.6585}{2.503} = 0.263 = \mathbf{26.3\%}$$

This efficiency may also be determined from

$$\eta_{II} = \frac{COP}{COP_{Carnot}}$$

where

$$COP_{Carnot} = \frac{T_L}{T_H - T_L} = \frac{(-10 + 273)\ K}{[22 - (-10)]\ K} = 8.22$$

Substituting,

$$\eta_{II} = \frac{COP}{COP_{Carnot}} = \frac{2.16}{8.22} = 0.263 = 26.3\%$$

The results are identical as expected.

(e) The total exergy destruction in the cycle is the difference between the exergy supplied (power input) and the exergy recovered (the exergy of the heat transferred from the low-temperature medium):

$$\dot{E}x_{dest,total} = \dot{W} - \dot{E}x_{\dot{Q}_L} = 2.503 - 0.6585 = \mathbf{1.845\,kW}$$

The total exergy destruction can also be determined by adding exergy destructions in each component:

$$\dot{E}x_{dest,total} = \dot{E}x_{dest,1-2} + \dot{E}x_{dest,2-3} + \dot{E}x_{dest,3-4} + \dot{E}x_{dest,4-1}$$
$$= 0.4432 + 0.6308 + 0.5651 + 0.2054 = 1.845\ kW$$

The results are identical as expected.

4.5 Practical Vapor-Compression Refrigeration Cycle

There are some clear differences between the practical (actual) cycle and the theoretical cycle (standard ideal cycle) primarily because of the pressure and temperature drops associated with refrigerant flow and heat transfer to or from the surroundings. Figure 4.5 shows an actual vapor-compression

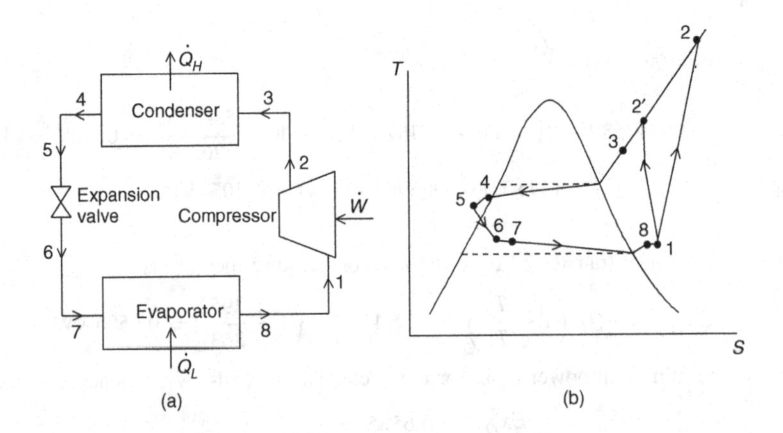

Figure 4.5 An actual vapor-compression refrigeration system and its T–s diagram.

refrigeration cycle. The refrigerant vapor entering the compressor is normally superheated. During the compression process, because of the irreversibilities and heat transfer either to or from the surroundings, depending on the temperatures of the refrigerant and the surroundings, entropy may increase (for the irreversibility and heat transferred to the refrigerant) or decrease (for the irreversibility and heat transferred from the refrigerant), as shown by the two dashed lines 1–2 and 1–2'. The pressure of the liquid leaving the condenser becomes less than the pressure of the vapor entering, and the temperature of the refrigerant in the condenser is somewhat higher than that of the surroundings to which heat is being transferred. As always, the temperature of the liquid leaving the condenser is lower than the saturation temperature. It may drop even more in the piping between the condenser and expansion valve. This represents a gain, however, because as a result of this heat transfer the refrigerant enters the evaporator with a lower enthalpy, which permits more heat to be transferred to the refrigerant in the evaporator. There is also some pressure drop as the refrigerant flows through the evaporator. It may be slightly superheated as it leaves the evaporator, and through heat transferred from the surroundings its temperature increases in the piping between the evaporator and the compressor. This heat transfer represents a loss, because it increases the work of the compressor, since the fluid entering it has an increased specific volume.

A practical commercial mechanical vapor-compression refrigeration system is shown in Figure 4.6. In the system shown, it is possible to use properly the temperature, pressure, and

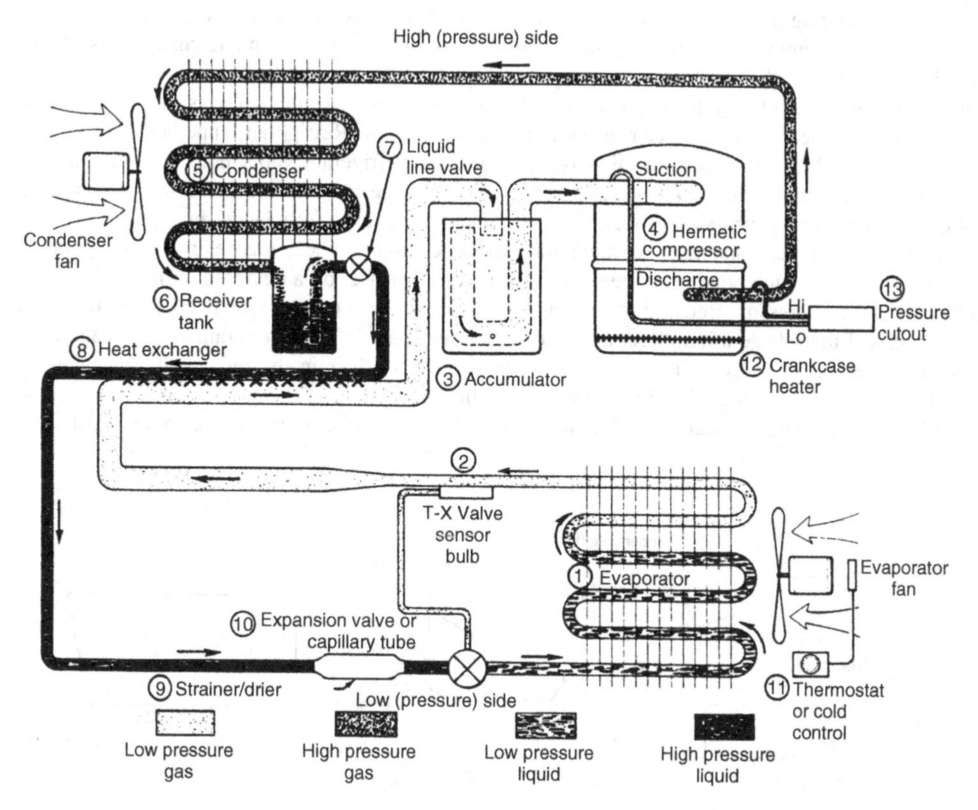

Figure 4.6 A typical commercial refrigerating unit. 1. Evaporator inlet. 2. Evaporator outlet. 3. Accumulator. 4. Compressor. 5. Condenser inlet. 6. Condenser outlet. 7. Receiver outlet. 8. Heat exchanger. 9. Liquid line strainer/drier. 10. Expansion valve. 11. Thermostat. 12. Compressor crankcase heater. 13. High- and low-pressure cutout (*Courtesy of Tecumseh Products Co.*).

latent heat of vaporization. This system utilizes a water-cooled condenser, and water removes heat from the hot refrigerant vapor to condense it. Therefore, the water carries away the heat that is picked up by the evaporator as the refrigerant boils. The refrigerant is then recirculated through the system again to carry out its function to absorb heat in the evaporator.

4.5.1 Superheating and Subcooling

Superheating (referring to superheating of the refrigerant vapor leaving evaporator) and subcooling (referring to subcooling of refrigerant liquid leaving the condenser) are apparently two significant processes in practical vapor-compression refrigeration systems and are applied to provide better efficiency (COP) and to avoid some technical problems, as will be explained below.

4.5.1.1 Superheating

During the evaporation process, the refrigerant is completely vaporized partway through the evaporator. As the cool refrigerant vapor continues through the evaporator, additional heat is absorbed to superheat the vapor. Under some conditions such pressure losses caused by friction increase the amount of superheat. If the superheating takes place in the evaporator, the enthalpy of the refrigerant is raised, extracting additional heat and increasing the refrigeration effect of the evaporator. If it is provided in the compressor suction piping, no useful cooling occurs. In some refrigeration systems, liquid–vapor heat exchangers can be employed to superheat the saturated refrigerant vapor from the evaporator with the refrigerant liquid coming from the condenser (Figure 4.7). As can be seen from Figure 4.7, the heat exchanger can provide high system COP. Refrigerant superheating can also be obtained in the compressor. In this case, the saturated refrigerant vapor enters the compressor and is superheated by increasing the pressure, leading to the temperature increase. Superheating obtained from the compression process does not improve the cycle efficiency, but results in larger condensing equipment and large compressor discharge piping. The increase in the refrigeration effect obtained by superheating in the evaporator is usually offset by a decrease in the refrigeration effect in the compressor. Because the volumetric flow rate of a compressor is constant, the mass flow rate and the refrigeration effect are reduced by decreases in the refrigerant density caused by the superheating. In practice, it is well known that there is a loss in the refrigerating capacity of 1% for every 2.5 °C of superheating in the suction line. Insulation of the suction lines is a solution to minimize undesirable heat gain. The desuperheating is a process to remove excess heat from

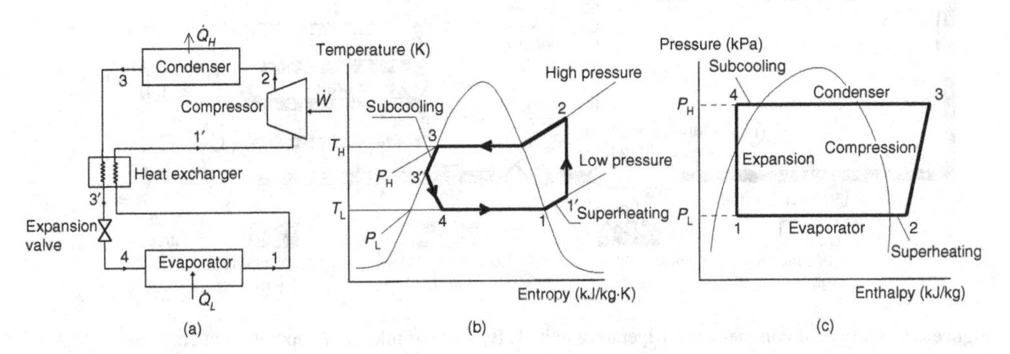

Figure 4.7 (a) A vapor-compression refrigeration system with a heat exchanger for superheating and subcooling, (b) its T–s diagram, and (c) its log P–h diagram.

superheated refrigerant vapor, and if accomplished by using an external effect it will be more useful to the COP. Desuperheating is often considered impractical, owing to the low temperatures (less than 10 °C) and small amount of available energy.

4.5.1.2 Subcooling

This is a process of cooling the refrigerant liquid below its condensing temperature at a given pressure (Figure 4.7). Subcooling provides 100% refrigerant liquid to enter the expansion device, preventing vapor bubbles from impeding the flow of refrigerant through the expansion valve. If the subcooling is caused by a heat-transfer method external to the refrigeration cycle, the refrigerant effect of the system is increased, because the subcooled liquid has less enthalpy than the saturated liquid. Subcooling is accomplished by refrigerating the liquid line of the system, using a higher temperature system. Simply we can state, subcooling cools the refrigerant more and provides the following accordingly:

- increase in energy loading,
- decrease in electrical usage,
- reducing pulldown time,
- more uniform refrigerating temperatures, and
- reduction in the initial cost.

Note that the performance of a simple vapor-compression refrigeration system can be significantly improved by further cooling the liquid refrigerant leaving the condenser coil. This subcooling of the liquid refrigerant can be accomplished by adding a mechanical-subcooling loop in a conventional vapor-compression cycle. The subcooling system can be either a dedicated mechanical-subcooling system or an integrated mechanical-subcooling system (Khan and Zubair, 2000). In a dedicated mechanical-subcooling system, there are two condensers, one for each of the main cycle and the subcooler cycle, whereas, for an integrated mechanical-subcooling system, there is only one condenser serving both the main cycle and the subcooler cycle.

For example, subcooling of R-22 by 13 °C increases the refrigeration effect by about 11%. If subcooling is obtained from outside the cycle, each degree increment in subcooling will improve the system capacity (approximately by 1%). Subcooling from within the cycle may not be as effective because of offsetting effects in other parts of the cycle. Mechanical subcooling can be added to existing systems or designed into new ones. It is ideal for any refrigeration process in which more capacity may be necessary or operating costs must be lowered. It has proved cost efficient in a variety of applications and is recommended for large supermarkets, warehouses, plants, and so on. Figure 4.8 shows a typical subcooler for commercial refrigeration applications.

4.5.2 Defrosting

One of the most common applications of refrigeration systems is to produce and maintain space temperatures by circulating air through a refrigerated coil. If the temperature of the refrigerant in the coil is below 0 °C, water in the air freezes and accumulates on the coil. The ice blocks airflow and acts as an insulator, penalizing coil performance. For efficient performance, the coil must be defrosted periodically. The defrost cycle is a necessary and important part of the design of the refrigeration system.

Over the years, various defrost methods have been used. One of the first methods was to arrange the coil in such a manner that it could be isolated from the cold room. Warm air was circulated over it until the ice melted. Another method is to run water over the coil. Careful design of the water lines into and out of the cold room prevents freezing of the defrost water. Electric heater rods

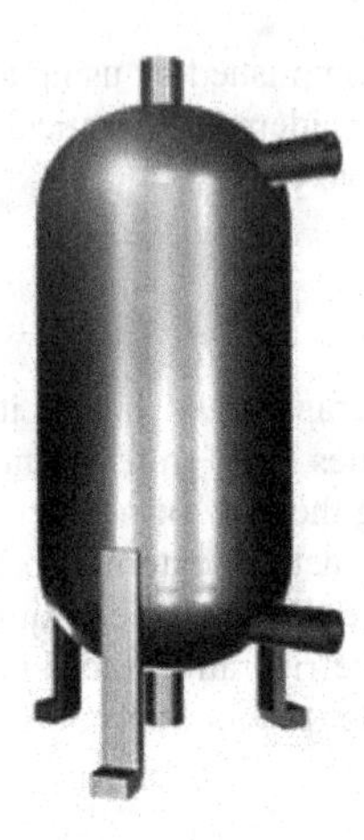

Figure 4.8 A subcooler (*Courtesy of Standard Refrigeration Company*).

inserted into formed holes through aluminum fins work effectively, and this type is common for halocarbon systems. All of these have been used for ammonia coils, but the most common method is hot gas from the compressor discharge. Hot gas defrost is simple and effective, removes ice rapidly, and is relatively inexpensive to install. However, the control valves selection and the sequence of operation must be correct for reliable and efficient defrosts (for details, see Norton, 2000).

Defrost systems vary with the size and type of evaporator, with some choices possible for the larger size coils. Electrical heating defrost via elements in the drip trays under the evaporator or as elements through the coil fins are the most common and economical for small evaporators. Hot gas systems that pump hot refrigerant gas through the coils or defrosting by running ambient water over the coils are more common on larger systems.

Auto cycle defrost is not as complicated as it sounds. In fact, cycle defrost systems are the least complex in operation and most effective defrost systems available. Cycle defrost systems are feasible only on all-refrigerator units because these units do not contain a freezer compartment. Cycle defrost units contain a special thermostat which senses the evaporator plate temperature. At the completion of each compressor run cycle, the thermostat disconnects the electrical power and turns off the compressor. The thermostat will not connect the electrical power again to initiate the next compressor run cycle until the evaporator plate reaches a preset temperature well above freezing.

During this evaporator warm-up period, the frost which has accumulated during the previous compressor run cycle melts and becomes water droplets. These water droplets run down the vertical surface of the evaporator and drop off into the drip tray located just underneath the evaporator, which then empties into a drain tube. The drain tube discharges the water droplets into the condensation pan located in the mechanical assembly under and outside the refrigerator compartment. There, the compressor heat and the airflow from the condenser fan evaporate the moisture.

4.5.3 Purging Air in Refrigeration Systems

Air is known as the enemy of any refrigeration system. Purging, whether manual or automatic, removes air and maximizes refrigeration system performance (Rockwell and Quake, 2001).

Air in a refrigeration system robs it of its capacity to function, and failure to remove such air can be costly in terms of operating efficiency and equipment damage. Such damage is especially notable in the industrial-sized refrigeration systems commonly used in major cold storage facilities, food processing plants, and some chemical plants.

Regardless of whether a system is charged with ammonia or a Freon refrigerant, the thermal efficiency of such systems will greatly improve when undesirable, noncondensable gas (air) is removed. The process of removing air, which is colorless and odorless, is called purging. Over time, this process has become increasingly automatic. But, it is important to understand why, where, and how to purge the system before attempting to rely on an automatic purging system. Figure 4.9 shows an industrial air purger unit.

Air can enter a refrigeration system through several places (Rockwell and Quake, 2001):

- When suction pressure is below atmospheric conditions, air can enter through seals and valve packing.
- Air can rush in when the system is open for repair, coil cleaning, or adding equipment.
- Air can enter when the refrigerant truck is charging the system or when oil is being added.

Therefore, the accumulated air has negative impact on the system performance, which can be summarized as follows:

- Accumulated air insulates the transfer surface and effectively reduces the size of the condenser. To offset this size reduction, the system must work harder by increasing the pressure and temperature of the refrigerant. Therefore, removal of air, as quickly and as efficiently as possible, is essential.
- Air in the system can result in excess wear and tear on bearings and drive motors and contribute to a shorter service life for seals and belts. Also, the added head pressure increases the likelihood of premature gasket failures. It can also decrease the power cost to operate the compressor by about 2% for each 1% reduction in compressor capacity. Thus, it is essential to choose the proper size and type of purger for the job.

The easiest way to determine the amount of air in a refrigeration system is to check the condenser pressure and the temperature of the refrigerant leaving the condenser. Then, these findings should be compared with the standard temperature–pressure for that particular refrigerant.

Figure 4.9 An industrial air purger (*Courtesy of Hansen Technologies Corporation*).

Example 4.3

If, for example, the ammonia temperature is 30 °C, the theoretical condenser pressure should be 1065.2 kPa. If your gauge reads 1199.7 kPa, the excess pressure is 134.5 kPa. Under this condition, the power costs increase by 10% and the compressor capacity decreases by 5%, as determined by the per kWh cost of energy. As an example, if the pressure is reduced by 20 psi (138 kPa) and the cost of electricity is $0.05 per kWh, the annual savings will be more than $2600 per 100 tons (for details, see Rockwell and Quake, 2001).

4.5.3.1 Air Purging Methods

Basically, there are two ways to purge a system of air: manual or automatic. To purge manually, a properly positioned valve is opened by hand, allowing the air to escape. It is a common misconception that when a cloud of refrigerant gas is seen being discharged to atmosphere, the system has been purged of air. Air can still be trapped in the system.

Therefore, many refrigeration system users prefer automatic purging. Refrigeration systems include the compressor, condenser, receiver, evaporator, and purger (Figure 4.10). Of these components, the purger is perhaps the least understood and appreciated. The purger's job is to remove air from the system, thus improving compressor and condenser operating efficiency.

Two types of automatic purgers are used as follows (Rockwell and Quake, 2001): (i) nonelectrical mechanical and (ii) automatic electronic purgers. Determining the type of automatic purger to use

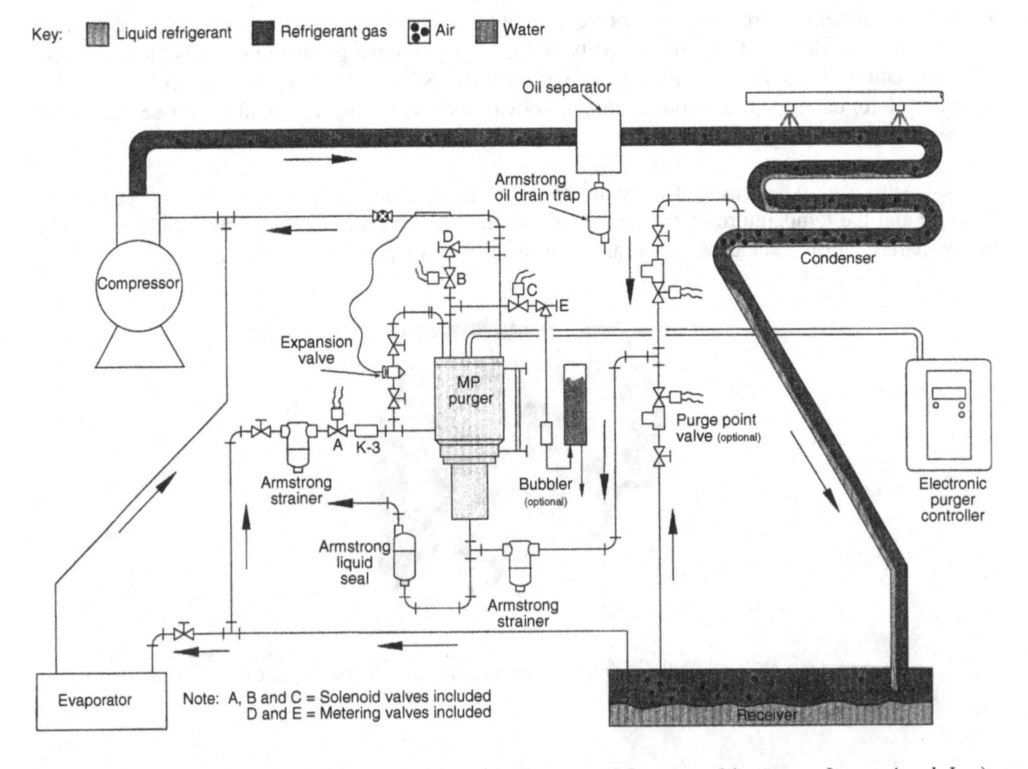

Figure 4.10 A basic refrigeration system with multipoint purger (*Courtesy of Armstrong International, Inc.*).

is a matter of whether electricity is available at the purger location and if it safe to allow electrical components to be used. The nonelectrical mechanical units are used primarily in applications where electricity is not available at the point of use or in hazardous applications where electric components are not allowed. They remove air by sensing the density difference between the liquid refrigerant and gases. An operator opens and closes valves to start and stop the purging operation and ensure its efficiency. Electronic automatic refrigeration purgers are classed as single-point and multipoint purgers. The single-point electronic refrigerated purger has a mechanical-purge operation with a temperature/gas level monitor that controls the discharge to atmosphere. The purging sequence is performed manually. A multipoint refrigerated purger will purge a number of points using the same unit. However, each purge point is purged individually, and the multipoint purger offers total automation, including start-up, shutdown, and alarm features. With this purger, it is important to choose a purger designed for the total tonnage of your system. Undersized purgers may cost less initially but may adversely impact the system's efficiencies and payback period. Some multipoint purgers include a microprocessor-based programmable controller rather than a clock timer. The fuzzy logic controller can "learn" as it cycles through the system. As the purger accumulates air and purges, the controller records and prioritizes each purge point in its memory, thus removing air more efficiently.

Example 4.4

A practical refrigerator operates on the vapor-compression refrigeration cycle with refrigerant-22 as the working fluid. The pressure of R-22 is 300 psia at the compressor exit, and 50 psia at the evaporator inlet. The isentropic efficiency of the compressor is 80%. The refrigerant is superheated by 10 °F at the compressor inlet and subcooled by 10 °F at the exit of the condenser. There is a pressure drop of 10 psia in the condenser and 5 psia in the evaporator. Determine (a) the heat absorption in the evaporator per unit mass of R-22, the work input, and the COP. (b) Determine the refrigeration load, the work input, and the COP if the cycle operated on the ideal vapor-compression refrigeration cycle between the pressure limits of 300 and 50 psia.

The properties of R-22 in the case of actual operation are obtained from R-22 tables to be

$$h_1 = 173.44 \text{ Btu/Ibm},\ h_2 = 200.37 \text{ Btu/Ibm},\ h_3 = 110.65 \text{ Btu/Ibm},\ h_4 = 110.65 \text{ Btu/Ibm}$$

The properties of R-22 in the case of ideal operation are obtained from R-22 tables to be

$$h_1 = 172.30 \text{ Btu/Ibm},\ h_2 = 191.99 \text{ Btu/Ibm},\ h_3 = 114.90 \text{ Btu/Ibm},\ h_4 = 114.90 \text{ Btu/Ibm}$$

Solution

(a) Temperature–entropy diagram of the cycle for the actual operating conditions is given in Figure 4.11.

The heat absorption in the evaporator per unit mass of R-22, the work input, and the COP are determined as follows:

$$q_L = h_1 - h_4 = 173.44 - 110.65 = \mathbf{62.8\,Btu/lbm}$$

$$q_H = h_2 - h_3 = 200.37 - 110.65 = \mathbf{89.7\,Btu/lbm}$$

$$w = h_2 - h_1 = 200.37 - 173.44 = \mathbf{26.9\,Btu/lbm}$$

$$\text{COP} = \frac{q_L}{w} = \frac{62.8 \text{ Btu/lbm}}{26.9 \text{ Btu/lbm}} = \mathbf{2.33}$$

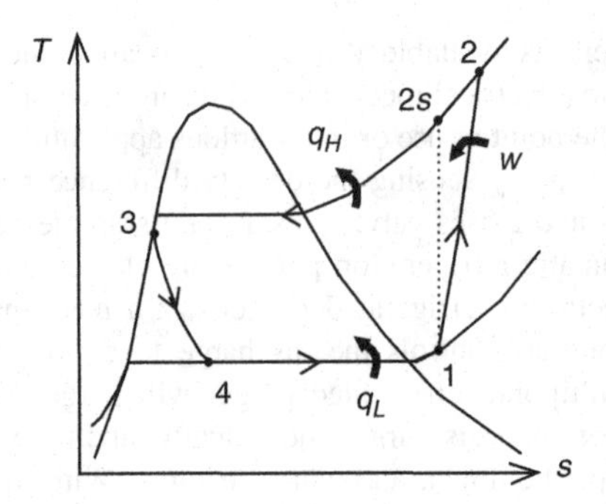

Figure 4.11 Temperature–entropy diagram of vapor-compression refrigeration cycle considered in the solution of Example 4.4a.

(b) Temperature–entropy diagram of the ideal cycle is given in Figure 4.12.
Ideal vapor-compression refrigeration cycle solution is as follows:

$$q_L = h_1 - h_4 = 172.30 - 114.90 = \mathbf{57.4\,Btu/lbm}$$

$$q_H = h_2 - h_3 = 191.99 - 114.90 = \mathbf{77.1\,Btu/lbm}$$

$$w = h_2 - h_1 = 191.99 - 172.30 = \mathbf{19.7\,Btu/lbm}$$

$$\text{COP} = \frac{q_L}{w} = \frac{57.4\text{ Btu/lbm}}{19.7\text{ Btu/lbm}} = \mathbf{2.91}$$

In the ideal operation, the refrigeration load decreases by 8.6% and the work input by 26.8% while the COP increases by 24.9%. Also, it can be shown that the cycle operation in part (a) with a compressor isentropic efficiency of 100% would give the following results: $q_L = 62.8$ Btu/lbm, $q_H = 84.3$ Btu/lbm, $w = 21.6$ Btu/lbm, COP = 2.91.

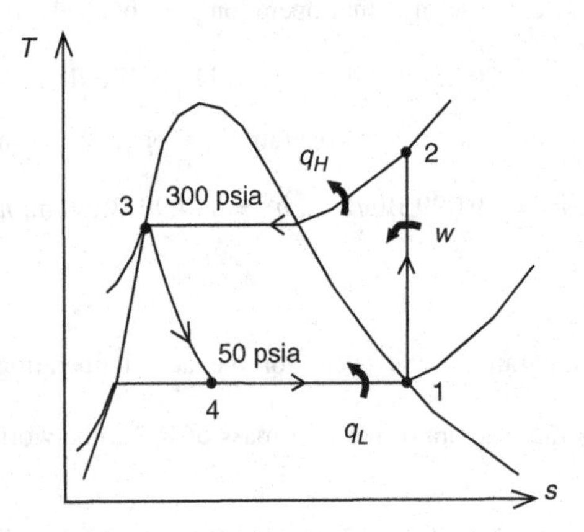

Figure 4.12 Temperature–entropy diagram of the ideal vapor-compression refrigeration cycle considered in Example 4.4b.

4.5.4 *Twin Refrigeration System*

The twin refrigeration system is a new refrigeration technology that solves the problems of conventional vapor-compression refrigerators. A no-frost cooling system is the latest craze, but conventional no-frost features reduce energy efficiency and humidity. To overcome this problem, a new refrigeration system, named the *Twin Refrigeration System* (Figure 4.13a) has been developed by Samsung. Here are the primary features of this new system:

- **Two evaporators and two fans.** The evaporators and fans of the freezer and the refrigerator operate independently to achieve the necessary temperature in each compartment. This minimizes unnecessary airflow from one compartment to another. It eliminates the need for a complicated air flow system which would lead to energy loss.
- **Turbo fans.** Newly developed turbo-fan and multiple-scroll air distribution duct system minimizes the air path.
- **Inverting compressor.** Variable compressor PRM according to the condition of the refrigerator 4-step control is utilized.

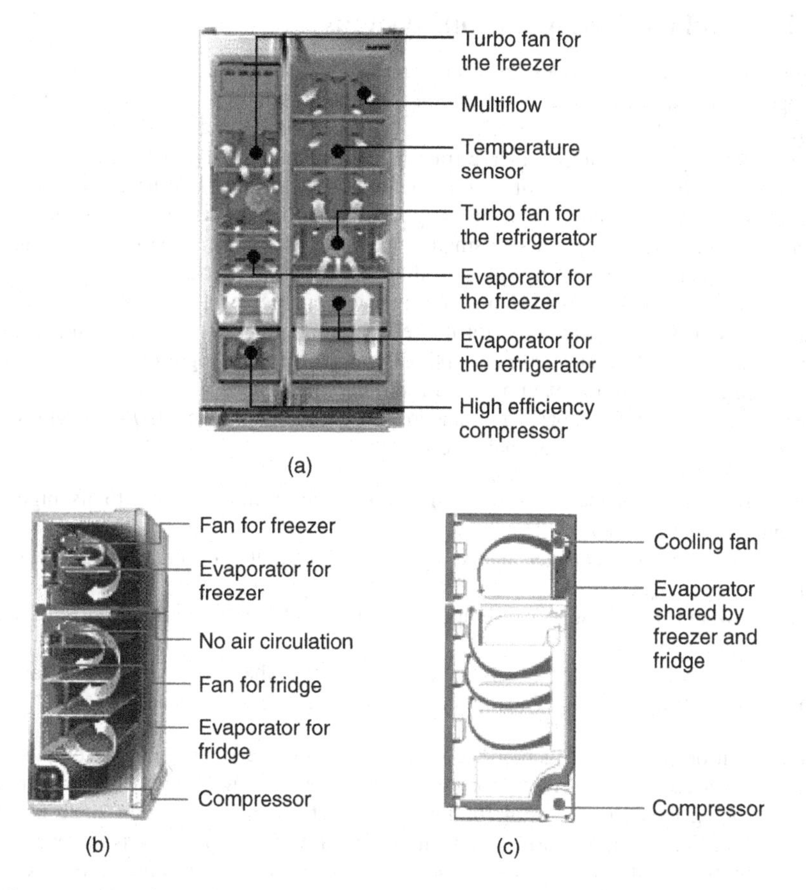

Figure 4.13 (a) A twin refrigeration system and its components. Comparison of (b) a twin refrigeration system with (c) a conventional no-frost system (*Courtesy of Samsung Electronics*).

- **High-efficiency fan motors.** Brushless DC variable motors are employed.
- **High-efficiency insulation.** The insulation material is cyclo-pentane. It helps minimize heat penetration, because of its low thermal conductivity.
- **CFC-free.** All these new refrigerators use R-134a and R-600a only, and are free of CFC and HCFC. Therefore, they are environmentally benign.

As seen in Figure 4.13a, the system has both freezer and refrigerator compartments which are controlled independently because of each compartment's separate evaporator and precise control unit. These features also eliminate inefficient air circulation between the compartments. The result is considered a technological ingenuity, because of the following:

- high humidity preservation,
- ideal constant temperature storage,
- high energy savings, and
- no mixed odors between compartments.

4.6 Air-Standard Refrigeration Systems

The air-standard refrigeration cycles are also known as the reverse Brayton cycles. In these systems, refrigeration is accomplished by means of a noncondensing gas (e.g., air) cycle rather than a refrigerant vapor cycle. While the refrigeration load per kilogram of refrigerant circulated in a vapor-compression cycle is equal to a large fraction of the enthalpy of vaporization, in an air cycle it is only the product of the temperature rise of the gas in the low-side heat exchanger and the specific heat of the gas. Therefore, a large refrigeration load requires a large mass rate of circulation. In order to keep the equipment size smaller, the complete unit may be under pressure, which requires a closed cycle. The throttling valve used for the expansion process in a vapor-compression refrigeration cycle is usually replaced by an expansion engine (e.g., expander) for an air cycle refrigeration system. The work required for the refrigeration effect is provided by the gas refrigerant. These systems are of great interest in applications where the weight of the refrigerating unit must be kept to a minimum, for example, in aircraft cabin cooling.

A schematic arrangement of a basic air-standard refrigeration cycle and its T–s diagram is shown in Figure 4.14. This system has four main elements:

- a compressor that raises the pressure of the refrigerant from its lowest to its highest value (e.g., isentropic compression: 1–2),
- an energy output heat exchanger where the high temperature of the refrigerant is lowered (e.g., isobaric heat rejection: 2–3),
- an expander where the pressure and temperature of the refrigerant are reduced (e.g., isentropic expansion: 3–4), and
- an energy input heat exchanger that raises the temperature of the refrigerant at a constant pressure (e.g., isobaric heat input: 4–1). This input is known as refrigeration load.

The utilization of air as a refrigerant becomes more attractive when a double purpose is to be met. This is so in the case of air conditioning, when the air can be both the refrigerating and the air conditioning medium. Figure 4.15 shows an air-standard refrigeration cycle using a heat exchanger and its T–s diagram. Furthermore, air-standard refrigeration cycle is commonly used in the liquefaction of air and other gases and also in certain cases where refrigeration is needed such as aircraft cooling systems.

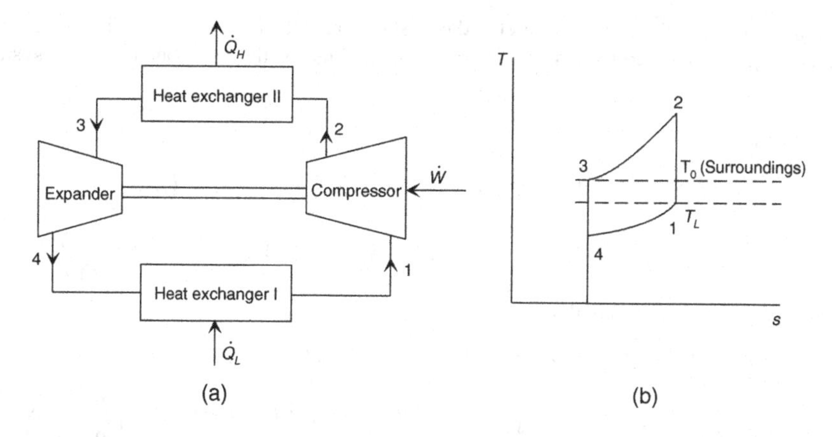

Figure 4.14 (a) A basic air-standard refrigeration cycle and (b) its T–s diagram.

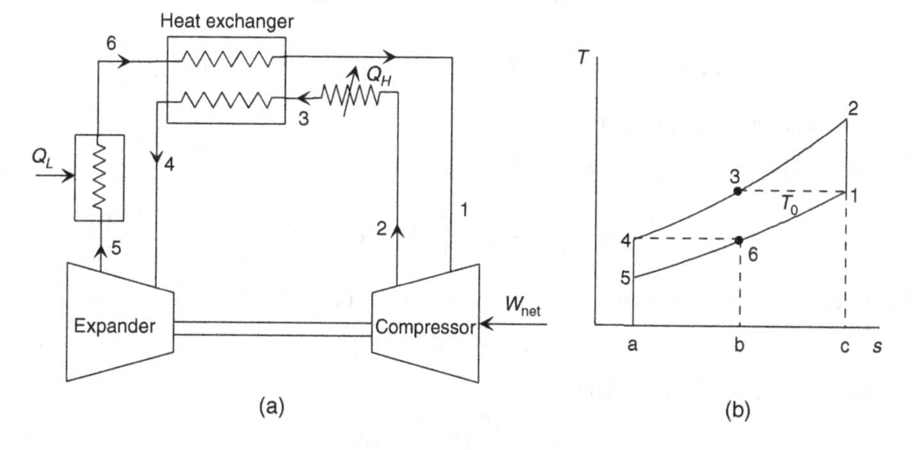

Figure 4.15 (a) An air-standard refrigeration cycle using a heat exchanger and (b) its T–s diagram.

4.6.1 Energy and Exergy Analyses of a Basic Air-Standard Refrigeration Cycle

Here, in energy analysis of a basic air-standard refrigeration cycle as shown in Figure 4.14, we follow the same methodology that we used in energy analysis of a vapor-compression refrigeration cycle. The only difference is that we can treat the gaseous working fluid (i.e., air) as an ideal gas. Therefore, we can write the following for enthalpy and entropy difference equations:

$$\Delta h = (h_e - h_i) = c_p \Delta T = c_p(T_e - T_i) \tag{4.28}$$

$$\Delta s = (s_e - s_i) = c_p \ln \frac{T_e}{T_i} - R \ln \frac{P_e}{P_i} \tag{4.29}$$

where the subscripts i and e represent inlet and exit states, respectively. On the basis of Figure 4.14, we list the energy balance equations and exergy destructions for the components of the system as follows:

- For compressor:

$$\dot{m}h_1 + \dot{W}_{Comp} = \dot{m}h_2 \Rightarrow \dot{W}_{Comp} = \dot{m}(h_2 - h_1) = \dot{m}c_p(T_2 - T_1) \tag{4.30}$$

$$\dot{E}x_{dest,1-2} = T_0\dot{S}_{gen,1-2} = \dot{m}T_0(s_2 - s_1) = \dot{m}T_0\left(c_p \ln\frac{T_2}{T_1} - R\ln\frac{P_2}{P_1}\right)$$

- For heat exchanger II (i.e., condenser):

$$\dot{m}h_2 = \dot{m}h_3 + \dot{Q}_H \Rightarrow \dot{Q}_H = \dot{m}(h_2 - h_3) = \dot{m}c_p(T_2 - T_3) \tag{4.31}$$

$$\dot{E}x_{dest,2-3} = T_0\dot{S}_{gen,2-3} = \dot{m}T_0\left(s_3 - s_2 + \frac{q_H}{T_H}\right) = \dot{m}T_0\left[\left(c_p \ln\frac{T_3}{T_2} - R\ln\frac{P_3}{P_2}\right) + \frac{q_H}{T_H}\right]$$

- For expander (turbine):

$$\dot{m}h_3 = \dot{m}h_4 + \dot{W}_{Turb} \Rightarrow \dot{W}_{Turb} = \dot{m}(h_3 - h_4) = \dot{m}c_p(T_3 - T_4) \tag{4.32}$$

$$\dot{E}x_{dest,3-4} = T_0\dot{S}_{gen,3-4} = \dot{m}T_0(s_4 - s_3) = \dot{m}T_0\left(c_p \ln\frac{T_4}{T_3} - R\ln\frac{P_4}{P_3}\right)$$

- For heat exchanger I (i.e., evaporator):

$$\dot{m}h_4 + \dot{Q}_L = \dot{m}h_1 \Rightarrow \dot{Q}_L = \dot{m}(h_1 - h_4) = \dot{m}c_p(T_1 - T_4) \tag{4.33}$$

$$\dot{E}x_{dest,4-1} = T_0\dot{S}_{gen,4-1} = \dot{m}T_0\left(s_1 - s_4 - \frac{q_L}{T_L}\right) = \dot{m}T_0\left[\left(c_p \ln\frac{T_1}{T_4} - R\ln\frac{P_1}{P_4}\right) - \frac{q_L}{T_L}\right]$$

For the entire refrigeration system, the energy balance can be written as

$$\dot{W}_{Comp} + \dot{Q}_L = \dot{W}_{Turb} + \dot{Q}_H \tag{4.34}$$

The net work for the system becomes

$$\dot{W}_{net} = \dot{W}_{Comp} - \dot{W}_{Turb} \tag{4.35}$$

The COP of the air-standard refrigeration system is

$$\text{COP} = \frac{\dot{Q}_L}{\dot{W}_{net}} \tag{4.36}$$

The total exergy destruction in the cycle can be determined by adding exergy destructions in each component:

$$\dot{E}x_{dest,total} = \dot{E}x_{dest,1-2} + \dot{E}x_{dest,2-3} + \dot{E}x_{dest,3-4} + \dot{E}x_{dest,4-1} \tag{4.37}$$

It can also be expressed as

$$\dot{E}x_{dest,total} = \dot{W}_{net} - \dot{E}x_{\dot{Q}_L} \tag{4.38}$$

where the exergy of the heat transferred from the low-temperature medium is given by

$$\dot{E}x_{\dot{Q}_L} = -\dot{Q}_L\left(1 - \frac{T_0}{T_L}\right) \tag{4.39}$$

This is in fact the minimum power input to accomplish the required refrigeration load $\dot{Q}_L$:

$$\dot{W}_{\min} = \dot{E}x_{\dot{Q}_L} \tag{4.40}$$

The second-law efficiency (or exergy efficiency) of the cycle is defined as

$$\eta_{II} = \frac{\dot{E}x_{\dot{Q}_L}}{\dot{W}_{net}} = \frac{\dot{W}_{\min}}{\dot{W}_{net}} = 1 - \frac{\dot{E}x_{dest,total}}{\dot{W}_{net}} \tag{4.41}$$

Example 4.5

Air enters the compressor of a gas refrigeration system with a regenerator at −20 °C at a flow rate of 0.45 kg/s (Figure 4.16). The cycle has a pressure ratio of 4. The temperature of the air decreases from 16 to −30 °C in the regenerator. The isentropic efficiency of the compressor is 82% and that of the turbine is 84%. Determine (a) the rate of refrigeration and the COP of the cycle and (b) the minimum power input, the second-law efficiency of the cycle, and the total exergy destruction in the cycle. The temperature of the cooled space is −40 °C and heat is released to the ambient at 7 °C. (c) Determine the minimum power input, the second-law efficiency of the cycle, and the total exergy destruction in the cycle if the temperature of the cooled space is −15 °C. Also, determine (d) the refrigeration load and the COP if this system operated on the simple gas refrigeration cycle. In this cycle, take the compressor and turbine inlet temperatures to be −20 and 16 °C, respectively, and use the same compressor and turbine efficiencies. Use constant specific heat for air at room temperature with $c_p = 1.005$ kJ/kg·K and $k = 1.4$.

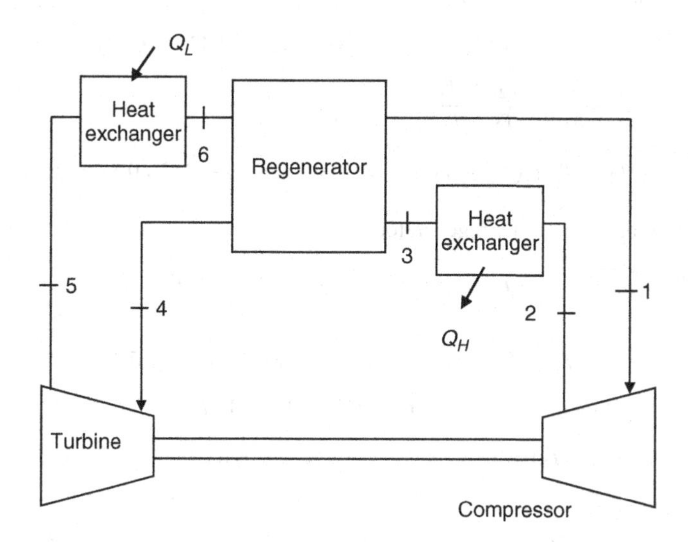

Figure 4.16 The schematic of gas refrigeration system with a regenerator considered in Example 4.5.

Solution

The T–s diagram of the cycle is given in Figure 4.17.

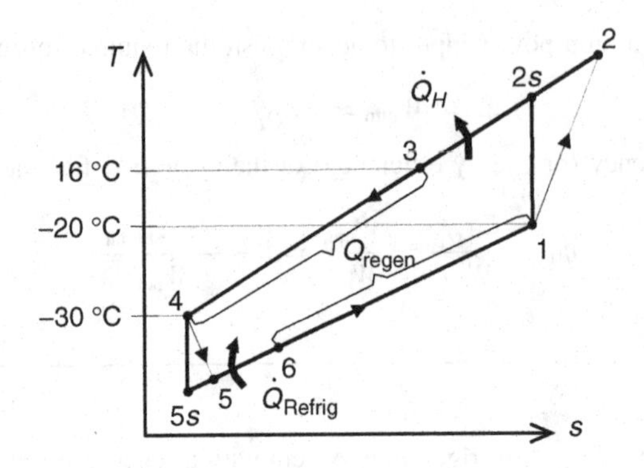

Figure 4.17 Temperature–entropy diagram of gas refrigeration cycle considered in Example 4.5.

(a) From the isentropic relations,

$$T_{2s} = T_1 \left(\frac{P_2}{P_1}\right)^{(k-1)/k} = (253\ \text{K})\,(4)^{0.4/1.4} = 376.0\ \text{K}$$

$$T_{5s} = T_4 \left(\frac{P_5}{P_4}\right)^{(k-1)/k} = (243\ \text{K})\left(\frac{1}{4}\right)^{0.4/1.4} = 163.5\ \text{K}$$

$$\eta_T = \frac{h_4 - h_5}{h_4 - h_{5s}} = \frac{T_4 - T_5}{T_4 - T_{5s}}$$

$$\longrightarrow T_5 = T_4 - \eta_T\,(T_4 - T_{5s}) = 243 - (0.84)\,(243 - 163.5) = 176.2\ \text{K}$$

$$\eta_C = \frac{h_{2s} - h_1}{h_2 - h_1} = \frac{T_{2s} - T_1}{T_2 - T_1}$$

$$\longrightarrow T_2 = T_1 + (T_{2s} - T_1)/\eta_C = 253 + (376.0 - 253)/0.82 = 402.9\ \text{K}$$

From an energy balance on the regenerator,

$$\dot{m}c_p\,(T_3 - T_4) = \dot{m}c_p\,(T_1 - T_6) \longrightarrow T_3 - T_4 = T_1 - T_6$$

or

$$T_6 = T_1 - T_3 + T_4 = 253 - 289 + 243 = 207\ \text{K}$$

The rate of refrigeration, the net power input and the COP are

$$\dot{Q}_L = \dot{m}c_p(T_6 - T_5) = (0.45\ \text{kg/s})(1.005\ \text{kJ/kg}\cdot\text{K})(207 - 176.2)\ \text{K} = \mathbf{13.91\ kW}$$

$$\dot{W}_{net} = \dot{m}c_p[(T_2 - T_1) - (T_4 - T_5)]$$

$$= (0.45\ \text{kg/s})(1.005\ \text{kJ/kg}\cdot\text{K})[(402.9 - 253) - (243 - 176.2)]\ \text{K}$$

$$= 37.62\ \text{kW}$$

$$\text{COP} = \frac{\dot{Q}_L}{\dot{W}_{net}} = \frac{13.91}{37.62} = \mathbf{0.370}$$

(b) The exergy of the heat transferred from the low-temperature medium is

$$\dot{E}x_{\dot{Q}_L} = -\dot{Q}_L\left(1 - \frac{T_0}{T_L}\right) = -(13.91\text{ kW})\left(1 - \frac{280}{233}\right) = 2.81\text{ kW}$$

This is the minimum power input:

$$\dot{W}_{\min} = \dot{E}x_{\dot{Q}_L} = \mathbf{2.81\,kW}$$

The second-law efficiency of the cycle is

$$\eta_{\text{II}} = \frac{\dot{E}x_{\dot{Q}_L}}{\dot{W}_{\text{net}}} = \frac{2.81}{37.62} = \mathbf{0.075 = 7.5\%}$$

The total exergy destruction in the cycle can be determined from

$$\dot{E}x_{\text{dest,total}} = \dot{W}_{\text{net}} - \dot{E}x_{\dot{Q}_L} = 37.62 - 2.81 = \mathbf{34.8\,kW}$$

(c) If the temperature of the cooled space is $T_L = -15\,°\text{C} = 258$ K

$$\dot{E}x_{\dot{Q}_L} = -\dot{Q}_L\left(1 - \frac{T_0}{T_L}\right) = -(13.91\text{ kW})\left(1 - \frac{280}{258}\right) = 1.19\text{ kW}$$

$$\dot{W}_{\min} = \dot{E}x_{\dot{Q}_L} = \mathbf{1.19\,kW}$$

$$\eta_{\text{II}} = \frac{\dot{E}x_{\dot{Q}_L}}{\dot{W}_{\text{net}}} = \frac{1.19}{37.62} = \mathbf{0.032 = 3.2\%}$$

$$\dot{E}x_{\text{dest,total}} = \dot{W}_{\text{net}} - \dot{E}x_{\dot{Q}_L} = 37.62 - 1.19 = \mathbf{36.4\,kW}$$

(d) The simple gas refrigeration cycle analysis is as follows (Figure 4.18):

$$T_{2s} = T_1\left(\frac{P_2}{P_1}\right)^{(k-1)/k} = (253\text{ K})\,(4)^{0.4/1.4} = 376.0\text{ K}$$

$$\eta_C = \frac{h_{2s} - h_1}{h_2 - h_1} = \frac{T_{2s} - T_1}{T_2 - T_1} \longrightarrow 0.82 = \frac{376.0 - 253}{T_2 - 253} \longrightarrow T_2 = 402.9\text{ K}$$

$$T_{4s} = T_3\left(\frac{1}{r}\right)^{(k+1)/k} = (289\text{ K})\left(\frac{1}{4}\right)^{0.4/1.4} = 194.5\text{ K}$$

$$\eta_T = \frac{T_3 - T_4}{T_3 - T_{4s}} \longrightarrow 0.84 = \frac{289 - T_4}{289 - 194.5} \longrightarrow T_4 = 209.6\text{ K}$$

$$\dot{Q}_L = \dot{m}c_p(T_1 - T_4) = (0.45\text{ kg/s})(1.005\text{ kJ/kg}\cdot\text{K})(253 - 209.6)\text{ kJ/kg} = \mathbf{19.63\,kW}$$

$$\begin{aligned}\dot{W}_{\text{net,in}} &= \dot{m}c_p(T_2 - T_1) - \dot{m}c_p(T_3 - T_4)\\ &= (0.45\text{ kg/s})(1.005\text{ kJ/kg}\cdot\text{K})\,[(402.9 - 253) - (289 - 209.6)\text{ K}]\\ &= 31.91\text{ kW}\end{aligned}$$

$$\text{COP} = \frac{\dot{Q}_L}{\dot{W}_{\text{net}}} = \frac{19.63}{31.91} = \mathbf{0.615}$$

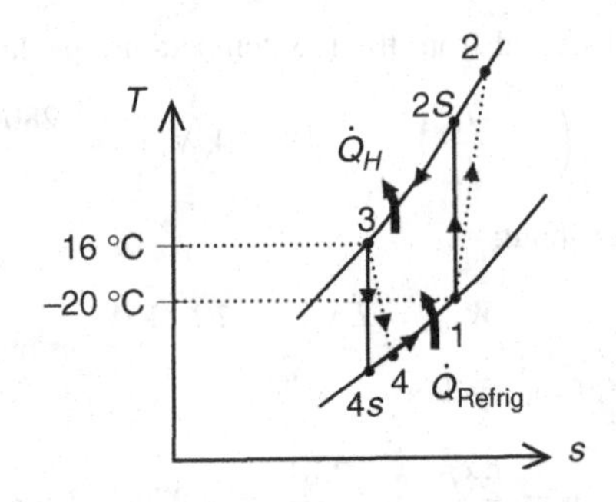

Figure 4.18 Temperature–entropy diagram of simple gas refrigeration cycle considered in Example 4.5, part (d).

4.7 Absorption–Refrigeration Systems (ARSs)

Although the principle of the absorption–refrigeration cycle has been known since the early 1800s, the first one was invented by French engineer Ferdinand P.E. Carre in 1860, an intermittent crude ammonia absorption apparatus based on the chemical affinity of ammonia for water, and produced ice on a limited scale. The first five Absorption–Refrigeration System (ARS) units Carre produced were used to make ice, up to 100 kg/hour. In the 1890s, many large ARS units were manufactured for chemical and petroleum industries. The development of ARSs slowed to a standstill by 1911 as vapor-compression refrigeration systems came to the forefront. After 1950, large ARSs gained in popularity. In 1970s, the market share of ARSs dropped rapidly because of the oil crisis and hence the government regulations. Because of the increasing energy prices and environmental impact of refrigerants, during the past decade ARSs have received increasing attention. So, many companies have concentrated on ARSs and now do research and development on these while the market demand increases dramatically.

ARSs have experienced many ups and downs. The system was the predecessor of the vapor-compression refrigeration system in the nineteenth century, and water–ammonia systems enjoyed a variety of applications in domestic refrigerators and large industrial installations in the chemical and process industries. They were energized by steam or hot water generated from natural gas, oil-fired boilers, and electrical heaters. In the 1970s, the shift from direct burning of oil and natural gas struck a blow at the application of the ARSs but at the same time opened up other opportunities, such as the use of heat derived from solar collectors to energize these systems.

The concept of absorption refrigeration developed well before the advent of electrically driven refrigerators. In the last decades, the availability of cheap electricity has made absorption systems less popular. Today, improvements in absorption technology, the rising cost, and the environmental impact of generating electricity are contributing to the increasing popularity of absorption systems. ARSs for industrial and domestic applications have been attracting increasing interest throughout the world because of the following advantages over other refrigeration systems:

- quiet operation,
- high reliability,
- long service life,
- efficient and economic use of low-grade energy sources (e.g., solar energy, waste energy, geothermal energy),
- easy capacity control,
- no cycling losses during on-off operations,

- simpler implementation, and
- meeting the variable load easily and efficiently.

Recently, there has been increasing interest in the industrial (Figure 4.19) and domestic use of the ARSs for meeting cooling and air conditioning demands as alternatives, because of a trend in the world for rational utilization of energy sources, protection of the natural environment, and prevention of ozone depletion, as well as reduction of pollution. There are a number of applications in various industries where ARSs are employed, including the following:

- food industry (meat, dairy, vegetables, and food freezing and storage, fish industry, freeze drying),
- chemical and petrochemical industry (liquefying if gases, separation processes),
- cogeneration units in combination with production of heat and cold (trigeneration plants),
- leisure sector (skating rinks),
- HVAC,
- refrigeration, and
- cold storage.

The absorption cycle is a process by which the refrigeration effect is produced through the use of two fluids and some quantity of heat input, rather than electrical input as in the more familiar vapor-compression cycle. In ARSs, a secondary fluid (i.e., absorbent) is used to circulate and absorb the primary fluid (i.e., refrigerant), which is vaporized in the evaporator. The success of the absorption process depends on the selection of an appropriate combination of refrigerant and absorbent. The most widely used refrigerant and absorbent combinations in ARSs have been ammonia–water and lithium bromide-water. The lithium bromide-water pair is available for air-conditioning and chilling applications (over 4 °C, because of the crystallization of water). Ammonia-water is used for cooling and low-temperature freezing applications (below 0 °C).

The absorption cycle uses a heat-driven concentration difference to move refrigerant vapors (usually water) from the evaporator to the condenser. The high concentration side of the cycle absorbs refrigerant vapors (which, of course, dilutes that material). Heat is then used to drive off these refrigerant vapors thereby increasing the concentration again.

(a)

(b)

(c)

Figure 4.19 (a) An ARS of 2500 kW at −15 °C installed in a meat factory in Spain. (b) An ARS of 2700 kW at −30 °C installed in a refinery in Germany. (c) An ARS of 1400 kW at −28 °C installed in a margarine factory in The Netherlands (*Courtesy of Colibri b.v.-Stork Thermeq b.v.*).

Both vapor-compression and absorption-refrigeration cycles accomplish the removal of heat through the evaporation of a refrigerant at a low pressure and the rejection of heat through the condensation of the refrigerant at a higher pressure.

Extensive studies to find suitable chemicals for ARSs were conducted using solubility measurements for given binary systems. Although this information is useful as a rough screening technique for suitable binary systems, more elaborate investigations now seem necessary to learn more of the fundamentals of the absorption phenomena.

During the last decade, numerous experimental and theoretical studies on ARSs have been undertaken to develop alternative working fluids, such as R22-dimethyl ether tetraethylene glycol (DMETEG), R21-DMETEG, R22-dimethylformamide (DMF), R12-dimethylacetamide, R22-dimethylacetamide, and R21-dimethyl ester. Previous studies indicated that ammonia, R21, R22, and methylamine hold promise as refrigerants, whereas the organic glycols, some amides, esters, and so on fulfill the conditions for good absorbents. Recently, environmental concerns have brought some alternative working fluids to the forefront, for example, R123a-ethyl tetrahydrofurfuryl ether (ETFE), R123a-DMETEG, R123a-DMF, and R123a-trifluoroethanol, because of the CFCs' ozone depletion effects.

The cycle efficiency and the operating characteristics of an ARS depend on the thermophysical properties of the refrigerant, the absorbent, and their combinations. The most important properties for the selection of the working fluids are vapor pressure, solubility, density, viscosity, and thermal stability. Knowledge of these properties is required to determine the other physical and chemical properties, as well as the parameters affecting performance, size, and cost.

Note that ammonia will quickly corrode copper, aluminum, zinc, and all alloys of these metals, therefore these metals cannot be used where ammonia is present. From common materials only steel, cast iron, and stainless steel can be used in ammonia ARSs. Most plastics are also resistant to chemical attack by ammonia, hence plastics are suitable for valve seats, pump parts, and other minor parts of the system.

4.7.1 Basic ARSs

It is considered that the ARS is similar to the vapor-compression refrigeration cycle (using the evaporator, condenser, and throttling valve as in a basic vapor-compression refrigeration cycle), except that the compressor of the vapor-compression system is replaced by three main elements – an absorber, a solution pump, and a generator. Three steps, absorption, solution pumping, and vapor release, take place in an ARS.

In Figure 4.20, a basic ARS, which consists of an evaporator, a condenser, a generator, an absorber, a solution pump, and two throttling valves, is schematically shown. The strong solution (a mixture strong in refrigerant), which consists of the refrigerant and absorbent, is heated in the high-pressure portion of the system (the *generator*). This drives refrigerant vapor off the solution. The hot refrigerant vapor is cooled in the condenser until it condenses. Then the refrigerant liquid passes through a throttling valve into the low-pressure portion of the system, the evaporator. The reduction in pressure through this valve facilitates the vaporization of the refrigerant, which ultimately effects the heat removal from the medium. The desired refrigeration effect is then provided accordingly. The weak solution (weak in refrigerant) flows down through a throttling valve to the absorber. After the evaporator, the cold refrigerant comes to the absorber and is absorbed by this weak solution (i.e., absorbent), because of the strong chemical affinity for each other. The strong solution is then obtained and is pumped by a solution pump to the generator, where it is again heated, and the cycle continues. It is significant to note that the system operates at high vacuum at an evaporator pressure of about 1.0 kPa; the generator and the condenser operate at about 10.0 kPa.

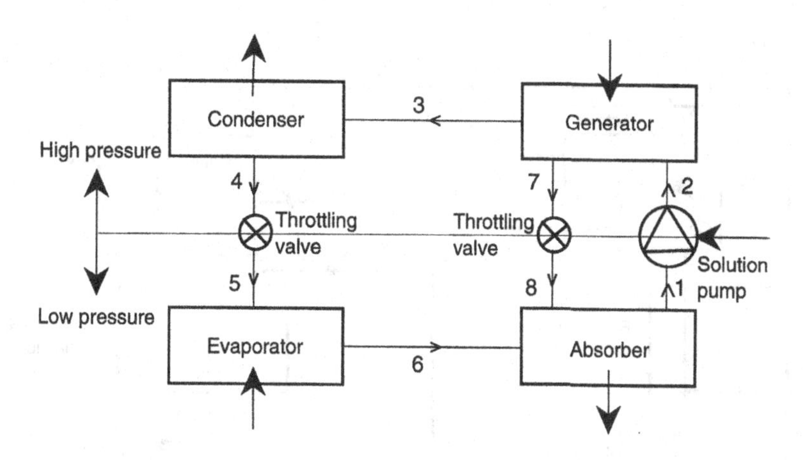

Figure 4.20 A basic ARS.

4.7.2 Ammonia–Water (NH_3–H_2O) ARSs

In practical ARSs, the utilization of one or two heat exchangers is very common. Figure 4.21 represents a practical ARS using a working fluid of ammonia as the refrigerant and water as the absorbent, with two exchangers. As can be seen from the figure, in addition to two heat exchangers, this system employs an analyzer and a rectifier. These devices are used to remove the water vapor that may have formed in the generator, so that only ammonia vapor goes to the condenser.

The system shown in Figure 4.21 utilizes the inherent ability of water to absorb and release ammonia as the refrigerant. The amount of ammonia vapor which can be absorbed and held in a water solution increases with rising pressure and decreases with rising temperature. Its operation is same as the system given in Figure 4.20, except for the analyzer, rectifier, and heat exchangers. In the absorber, the water absorbs the ammonia at the condenser temperature supplied by circulating water or air, and hence a strong solution (about 38% ammonia concentration) occurs.

Because of physical limitations, sometimes complete equilibrium saturation may not be reached in the absorber, and the strong solution leaving the absorber may not be as fully saturated with water as its pressure and temperature would require. This strong solution from the absorber enters the solution pump (the only moving part of the system), which raises its pressure and delivers the solution into the generator through the heat exchanger. Pumped strong solution passes into generator via heat exchanger where strong solution is preheated before being discharged into ammonia generator. Note that the pumping energy required is only a few percent of the entire refrigeration energy requirement. The generator, which is heated by an energy source (saturated steam or other heat source via heating coils or tube bundles), raises the temperature of the strong solution causing the ammonia to separate from it. The remaining weak solution (about 24% ammonia concentration) absorbs some of the water vapor coming from the analyzer/rectifier combination and flows down to the expansion valve through the heat exchanger. It is then throttled into the absorber for further cooling as it picks up a new charge of the ammonia vapor, thus becoming a strong solution. The hot ammonia in the vapor phase from the generator is driven out of solution and rises through the rectifier for possible separation of the remaining water vapor. Then it enters the condenser and is released to the liquid phase. Liquid ammonia enters the second heat exchanger and loses some heat to the cool ammonia vapor. The pressure of liquid ammonia significantly drops in the throttling valve before it enters the evaporator. The cycle is completed when the desired cooling load is achieved in the evaporator. Cool ammonia vapor obtained from the evaporator passes into

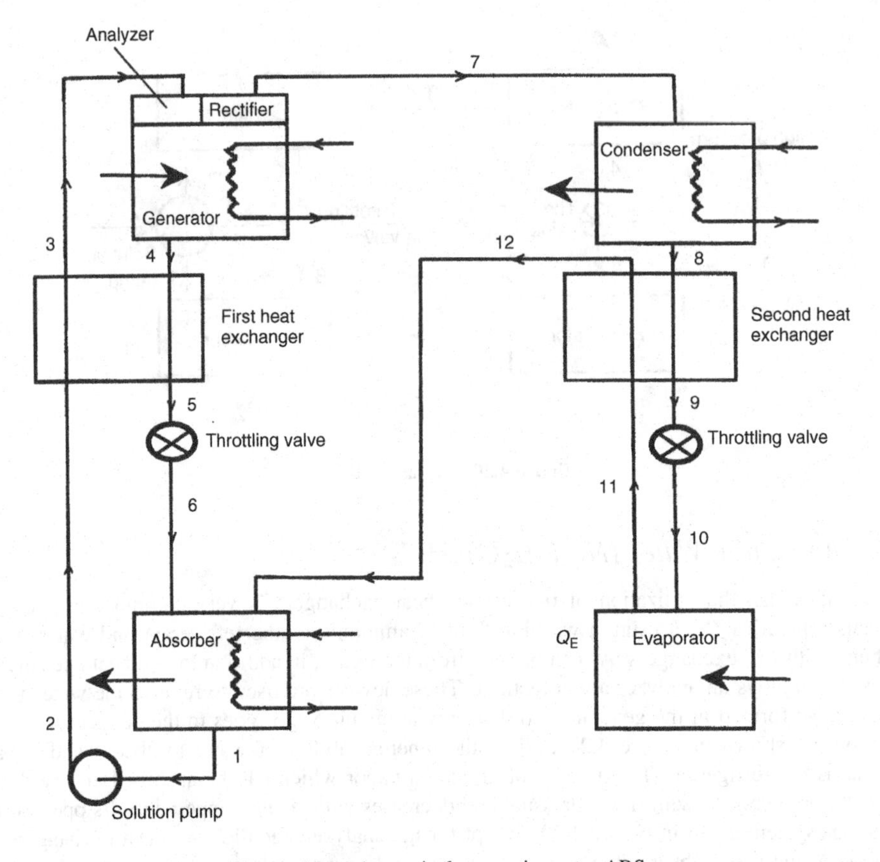

Figure 4.21 A practical ammonia–water ARS.

the absorber and is absorbed there. This absorption activity lowers the pressure in the absorber and causes the vapor to be taken off from the evaporator. When the vapor goes into liquid solution it releases both its latent heat and a heat of dilution. This energy release has to be continuously dissipated by the cooling water or air.

The heat introduced into the absorption system in the generator (from steam heat) and the evaporator (from actual refrigeration operation) has to be rejected to the outside. One heat ejection occurs in the ammonia condenser and other heat ejection occurs in the ammonia absorber. Reabsorption of ammonia into weak solution generates heat and unfortunately this heat has to be rejected so that the absorption process can function. Aqua ammonia consists of water and ammonia. Water can easily absorb ammonia and stay in solution under normal temperature; hence the absorber has to be cooled with cooling water or air. Evaporated ammonia in the generator is passed through the distilling column where the ammonia is concentrated into nearly pure ammonia vapor before going into the condenser. Once ammonia is turned into liquid it is let down into the evaporator, low-pressure side, where ammonia is again turned into vapor, by evaporation, while picking up heat from the confined refrigerated space. Ammonia vapor is then absorbed in the absorber to complete the cycle.

For ammonia–water ARSs, the most suitable absorber is the film-type absorber for the following reasons (Keizer, 1982):

- high heat and mass transfer rates,
- good overall performance, and
- large concentration rates.

Further information and detailed discussion of energy and mass balances and limiting conditions in the analyzer and rectifier, along with some examples, can be found in Gosney (1982).

4.7.3 *Energy Analysis of an ARS*

As mentioned earlier, energy analysis of an ARS refers to the first law of thermodynamic analysis of an open (control volume) system. Therefore, each component in the ARS is considered a steady-state steady-flow process, and we will write energy balance equations, equating that input energies (including work) to output energies. Note that in vapor-compression refrigeration systems, the mass flow rate of the refrigerant was constant throughout the cycle. However, here in ARS we have two fluids (making a working fluid) as refrigerant and absorbent and their composition at different points is different, particularly in the absorber and generator. Therefore, we also include mass balance equations for those two components in addition to energy balance equations. We refer to Figure 4.21 for the state points in the following equations.

- Absorber:

  Energy balance:
  $$\dot{m}_6 h_6 + \dot{m}_{12} h_{12} = \dot{m}_1 h_1 + \dot{Q}_{\mathrm{A}} \tag{4.42}$$

  Mass balance equation:
  $$\dot{m}_{\mathrm{ws}} X_{\mathrm{ws}} + \dot{m}_{\mathrm{r}} = \dot{m}_{\mathrm{ss}} X_{\mathrm{ss}} \tag{4.43}$$

  where $\dot{Q}_{\mathrm{A}}$ is the absorber head load in kW; X is the concentration; $\dot{m}_{\mathrm{ws}} = \dot{m}_6$ is the mass flow rate of the weak solution in kg/s; $\dot{m}_{\mathrm{ss}} = \dot{m}_1$ is the mass flow rate of the strong solution in kg/s; and $\dot{m}_{\mathrm{r}}$ is the mass flow rate of the refrigerant in kg/s. Here, state 1 is a saturated liquid at the lowest temperature in the absorber and is determined by the temperature of the available cooling water flow or air flow.
- Solution pump:
  $$\dot{m}_1 h_1 + \dot{W}_{\mathrm{P}} = \dot{m}_2 h_2 \tag{4.44}$$

  The compression is almost isothermal.
- First heat exchanger:
  $$\dot{m}_2 h_2 + \dot{m}_4 h_4 = \dot{m}_3 h_3 + \dot{m}_5 h_5 \tag{4.45}$$
- Generator:

  Energy balance:
  $$\dot{m}_3 h_3 + \dot{Q}_{\mathrm{gen}} = \dot{m}_4 h_4 + \dot{m}_7 h_7 \tag{4.46}$$

  Mass balance:
  $$\dot{m}_{\mathrm{ws}} X_{\mathrm{ws}} + \dot{m}_{\mathrm{r}} = \dot{m}_{\mathrm{ss}} X_{\mathrm{ss}} \tag{4.47}$$

  where $\dot{Q}_{\mathrm{G}}$ is the heat input to generator in kW; $\dot{m}_{\mathrm{ws}} = \dot{m}_4$ and $\dot{m}_{\mathrm{ss}} = \dot{m}_3$.
- Condenser:
  $$\dot{m}_7 h_7 = \dot{m}_8 h_8 + \dot{Q}_H \tag{4.48}$$
- Second heat exchanger:
  $$\dot{m}_8 h_8 + \dot{m}_{11} h_{11} = \dot{m}_9 h_9 + \dot{m}_{12} h_{12} \tag{4.49}$$
- Expansion (throttling) valves:
  $$\dot{m}_5 h_5 = \dot{m}_6 h_6 \Rightarrow h_5 = h_6 \tag{4.50}$$
  $$\dot{m}_9 h_9 = \dot{m}_{10} h_{10} \Rightarrow h_9 = h_{10} \tag{4.51}$$

  The process is isenthalpic pressure reduction.

- Evaporator:

$$\dot{m}_{10}h_{10} + \dot{Q}_L = \dot{m}_{11}h_{11} \tag{4.52}$$

For the entire system, the overall energy balance of the complete system can be written as follows, by considering that there is negligible heat loss to the environment:

$$\dot{W} + \dot{Q}_L + \dot{Q}_{gen} = \dot{Q}_A + \dot{Q}_H \tag{4.53}$$

The COP of the system then becomes

$$COP = \frac{\dot{Q}_L}{\dot{W}_P + \dot{Q}_{gen}} \tag{4.54}$$

where $\dot{W}_P$ is the pumping power requirement, and it is usually neglected in the COP calculation.

Example 4.6

Consider a basic ARS using ammonia–water solution as shown in Figure 4.22. Pure ammonia enters the condenser at 2.5 MPa and 60 °C at a rate of 0.022 kg/s. Ammonia leaves the condenser as a saturated liquid and is throttled to a pressure of 0.15 MPa. Ammonia leaves the evaporator as a saturated vapor. Heat is supplied to the generator by geothermal liquid water that enters at 135 °C at a rate of 0.35 kg/s and leaves at 120 °C. Determine (a) the rate of cooling provided by the system and (b) the COP of the system. (c) Also, determine the second-law efficiency of the system if the ambient temperature is 25 °C and the temperature of the refrigerated space is 2 °C. The enthalpies of ammonia at various states of the system are given as $h_3 = 1497.4$ kJ/kg, $h_4 = 482.5$ kJ/kg, $h_6 = 1430.0$ kJ/kg. Also, take the specific heat of water to be 4.2 kJ/kg·°C.

Solution

(a) The rate of cooling provided by the system is

$$\dot{Q}_L = \dot{m}_R(h_6 - h_5) = (0.022\ \text{kg/s})(1430.0 - 482.5)\ \text{kJ/kg} = \mathbf{20.9\,kW}$$

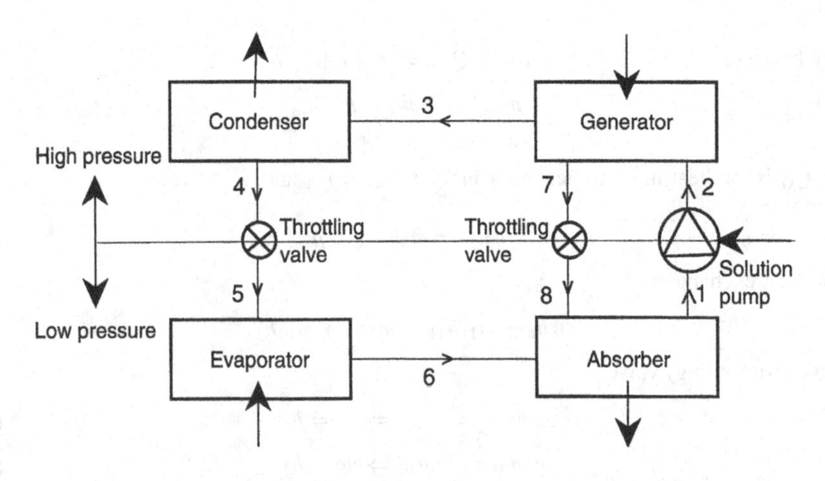

Figure 4.22 The basic ARS considered in Example 4.6.

(b) The rate of heat input to the generator is

$$\dot{Q}_{\text{gen}} = \dot{m}_{\text{geo}} c_p (T_{\text{geo,in}} - T_{\text{geo,out}}) = (0.35 \text{ kg/s})(4.2 \text{ kJ/kg} \cdot {}^\circ\text{C})(135 - 120)\ {}^\circ\text{C} = 22.1 \text{ kW}$$

Then the COP becomes

$$\text{COP} = \frac{\dot{Q}_L}{\dot{Q}_{\text{gen}}} = \frac{20.9 \text{ kW}}{22.1 \text{ kW}} = \mathbf{0.946}$$

(c) In order to develop a relation for the maximum (reversible) COP of an ARS, we consider a reversible heat engine and a reversible refrigerator as shown in Figure 4.23. Heat is absorbed from a source at T_s by a reversible heat engine and the waste heat is rejected to an environment T_0. Work output from the heat engine is used as the work input in the reversible refrigerator, which keeps a refrigerated space at T_L while rejecting heat to the environment at T_0. Using the definition of COP for an ARS, thermal efficiency of a reversible heat engine and the COP of a reversible refrigerator, we obtain

$$\text{COP}_{\text{abs,rev}} = \frac{\dot{Q}_L}{\dot{Q}_{\text{gen}}} = \frac{\dot{W}}{\dot{Q}_{\text{gen}}}\frac{\dot{Q}_L}{\dot{W}} = \eta_{\text{th,rev}}\text{COP}_{\text{R,rev}} = \left(1 - \frac{T_0}{T_s}\right)\left(\frac{T_L}{T_0 - T_L}\right)$$

Substituting,

$$\text{COP}_{\text{abs,rev}} = \left(1 - \frac{T_0}{T_s}\right)\left(\frac{T_L}{T_0 - T_L}\right) = \left(1 - \frac{(25 + 273) \text{ K}}{(127.5 + 273) \text{ K}}\right)\left(\frac{(2 + 273) \text{ K}}{(25 - 2) \text{ K}}\right) = 3.06$$

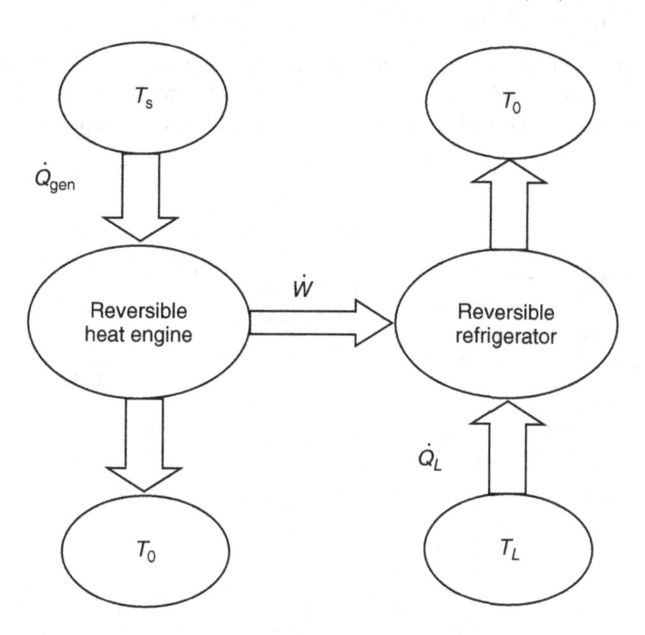

Figure 4.23 The system used to develop reversible COP of an absorption-refrigeration system.

The temperature of the heat source is taken as the average temperature of geothermal water. Then the second-law efficiency of this absorption system is determined to be

$$\eta_{\text{II}} = \frac{\text{COP}}{\text{COP}_{\text{abs,rev}}} = \frac{0.946}{3.06} = 0.309 = \mathbf{30.9\%}$$

4.7.4 Three-Fluid (Gas Diffusion) ARSs

The two-fluid ARS succeeded in replacing a compressor which requires a large amount of shaft work by a liquid pump with a negligible energy requirement compared to the refrigeration effect. By addition of a third fluid, the pump is removed, completely eliminating all moving parts. This system is also called the *von Platen–Munters system* after its Swedish inventors. This type of system is shown in Figure 4.24. The most commonly used fluids are ammonia (as refrigerant), water (as absorbent), and hydrogen, a neutral gas used to support a portion of the total pressure in part of the system. Hydrogen is called the *carrier gas*. The unit consists of four main parts: the boiler, condenser, evaporator, and absorber. In gas units, heat is supplied by a burner, and when the unit operates on electricity the heat is supplied by a heating element. The unit charge consists of a quantity of ammonia, water, and hydrogen at a sufficient pressure to condense ammonia at the room temperature for which the unit is designed. This method of absorption refrigeration is presently used in domestic systems where the COP is less important than quiet trouble-free operation. In the system shown in Figure 4.24, the cold ammonia vapor with hydrogen is circulated by natural convection through a gas–gas heat exchanger to the absorber, where the ammonia vapor comes in contact with the weak solution from the separator. At the low temperature of the ammonia and hydrogen, absorption of the ammonia occurs and hence hydrogen alone rises through the heat exchanger to the evaporator, while the strong solution flows down by gravity to the generator.

4.7.5 Water–Lithium Bromide (H_2O–LiBr) ARSs

These ARSs utilize a combination of water (as the refrigerant) and lithium bromide (as the absorbent), as the working fluid. These systems are also called *absorption chillers* and have a wide range of application in air conditioning and chilling or precooling operations and are manufactured

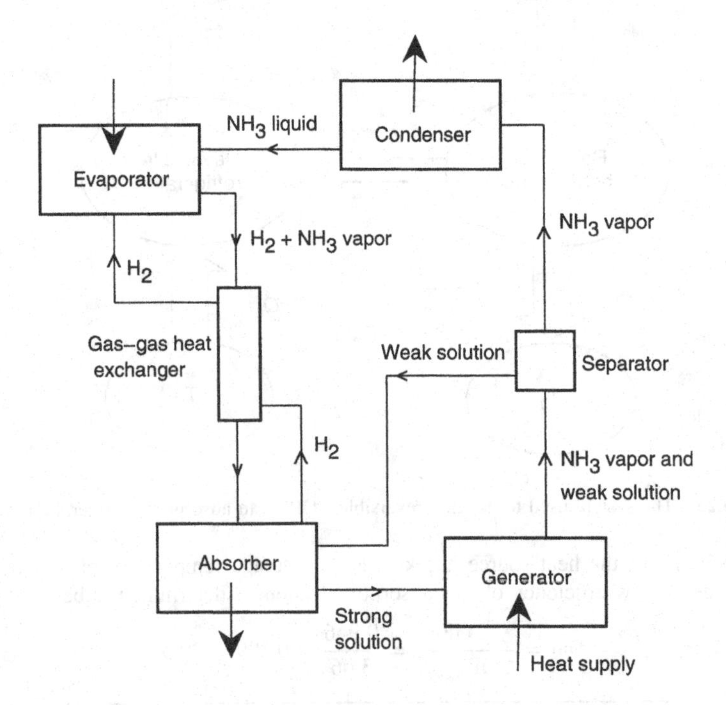

Figure 4.24 A three-fluid ARS.

in sizes from 10 to 1000 tons, leading to the lowest evaporation temperature of 4 °C (with a minimum pressure of 0.8 kPa) because the water is used as the refrigerant. In practical applications, the temperature is 5 °C. Low-pressure steam is the main energy source for these H_2O–LiBr absorption systems. Despite their COPs less than unity, cheap energy can make these systems economically competitive with much higher COP values for vapor-compression systems. In practical H_2O–LiBr ARSs, the evaporator and absorber are combined in a shell at the lower-pressure side and the condenser and generator are combined in another shell at the higher-pressure level. A liquid–liquid heat exchanger is arranged to increase system efficiency and hence to improve the COP. Its operating principle is the same as that of other ARSs. In the H_2O–LiBr ARS, crystallization (which is a solidification of the LiBr) appears to be a significant problem. The crystallization lines are shown on the pressure–temperature and enthalpy–concentration charts. Dropping into the crystallization region causes the formation of slush, resulting in blockage of the flow inside the pipe and interruption of the system operation. In order to prevent this problem, practical systems are designed with control devices to keep the condensation pressure artificially high. Note that absorption chillers and/or refrigeration systems are classified into three categories as follows:

- **Single-effect ARS.** Units using low pressure (135 kPa or less) as the driving force. These units typically have a COP of 0.7.
- **Double-effect ARS.** Units are available as gas-fired (either direct gas firing, or hot exhaust gas from a gas-turbine or engine) or steam-driven with high-pressure steam (270 to 950 kPa). These units typically have a COP of 1.0 to 1.2. To achieve this improved performance they have a second generator in the cycle and require a higher temperature energy source.
- **Triple-effect ARS.** Although the units are not fully available for commercial applications, the concept is well-developed and experiments are conducted for applications through various patents (e.g., Patent Storm, 2010) and some papers (e.g., Kaita, 2001). This triple-effect ARS can use any heat source from waste heat to renewable energy sources, including solar and geothermal heat. The pressure of steam further increases here due to the additional effect (stage) and may easily go beyond the double-effect ARS pressures. The COPs of these three-effect units may become 13 and higher for ammonia-water ARSs, and 1.6 and higher for water-LiBr ARSs. Such COPs are really encouraging for practical applications. The operation of this kind of triple-effect ARS may be described briefly as follows:
  An absorber provides strong solution to three generators, including a high-temperature generator, an intermediate-temperature generator, and a low-temperature generators in which all may be connected in parallel or inverse series. Each generator feeds refrigerant vapor to a corresponding condenser, including a high-temperature condenser, an intermediate-temperature condenser, and a low-temperature condenser. The higher-temperature condensers are essentially coupled with the lower temperature generators, respectively. Hence, the system is referred to as a double-coupled condenser triple effect absorption system. The three heat exchangers may be provided in the parallel or inverse series flowpath from the absorber. It is possible to configure this system differently which requires further research and development to find the best option for applications.

4.7.5.1 Single-Effect ARS

As stated earlier, in ARS, an absorber, generator, pump, and recuperative heat exchanger replace the compressor. Like mechanical refrigeration, as shown in Figure 4.25, the cycle begins when high-pressure liquid refrigerant from the condenser passes through a metering device (1) into the lower-pressure evaporator (2) and collects in the evaporator pan or sump. As before, the flashing that occurs at the entrance to the evaporator cools the remaining liquid refrigerant. Similarly, the transfer of heat from the comparatively warm system water to the now-cool refrigerant causes the latter to evaporate (2), and the resulting refrigerant vapor migrates to the lower-pressure absorber (3).

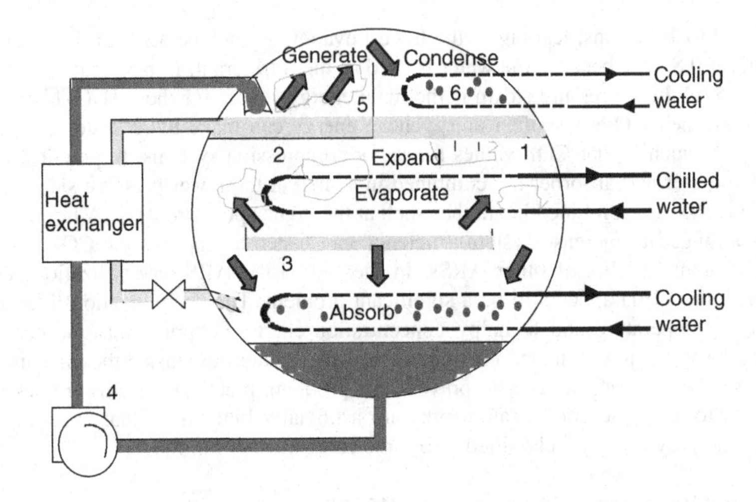

Figure 4.25 Schematic of a single-effect ARS.

There, it is soaked up by an absorbent lithium bromide solution. This process not only creates a low-pressure area that draws a continuous flow of refrigerant vapor from the evaporator to the absorber, but also causes the vapor to condense (3) as it releases the heat of vaporization picked up in the evaporator. This heat – along with the heat of dilution produced as the refrigerant condensate mixes with the absorbent – is transferred to the cooling water and is released in the cooling tower. Of course, assimilating refrigerant dilutes the lithium bromide solution and reduces its affinity for refrigerant vapor. To sustain the refrigeration cycle, the solution must be reconcentrated. This is accomplished by constantly pumping (4) dilute solution from the absorber to the generator (5), where the addition of heat boils the refrigerant from the absorbent. Once the refrigerant is removed, the reconcentrated lithium bromide solution returns to the absorber, ready to resume the absorption process. Meanwhile, the refrigerant vapor liberated in the generator migrates to the cooler condenser (6). There, the refrigerant returns to its liquid state as the cooling water picks up the heat of vaporization carried by the vapor. The liquid refrigerant's return to the metering device (1) completes the cycle.

4.7.5.2 Double-Effect ARS

The energy efficiency of absorption can be improved by recovering some of the heat normally rejected to the cooling tower circuit. A two-stage or two-effect ARS accomplishes this by taking vapors driven off by heating the first-stage concentrator (or generator) to drive off more water in a second stage. Many ARS manufacturers offer this higher efficiency alternative.

The double-effect ARS takes absorption to the next level. The easiest way to picture a double-effect cycle is to think of two single-effect cycles stacked on top of each other (as shown in Figure 4.26). Note that two separate shells are used. The smaller is the first-stage concentrator. The second shell is essentially the single-effect ARS from before, containing the concentrator, condenser, evaporator, and ARS. The temperatures, pressures, and solution concentrations within the larger shell are similar to the single-effect ARS as well. The cycle on top is driven either directly by a natural gas or oil burner, or indirectly by steam. Heat is added to the generator of the topping cycle (primary generator), which generates refrigerant vapor at a relatively higher temperature and pressure. The vapor is then condensed at this higher temperature and pressure and the heat of condensation is used to drive the generator of the bottoming cycle (secondary generator), which is at a lower temperature. If the heat added to the generator is thought to be equivalent to the heat of condensation of the refrigerant, it becomes clear where the efficiency improvement comes from.

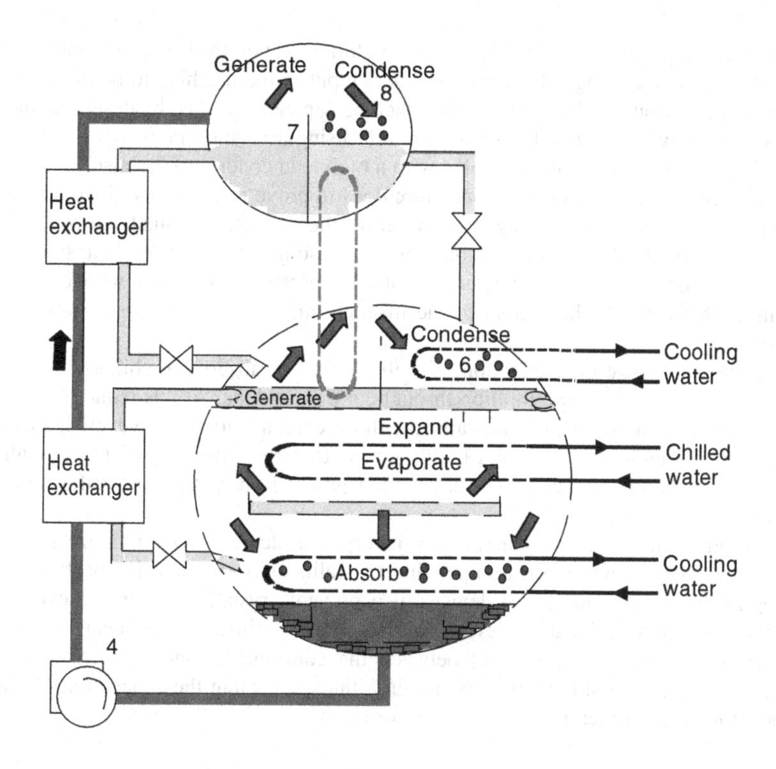

Figure 4.26 Schematic of a double-effect ARS.

For every unit of heat into the primary generator, two masses of refrigerant are boiled out of solution, or generated: one in the primary generator and one in the secondary generator. In a single-effect cycle only one mass is generated. Therefore, in a double-effect system, twice the mass flow of refrigerant is sent through the refrigerant loop per unit of heat input, so twice the cooling is delivered per unit of heat input. Using this approach a double-effect system has a COP that is roughly twice that of a single-effect cycle. However, this simplifying assumption does not account for cycle inefficiencies and losses. In actuality, a single-effect system has a COP of about 0.65 and a double-effect system has a COP of about 1.0. Note that the reuse of the vapors from the first-stage generator makes this machine more efficient than single-stage absorption chillers, typically by about 30%.

4.7.5.3 Crystallization

Some absorption chillers are notorious for "freezing up" or crystallizing. The basic mechanism of failure is simple enough – the lithium bromide solution becomes so concentrated that crystals of lithium bromide form and plug the machine (usually the heat exchanger section). The most frequent causes are as follows:

- air leakage into the machine,
- low-temperature condenser water, and
- electric power failures.

The first two are actually very similar since they both drive the heat input up to the point that crystallization can occur. Whether air leaks into the machine or the condenser water temperature is too low, the water vapor pressure in the absorption chiller evaporator has to be lower than normal to produce the required cooling. This forces the heat input to the machine to be higher to increase the solution concentration. Air leakage into the machine can be controlled by designing the machine with hermetic integrity and routinely purging the unit using a vacuum pump.

Excessively cold condenser water (coupled with a high load condition) can also cause crystallization. While reducing condenser water temperature does improve performance, it could cause a low enough temperature in the heat exchanger to crystallize the concentrate. Sudden drops in condenser water temperature could cause crystallization. For this reason, some of the early absorption chillers were designed to produce a constant condenser water temperature. Modern absorption chillers have special controls that limit the heat input to the machine during these periods of lower condenser water temperatures.

Power failures can cause crystallization as well. A normal absorption chiller shutdown uses a dilution cycle that lowers the concentration throughout the machine. At this reduced concentration, the machine may cool to ambient temperature without crystallization. However, if power is lost when the machine is under full load and highly concentrated solution is passing through the heat exchanger, crystallization can occur. The longer the power is out, the greater the probability of crystallization.

Major absorption chiller manufacturers now incorporate devices that minimize the possibility of crystallization. These devices sense impending crystallization and shut the machine down after going through a dilution cycle. These devices also prevent crystallization in the event of power failure. A typical anti-crystallization device consists of two primary components: (i) a sensor in the concentrated solution line at a point between the concentrator and the heat exchanger and (ii) a normally open, two-position valve located in a line connecting the concentrated solution line and the line supplying refrigerant to the evaporator sprays.

4.7.6 The Steam Ejector Recompression ARS

The ejector recompression absorption cycle, which was developed by Eames and Wu (2000), is similar to the conventional single-effect lithium bromide absorption cycle. The difference between them is that there is a steam ejector in this novel cycle for enhancing the concentration process. Because of the use of the steam ejector, the performance and the operating characteristics of the novel cycle are different from the conventional cycle.

The steam ejector recompression absorption cycle is shown schematically in Figure 4.27a. In this figure, the expansion of the high-pressure steam causes a low pressure at the exit of the primary nozzle of the steam ejector; therefore, the vapor at point 8 in the concentrator is entrained by the primary flow. The two streams are mixed in the steam ejector and condensed in the heat exchanger of the concentrator. The condensation heat is used to heat the solution in the concentrator. Obviously, the heat of the entrained vapor is recovered by the steam ejector in this process. Water at point 3 splits into two streams; one flows back to the steam generator and the other flows into the condenser. In stable operation, the mass flow rate of the first stream equals that of primary flow, while the mass flow rate of the second stream equals that of the entrained vapor. The rest of the cycle is similar to that of the conventional single-effect lithium bromide absorption cycle. Figure 4.27b shows the novel cycle on a P–T–C diagram. As shown in Figure 4.27b, the cycle 6–7–9–10–6 takes up water at the absorber (10–6) and releases it as vapor at the concentrator (7–9). In the conventional absorption cycle, the vapor is condensed at 8′ and the condensation heat is rejected to the surroundings. In the novel cycle, this vapor undergoes a compression process through the ejector to point 2. Since the vapor temperature is greater than the solution temperature in the concentrator, this vapor is used to heat the solution by condensation to point 3. Therefore the heat otherwise wasted is recovered and the energy efficiency is improved.

Eames and Wu (2000) investigated the energy efficiency and the performance characteristics of the novel cycle and the theoretical results showed that the COP of the novel cycle is better than

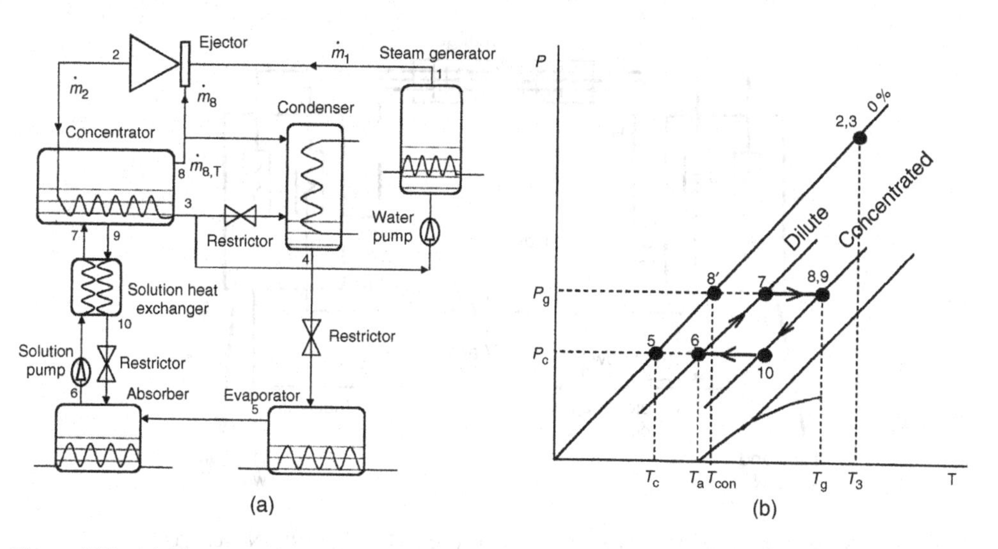

Figure 4.27 (a) The steam ejector recompression ARS and (b) its P–T–C diagram (Eames and Wu, 2000) (Reprinted with permission from Elsevier Science).

that of the conventional single-effect absorption cycle. The characteristics of the cycle performance show its promise in using high-temperature heat source at low cost.

In the past, Kang *et al.* (2000) undertook a study to propose and evaluate advanced absorption cycles for the COP improvement and temperature lift enhancement applications. The characteristics of each cycle are assessed from the viewpoints of the ideal cycle COP and its applications. The advanced cycles for the COP improvement are categorized according to their heat recovery method: condensation heat recovery, absorption heat recovery, and condensation/absorption heat recovery. In H_2O–LiBr systems, the number of effects and the number of stages can be improved by adding a third or a fourth component to the solution pairs. The performance of NH_3–H_2O systems can be improved by internal heat recovery because of their thermal characteristics such as temperature gliding. NH_3–H_2O cycles can be combined with adsorption cycles and power generation cycles for waste heat utilization, performance improvement, panel heating, and low-temperature applications. The H_2O–LiBr cycle is better from the high COP viewpoint for evaporation temperature over 0 °C while the NH_3–H_2O cycle is better from the viewpoint of low-temperature applications. This study suggests that the cycle performance would be significantly improved by combining the advanced H_2O–LiBr and NH_3–H_2O cycles.

4.7.7 *The Electrochemical ARS*

In another study, Newell (2000) proposed a new electrochemical ARS as shown in Figure 4.28, which consists of four main components. An electrochemical cell is the heat absorber, equivalent to an evaporator in a conventional vapor-compression refrigeration system. A fuel cell rejects heat in a manner similar to a condenser in a common vapor-compression refrigeration cycle. The third component is a heat exchanger between gas streams and water flow stream. The fourth component is a current pump for elevating the fuel cell's voltage output to a level sufficient for driving the electrochemical cell. The voltage required is sufficiently low such that the cycle may be one that is conveniently matched for solar photovoltaic cells or other direct current electric energy conversion systems. In fact, the system shown in Figure 4.28 can be used as a thermally driven power cycle by operating the fuel cell at a temperature lower than the electrochemical cell. The voltage supply

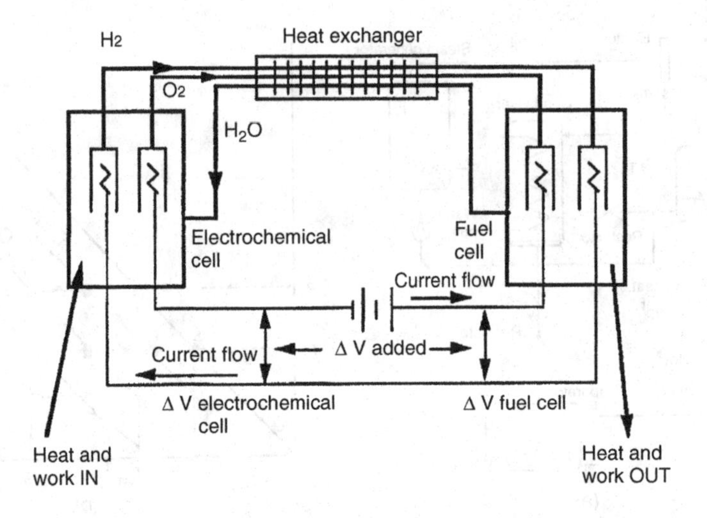

Figure 4.28 Schematic representation of the electrochemical ARS (Newell, 2000).

becomes a load driven by the electric circuit. Lowering component irreversibilities is essential to reach a breakeven operating condition where the fuel cell is generating sufficient power for operation of the electrochemical cell. Newell's system is based on a water/hydrogen/oxygen fuel cell and electrochemical cell combination. Other combinations are also considered. Each one has its own advantages or disadvantages. The configuration envisioned for the system operates near atmospheric pressure. The components could be operated at nearly uniform pressures with gravitation, surface tension, or low head pumping used for transporting the working fluids within and between components. Water may be moved from the electrochemical cell and fuel cell to external heat exchange surfaces, or the cells could be configured for direct heat exchange with their surroundings.

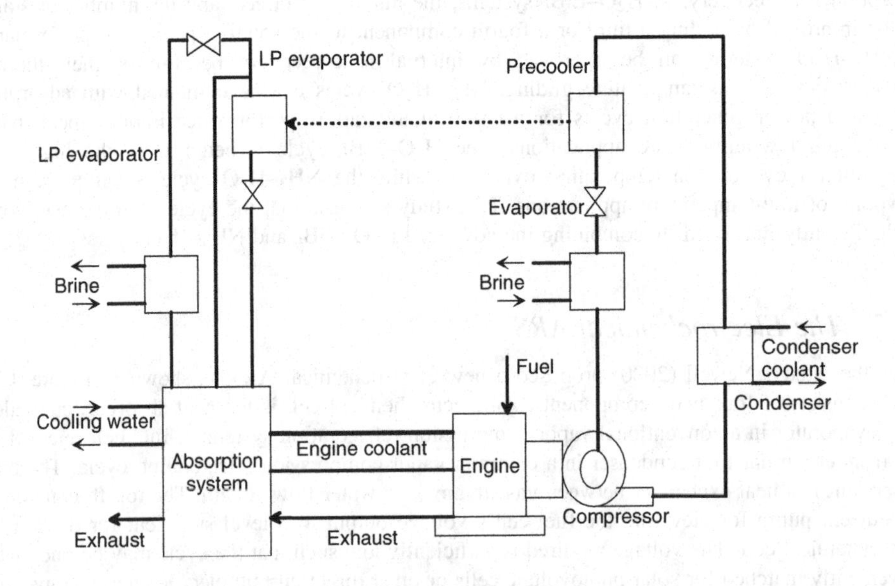

Figure 4.29 Absorption-augmented engine-driven refrigeration system (Turpin, 2000).

4.7.8 The Absorption-Augmented Refrigeration System

Recently, a new absorption-augmented refrigeration system has been under development. The system is based on another development called the generator absorber heat exchange (GAX) cycle. These heat-activated absorption cycles excel at using low-temperature waste heat and turning it into refrigeration or air conditioning. In the absorption-augmented refrigeration system, the prime mover is a gas-fired engine. Gas-fired engines are quite efficient at using high-temperature heat; however, they leave a lot of their energy (approximately 65 or 70%) behind as low-temperature waste heat, which is ideal for absorption (Turpin, 2000). The total system combines an internal combustion engine with a mechanical compression refrigeration system powered by the engine shaft power and the waste heat driven ARS (Figure 4.29).

Example 4.7

In this example, we present one of our earlier works (Dincer and Ture, 1993; Dincer, Edin and Ture, 1996) on the design and construction of a solar powered ARS (Figure 4.30) using a mixture of R-22 and DMETEG as the working fluid. In this project, a combined water-heating and cooling system based on absorption refrigeration was designed and constructed. The system consists of four plate collectors, an evaporator, an absorber, a generator, a condenser, a solution pump, and two heat exchangers. Each part was custom-designed to provide 4000 kcal/h cooling load, although R-22 which has a less damaging effect on the ozone layer compared to other CFCs was employed as a refrigerant. The energy analysis results of the experimental system were compared with the theoretical calculations and a reasonably good agreement was found. The results show that the ARS appeared to be efficient and effective.

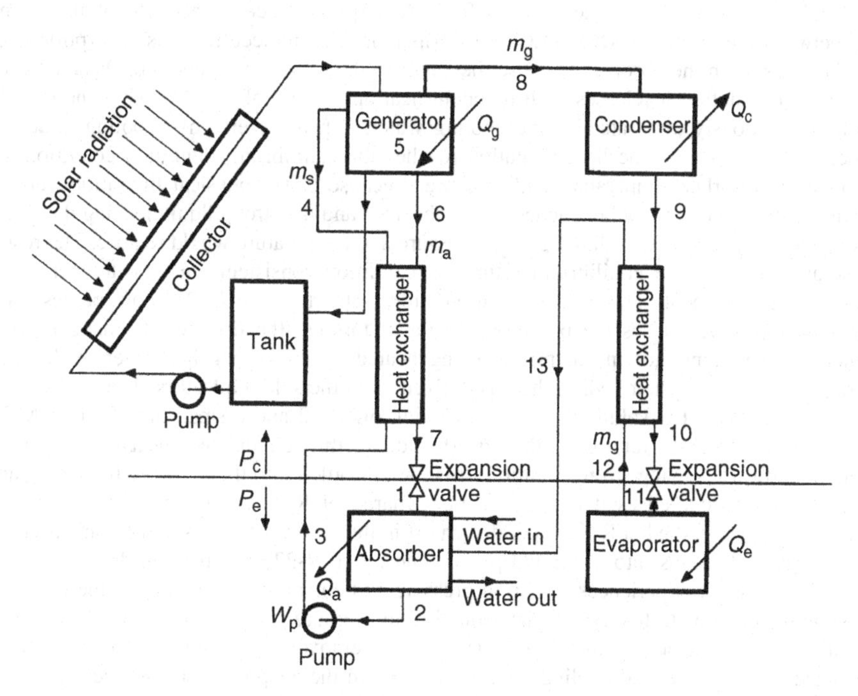

Figure 4.30 An R-22 and DMETEG ARS (Dincer *et al.*, 1996) (Reprinted with permission from Elsevier Science).

Solution

In the energy analysis, we used the energy balance equations presented earlier in this section.

Experimental Apparatus and Procedure. The experimental system used a working fluid combining R-22 as refrigerant and DMETEG as absorbent. The cycle efficiency and the operational characteristics of an ARS are dependent on the properties of the refrigerant, the absorbent, and their relative mixtures. The combination of R-22 and DMETEG was suggested as one of the new alternative combinations and was employed in the present system.

A schematic diagram of the solar powered ARS built in the Solar Energy Laboratory of the Energy Systems Department at Marmara Research Centre in Gebze, Turkey, is depicted in Figure 4.30. Generally, the system is considered as one which incorporates solar energy equipment with conventional ARS. The basic elements of the system were four flat-plate collectors, an evaporator, an absorber, a generator, a condenser, a solution pump, and two heat exchangers. The system has been designed specifically to provide 4000 kcal/h cooling load in the evaporator. Although the construction of the components of the system followed closely that of Van Den Bulck, Trommelmans and Berghmans 1982, some modifications were introduced into the individual components of the system, not only to achieve the desired design parameters, but also to improve the operation of the ARS. For instance, a 1.2 m long and 1.3 cm diameter copper pipe with 2 mm holes in every 3 cm of length was employed in the absorber to provide homogeneous, fast, and effective absorption of DMETEG and R-22 vapor.

The operation of the system is described as follows. Starting from the change of R-22 from liquid to gas via the throttling effect of the expansion valve, the resulting R-22 vapor begins to absorb heat from its immediate surroundings in a conventional natural convection type evaporator. Cool vapor leaving the evaporator passes through the second heat exchanger into the absorber where it combines with DMETEG which absorbs the gaseous R22. Absorption proceeds because of the chemical affinity between the absorbent DMETEG and refrigerant R-22 molecules. This absorption activity lowers the pressure in the absorber to cause the vapor to flow from the evaporator. When the vapor goes into liquid solution it releases both its latent heat and a heat of dilution. This energy release has to be continuously dissipated by the cooling water. When the effective cooling is achieved, the process continues until the liquid solution reaches the equilibrium saturation condition which exists for each absorber temperature and pressure. Because of the physical limitations, complete equilibrium saturation may not be reached in the absorber and the strong liquid leaving the absorber may not be as fully saturated with R-22 as its pressure and temperature would require. The resulting liquid solution reaches the equilibrium saturation condition consistent with the temperature and pressure of the absorbent. This strong solution, fully saturated with R-22, now passes through a solution pump which raises the pressure, passing it through the first heat exchanger and into the generator. The generator meanwhile is being heated by circulating hot water in the higher-pressure portion of the system, whose heat is derived from the solar collectors. The temperature of the strong R-22/DMETEG solution increases, driving off R-22 and a small amount of DMETEG vapor. The weak solution returns to the absorber down through the first heat exchanger while warming the upward flowing strong solution. It is then throttled into the absorber by the expansion valve, to be further cooled as it picks up a new charge of R-22 coming from the second heat exchanger. Meanwhile, the hot R-22 vapor driven off in the generator passes to the condenser where it loses energy and passes into the liquid phase. The liquid R-22 after passing down through the second heat exchanger experiences a drop in pressure and enters the evaporator as the low portion of the system to complete the cycle. The reduction in pressure through this valve 2 facilities the vaporization of R-22 which ultimately effects the heat removal from the environment. The cycle is completed when the desired cooling load is achieved in the evaporator. Consequently, it can be seen that there are essentially three circuits for the absorption cooling system; (a) the almost pure R22 circuit – condenser, heat exchanger 2, evaporator, and heat exchanger 2 to the absorber, (b) the strong solution circuit – absorber, pump, and heat exchanger 1 to the generator, and (c) the weak solution circuit – from the generator through the heat exchanger 1 and into the absorber.

Results and Discussion. All the required parameters for the design of a solar powered absorption cooling system were obtained by using the calculation techniques for the individual components. In the theoretical calculations for the design of an ARS, an enthalpy–concentration diagram for the R-22 and DMETEG pair (Figure 4.30) was used. Some obtained results are as follows: $T_1 \ldots T_{13}$ (°C) =39, 30, 30, 65, 90, 87, 53, 82, 40, 27, −5, −5, 20; $T_{cw} = 20\,°C$; $P_e = 4.8$ bar; $P_c = 16$ bar; $Q_e = 4.65$ kW; $Q_c = 5.0$ kW; $Q_a = 7.5$ kW; $Q_g = 7.6$ kW; $W_p = 0.25$ kW; COP = 0.6; $m_s = 290.6$ kg/h; $m_g = 90.0$ kg/h; $m_a = 200.6$ kg/h. In addition, some results related to the collector system are as follows: $L = 40°46'$; $S = 36°$; $Q_{th} = 4313$ kcal/m^2·day; $R = 1.08$; $E_c = 0.65$; $E_m = 0.60$; $E_g = 0.80$; $T_w = 45\,°C$; $T_s = 18.7\,°C$; $Q_u = 1450$ kcal/m^2·day; $m = 420$ kg/day; $C_p = 1$ kcal/kg·°C; $F_c = 8$ m^2; $n = 4$; $V = 0.48$ m^3; $G = 8$ L/min; $\varepsilon_1 = 0.06$ m^3/m^2; $f_2 = 1$ L/min·m^2; $Q_r = 1{,}1046$ kcal/day. Note that Figure 4.31, known as the enthalpy–concentration diagram of the R-22/DMETEG pair, was used to find the enthalpy and other relevant data. This graph was originally developed by Jelinek, Yaron and Borde 1980.

Under a working regime, the measured experimental and theoretical values were used to determine the values of COP which were plotted against the variations in evaporator temperature as shown as Figure 4.32a. In this graph, in order to eliminate the negative evaporator temperatures, the term $(T_a - T_e)$ was used where T_a is the initial evaporator temperature whose average value

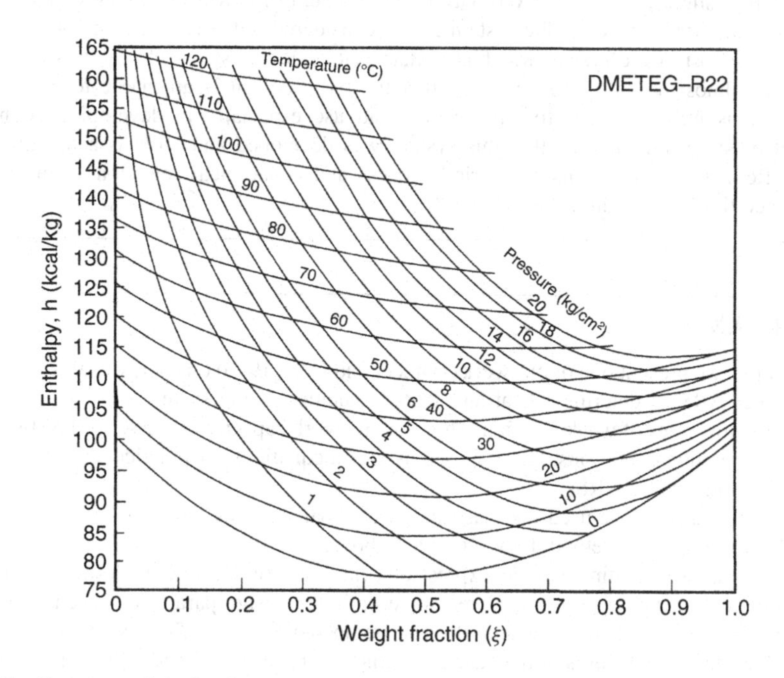

Figure 4.31 Enthalpy-weight fraction (concentration) diagram for the pair of R-22 and DMETEG (Dincer *et al.*, 1996) (Reprinted with permission from Elsevier Science).

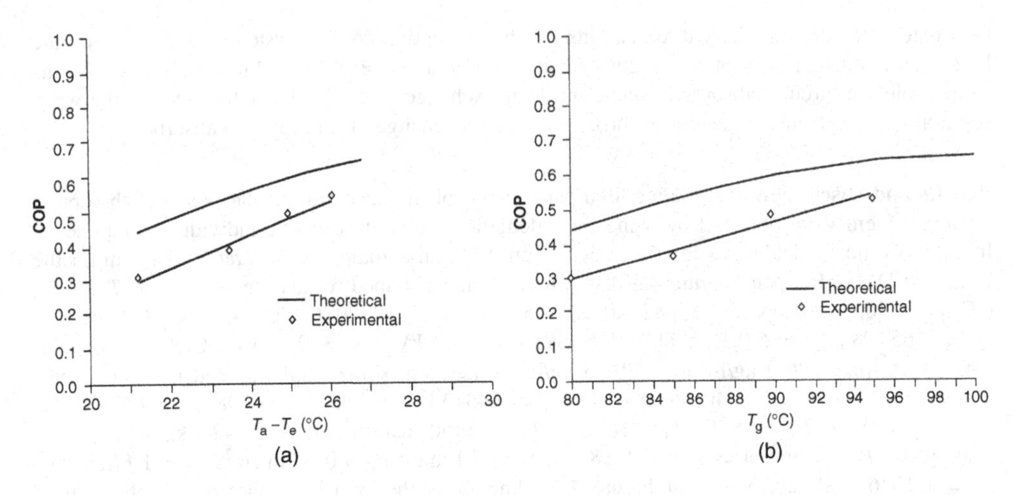

Figure 4.32 Variation of COP versus (a) evaporator temperature and (b) generator temperature (Dincer *et al.*, 1996) (Reprinted with permission from Elsevier Science).

was around 20 °C. The experimental evaporator temperature values which were found to be lower than that of the theoretical ones clearly indicated some heat losses in the system. Similarly, when the actual and theoretical values of COP were plotted against the variation of generator temperature (Figure 4.21b) a slight increase in COP with increasing temperature was observed. This indicated that the operating performance of the system can be considered stable in that range of temperatures.

Although the cost of the system was higher than that of compression refrigerators of equivalent performance (almost double), it is believed that the basic design is amenable to low cost mass production. This makes it attractive for widespread use especially in developing countries. In addition, it must be remembered that this system was developed and built as a one-off research prototype. Refinement of the manufacturing process and economical selection of materials will further reduce the cost per unit.

Example 4.8

In this example, we present one of our works (Dincer and Dost, 1996) on energy analysis of a lithium bromide–water ARS to determine heat and work capacities of the system's components varying with the mass flow rates of weak solution. The heat and work capacity expressions developed, based on optimum operation conditions, are proposed as useful equations for practical design calculations of lithium bromide–water ARSs.

In the energy analysis, we used the mass and energy balance equations for the system shown in Figure 4.33, based on the methodology presented above.

The main goal was to find simple expressions that can be utilized in design calculation of lithium bromide–water ARS. We considered some optimum design parameters, such as evaporator temperature $T_E = 4.5\,°C$, condenser and absorber temperatures $T_C = T_A = 30\,°C$, heat exchanger efficiency $E = 0.9$ for two cases of the generator temperatures $T_G = 90$ and $100\,°C$. For comparison purposes, two temperature values of the generator were investigated. In addition, the concentration values of LiBr on a mass basis were taken as $X_a = 0.685$ and 0.695 for weak solution, and $X_s = 0.5$ for strong solution to avoid crystallization at the two generator temperatures given above, using Figure 4.34. Using the optimum conditions, the following temperature and enthalpy values for two

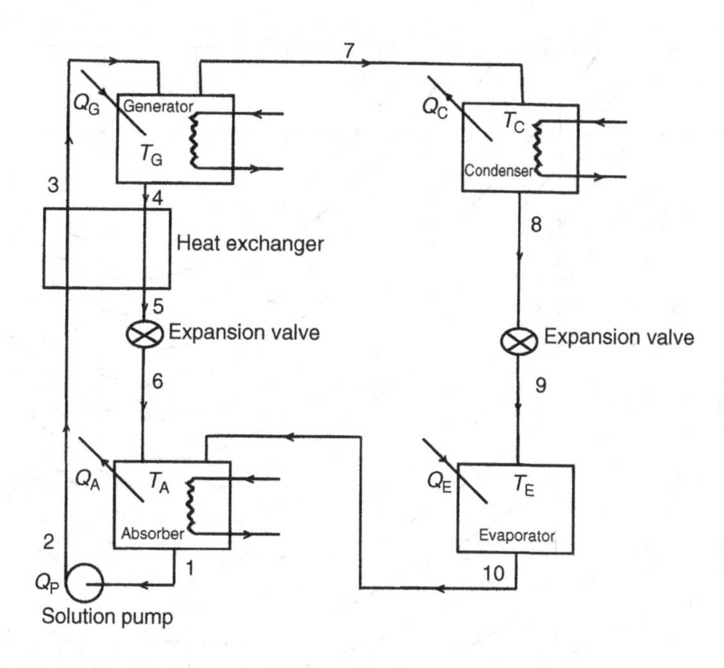

Figure 4.33 The lithium bromide–water ARS used in the model development (Dincer and Dost, 1996).

generator temperatures were computed using a computer program which is partly based on some equations given in ASHRAE 1997: $T_1 = T_2 = T_8 = 30\,°C$, $T_3 = 32.3\,°C$, $T_4 = 90\,°C$, $T_5 = 88\,°C$, $T_6 = 75.5\,°C$, $T_7 = 90\,°C$, $T_9 = T_{10} = 4.5\,°C$; $h_1 = 60.5$ kJ/kg, $h_2 = 95.6$ kJ/kg, $h_3 = 102.1$ kJ/kg, $h_4 = 256.1$ kJ/kg, $h_5 = h_6 = 251.19$ kJ/kg, $h_7 = 2660.1$ kJ/kg, $h_8 = h_9 = 125.66$ kJ/kg, $h_{10} =$ 2509.9 kJ/kg; and $P_E = 1.23$ kPa, for $T_G = 90\,°C$; and $T_1 = T_2 = T_8 = 30\,°C$, $T_3 = 34.4\,°C$, $T_4 =$ $100\,°C$, $T_5 = 96\,°C$, $T_6 = 80.1\,°C$, $T_7 = 100\,°C$, $T_9 = T_{10} = 4.5\,°C$; $h_1 = 60.5$ kJ/kg, $h_2 = 104.12$ kJ/kg, $h_3 = 111.3$ kJ/kg, $h_4 = 278.08$ kJ/kg, $h_5 = h_6 = 268.13$ kJ/kg, $h_7 = 2676.0$ kJ/kg, $h_8 =$ $h_9 = 125.66$ kJ/kg, $h_{10} = 2509.9$ kJ/kg; and $P_E = 1.23$ kPa, $P_C = 5.45$ kPa for $T_G = 100\,°C$.

Using the above temperature, enthalpy, and pressure values in energy balance equations we obtained the following absorber heat capacity, solution pump work, generator heat capacity, condenser heat capacity, and evaporator heat capacity with the mass flow rate of weak solution for two cases of generator temperatures of 90 and 100 °C:

$$\dot{Q}_A = 1093.65\dot{m}_a \text{ and } \dot{Q}_A = 1162.89\dot{m}_a$$

$$\dot{W}_P = 48.09\dot{m}_a \text{ and } \dot{W}_P = 60.63\dot{m}_a$$

$$\dot{Q}_G = 1101.13\dot{m}_a \text{ and } \dot{Q}_G = 1167.89\dot{m}_a$$

$$\dot{Q}_C = 937.76\dot{m}_a \text{ and } \dot{Q}_C = 994.63\dot{m}_a$$

$$\dot{Q}_E = 882.19\dot{m}_a \text{ and } \dot{Q}_C = 929.85\dot{m}_a$$

with COP = 0.7676 at the generator temperature of 90 °C and COP = 0.7574 at the generator temperature of 100 °C. These COP values indicate that the lithium bromide–water ARS at the generator temperature of 90 °C becomes more efficient in providing the evaporator temperature of 4.5 °C, because of the fact that the lower generator temperature provides the same evaporator temperature. In this respect, this generator temperature is recommended for practical applications.

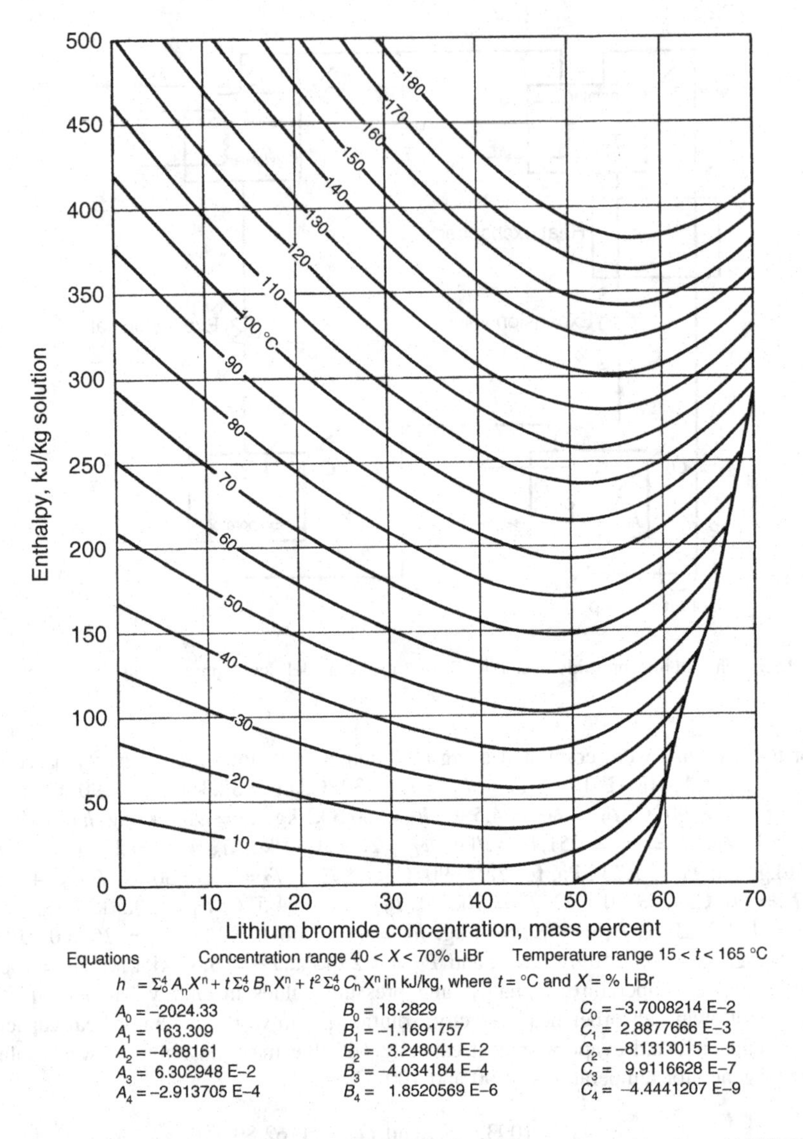

Equations Concentration range $40 < X < 70\%$ LiBr Temperature range $15 < t < 165$ °C

$h = \Sigma_0^4 A_n X^n + t \Sigma_0^4 B_n X^n + t^2 \Sigma_0^4 C_n X^n$ in kJ/kg, where t = °C and X = % LiBr

$A_0 = -2024.33$	$B_0 = 18.2829$	$C_0 = -3.7008214$ E–2
$A_1 = 163.309$	$B_1 = -1.1691757$	$C_1 = 2.8877666$ E–3
$A_2 = -4.88161$	$B_2 = 3.248041$ E–2	$C_2 = -8.1313015$ E–5
$A_3 = 6.302948$ E–2	$B_3 = -4.034184$ E–4	$C_3 = 9.9116628$ E–7
$A_4 = -2.913705$ E–4	$B_4 = 1.8520569$ E–6	$C_4 = -4.4441207$ E–9

Figure 4.34 Enthalpy–concentration diagram for lithium bromide–water combinations (ASHRAE, 1997) (Reprinted with permission from ASHRAE).

The above equations were checked by inserting them into energy balance equations and were found to be correct. The variations in the capacities of absorber, pump, generator, condenser, and evaporator against the flow rate of the weak solution (for the range of 0 and 0.2 kg/s) were computed from the above equations and are shown in Figure 4.35a and b. As seen, the capacities increase with the flow rate linearly. Thus, one can claim that the above equations (i.e., heat load-mass flow rate equations) under the optimum conditions will lead to a simple solution for the design of a practical lithium bromide–water ARS.

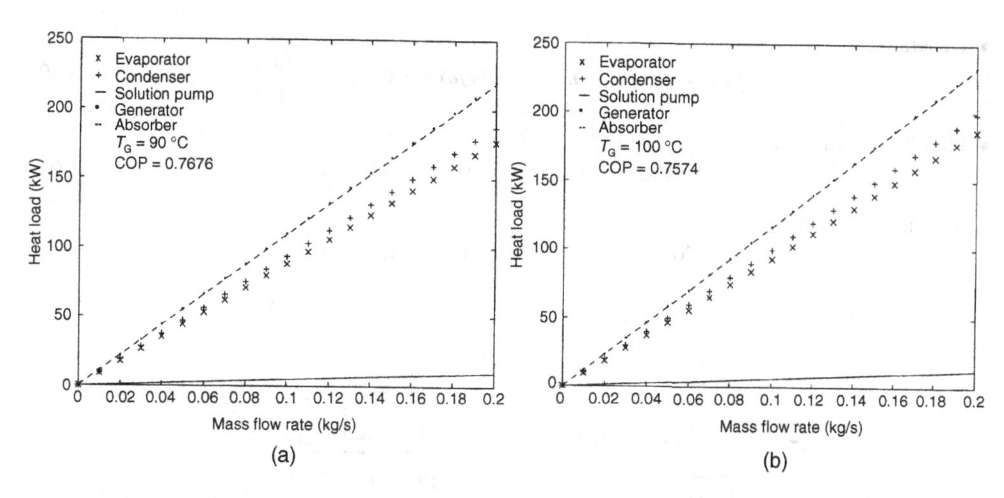

Figure 4.35 Variations of absorber, pump, generator, condenser, and evaporator capacities versus mass flow rate of weak solution (a) at $T_G = 90\,°C$ and (b) at $T_G = 100\,°C$ (Dincer and Dost, 1996).

4.7.9 Exergy Analysis of an ARS

As given earlier, the change in exergy rate or the rate of exergy loss can be defined in terms of physical terms as follows:

$$\Delta \dot{E}x = \sum \dot{m}_i ex_i - \sum \dot{m}_e ex_e + \dot{Q}\left(1 - \frac{T_0}{T}\right) + \dot{W}_i \tag{4.55}$$

where the first terms are the sum of exergy input and output rates of the flow, respectively. The third term is the heat of exergy (+ if it is heat input; – if it is heat output). The last term is the work given to the system (e.g., pump work).

The exergy balance equations for the components of the ARS can be written with respect to Figure 4.36 as follows:

- Condenser:

$$\Delta \dot{E}x_C = \dot{m}_7(ex_{7'} - ex_8) \tag{4.56}$$

since there is a heat rejection to the surroundings ($T_C = T_0$) resulting in $\dot{Q}_C(1 - T_0/T_C) = 0$ and no work input makes $\dot{W}_i = 0$.

- Evaporator:

$$\Delta \dot{E}x_E = \dot{m}_7(ex_{10} - ex_{11}) + \dot{Q}_E\left(1 - \frac{T_0}{T_E}\right) \tag{4.57}$$

- Absorber:

$$\Delta \dot{E}x_A = \dot{m}_7 ex_{12'} + \dot{m}_6 ex_6 - \dot{m}_1 ex_1 \tag{4.58}$$

since there is a heat rejection to the surroundings ($T_A = T_0$) resulting in $\dot{Q}_A(1 - T_0/T_A) = 0$ and no work input makes $\dot{W}_i = 0$.

- Solution pump:

$$\Delta \dot{E}x_P = \dot{m}_1(ex_1 - ex_2) + \dot{W}_P \tag{4.59}$$

- Generator:

$$\Delta \dot{E}_G = \dot{m}_3 ex_3 - \dot{m}_4 ex_4 - \dot{m}_7 ex_7 + \dot{Q}_G \left(1 - \frac{T_0}{T_G}\right) \tag{4.60}$$

- First heat exchanger:

$$\Delta \dot{E}x_{HE1} = \dot{m}_2(ex_2 - ex_3) + \dot{m}_4(ex_4 - ex_5) \tag{4.61}$$

- Second heat exchanger:

$$\Delta \dot{E}x_{HE2} = \dot{m}_7(ex_8 - ex_9 + ex_{11} - ex_{12}) \tag{4.62}$$

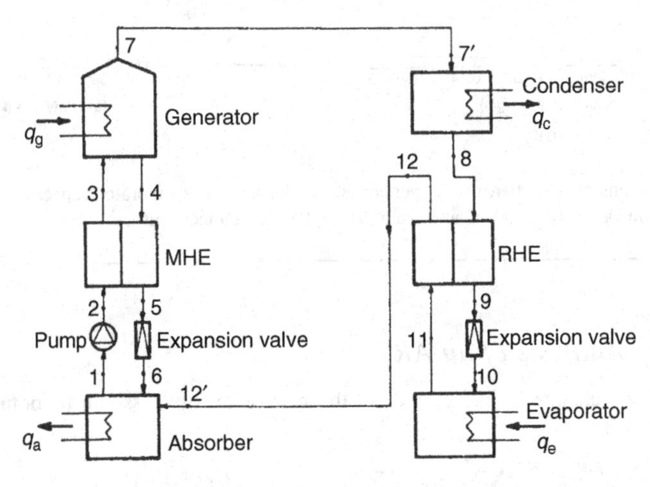

Figure 4.36 Schematic of the ammonia–water ARS (Ataer and Gogus, 1991) (Reprinted with permission from Elsevier Science).

Note that the exergy losses in expansion valves are neglected because of the fact that their magnitudes are comparatively small. Therefore, the total exergy loss of the ARS system becomes the sum of the exergy losses in the components as listed above:

$$\Delta \dot{E}x_T = \Delta \dot{E}x_C + \Delta \dot{E}x_E + \Delta \dot{E}x_P + \Delta \dot{E}x_G + \Delta \dot{E}x_{HE1} + \Delta \dot{E}x_{HE2} \tag{4.63}$$

Consequently, the exergetic COP (ECOP)(i.e., exergy efficiency) for the entire system can be defined as follows:

$$\eta_{COP,ex} = \frac{\dot{Q}_E \left(1 - \frac{T_0}{T_E}\right)}{\dot{Q}_G \left(1 - \frac{T_0}{T_G}\right) + \dot{W}_P} \tag{4.64}$$

Note that the heat-transfer rates for condenser, evaporator, absorber, and generator can be calculated through the energy balance equations as given earlier in this section.

Example 4.9

In this example, we present work done by Ataer and Gogus 1991 on exergy analysis of an ammonia–water ARS which is similar to that shown in Figure 4.21 and the performance results. These authors determined irreversibilities in components (absorber, generator, pump, expansion valves, mixture heat exchanger (MHE), and refrigerant heat exchanger (RHE)) of the ARS by exergy analysis. It was assumed that the ammonia concentration at the generator exit is independent of the other parameters, equal to 0.999, and at the evaporator exit the gas was saturated vapor. Pressure losses between the generator and condenser, and the evaporator and absorber, were taken into consideration. For each condenser, evaporator, absorber, and generator temperature it was assumed that a separate ARS design capacity is a 1 kW cooling load. The assumptions were that the mixture leaves the condenser at the condenser temperature and in a saturated liquid state; the weak solution leaves the generator at the generator temperature; the strong solution leaves the absorber at the absorber temperature; the mixture at the evaporator exit is saturated vapor at the evaporator temperature; and the rectifier of the ARS and its effect is ignored. In the results, the exergy values of each component, the COP, and the ECOP were given graphically for different generator temperatures.

Solution

The mass and exergy balance equations for each component of the system were written, based on the methodology presented earlier in this section. Ataer and Gogus 1991 conducted an analysis to determine COP and ECOP values varying with generator temperature. They obtained Table 4.1, presenting the temperature, pressure, concentration (mass fraction), mass flow rate, enthalpy, entropy, and exergy data for each point of the system as shown in Figure 4.36.

As shown in Figure 4.38a, typical COP values of the ARS are in the range 0.2–0.9. For a given evaporator, absorber, and condenser temperature there is a minimum generator temperature which

Table 4.1 ARS data obtained from the analysis.

Point	T (°C)	P (bar)	X	m (g/s)	h (kJ/kg)	S (kJ/kg·K)	e (kJ/kg)
1	24.00	2.571	0.459	2.37	−138.65	2.032	−734.43
2	24.04	9.094	0.459	2.37	−137.80	2.031	−733.09
3	81.53	9.094	0.459	2.37	127.49	2.847	−707.07
4	130.0	9.094	0.170	1.54	444.43	4.532	−884.26
5	45.41	9.094	0.170	1.54	66.22	3.479	−953.71
6	45.50	2.571	0.170	1.54	66.22	3.482	−954.54
7	130.00	9.094	0.999	0.83	1562.43	5.150	52.82
7′	129.59	8.661	0.999	0.83	1562.43	5.173	45.97
8	22.00	8.661	0.990	0.83	102.60	0.378	−8.15
9	11.02	8.661	0.999	0.83	50.64	0.198	−7.52
10	11.06	2.764	0.999	0.83	50.64	0.201	−8.41
11	−10.00	2.764	0.999	0.83	1255.82	4.777	−144.58
12	15.59	2.764	0.999	0.83	1316.44	4.997	−148.44
12′	15.00	2.571	0.999	0.83	1316.44	5.031	−158.53

Note that the temperature–concentration diagram of ammonia–water mixture is utilized to get enthalpy and other relevant data. Such a diagram is shown in Figure 4.37.
Source: Ataer and Gogus (1991) (Reprinted with permission from Elsevier Science).

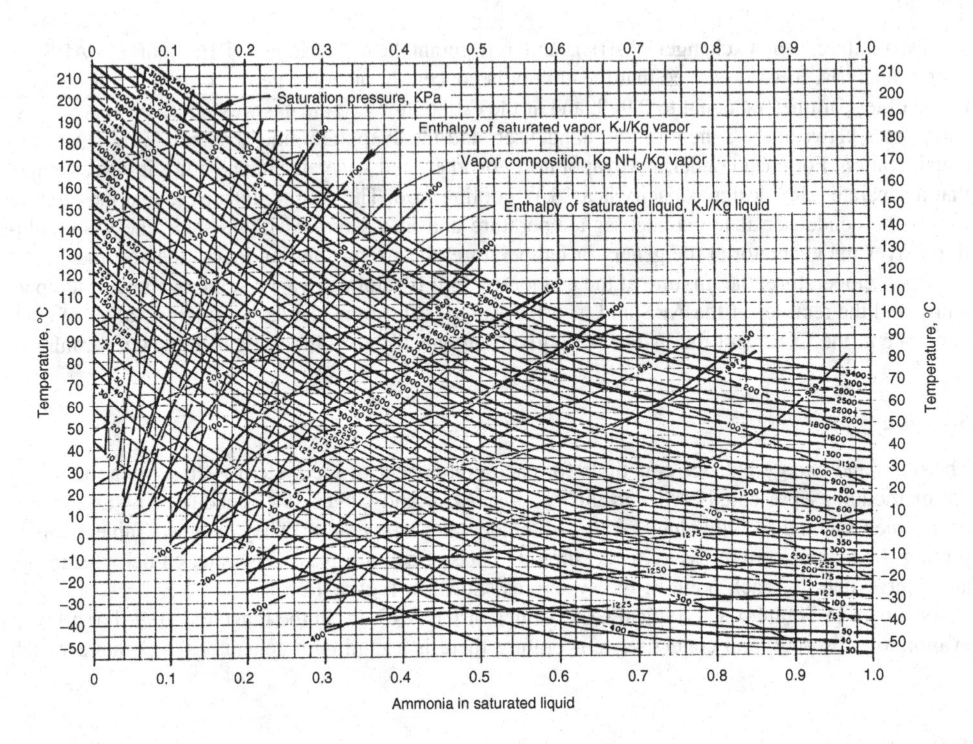

Figure 4.37 Temperature–concentration diagram of ammonia–water mixture (ASHRAE, 1997) (Reprinted with permission from ASHRAE).

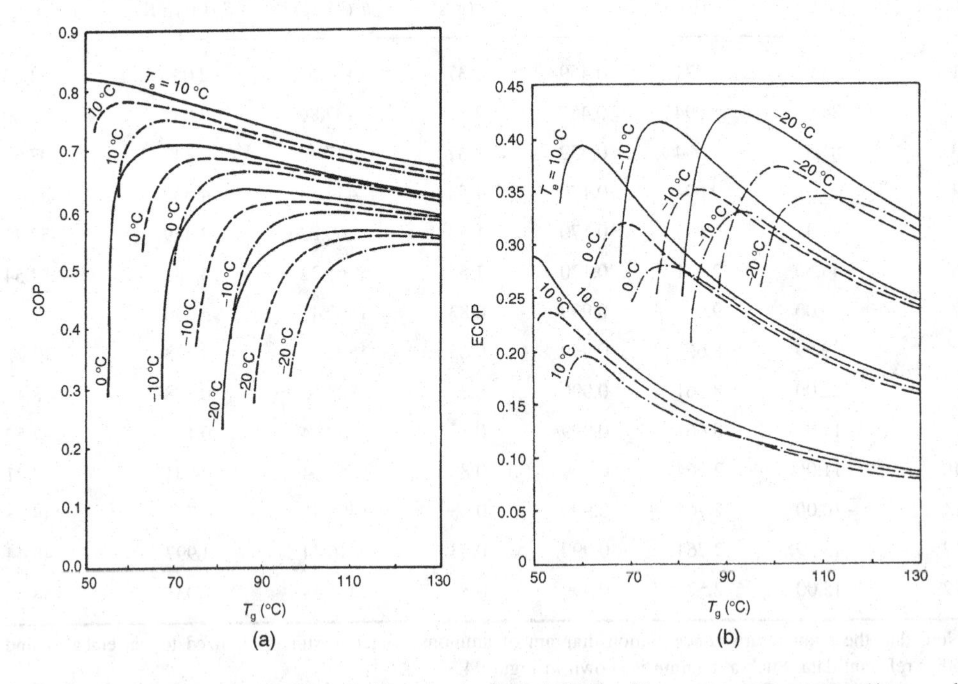

Figure 4.38 Variation of (a) COP and (b) exergetic COP (ECOP) with generator temperature (Ataer and Gogus, 1991) (Reprinted with permission from Elsevier Science).

corresponds to equalization of ammonia concentrations of the solution flowing into and out of the generator. This temperature is called the cut-in temperature. For generator temperatures above this value, the COP increases until it reaches a maximum. The value of ECOP is at a maximum at the points where the COP values are at a maximum, as shown in Figure 4.38b. However, ECOP values are relatively much smaller than the corresponding COP values, because of the fact that there are considerable irreversibilities occurring in the system.

Consequently, for each condenser, absorber, and evaporator temperature, there is a generator temperature at which the dimensionless total exergy loss of the ARS is a minimum. At this point the COP and ECOP of the system are at a maximum. It can be noted that the results of the second-law analysis can be used to identify the less efficient components of the system and also to modify them. Moreover, the suitability of the selected components can be judged by exergy analysis. Furthermore, the exergy analysis appears to be a significant tool for the determination of the optimum working conditions of such systems.

4.7.10 Performance Evaluation of an ARS

The efficiency of ARS is defined by COP as in a vapor-compression refrigeration system. Figure 4.39 presents the change of COP for a single-stage NH_3–H_2O ARS with evaporation temperature at different condensation temperature ranges of 10–20 °C, 20–30 °C, and 30–40 °C, respectively. As shown in the figure, increasing evaporation temperature will decrease the COP, and for the same evaporation temperature a lower condensation temperature will give better COP. These trends should be taken into consideration carefully when designing an ARS for any particular application.

4.8 Concluding Remarks

This chapter has dealt with a large number of theoretical and practical topics in refrigeration systems, covering refrigeration cycles/systems and their energy and exergy analyses along with the representative examples. In addition to conventional vapor-compression cycles, air-standard refrigeration cycle and absorption–refrigeration cycles are studied in a greater detail.

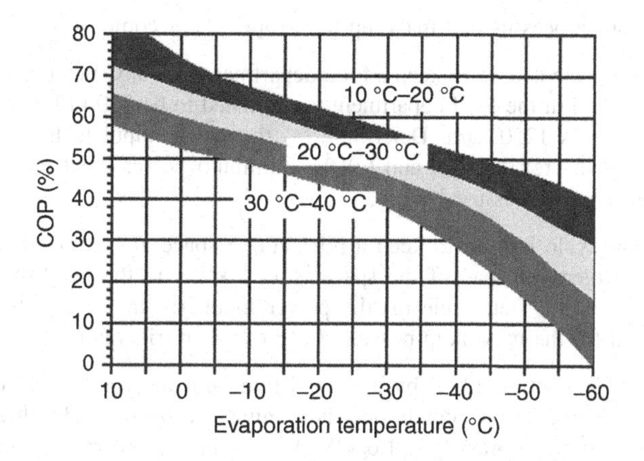

Figure 4.39 Variation of COP with evaporation temperature at various condensation temperature ranges (*Courtesy of Colibri b.v.-Stork Thermeq b.v.*).

Nomenclature

COP coefficient of performance
c_p constant pressure specific heat, kJ/kg·K
ex specific exergy, kJ/kg
$\dot{E}x$ exergy rate, kW
h enthalpy, kJ/kg
$\dot{m}$ mass flow rate, kg/s
P pressure, kPa
q specific heat, kJ/kg
$\dot{Q}$ heat load; power, kW
s entropy, kJ/kg
$\dot{S}_{gen}$ entropy generation rate, kW/K
T temperature, °C or K
v specific volume, m^3/kg
$\dot{V}$ volumetric flow rate, m^3/s
w specific work, kJ/kg
$\dot{W}$ work input to compressor or pump, kW
X concentration of refrigerant in solution, kg/kg

Greek Letters

η efficiency

Study Problems

Vapor-Compression Refrigeration Systems

4.1 Draw temperature–entropy and pressure–enthalpy diagrams of simple vapor-compression refrigeration cycle.

4.2 Explain the four processes that make up the simple vapor-compression refrigeration cycle.

4.3 A refrigeration cycle is used to keep a food department at −5 °C in an environment at 20 °C. The total heat gain in the food department is estimated to be 750 kJ/h and the heat rejection in the condenser is 1250 kJ/h. Determine (a) the power input to the compressor in kW, (b) the COP of the refrigerator, and (c) the minimum power input to the compressor if a reversible refrigerator was used.

4.4 A refrigeration cycle is used to keep a refrigerated space at −25 °C in an environment at 27 °C. The refrigeration load of the space is 11.5 kW and the COP of the refrigerator is estimated to be 0.90. Determine (a) the power input, (b) the rate of heat rejected in the condenser, and (c) the maximum possible COP of this refrigerator.

4.5 A small room is kept at 23 °C by a 9000 Btu/h split air conditioner when the ambient temperature is 35 °C. The air conditioner is running at full load under these conditions. The power input to the compressor is 1.6 kW. Determine (a) the rate of heat rejected in the condenser in Btu/h, (b) the COP of the air conditioner, and (c) the rate of cooling in Btu/h if the air conditioner operated as a Carnot refrigerator for the same power input.

4.6 A commercial refrigerator is to cool eggplants from 28 to 12 °C at a rate of 660 kg/h. The power input to the refrigerator is 10 kW. Determine the rate of cooling and the COP of the refrigerator. The specific heat of eggplant above freezing is 3.92 kJ/kg·°C.

4.7 Water is continuously cooled in a refrigerator from 17 to 3 °C. The heat rejected in the condenser is 380 kJ/min and the power input is 2.2 kW. Determine the rate at which water is cooled in L/min and the COP of the refrigerator. The specific heat of water is 4.18 kJ/kg·°C.

4.8 A refrigeration system absorbs heat from a space at 5 °C at a rate of 25 kW and rejects heat to water in the condenser. Water enters the condenser at 15 °C at a rate of 0.84 kg/s. The COP of the system is estimated to be 1.75. Determine (a) the power input to the system, (b) the temperature of the water at the exit of the condenser, and (c) the maximum possible COP of the system. The specific heat of water is 4.18 kJ/kg·°C.

4.9 A refrigeration system absorbs heat from a space at 40 °F at a rate of 17 Btu/s and rejects heat to water in the condenser. Water enters the condenser at 60 °F at a rate of 1.75 lbm/s. The COP of the system is estimated to be 1.85. Determine (a) the power input to the system in kW, (b) the temperature of the water at the exit of the condenser, and (c) the maximum possible COP of the system. The specific heat of water is 1.0 Btu/lbm·°F.

4.10 Refrigerant-134a enters the compressor of a refrigeration system at 140 kPa as a saturated vapor and leaves at 800 kPa and 70 °C. The refrigerant leaves the condenser as a saturated liquid. The rate of cooling provided by the system is 900 W. Determine the mass flow rate of R-134a and the COP of the system.

4.11 Refrigerant-134a enters the evaporator coils of a household refrigerator placed at the back of the freezer section at 120 kPa with a quality of 20% and leaves at 120 kPa and −20 °C. If the compressor consumes 620 W of power and the COP of the refrigerator is 1.3, determine (a) the mass flow rate of the refrigerant and (b) the rate of heat rejected to the kitchen air.

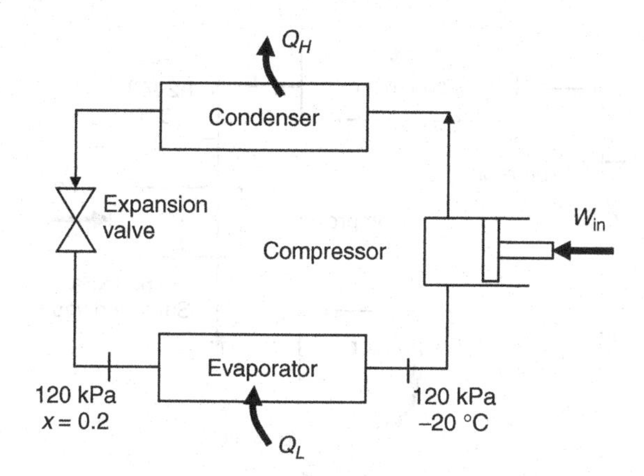

4.12 A refrigerated room is kept at −35 °C by a vapor-compression cycle with R-134a as the refrigerant. Heat is rejected to cooling water that enters the condenser at 16 °C at a rate of 0.45 kg/s and leaves at 26 °C. The refrigerant enters the condenser at 1.2 MPa and 50 °C and leaves at the same pressure subcooled by 5 °C. If the compressor consumes 7.4 kW of power, determine (a) the mass flow rate of the refrigerant, (b) the refrigeration load, (c) the COP, and (d) the minimum power input to the compressor for the same refrigeration load. Take specific heat of water to be 4.18 kJ/kg·°C.

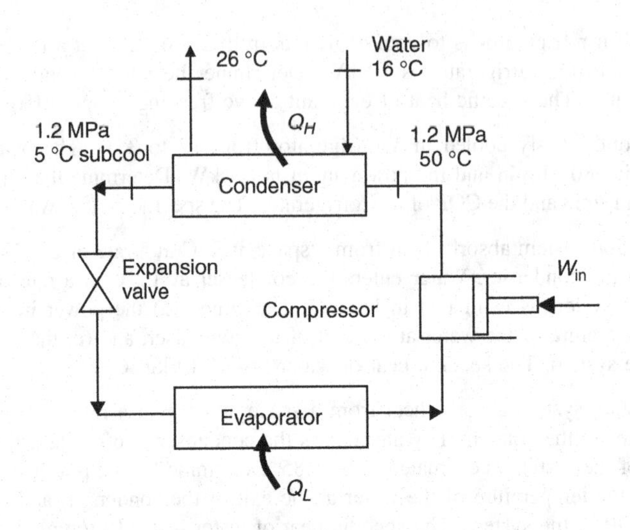

4.13 An air conditioner with refrigerant-134a as the refrigerant is used to keep a room at 24 °C by rejecting the waste heat to the outside air at 36 °C. The room is gaining heat through the walls and the windows at a rate of 125 kJ/min, while the heat generated by the computer, TV, and lights amounts to 800 W. The refrigerant enters the compressor at 500 kPa as a saturated vapor at a rate of 100 L/min and leaves at 1200 kPa and 50 °C. Determine (a) the actual COP, (b) the maximum COP, and (c) the minimum volume flow rate of the refrigerant at the compressor inlet for the same compressor inlet and exit conditions.

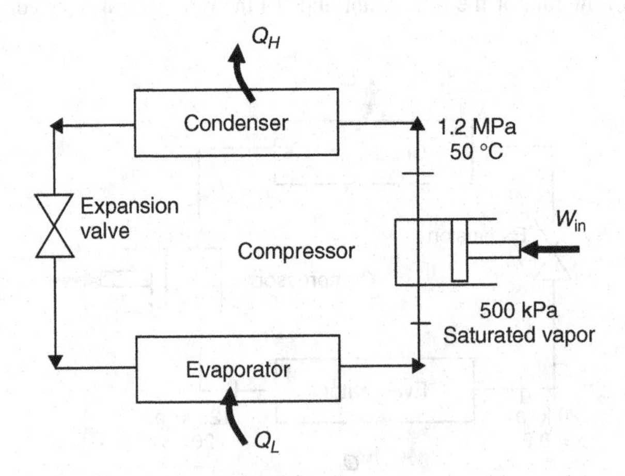

4.14 An ideal vapor-compression refrigeration cycle with refrigerant-134a as the working fluid operates between pressure limits of 240 kPa and 1600 kPa. Determine (a) the heat absorption in the evaporator, (b) the heat rejection in the condenser, (c) the work input, and (d) the COP.

4.15 An ideal vapor-compression refrigeration cycle with refrigerant-134a as the working fluid operates between pressure limits of 40 and 300 psia. Determine (a) the heat absorption in the evaporator, (b) the heat rejection in the condenser, (c) the work input, and (d) the COP.

4.16 A refrigerator operates on the ideal vapor-compression refrigeration cycle with refrigerant-134a as the working fluid. The evaporator pressure is 160 kPa and the temperature at the exit of the condenser is 35.5 °C. The flow rate at the compressor inlet is 0.85 m^3/min. Determine (a) the rate of heat absorption in the evaporator, (b) the rate of heat rejection in the condenser, (c) the power input, and (d) the COP.

4.17 An 18,000 Btu/h air conditioner operates on the ideal vapor-compression refrigeration cycle with R-22 as the refrigerant. The evaporator pressure is 125 kPa and the condenser pressure is 1750 kPa. If the air conditioner operates at full cooling load, determine (a) the mass flow rate of R-22, (b) the power input, and (c) the COP. The enthalpies of R-22 at various states are given as (R-22 tables are not available in the text): $h_1 = 389.67$ kJ/kg, $h_2 = 458.68$ kJ/kg, $h_3 = 257.20$ kJ/kg, $h_4 = 257.20$ kJ/kg.

4.18 An ideal vapor-compression refrigeration cycle uses R-134a as the refrigerant. The refrigerant enters the evaporator at 160 kPa with a quality of 25% and leaves the compressor at 65 °C. If the compressor consumes 800 W of power, determine (a) the mass flow rate of the refrigerant, (b) the condenser pressure, and (c) the COP of the refrigerator.

4.19 A vapor-compression refrigeration cycle with refrigerant-134a as the working fluid operates between pressure limits of 240 and 1600 kPa. The isentropic efficiency of the compressor is 78%. Determine (a) the heat absorption in the evaporator, (b) the heat rejection in the condenser, (c) the work input, and (d) the COP.

4.20 A refrigerator operates on the vapor-compression refrigeration cycle with refrigerant-134a as the working fluid. The evaporator pressure is 100 kPa and the condenser pressure is 1400 kPa. The flow rate at the compressor inlet is 0.22 m^3/min. The isentropic efficiency of the compressor is 84%. Determine (a) the rate of heat absorption in the evaporator, (b) the rate of heat rejection in the condenser, (c) the power input, and (d) the COP.

4.21 A refrigerator operates on the vapor-compression refrigeration cycle with refrigerant-134a as the working fluid. The evaporator pressure is 20 psia and the condenser pressure is 180 psia. The flow rate at the compressor inlet is 8.4 ft^3/min. The isentropic efficiency of the compressor is 86%. Determine (a) the rate of heat absorption in the evaporator, (b) the rate of heat rejection in the condenser, (c) the power input, and (d) the COP.

4.22 An automotive air conditioner operates on the vapor-compression refrigeration cycle with refrigerant-134a as the working fluid. The refrigerant enters the compressor at 180 kPa superheated by 2.7 °C at a rate of 0.007 kg/s and leaves the compressor at 1200 kPa and 60 °C. R-134a is subcooled by 6.3 °C at the exit of the condenser. Determine (a) the isentropic efficiency of the compressor, (b) the rate of cooling, and (c) the COP.

4.23 A vapor-compression refrigeration cycle with refrigerant-134a as the working fluid operates between pressure limits of 240 and 1600 kPa. The isentropic efficiency of the compressor is 78%. The refrigerant is superheated by 5.4 °C at the compressor inlet and subcooled by 5.9 °C at the exit of the condenser. Determine (a) the heat absorption in the evaporator, (b) the heat rejection in the condenser, (c) the work input, and (d) the COP. (e) Also determine all parameters if the cycle operated on the ideal vapor-compression refrigeration cycle between the same pressure limits.

4.24 A practical refrigerator operates on the vapor-compression refrigeration cycle with refrigerant-22 as the working fluid. The pressure of R-22 at the compressor exit is 2000 and 300 kPa at the inlet of the evaporator. The isentropic efficiency of the compressor is 75%. The refrigerant is superheated by 5 °C at the compressor inlet and subcooled by 5 °C at the exit of the condenser. There is a pressure drop of 50 kPa in the condenser and

25 kPa in the evaporator. Determine (a) the heat absorption in the evaporator per unit mass of R-22, (b) the work input, and the COP. (c) Determine the refrigeration load, the work input, and the COP if the cycle operated on the ideal vapor-compression refrigeration cycle between the pressure limits of 2000 and 300 kPa.

The properties of R-22 in the case of actual operation are obtained from R-22 tables to be

$$h_1 = 401.62 \text{ kJ/kg}, h_2 = 487.29 \text{ kJ/kg}, h_3 = 283.76 \text{ kJ/kg}, h_4 = 283.76 \text{ kJ/kg}$$

The properties of R-22 in the case of ideal operation are obtained from R-22 tables to be

$$h_1 = 399.18 \text{ kJ/kg}, h_2 = 459.30 \text{ kJ/kg}, h_3 = 293.30 \text{ kJ/kg}, h_4 = 293.30 \text{ kJ/kg}$$

4.25 An air conditioner with refrigerant-134a as the refrigerant is used to keep a large space at 20 °C by rejecting the waste heat to the outside air at 37 °C. The room is gaining heat through the walls and the windows at a rate of 125 kJ/min while the heat generated by the computer, TV, and lights amounts to 0.7 kW. Unknown amount of heat is also generated by the people in the room. The condenser and evaporator pressures are 1200 and 500 kPa, respectively. The refrigerant is saturated liquid at the condenser exit and saturated vapor at the compressor inlet. If the refrigerant enters the compressor at a rate of 65 L/min and the isentropic efficiency of the compressor is 70%, determine (a) the temperature of the refrigerant at the compressor exit, (b) the rate of heat generated by the people in the room, (c) the COP of the air conditioner, and (d) the minimum volume flow rate of the refrigerant at the compressor inlet for the same compressor inlet and exit conditions.

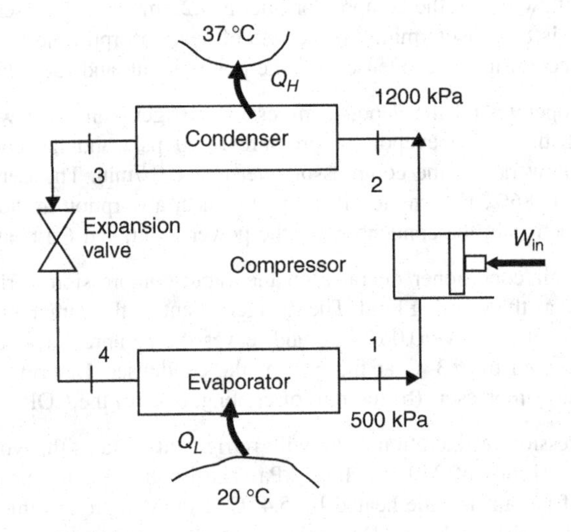

4.26 A refrigerated room is kept at −27 °C by a vapor-compression cycle with R-134a as the refrigerant. Heat is rejected to cooling water that enters the condenser at 16 °C at a rate of 0.22 kg/s and leaves at 23 °C. The refrigerant enters the condenser at 1.2 MPa and 65 °C and leaves at 42 °C. The inlet state of the compressor is 60 kPa and −34 °C and the compressor is estimated to gain a net heat of 150 W from the surroundings. Determine (a) the quality of the refrigerant at the evaporator inlet, (b) the refrigeration load, (c) the COP of the refrigerator, and (d) the theoretical maximum refrigeration load for the same power input to the compressor.

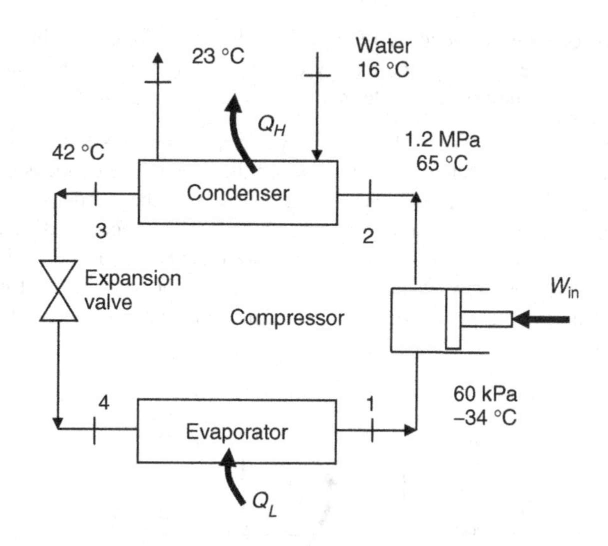

Exergy Analysis of Refrigeration Cycles

4.27 A refrigeration cycle is used to keep a food department at −15 °C in an environment at 22 °C. The total heat gain to the food department is estimated to be 1750 kJ/h and the heat rejection in the condenser is 3250 kJ/h. Determine (a) the power input to the compressor, (b) the COP of the refrigerator, and (c) the second-law efficiency of the cycle.

4.28 A refrigeration cycle is used to keep a food department at −22 °F in an environment at 73 °F. The total heat gain by the food department is estimated to be 7900 Btu/h and the heat rejection in the condenser is 4150 Btu/h. Determine (a) the power input to the compressor, (b) the COP of the refrigerator, and (c) the second-law efficiency of the cycle.

4.29 A commercial refrigerator is to cool eggplants from 26 to 5 °C at a rate of 380 kg/h. The power input to the refrigerator is 4.5 kW. Determine (a) the rate of cooling, (b) the COP, (c) the exergy of the heat transferred from the low-temperature medium, and (d) the second-law efficiency and the exergy destruction for the cycle. The specific heat of eggplant above freezing is 3.92 kJ/kg·°C.

4.30 A refrigeration system absorbs heat from a space at 2 °C at a rate of 6.9 kW and rejects heat to water in the condenser. Water enters the condenser at 16 °C at a rate of 0.27 kg/s. The COP of the system is estimated to be 1.85. Determine (a) the power input to the system, (b) the temperature of the water at the exit of the condenser, and (c) the second-law efficiency and the exergy destruction for the refrigerator. Take the dead-state temperature to be the inlet temperature of water in the condenser. The specific heat of water is 4.18 kJ/kg·°C.

4.31 A refrigerator using R-134a as the refrigerant is used to keep a space at −10 °C by rejecting heat to ambient air at 22 °C. R-134a enters the compressor at 140 kPa as a saturated vapor and leaves at 800 kPa and 70 °C. The refrigerant leaves the condenser as a saturated liquid. The rate of cooling provided by the system is 2600 W. Determine (a) the mass flow rate of R-134a, (b) the COP, (c) the exergy destruction in each component of the cycle, (d) the second-law efficiency of the cycle, and (e) the total exergy destruction in the cycle.

4.32 A refrigerator using R-134a as the refrigerant is used to keep a space at 15 °F by rejecting heat to ambient air at 75 °F. R-134a enters the compressor at 25 psia as a saturated vapor and leaves at 140 psia and 160 °F. The refrigerant leaves the condenser as a saturated liquid.

The rate of cooling provided by the system is 9000 Btu/h. Determine (a) the mass flow rate of R-134a, (b) the COP, (c) the exergy destruction in each component of the cycle, (d) the second-law efficiency of the cycle, and (e) the total exergy destruction in the cycle.

4.33 A refrigerated room is kept at −18 °C by a vapor-compression cycle with R-134a as the refrigerant. Heat is rejected to cooling water that enters the condenser at 14 °C at a rate of 0.35 kg/s and leaves at 22 °C. The refrigerant enters the condenser at 1.2 MPa and 50 °C and leaves at the same pressure subcooled by 5 °C. If the compressor consumes 5.5 kW of power, determine (a) the mass flow rate of the refrigerant, (b) the refrigeration load and the COP, (c) the second-law efficiency of the refrigerator and the total exergy destruction in the cycle, and (d) the exergy destruction in the condenser. Take specific heat of water to be 4.18 kJ/kg·°C.

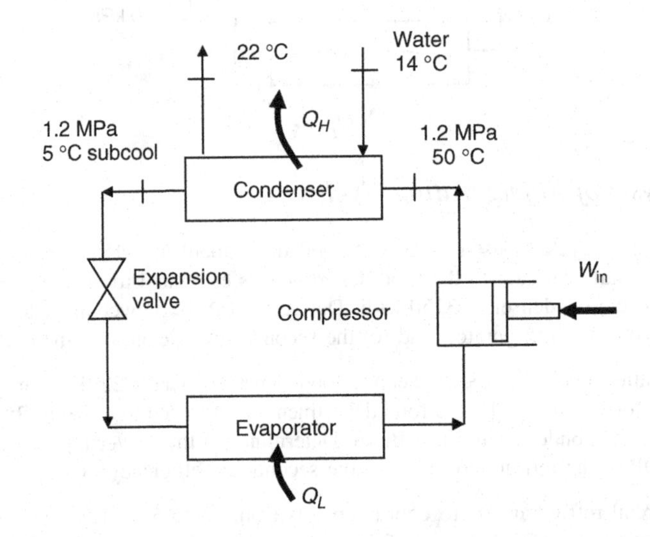

4.34 An ideal vapor-compression refrigeration cycle with refrigerant-134a as the working fluid operates between pressure limits of 200 and 1600 kPa. The refrigerant absorbs heat from a space at 3 °C and rejects heat to ambient air at 27 °C. Determine (a) the heat absorbed in the evaporator and the work input, (b) the COP, (c) the exergy destruction in each component of the cycle and the total exergy destruction in the cycle, (d) the second-law efficiency of the cycle.

4.35 A 45,000 Btu/h refrigeration system operates on the ideal vapor-compression refrigeration cycle with R-22 as the refrigerant. The evaporator pressure is 150 kPa and the condenser pressure is 1500 kPa. The refrigerant exchanges heat with air at −5 °C in the evaporator and with air at 19 °C in the condenser. If the air conditioner operates at full cooling load, determine (a) the mass flow rate of R-22, (b) the power input and the COP, and (c) the second-law efficiency of the cycle and the total exergy destruction in the cycle. The enthalpies of R-22 at various states are given as (R-22 tables are not available in the text): $h_1 = 391.58$ kJ/kg, $s_1 = 1.8051$ kJ/kg·K, $h_2 = 450.98$ kJ/kg, $h_3 = 248.58$ kJ/kg, $s_3 = 1.1632$ kJ/kg·K, $h_4 = 248.58$ kJ/kg, $s_4 = 1.2119$ kJ/kg·K. Take $T_0 = 19$ °C.

4.36 An automotive air conditioner operates on the vapor-compression refrigeration cycle with refrigerant-134a as the working fluid. The refrigerant absorbs heat from the air inside the car at 23 °C and rejects heat to ambient air at 36 °C. The refrigerant enters the compressor at

180 kPa superheated by 2.7 °C at a rate of 0.0095 kg/s and leaves the compressor at 1200 kPa and 60 °C. R-134a is subcooled by 6.3 °C at the exit of the condenser. Determine (a) the rate of cooling and the COP, (b) the isentropic efficiency and the exergetic efficiency of the compressor, (c) the exergy destruction in each component of the cycle and the total exergy destruction in the cycle, (d) the minimum power input and the second-law efficiency of the cycle. Take $T_0 = 36\,°C$.

Air-Standard Refrigeration Systems

4.37 An ideal gas refrigeration cycle with a pressure ratio of four uses air as the working fluid. Air enters the compressor at 100 kPa and 0 °C and the turbine at 50 °C. Determine (a) the temperature at the turbine exit, (b) the heat removed per unit mass of the air, and (c) the COP of the cycle. Use constant specific heat for air at room temperature with $c_p = 1.005$ kJ/kg·K and $k = 1.4$.

4.38 A gas refrigeration cycle with a pressure ratio of four uses air as the working fluid. Air enters the compressor at 100 kPa and 0 °C and the turbine at 50 °C. The isentropic efficiencies of the compressor and turbine are 84%. Determine (a) the temperature at the turbine exit, (b) the heat removed per unit mass of the air, and (c) the COP of the cycle. Use constant specific heat for air at room temperature with $c_p = 1.005$ kJ/kg·K and $k = 1.4$.

4.39 Argon gas enters the compressor of a gas refrigeration cycle at 40 kPa and −35 °C at a flow rate of 7500 L/min and leaves at 130 kPa and 125 °C. The argon enters the turbine at 40 °C. The isentropic efficiency of the turbine is 88%. Determine (a) the minimum temperature in the cycle, (b) the isentropic efficiency of the compressor, (c) the net power input to the cycle, (d) the rate of refrigeration, and (e) the COP of the cycle. Use constant specific heat for argon with $c_p = 0.5203$ kJ/kg·K, $R = 0.2081$ kJ/kg·K, and $k = 1.667$.

4.40 Argon gas enters the compressor of a gas refrigeration cycle at 6 psia and −30 °F at a flow rate of 265 ft^3/min and leaves at 19 psia and 255 °F. The argon enters the turbine at 105 °F. The isentropic efficiency of the turbine is 83%. Determine (a) the minimum temperature in the cycle, (b) the isentropic efficiency of the compressor, (c) the net power input to the cycle in kW, (d) the rate of refrigeration in Btu/h, and (e) the COP of the cycle. Use constant specific heat for argon with $c_p = 0.1253$ Btu/lbm·R, $R = 0.04971$ Btu/lbm·R, and $k = 1.667$.

4.41 Air enters the compressor of an ideal gas refrigeration system with a regenerator at −20 °C at a flow rate of 0.12 kg/s. The cycle has a pressure ratio of 4.5. The temperature of the air decreases from 15 to −28 °C in the regenerator. Both the turbine and compressor are assumed to be isentropic. Determine (a) the rate of refrigeration, (b) the power input, and (c) the COP of the cycle. Use constant specific heat for air at room temperature with $c_p = 1.005$ kJ/kg·K and $k = 1.4$.

4.42 Air enters the compressor of a gas refrigeration system with a regenerator at −20 °C at a flow rate of 0.12 kg/s. The cycle has a pressure ratio of 4.5. The temperature of the air decreases from 15 to −28 °C in the regenerator. The isentropic efficiency of the compressor is 85% and that of the turbine is 80%. Determine (a) the rate of refrigeration, (b) the power input, and (c) the COP of the cycle. Use constant specific heat for air at room temperature with $c_p = 1.005$ kJ/kg·K and $k = 1.4$.

4.43 Consider a gas refrigeration system with air as the working fluid. The pressure ratio is 5.5. Air enters the compressor at 0 °C. The high-pressure air is cooled to 35 °C by rejecting heat to the surroundings. The refrigerant leaves the turbine at −95 °C and then it absorbs heat from the refrigerated space before entering the regenerator. The mass flow rate of air is

0.55 kg/s. Assuming isentropic efficiencies of 90% for both the compressor and the turbine, determine (a) the effectiveness of the regenerator, (b) the rate of heat removal from the refrigerated space, and (c) the COP of the cycle. Also, determine (d) the refrigeration load and the COP if this system operated on the simple gas refrigeration cycle. In this cycle, take the compressor and turbine inlet temperatures to be 0 and 35 °C, respectively, and use the same compressor and turbine efficiencies. Use constant specific heat for air at room temperature with $c_p = 1.005$ kJ/kg·K and $k = 1.4$.

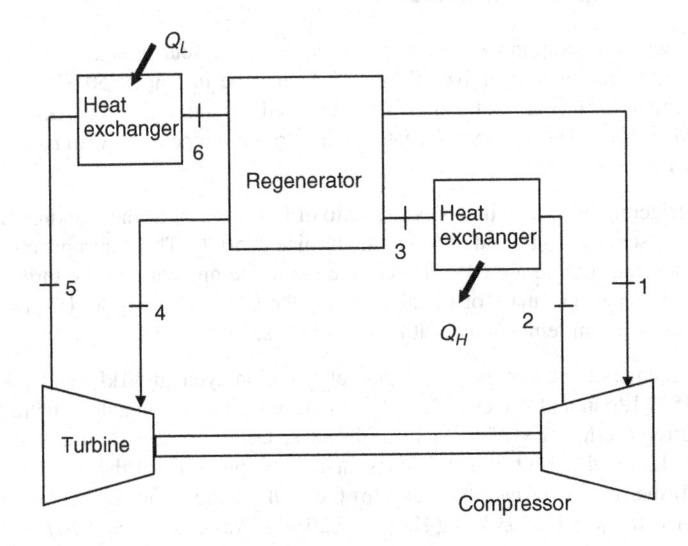

4.44 A gas refrigeration cycle with a pressure ratio of six operates between a cooled space temperature of 3 °C and an ambient temperature of 22 °C. Air enters the compressor at 90 kPa and 10 °C and the turbine at 65 °C. Both the compressor and turbine are isentropic. Determine (a) the heat removed per unit mass of the air, (b) the COP of the cycle, (c) the exergy destruction in each component of the cycle and the total exergy destruction in the cycle, and (d) the second-law efficiency of the cycle. Use constant specific heat for air at room temperature with $c_p = 1.005$ kJ/kg·K and $k = 1.4$.

4.45 Helium gas enters the compressor of a gas refrigeration cycle at 65 kPa and −25 °C at a flow rate of 9000 L/min and leaves at 240 kPa and 160 °C. The helium enters the turbine at 60 °C. The isentropic efficiency of the turbine is 85%. Determine (a) the isentropic efficiency and the exergetic efficiency for the compressor, (b) the rate of refrigeration and the COP of the cycle, (c) the exergy destruction in each component of the cycle and the total exergy destruction in the cycle, and (d) the exergy efficiency of the compressor, the minimum power input, and the second-law efficiency of the cycle. The temperature of the cooled space is −15 °C and heat is rejected to the ambient at −5 °C. Use constant specific heat for argon with $c_p = 5.1926$ kJ/kg·K, $R = 2.0769$ kJ/kg·K, and $k = 1.667$. Take $T_0 = -5$ °C.

4.46 Air enters the compressor of a gas refrigeration system with a regenerator at −28 °C at a flow rate of 0.75 kg/s. The cycle has a pressure ratio of four. The temperature of the air decreases from 12 to −35 °C in the regenerator. The isentropic efficiency of the compressor is 85% and that of the turbine is 80%. Determine (a) the rate of refrigeration and the COP of the cycle and (b) the minimum power input, the second-law efficiency of the cycle, and the total exergy destruction in the cycle. The temperature of the cooled space is −40 °C and heat is rejected to the ambient at 7 °C. Use constant specific heat for air at room temperature with $c_p = 1.005$ kJ/kg·K and $k = 1.4$. (c) Determine the minimum power input, the second-law efficiency of the cycle, and the total exergy destruction in the cycle if the temperature of the cooled space is −20 °C.

Absorption-Refrigeration Systems (ARSs)

4.47 An ARS removes heat from a cooled space at 2 °C at a rate of 66 kW. The system operates in an environment at 20 °C. If the heat is supplied to the cycle by condensing saturated steam at 200 °C at a rate of 0.04 kg/s. Determine (a) the COP of the system and (b) the COP of a reversible ARS operating between the same temperatures. The enthalpy of vaporization of water at 200 °C is $h_{fg} = 1940.34$ kJ/kg.

4.48 Consider a reversible ARS that can be modeled to consist of a reversible heat engine and a reversible refrigerator. The system removes heat from a cooled space at −7 °C at a rate of 22 kW. The refrigerator operates in an environment at 25 °C. If the heat is supplied to the cycle by condensing saturated steam at 175 °C, determine (a) the rate at which the steam condenses and (b) the power input to the reversible refrigerator. The enthalpy of vaporization of water at 175 °C is $h_{fg} = 2031.7$ kJ/kg.

4.49 Consider a basic ARS using ammonia–water solution as shown in the figure. Pure ammonia enters the condenser at 400 psia and 150 °F at a rate of 0.013 lbm/s. Ammonia leaves the condenser as a saturated liquid and is throttled to a pressure of 25 psia. Ammonia leaves the evaporator as a saturated vapor. Heat is supplied to the generator by geothermal liquid water that enters at 260 °F at a rate of 0.18 lbm/s and leaves at 220 °F. Determine (a) the rate of cooling provided by the system in Btu/h and tons of refrigeration and (b) the COP of the system. (c) Also, determine the second-law efficiency of the system if the ambient temperature is 77 °F and the temperature of the refrigerated space is 32 °F. The enthalpies of ammonia at various states of the system are given as follows: $h_3 = 646.0$ Btu/lbm, $h_4 = 216.5$ Btu/lbm, $h_6 = 616.8$ Btu/lbm. Also, take the specific heat of water to be 1.0 Btu/lbm·°F.

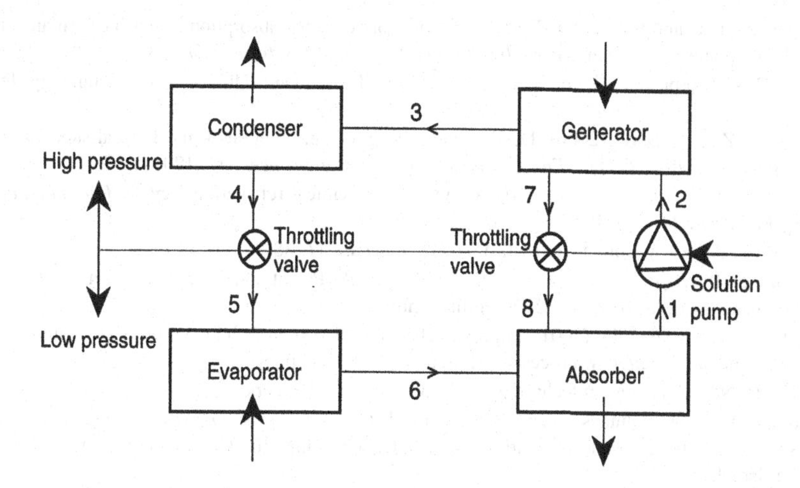

4.50 Consider a basic ARS using ammonia–water solution as shown in the figure of the previous problem. Pure ammonia enters the condenser at 3200 kPa and 70 °C. Ammonia leaves the condenser as a saturated liquid and is throttled to a pressure of 220 kPa. Ammonia leaves the evaporator as a saturated vapor. Heat is supplied to the solution in the generator by solar energy, which is incident on the collector at a rate of 550 W/m^2. The total surface area of the collectors is 31.5 m^2 and the efficiency of the collectors is 75% (i.e., 75% of the solar energy input is transferred to the solution). If the COP of the system is estimated to be 0.8, determine the mass flow rate of ammonia through the evaporator. The enthalpies of ammonia at various states of the system are given as $h_3 = 1491.9$ kJ/kg, $h_4 = 537.0$ kJ/kg, and $h_6 = 1442.0$ kJ/kg.

References

ASHRAE (1997) *Handbook of Fundamentals*, American Society of Heating, Refrigerating and AirConditioning Engineers, Atlanta, GA.

Ataer, O.E. and Gogus, Y. (1991) Comparative study of irreversibilities in an aqua-ammonia absorption refrigeration system. *International Journal of Refrigeration*, 14, 86–92.

Dincer, I. (2003) *Refrigeration Systems and Applications*, 1st edn, John Wiley & Sons, Ltd., New York.

Dincer, I. and Dost, S. (1996) A simple model for heat and mass transfer in absorption cooling systems (ACSs). *International Journal of Energy Research*, 20, 237–243.

Dincer, I., Edin, M. and Ture, I.E. (1996) Investigation of thermal performance of a solar powered absorption refrigeration system. *Energy Conversion and Management*, 37, 51–58.

Dincer, I. and Ture, I.E. (1993) *Design and Construction of a Solar Powered Absorption Cooling System*, Proceedings of the International Symposium on Energy Saving and Energy Efficiency, 16–18 November, Ankara, pp. 198–203.

Eames, I.W. and Wu, S. (2000) A theoretical study of an innovative ejector powered absorption-recompression cycle refrigerator. *International Journal of Refrigeration*, 23, 475–484.

Gosney, W.B. (1982) *Principles of Refrigeration*, Cambridge University Press, Cambridge, UK.

Jelinek, M., Yaron, I. and Borde, I. (1980) *Measurement of Vapour–Liquid Equilibria and Determination of Enthalpy-Concentration Diagrams of Refrigerant-Absorbent Combinations*, Proceedings of IIR, Commissions B1, B2, E1, E2, Mons (Belgium), pp. 57–65.

Kaita, Y. (2001) Simulation results of triple-effect absorption cycles. *International Journal of Refrigeration*, 25, 999–1007.

Kang, Y.T., Kunugi, Y. and Kashiwagi, T. (2000) Review of advanced absorption cycles: performance improvement and temperature lift enhancement. *International Journal of Refrigeration*, 23, 388–401.

Keizer, C. (1982) Absorption refrigeration machines, Ph.D. Thesis, Delft University of Technology, Delft, The Netherlands.

Khan, J.R. and Zubair, S.M. (2000) Design and rating of an integrated mechanical-subcooling vapor-compression refrigeration system. *Energy Conversion and Management*, 41, 1201–1222.

Newell, T.A. (2000) Thermodynamic analysis of an electrochemical refrigeration cycle. *International Journal of Energy Research*, 24, 443–453.

Norton, E. (2000) A look at hot gas defrost. *ASHRAE Journal*, 42, 88.

Patent Storm (2010) *Triple Effect Refrigeration System*, US Patent 5727397 issued on 17 March 1998, http://www.patentstorm.us/patents/5727397/fulltext.html.

Rockwell, T.C. and Quake, T.D. (2001) Improve refrigeration system efficiency, process cooling equipment (online magazine at: http://www.process-cooling.com) October, p. 5.

Turpin, J. (2000) New refrigeration technology poised to heat up the market. *Engineered Systems*, **October**, 4.

Van Den Bulck, E., Trommelmans, J. and Berghmans, J. (1982) *Solar Absorption Cooling Installation*, Proceedings of Solar Energy for Refrigeration and Air Conditioning, IIR Commission E1-E2, March 14–15, Jeruzalem, Israel, pp. 83–87.

5

Advanced Refrigeration Cycles and Systems

5.1 Introduction

Refrigeration cycles covered in Chapter 4 are simple and extensively used in most of the refrigeration needs encountered in practice. Household refrigerators, small coolers, and air-conditioning systems are some examples. For other refrigeration applications, the simple vapor-compression cycle may not be suitable and more advanced and innovative refrigeration systems may have to be used. Other motivations include the search for improved performance and efficiency as well as requirements to achieve very low temperatures. In this chapter, we cover some advanced refrigeration systems as well as special systems used in certain applications.

5.2 Multistage Refrigeration Cycles

Multistage refrigeration systems are widely used where ultralow temperatures are required, but cannot be obtained economically through the use of a single-stage system. This is due to the fact that the compression ratios are too large to attain the temperatures required to evaporate and condense the vapor. There are two general types of such systems: cascade and multistage. The multistage system uses two or more compressors connected in series in the same refrigeration system. The refrigerant becomes a denser vapor while it passes through each compressor. Note that a two-stage system (Figure 5.1) can attain a temperature of approximately −65 °C and a three-stage one about −100 °C.

Single-stage vapor-compression refrigerators are used by cold storage facilities with a range of +10 to −30 °C. In this system, the evaporator installed within the refrigeration system and the ice-making unit, as the source of low temperature, absorbs heat. Heat is released by the condenser at the high-pressure side.

In cases where large temperature and pressure differences exist between the evaporator and the condenser, multistage vapor-compression systems are employed accordingly. For example, if the desired temperature of a refrigerator (i.e., freezer) is below −30 °C, a several-stage compression system is required in order to prevent the occurrence of high compression ratios. The following are some of the disadvantages of a high compression ratio:

- decrease in the compression efficiency,
- increase in the temperature of the refrigerant vapor from the compressor, and
- increase in energy consumption per unit of refrigeration production.

Refrigeration Systems and Applications İbrahim Dinçer and Mehmet Kanoğlu

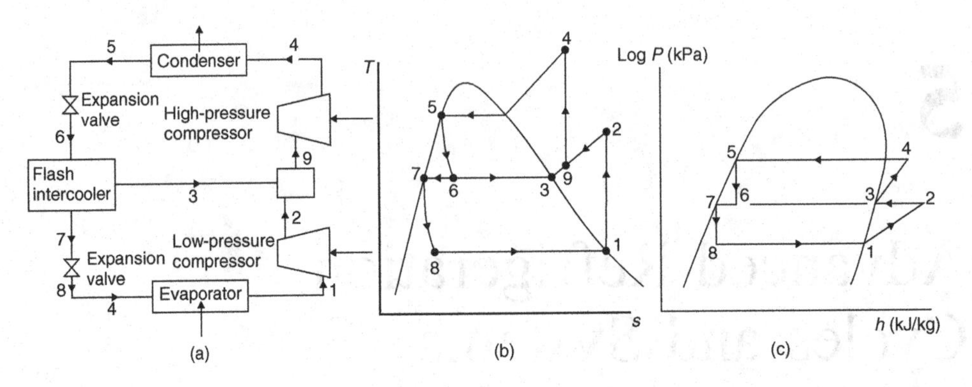

Figure 5.1 (a) A two-stage vapor-compression refrigeration system, (b) its $T-s$ diagram, and (c) its log $P-h$ diagram.

Figure 5.1 shows a schematic diagram of a two-stage vapor-compression refrigeration unit that can provide temperatures below $-30\,°C$ (approximately to $-50\,°C$), and its $T-s$ diagram. This system also uses an intercooler with air.

As an example, three-stage refrigeration systems can provide an evaporator temperature of $-100\,°C$. In the two-stage unit shown, the refrigerant is compressed in the first stage and, after being de-superheated by an intercooler, is further compressed in the second stage. An intercooler is used in between the two compression stages for reducing the compression work. In other words, a booster (first-stage) compressor and a gas–liquid intercooler are attached to the single-stage cycle. The intercooler subcools the refrigerant liquid supplied to the evaporator by vaporizing a portion of the refrigerant after the first throttling stage. The flash gas returns at an intermediate point in the compression process in order to improve the compression efficiency by cooling the superheated gas. Not only a single compressor but a set of compressors is also required to be used in each stage, depending on the capacity and temperature. In large systems with a number of evaporators and large compression (temperature) ratios, the number of intercoolers and compression stages yields increased system efficiency and hence increased coefficient of performance (COP).

5.3 Cascade Refrigeration Systems

For some industrial applications that require moderately low temperatures (with a considerably large temperature and pressure difference), single vapor-compression refrigeration cycles become impractical. One of the solutions for such cases is to perform the refrigeration in two or more stages (i.e., two or more cycles) which operate in series. These refrigeration cycles are called *cascade refrigeration cycles*. Therefore, cascade systems are employed to obtain high-temperature differentials between the heat source and heat sink and are applied for temperatures ranging from -70 to $100\,°C$. Application of a three-stage compression system for evaporating temperatures below $-70\,°C$ is limited, because of difficulties with refrigerants reaching their freezing temperatures. Impropriety of multi-stage vapor-compression systems can be avoided by applying a cascade vapor-compression refrigeration system.

Cascade refrigeration systems are commonly used in the liquefaction of natural gas and some other gases. A large-capacity industrial cascade refrigeration system is shown in Figure 5.2.

The most important advantage of these cascade systems is that refrigerants can be chosen with the appropriate properties, avoiding large dimensions for the system components. In these systems multiple evaporators can be utilized in any one stage of compression. Refrigerants used in each stage may be different and are selected for optimum performance at the given evaporator and condenser temperatures.

Figure 5.2 A cascade refrigeration system utilizing CO_2 as low-pressure stage refrigerant and ammonia as the high-pressure stage refrigerant (operating at about −50 °C), skid for a 100 ton/day CO_2 liquefaction plant (*Courtesy of Salof Refrigeration Co., Inc.*).

Conventional single compressor, mechanical refrigeration system condensing units are capable of achieving temperatures of about −40 °C. If lower temperatures are required then cascade refrigeration systems must be used. A two-stage cascade system uses two refrigeration systems connected in series to achieve temperatures of around −85 °C. There are single compressor systems that can achieve temperatures colder than −100 °C but they are not widely used. These systems are sometimes referred to as *autocascading* systems. The main disadvantage of such systems is that it requires the use of a proprietary blend of refrigerant. This characteristic results in three service-related problems:

- A leak in the system can easily cause the loss of only some of the refrigerant making up the blend (since the refrigerant blend is made up of different types of refrigerant with different boiling points), resulting in an imbalance in the ratio of the remaining refrigerants. To return the system to proper operation, all of the remaining refrigerant must be replaced with a new and potentially costly charge to ensure a proper blend ratio.
- The blend is proprietary and may not be readily available from the traditional refrigerant supply sources and therefore may be hard to obtain and costly.
- These types of cascade systems are not widely used; it is hard to find well-qualified field service staff who are familiar with repair and maintenance procedures.

Of course, these and other issues can cause undesirable expense and downtime.

5.3.1 Two-Stage Cascade Systems

A two-stage cascade system employs two vapor-compression units working separately with different refrigerants and interconnected in such a way that the evaporator of one system is used to serve as condenser to a lower temperature system (i.e., the evaporator from the first unit cools the condenser of the second unit). In practice, an alternative arrangement utilizes a common condenser with a booster circuit to provide two separate evaporator temperatures.

In fact, the cascade arrangement allows one of the units to be operated at a lower temperature and pressure than would otherwise be possible with the same type and size of single-stage system. It also allows two different refrigerants to be used, and it can produce temperatures below −150 °C. Figure 5.3 shows a two-stage cascade refrigeration system, where condenser B of system 1 is being cooled by evaporator C of system 2. This arrangement enables reaching ultralow temperatures in evaporator A of the system.

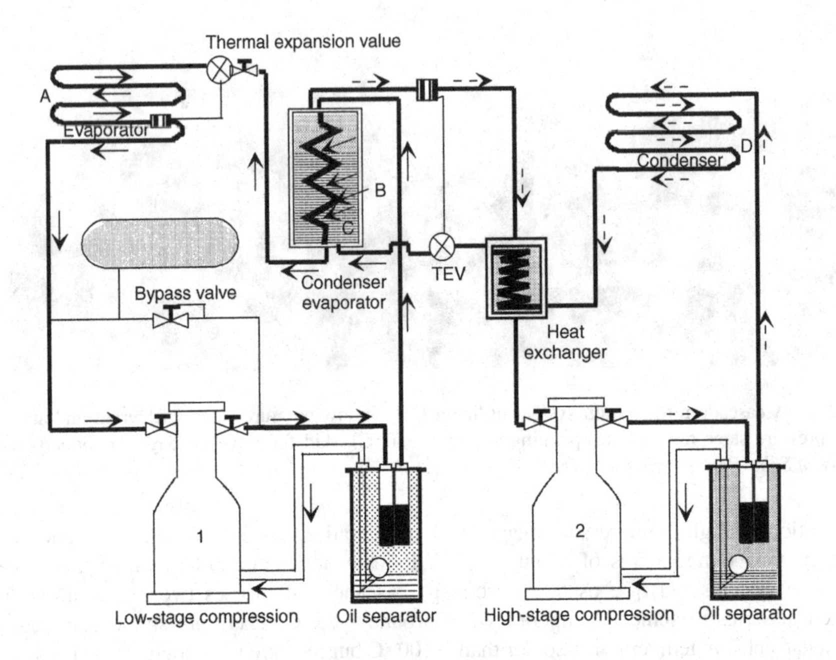

Figure 5.3 A practical two-stage cascade refrigeration system.

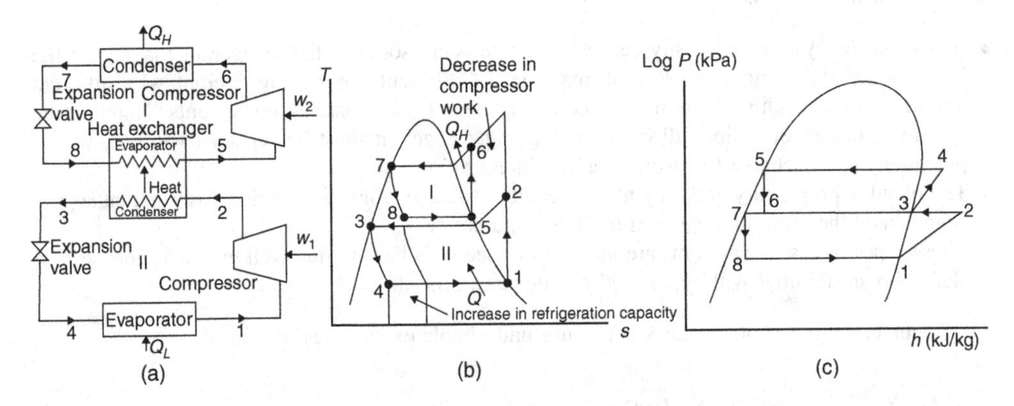

Figure 5.4 (a) Schematic of a two-stage (binary) cascade refrigeration system, (b) its T–s diagram, and (c) its log P–h diagram. [Adapted from Cengel and Boles (2008).]

For a schematic system shown in Figure 5.4, the condenser of system I, called the *first or high-pressure stage*, is usually fan cooled by the ambient air. In some cases a water supply may be used, but air cooling is much more common. The evaporator of system I is used to cool the condenser of system II called the *second or low-pressure stage*. The unit that makes up the evaporator of system I and the condenser of system II is often referred to as the *inter-stage or cascade condenser*. As stated earlier, cascade systems generally use two different refrigerants (i.e., one in each stage). One type is used for the low stage and a different one for the high stage. The reason why two refrigeration systems are used is that a single system cannot economically achieve the high compression ratios necessary to obtain the proper evaporating and condensing temperatures. It is clear from the $T-s$ diagram of the two-stage cascade refrigeration system, as shown in Figure 5.4, that the compressor work decreases and the amount of refrigeration load (capacity) in the evaporator increases as a result of cascading (Cengel and Boles, 2008). Therefore, cascading improves the COP.

Example 5.1

Consider a two-stage cascade refrigeration system operating between the pressure limits of 1.6 MPa and 180 kPa with refrigerant-134a as the working fluid (Figure 5.5). Heat rejection from the lower cycle to the upper cycle takes place in an adiabatic counter-flow heat exchanger where the pressure in the upper and lower cycles are 0.4 and 0.5 MPa, respectively. In both cycles, the refrigerant is a saturated liquid at the condenser exit and a saturated vapor at the compressor inlet, and the isentropic efficiency of the compressor is 85%. If the mass flow rate of the refrigerant through the lower cycle is 0.07 kg/s, (a) draw the temperature–entropy diagram of the cycle indicating pressures; determine (b) the mass flow rate of the refrigerant through the upper cycle, (c) the rate of heat removal from the refrigerated space, and (d) the COP of this refrigerator; and (e) determine the rate of heat removal and the COP if this refrigerator operated on a single-stage cycle between the same pressure limits with the same compressor efficiency. Also, take the mass flow rate of R-134a through the cycle to be 0.07 kg/s.

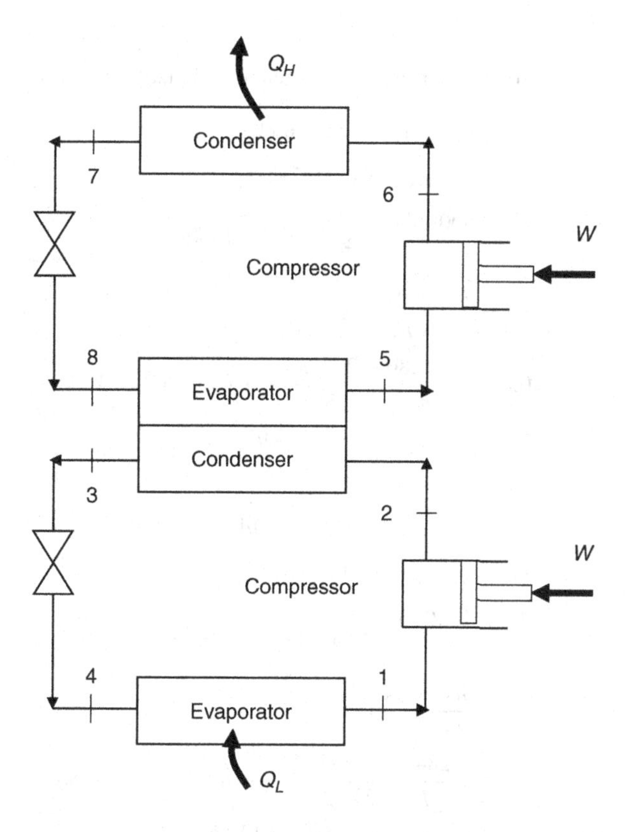

Figure 5.5 Schematic of two-stage cascade refrigeration system considered in Example 5.1.

Solution

(a) Noting that compression processes are not isentropic, the temperature–entropy diagram of the cycle can be drawn as shown in Figure 5.6.

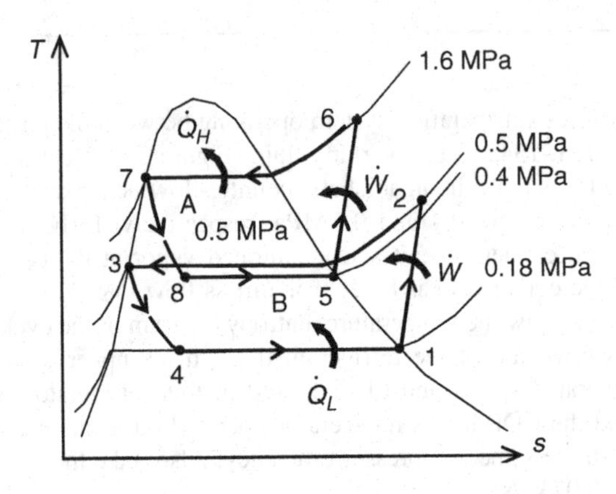

Figure 5.6 *T–s* diagram of the system considered in Example 5.1.

(b) The properties are to be obtained from the refrigerant-134a tables (Tables B.3 through B.5):

$$h_1 = h_{g@180\,\text{kPa}} = 242.86\,\text{kJ/kg}$$

$$s_1 = s_{g@180\,\text{kPa}} = 0.9397\,\text{kJ/kg}\cdot\text{K}$$

$$\left.\begin{aligned} P_2 &= 500\ \text{kPa} \\ s_2 &= s_1 \end{aligned}\right\} h_{2s} = 263.86\ \text{kJ/kg}$$

$$\eta_C = \frac{h_{2s} - h_1}{h_2 - h_1}$$

$$0.85 = \frac{263.86 - 242.86}{h_2 - 242.86} \longrightarrow h_2 = 267.57\ \text{kJ/kg}$$

$$h_3 = h_{f@500\,\text{kPa}} = 73.33\ \text{kJ/kg}$$

$$h_4 = h_3 = 73.33\ \text{kJ/kg}$$

$$h_5 = h_{g@400\,\text{kPa}} = 255.55\ \text{kJ/kg}$$

$$s_5 = s_{g@400\,\text{kPa}} = 0.9269\ \text{kJ/kg}\cdot\text{K}$$

$$\left.\begin{aligned} P_6 &= 1600\ \text{kPa} \\ s_6 &= s_5 \end{aligned}\right\} h_{6s} = 284.22\ \text{kJ/kg}$$

$$\eta_C = \frac{h_{6s} - h_5}{h_6 - h_5}$$

$$0.85 = \frac{284.22 - 255.55}{h_6 - 255.55} \longrightarrow h_6 = 289.28\ \text{kJ/kg}$$

$$h_7 = h_{f@1600\,\text{kPa}} = 135.93\ \text{kJ/kg}$$

$$h_8 = h_7 = 135.93\ \text{kJ/kg}$$

The mass flow rate of the refrigerant through the upper cycle is determined from an energy balance on the heat exchanger.

$$\dot{m}_A(h_5 - h_8) = \dot{m}_B(h_2 - h_3)$$

$$\dot{m}_A(255.55 - 135.93)\ \text{kJ/kg} = (0.07\ \text{kg/s})(267.57 - 73.33)\ \text{kJ/kg} \longrightarrow \dot{m}_A = \mathbf{0.1137\ kg/s}$$

(c) The rate of heat removal from the refrigerated space is

$$\dot{Q}_L = \dot{m}_B(h_1 - h_4) = (0.07\ \text{kg/s})(242.86 - 73.33)\ \text{kJ/kg} = \mathbf{11.87\ kW}$$

(d) The power input and the COP are

$$\dot{W} = \dot{m}_A(h_6 - h_5) + \dot{m}_B(h_2 - h_1)$$

$$= (0.1137\ \text{kg/s})(289.28 - 255.55)\ \text{kJ/kg} + (0.07\ \text{kg/s})(267.57 - 242.86)\ \text{kJ/kg} = 5.56\ \text{kW}$$

$$\text{COP} = \frac{\dot{Q}_L}{\dot{W}} = \frac{11.87}{5.56} = \mathbf{2.13}$$

(e) If this refrigerator operated on a single-stage cycle (Figure 5.7) between the same pressure limits, we would have

$$h_1 = h_{g@180\,\text{kPa}} = 242.86\ \text{kJ/kg}$$

$$s_1 = s_{g@180\,\text{kPa}} = 0.9397\ \text{kJ/kg}\cdot\text{K}$$

$$\left.\begin{array}{l} P_2 = 1600\ \text{kPa} \\ s_2 = s_1 \end{array}\right\} h_{2s} = 288.52\ \text{kJ/kg}$$

$$\eta_C = \frac{h_{2s} - h_1}{h_2 - h_1}$$

$$0.85 = \frac{288.52 - 242.86}{h_2 - 242.86} \longrightarrow h_2 = 296.58\ \text{kJ/kg}$$

$$h_3 = h_{f@1600\,\text{kPa}} = 135.93\ \text{kJ/kg}$$

$$h_4 = h_3 = 135.93\ \text{kJ/kg}$$

$$\dot{Q}_L = \dot{m}(h_1 - h_4) = (0.07\ \text{kg/s})(242.86 - 135.93)\ \text{kJ/kg} = \mathbf{7.49\ kW}$$

$$\dot{W} = \dot{m}(h_2 - h_1) = (0.07\ \text{kg/s})(296.58 - 242.86)\ \text{kJ/kg} = 3.76\ \text{kW}$$

$$\text{COP} = \frac{\dot{Q}_L}{\dot{W}} = \frac{7.49}{3.76} = \mathbf{1.99}$$

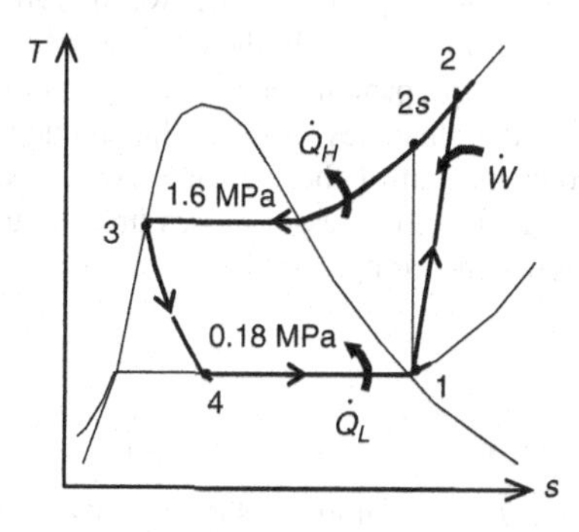

Figure 5.7 *T–s* diagram of the single-stage cycle considered in part (d) of Example 5.1.

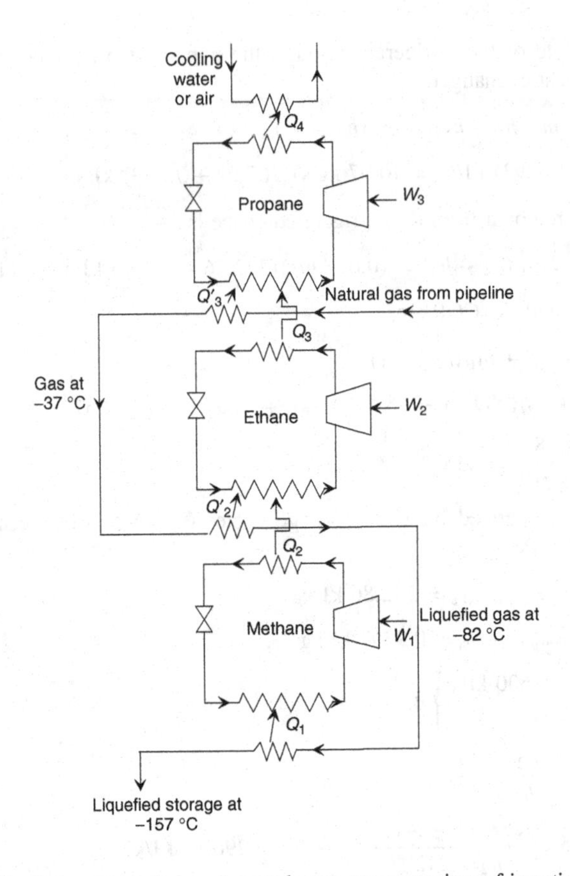

Figure 5.8 A three-stage (ternary) cascade vapor-compression refrigeration system.

5.3.2 Three-Stage (Ternary) Cascade Refrigeration Systems

Cascade refrigeration cycles are commonly used in the liquefaction of natural gas, which consists basically of hydrocarbons of the paraffin series, of which methane has the lowest boiling point at atmospheric pressure. Refrigeration down to that temperature can be provided by a ternary cascade refrigeration cycle using propane, ethane, and methane, whose boiling points at standard atmospheric pressure are 231.1, 184.5, and 111.7 K, respectively (Haywood, 1980). A simplified basic diagram for such a cascade cycle is shown in Figure 5.8. In the operation, the compressed methane vapor is first cooled by heat exchange with the propane in the propane evaporator before being condensed by heat exchange with the ethane in the ethane evaporator, thus reducing the degree of irreversibility involved in the cooling and condensation of the methane. Also, because of the high temperature after compression, the gas leaving each compressor passes first through a water-cooled intercooler. In a large-scale plant of this type, the compressors become rotary turbomachines instead of the reciprocating types.

5.4 Liquefaction of Gases

Cryogenics is associated with low temperatures, usually defined to be below −100 °C (173 K). The general scope of cryogenic engineering is the design, development, and improvement of low-temperature systems and components. The applications of cryogenic engineering include liquefaction of gases, separation of gases, high-field magnets, and sophisticated electronic devices that use the superconductivity property of materials at low temperatures, space simulation, food freezing, medical procedures such as cryogenic surgery, and various chemical processes (ASHRAE, 2006; Dincer, 2003).

The liquefaction of gases has always been an important area of refrigeration since many important scientific and engineering processes at cryogenic temperatures depend on liquefied gases. Some examples of such processes are the separation of oxygen and nitrogen from air, preparation of liquid propellants for rockets, study of material properties at low temperatures, and study of some exciting phenomena such as superconductivity. At temperatures above the critical-point value, a substance exists in the gas phase only. The critical temperatures of helium, hydrogen, and nitrogen (three commonly used liquefied gases) are −268, −240, and −147 °C, respectively (Cengel and Boles, 2008). Therefore, none of these substances will exist in liquid form at atmospheric conditions. Furthermore, low temperatures of this magnitude cannot be obtained with ordinary refrigeration techniques.

The general principles of various gas liquefaction cycles, including the Linde–Hampson cycle, and their general thermodynamic analyses are presented elsewhere, for example, Timmerhaus and Flynn (1989), Barron (1985), and Walker (1983).

Here we present the methodology for the first- and second-law-based performance analyses of the simple Linde–Hampson cycle, and investigate the effects of gas inlet and liquefaction temperatures on various cycle performance parameters.

5.4.1 Linde–Hampson Cycle

Several cycles, some complex and others simple, are used successfully for the liquefaction of gases. Here, we consider the simple Linde–Hampson cycle, which is shown schematically and on a $T-s$ diagram in Figure 5.9, in order to describe energy and exergy analyses of liquefaction cycles. See Kanoglu *et al.* (2008) for details of the analysis in this section. Makeup gas is mixed with the uncondensed portion of the gas from the previous cycle, and the mixture at state 1 is compressed by an isothermal compressor to state 2. The temperature is kept constant by rejecting compression heat to a coolant. The high-pressure gas is further cooled in a regenerative counter-flow heat exchanger by the uncondensed portion of gas from the previous cycle to state 3, and is then throttled to state 4, where it is a saturated liquid–vapor mixture. The vapour (state 5) is routed through the heat exchanger and the liquid (state 6) is collected as the desired product, to cool the high-pressure gas approaching the throttling valve. Finally, the gas is mixed with fresh makeup gas, and the cycle is repeated.

The refrigeration effect for this cycle can be defined as the heat removed from the makeup gas in order to turn it into a liquid at state 6. Assuming ideal operation for the heat exchanger (i.e., the gas leaving the heat exchanger and the makeup gas are at the same state as state 1, which is the compressor inlet state; this is also the dead state: $T_1 = T_0$), the refrigeration effect per unit mass of the liquefied gas is given by

$$q_L = h_1 - h_6 = h_1 - h_f \qquad \text{(per unit mass of liquefaction)} \tag{5.1}$$

where h_f is the enthalpy of saturated liquid that is withdrawn. From an energy balance on the cycle, the refrigeration effect per unit mass of the gas in the cycle prior to liquefaction may be expressed as

$$q_L = h_1 - h_2 \qquad \text{(per unit mass of gas in the cycle)} \tag{5.2}$$

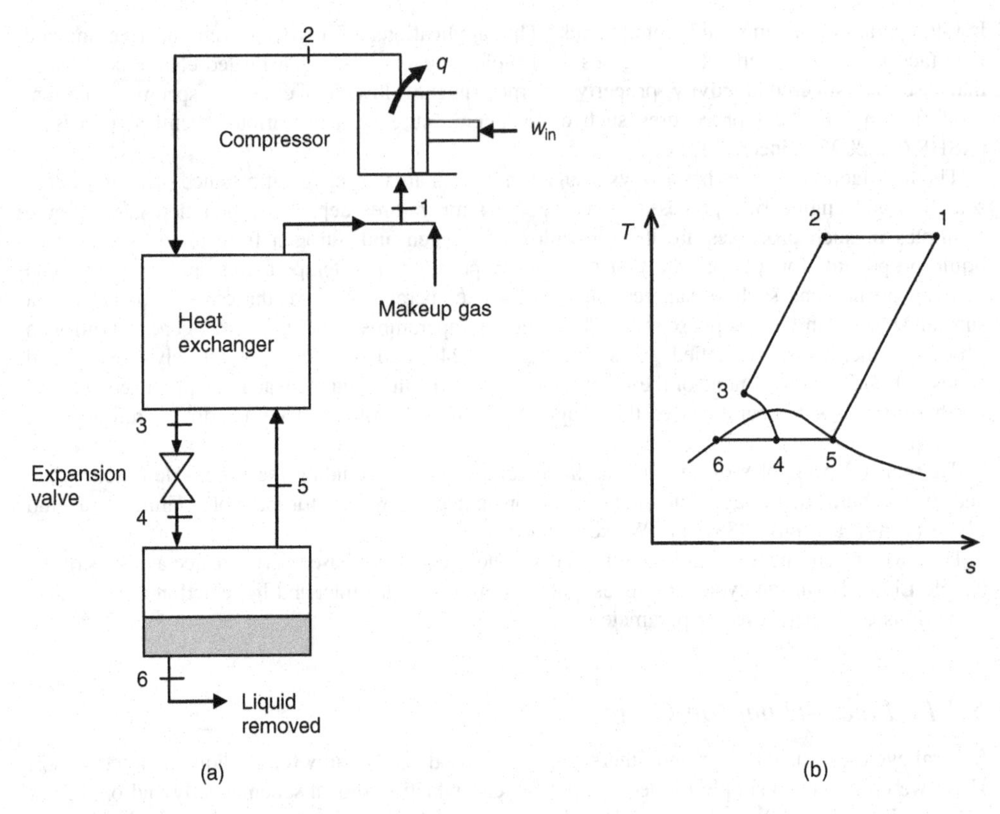

Figure 5.9 (a) Schematic and (b) temperature–entropy diagram for a simple Linde–Hampson liquefaction cycle (Kanoglu *et al.*, 2008).

Maximum liquefaction occurs when the difference between h_1 and h_2 (i.e., the refrigeration effect) is maximized. The ratio of Equations 5.2 and 5.1 is the fraction of the gas in the cycle that is liquefied. That is,

$$y = \frac{h_1 - h_2}{h_1 - h_f} \tag{5.3}$$

An energy balance on the heat exchanger gives

$$h_2 - h_3 = x(h_1 - h_5) \tag{5.4}$$

where x is the quality of the mixture at state 4. The fraction of the gas that is liquefied may also be determined from

$$y = 1 - x \tag{5.5}$$

An energy balance on the compressor gives the work of compression per unit mass of the gas in the cycle as

$$w_{\text{actual}} = h_2 - h_1 - T_1(s_2 - s_1) \qquad \text{(per unit mass of gas in the cycle)} \tag{5.6}$$

Note that $T_1 = T_0$. The last term in this equation is the isothermal heat rejection from the gas as it is compressed. Considering that the gas generally behaves as an ideal gas during this isothermal compression process, the compression work may also be determined from

$$w_{actual} = RT_1 \ln\left(\frac{P_2}{P_1}\right) \quad (5.7)$$

The COP of this cycle is given by

$$COP_{actual} = \frac{q_L}{w_{actual}} = \frac{h_1 - h_2}{h_2 - h_1 - T_1(s_2 - s_1)} \quad (5.8)$$

In liquefaction cycles, a performance parameter used is the work consumed in the cycle for the liquefaction of a unit mass of the gas. This is expressed as

$$w_{actual} = \frac{h_2 - h_1 - T_1(s_2 - s_1)}{y} \quad \text{(per unit mass of liquefaction)} \quad (5.9)$$

As the liquefaction temperature decreases work consumption increases. Noting that different gases have different thermophysical properties and require different liquefaction temperatures, this work parameter should not be used to compare work consumptions for the liquefaction of different gases. A reasonable use is to compare different cycles used for the liquefaction of the same gas.

An important object of exergy analysis for systems that consume work such as liquefaction of gases is finding the minimum work required for a certain desired result and comparing it to the actual work consumption. The ratio of these two quantities is often considered the exergy efficiency of such a liquefaction process (Kanoglu, 2002). Engineers are interested in comparing the actual work used to obtain a unit mass of liquefied gas to the minimum work required to obtain the same output. Such a comparison may be performed using the second law of thermodynamics. For instance, the minimum work input requirement (reversible work) and the actual work for a given set of processes may be related to each other by

$$w_{actual} = w_{rev} + T_0 s_{gen} = w_{rev} + ex_{dest} \quad (5.10)$$

where T_0 is the environment temperature, s_{gen} is the specific entropy generation, and ex_{dest} is the specific exergy destruction during the processes. The reversible work for the simple Linde–Hampson cycle shown in Figure 5.10 may be expressed by the stream exergy difference of states 1 and 6 as

$$w_{rev} = ex_6 - ex_1 = h_6 - h_1 - T_0(s_6 - s_1) \quad (5.11)$$

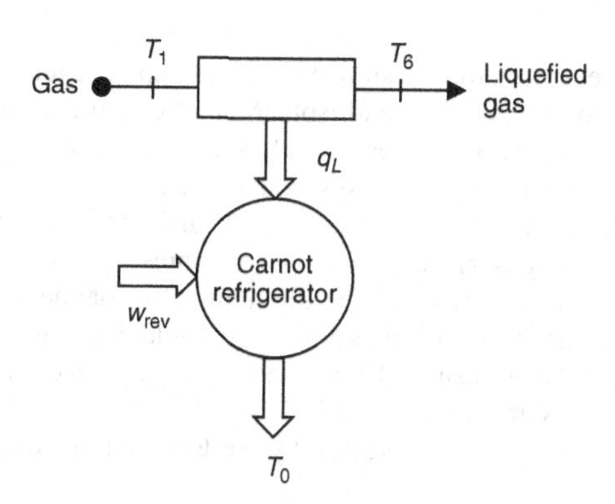

Figure 5.10 A Carnot refrigerator that uses a minimum amount of work for a liquefaction process.

where state 1 has the properties of the makeup gas, which is usually the dead state. As this equation clearly shows, the minimum work required for liquefaction depends only on the properties of the incoming and outgoing gas being liquefied and the ambient temperature T_0. An exergy efficiency may be defined as the reversible work input divided by the actual work input, both per unit mass of the liquefaction:

$$\eta_{ex} = \frac{w_{rev}}{w_{actual}} = \frac{h_6 - h_1 - T_0(s_6 - s_1)}{(1/y)\,[h_2 - h_1 - T_1(s_2 - s_1)]} \tag{5.12}$$

The exergy efficiency may also be defined using actual and reversible COPs of the system as

$$\eta_{ex} = \frac{COP_{actual}}{COP_{rev}} \tag{5.13}$$

where the reversible COP is given by

$$COP_{rev} = \frac{q_L}{w_{rev}} = \frac{h_1 - h_6}{h_6 - h_1 - T_0(s_6 - s_1)} \tag{5.14}$$

The minimum work input for the liquefaction process is simply the work input required for the operation of a Carnot refrigerator for a given heat removal, which can be expressed as

$$w_{rev} = \int \delta q \left(1 - \frac{T_0}{T}\right) \tag{5.15}$$

where δq is the differential heat transfer and T is the instantaneous temperature at the boundary where the heat transfer takes place. Note that T is smaller than T_0 for the liquefaction process, and to get a positive work input we have to take the sign of heat transfer to be negative since it is a heat output. The evaluation of Equation 5.15 requires knowledge of the functional relationship between the heat transfer δq and the boundary temperature T, which is usually not available. Equation 5.15 is also an expression of the exergy flow associated with the heat removal from the gas being liquefied.

Liquefaction process is essentially the removal of heat from the gas. Therefore, the minimum work can be determined by utilizing a reversible or Carnot refrigerator as shown in Figure 5.10. The Carnot refrigerator receives heat from the gas and supplies it to the heat sink at T_0 as the gas is cooled from T_1 to T_6. The amount of work that needs to be supplied to this Carnot refrigerator is given by Equation 5.11

Example 5.2

We present an illustrative example for the simple Linde–Hampson cycle shown in Figure 5.9. It is assumed that the compressor is reversible and isothermal; the heat exchanger has an effectiveness of 100% (i.e., the gas leaving the liquid reservoir is heated in the heat exchanger to the temperature of the gas leaving the compressor) the expansion valve is isenthalpic; and there is no heat leak to the cycle. Furthermore, the gas is taken to be air at 25 °C and 1 atm (0.101 MPa) at the compressor inlet and the pressure of the gas is 20 MPa at the compressor outlet. With these assumptions and specifications, the various properties at the different states of the cycle and the performance parameters discussed above are determined and listed in Table 5.1. The properties of air and other substances considered are obtained using EES software (Klein, 2006). This analysis is repeated for different fluids, and the results are listed in Table 5.2.

The COP of a Carnot refrigerator is expressed by the temperatures of the heat reservoirs as

$$COP_{rev} = \frac{1}{T_0/T - 1} \tag{5.16}$$

Table 5.1 Various properties and performance parameters of the cycle in Figure 5.9 for $T_1 = 25\,°C$, $P_1 = 1$ atm (0.101 MPa), and $P_2 = 20$ MPa. The fluid is air.

$h_1 = 298.4$ kJ/kg	$s_1 = 6.86$ kJ/kg · K	$q_L = 424$ kJ/kg liquid
$h_2 = 263.5$ kJ/kg	$s_2 = 5.23$ kJ/kg · K	$w_{actual} = 451$ kJ/kg gas
$h_3 = 61.9$ kJ/kg	$s_f = 2.98$ kJ/kg · K	$w_{actual} = 5481$ kJ/kg liquid
$h_4 = 61.9$ kJ/kg	$T_4 = -194.2$ °C	$w_{rev} = 733$ kJ/kg liquid
$h_5 = 78.8$ kJ/kg	$x_4 = 0.9177$	$COP_{actual} = 0.0775$
$h_6 = -126.1$ kJ/kg	$y = 0.0823$	$COP_{rev} = 0.578$
$h_f = -126.1$ kJ/kg	$q_L = 34.9$ kJ/kg gas	$\eta_{ex} = 0.134$

Table 5.2 Performance parameters of a simple Linde–Hampson cycle for various fluids.

Item	Air	Nitrogen	Oxygen	Argon	Methane	Fluorine
Liquefaction temperature T_4 (°C)	−194.2	−195.8	−183.0	−185.8	−161.5	−188.1
Fraction liquefied y	0.0823	0.0756	0.107	0.122	0.199	0.0765
Refrigeration effect q_L (kJ/kg gas)	34.9	32.6	43.3	33.2	181	26.3
Refrigeration effect q_L (kJ/kg liquid)	424	431	405	272	910	344
Work input w_{in} (kJ/kg gas)	451	468	402	322	773	341
Work input w_{in} (kJ/kg liquid)	5481	6193	3755	2650	3889	4459
Minimum work input w_{rev} (kJ/kg liquid)	733	762	629	472	1080	565
COP_{actual}	0.0775	0.0697	0.108	0.103	0.234	0.0771
COP_{rev}	0.578	0.566	0.644	0.576	0.843	0.609
Exergy efficiency η_{ex} (%)	13.4	12.3	16.8	17.8	27.8	12.7

Here, T represents the temperature of the gas being liquefied in Figure 5.10, which changes between T_1 and T_6 during the liquefaction process. An average value of T may be obtained using Equation 5.16 with $COP_{rev} = 0.578$ and $T_0 = 25\,°C$, yielding $T = -156\,°C$. This is the temperature a heat reservoir would have if a Carnot refrigerator with a COP of 0.578 operated between this reservoir at $-156\,°C$ and another reservoir at $25\,°C$. Note that the same reservoir temperature T could be obtained by writing Equation 5.15 in the form

$$w_{rev} = -q_L\left(1 - \frac{T_0}{T}\right) \tag{5.17}$$

where $q_L = 424$ kJ/kg, $w_{rev} = 733$ kJ/kg, and $T_0 = 25\,°C$.

As part of the analysis, the effects of liquefaction temperature and gas inlet temperature on various energy- and exergy-based performance parameters are investigated considering oxygen as the gas being liquefied. The results of these studies are given in Figures 5.11 through 5.18. The results involving various gases are shown in Figures 5.14 and 5.18.

The data obtained for various fluids in Table 5.2 show that different gases exhibit different behaviors in terms of performance parameters. The differences are due to the thermophysical properties of fluids and the liquefaction temperatures. Figures 5.11 through 5.18 show that as

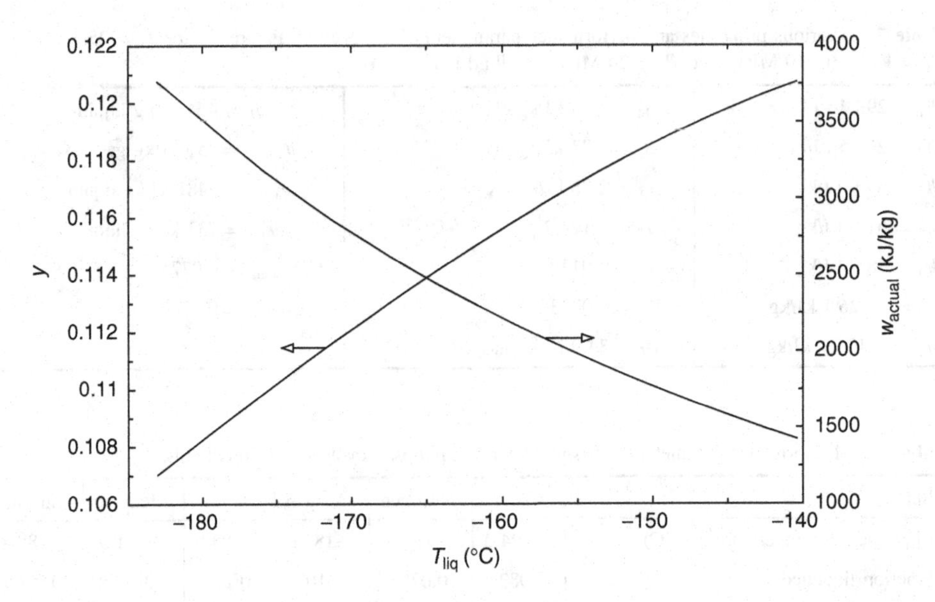

Figure 5.11 The liquefied mass fraction y and the actual work input versus liquefaction temperature for oxygen.

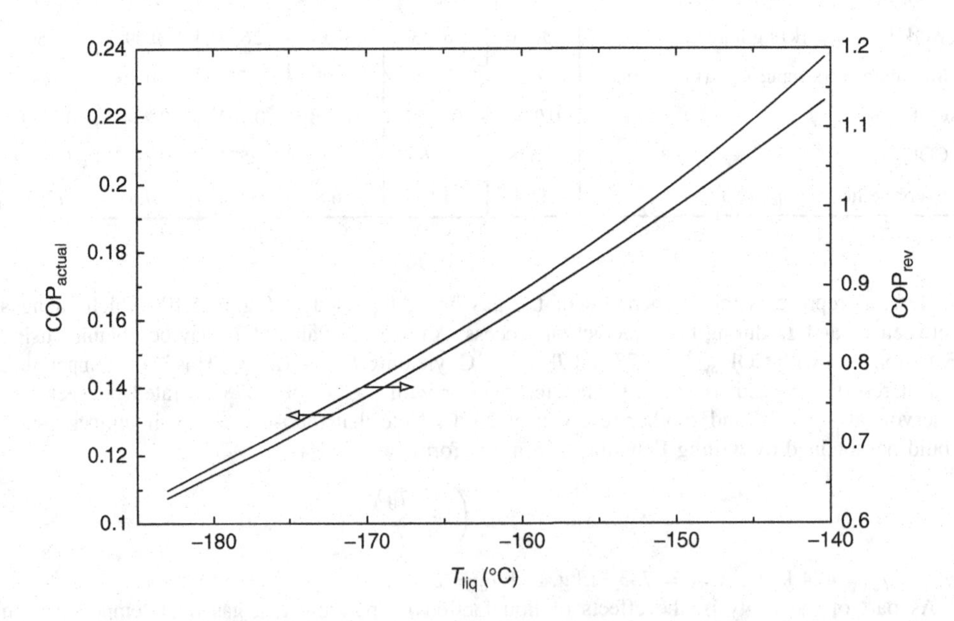

Figure 5.12 The actual and reversible COPs versus liquefaction temperature for oxygen.

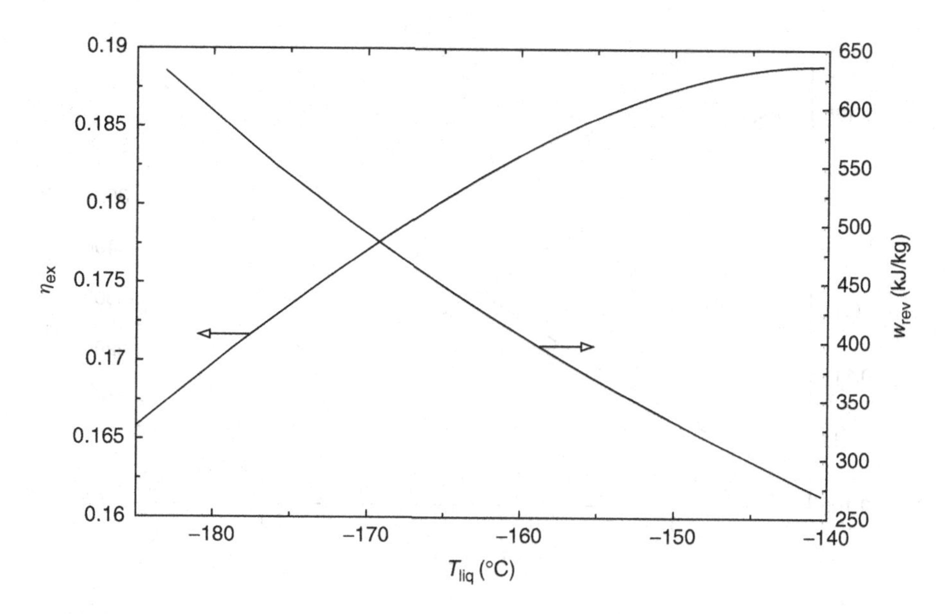

Figure 5.13 The exergy efficiency and reversible work versus liquefaction temperature for oxygen.

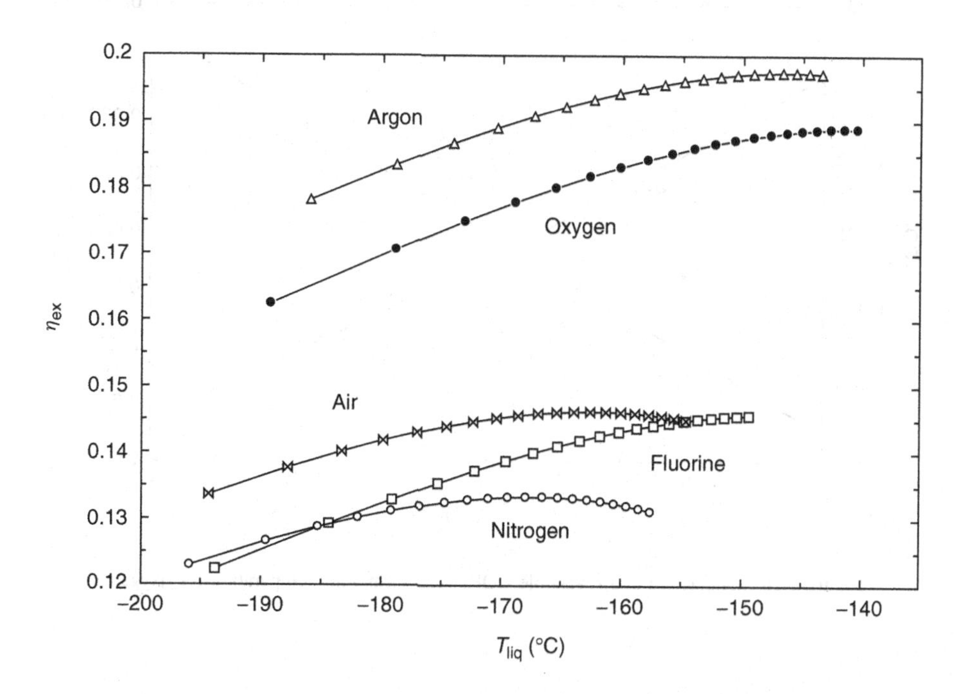

Figure 5.14 The exergy efficiency versus liquefaction temperature for various gases.

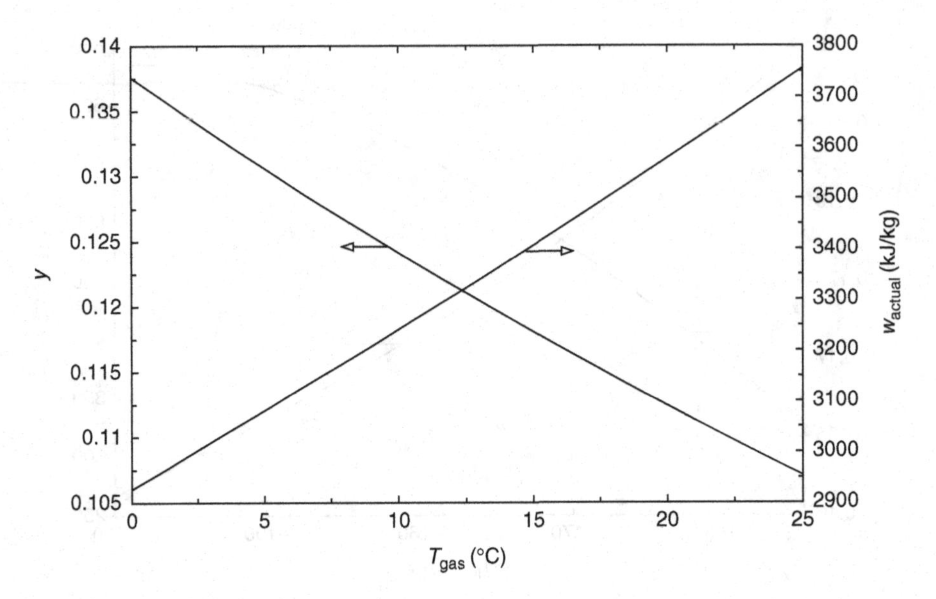

Figure 5.15 The liquefied mass fraction y and the actual work input versus gas inlet temperature for oxygen.

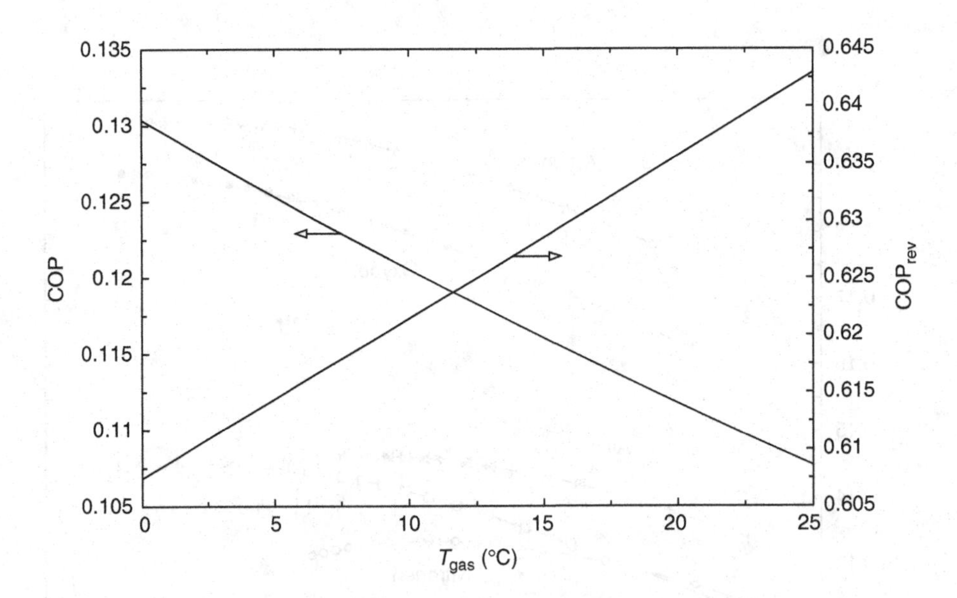

Figure 5.16 The actual and reversible COPs versus gas inlet temperature for oxygen.

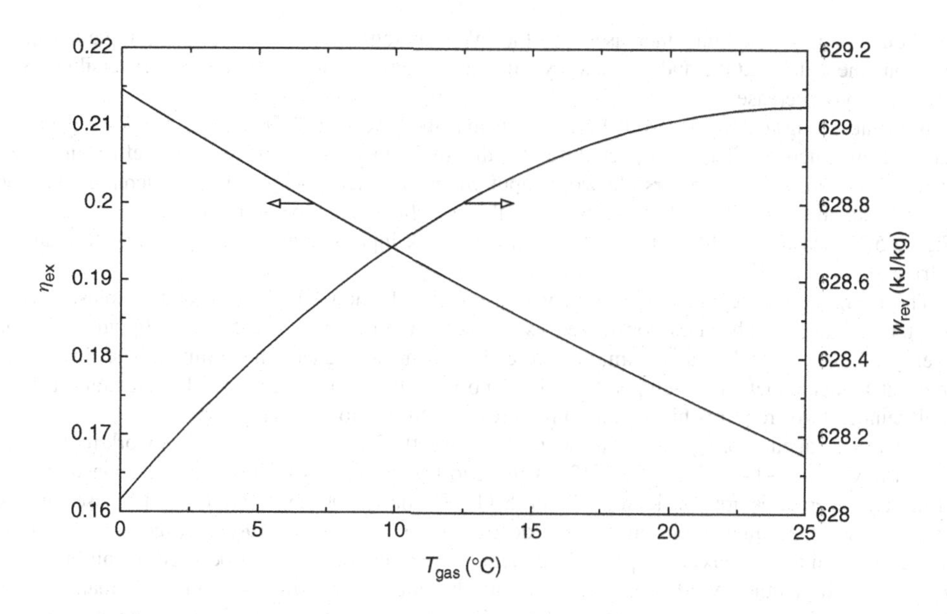

Figure 5.17 The exergy efficiency and reversible work versus gas inlet temperature for oxygen.

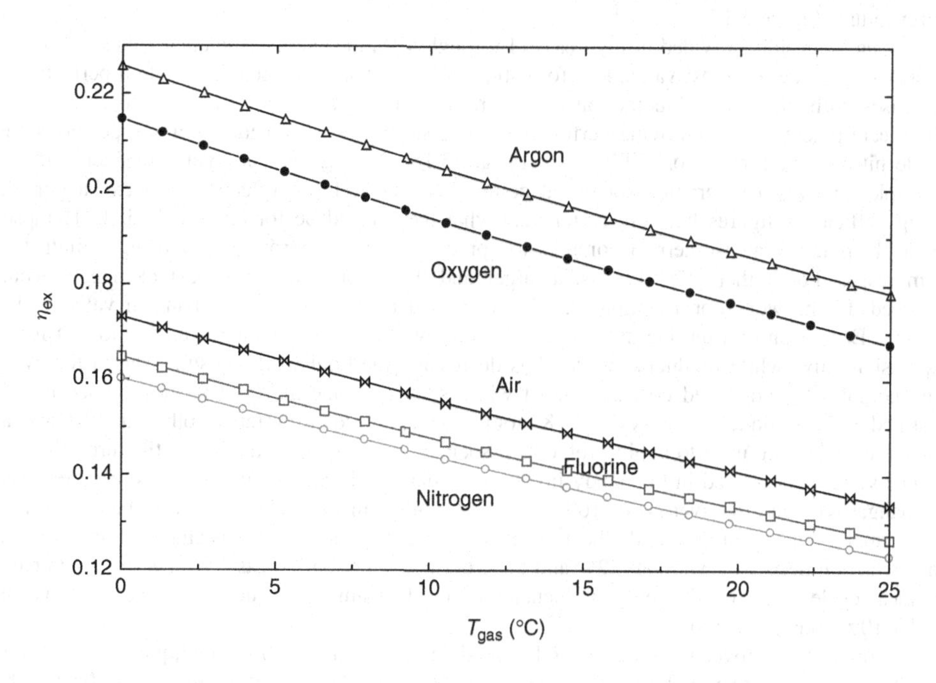

Figure 5.18 The exergy efficiency versus gas inlet temperature for various gases.

the liquefaction temperature increases and the inlet gas temperature decreases the liquefied mass fraction, the actual COP, and the exergy efficiency increase, while actual and reversible work consumptions decrease.

It is interesting to observe from Figure 5.16 that the reversible COP increases as the gas inlet temperatures increase. This unexpected trend is due to the fact that the refrigeration effect increases at a greater rate than the reversible work input when the inlet gas temperature increases. On the other hand, the reversible COP increases as the liquefaction temperature increases as shown in Figure 5.12 because of the fact that the reversible work input decreases at a greater rate than the refrigeration effect.

The exergy efficiency increases with increasing liquefaction temperature and decreasing inlet gas temperature for all gases considered as shown in Figures 5.14 and 5.18. In Figure 5.14, the exergy efficiency reaches a maximum before decreasing at higher temperatures. The decreasing trend at higher liquefaction temperatures is of no practical importance since liquefaction at these high temperatures requires higher inlet pressures, which are not normally used.

Obtaining liquefied oxygen at $-183\,°C$ requires exactly 2.1 times the minimum work required to obtain oxygen at $-145\,°C$ (Figure 5.13). This ratio becomes 2.4 when actual work consumptions at these temperatures are considered (Figure 5.11). Similarly, the reversible COP decreases almost by half when the liquefaction temperature decreases from -140 to $-190\,°C$ (Figure 5.12). These figures show that the maximum possible liquefaction temperature should be used to minimize the work input. In another words, the gas should not be liquefied to lower temperatures than needed. As the inlet gas temperature decreases from 25 to $0\,°C$, the actual specific work input decreases from 3755 to 2926 kJ/kg (Figure 5.15). The reversible work is not notably affected by the inlet gas temperature (Figure 5.17).

Among the results provided in Figures 5.11 through 5.18, the exergy efficiency values and trends appear to provide the most valuable information by clearly showing that the system performance increases with increasing liquefaction and decreasing inlet gas temperatures and that there is a significant potential for improving performance. Among the gases considered, argon performs best while nitrogen performs worst (Figures 5.14 and 5.18). Noting that the cycle considered in this example involves a reversible isothermal compressor and a 100% effective heat exchanger, the exergy efficiency figures here are better than what they would be for an actual Linde–Hampson cycle. In practice, an isothermal compression process may be approached by using a multistage compressor. For higher effectiveness, a larger and thus more expensive heat exchanger would be needed. The work consumption may be decreased by replacing the expansion valve with a turbine. Expansion in a turbine usually results in a lower outlet temperature relative to that for an expansion valve while producing work, thus decreasing the total work consumption in the cycle. The complexity and added cost associated with using a turbine as an expansion device is only justified in large liquefaction systems (Kanoglu, 2001). In some systems both a turbine and an expansion valve are used to avoid problems associated with liquid formation in the turbine.

The system considered in this study involves an ideal isothermal compressor and a perfect heat exchanger with an effectiveness of 100%. When a more realistic cycle for air liquefaction with an isothermal efficiency of 70% and a heat exchanger effectiveness of 96.5% is analyzed, the liquefied mass fraction decreases by about 22% and the work consumption increases by 1.8 times compared to ideal cycle. The actual exergy efficiencies of Linde–Hampson liquefaction cycle are usually under 10% (Barron, 1985).

The difference between the actual and reversible work consumptions in liquefaction systems are because of the exergy losses that occur during various processes in the cycle. Irreversible compression in the compressor, heat transfer across a finite temperature difference in heat exchangers (e.g., regenerator, evaporator, compressor), and friction are major sources of exergy losses in these systems. In actual refrigeration systems, these irreversibilities are normally reduced by applying modifications to the simple Linde–Hampson cycle, such as utilizing multistage compression and using a turbine in place of an expansion valve or in conjuction with an expansion valve

(Claude cycle). Other modified cycles that have resulted in greater efficiency are known as the *dual-pressure Claude cycle* and the *Collins helium cycle*. For natural gas liquefaction, mixed-refrigerant, cascade, and gas-expansion cycles are used (Kanoglu, 2002). In most large natural gas liquefaction plants, the mixed-refrigerant cycle is used in which the natural gas stream is cooled by the successive vaporization of propane, ethylene, and methane. Each refrigerant may be vaporized at two or three pressure levels to minimize the irreversibilities and thus increase the exergy efficiency of the system. This requires a more complex and costly system but the advantages usually more than offset the extra cost in large liquefaction plants.

5.4.2 Precooled Linde–Hampson Liquefaction Cycle

The precooled Linde–Hampson cycle is a well-known and relatively simple system used for the liquefaction of gases including hydrogen (Figure 5.19). Makeup gas is mixed with the uncondensed portion of the gas from the previous cycle and the mixture at state 1 is compressed to state 2. Heat is rejected from the compressed gas to a coolant. The high-pressure gas is cooled to state 3 in a regenerative counter-flow heat exchanger (I) by the uncondensed gas and is cooled further by flowing through two nitrogen baths (II and IV) and two regenerative heat exchangers (III and V) before being throttled to state 8, where it is a saturated liquid–vapor mixture. The liquid is collected as the desired product and the vapor is routed through the bottom half of the cycle. Finally, the gas is mixed with fresh makeup gas and the cycle is repeated.

Using an energy balance of heat exchanger V and the throttling valve taken together, the fraction of the liquefied gas can be determined to be

$$f_{\text{liq}} = \frac{h_9 - h_6}{h_9 - h_f} \tag{5.18}$$

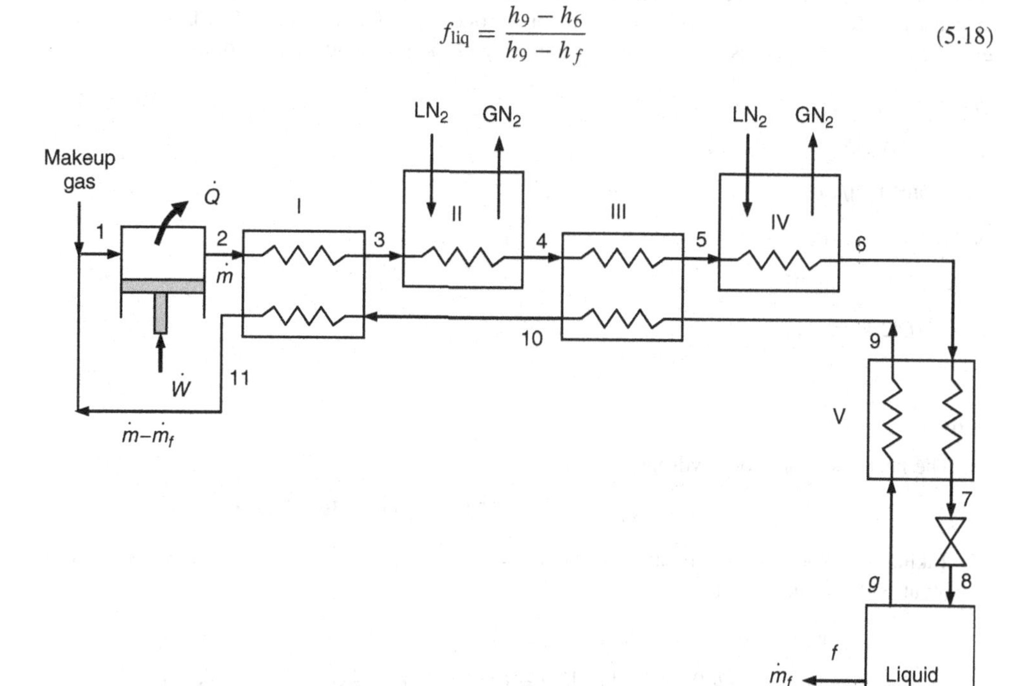

Figure 5.19 Precooled Linde–Hampson liquefaction cycle.

Energy balances for the heat exchangers can be written as

$$h_2 - h_3 = (1 - f_{liq})(h_{11} - h_{10}) \tag{5.19}$$

$$h_4 - h_5 = (1 - f_{liq})(h_{10} - h_9) \tag{5.20}$$

$$h_6 - h_7 = (1 - f_{liq})(h_9 - h_g) \tag{5.21}$$

Since the gas behaves ideally during compression, the specific compression work may be determined from

$$w_{in} = \frac{RT_0 \ln(P_2/P_1)}{\eta_{comp}} \quad \text{(per unit mass of gas in the cycle)} \tag{5.22}$$

where η_{comp} is the isothermal efficiency of the compressor, R is the gas constant, and P is the pressure. The numerator of the right side represents the work input for a corresponding isothermal process. The specific work input to the liquefaction cycle per unit mass of liquefaction is

$$w_{in,liq} = \frac{w_{in}}{f_{liq}} \quad \text{(per unit mass of liquefaction)} \tag{5.23}$$

Example 5.3

Hydrogen gas at 25 °C and 1 atm (101.325 kPa) is to be liquefied in a precooled Linde–Hampson cycle. Hydrogen gas is compressed to a pressure of 10 MPa in the compressor which has an isothermal efficiency of 65%. The effectiveness of heat exchangers is 90%. Determine (a) the heat removed from hydrogen and the minimum work input, (b) the fraction of the gas liquefied, (c) the work input in the compressor per unit mass of liquefied hydrogen, and (d) the second-law efficiency of the cycle if the work required for nitrogen liquefaction is 15,200 kJ/kg of hydrogen gas in the cycle. Properties of hydrogen in the cycle at various states are as follows:

$h_f = 271.1$ kJ/kg

$h_0 = 4200$ kJ/kg

$h_6 = 965.4$ kJ/kg

$h_9 = 1147.7$ kJ/kg

$s_f = 17.09$ kJ/kg · K

$s_0 = 70.42$ kJ/kg · K

Solution

(a) The heat rejection from hydrogen gas is

$$q_L = h_0 - h_f = (4200 - 271.1)\ \text{kJ/kg} = \mathbf{3929\,kJ/kg}$$

Taking the dead state temperature to be $T_0 = T_1 = 25\,°C = 298.15$ K, the minimum work input is determined from

$$\begin{aligned} w_{min} &= h_0 - h_f - T_0(s_0 - s_f) \\ &= (4200 - 271.1)\ \text{kJ/kg} - (298.15\ \text{K})(70.42 - 17.09)\ \text{kJ/kg} \cdot \text{K} \\ &= \mathbf{11{,}963\,kJ/kg} \end{aligned}$$

(b) The fraction of the gas liquefied is

$$f_{\text{liq}} = \frac{h_9 - h_6}{h_9 - h_f} = \frac{1147.7 - 965.4}{1147.7 - 271.1} = \mathbf{0.208}$$

(c) The work input in the compressor per unit mass of hydrogen gas compressed is

$$w_{\text{in}} = \frac{RT_0 \ln(P_2/P_1)}{\eta_{\text{comp}}} = \frac{(4.124)(298.15)\ln(10{,}000/101.325)}{0.65} = 8682 \text{ kJ/kg}$$

Per unit mass of liquefaction,

$$w_{\text{in,liq}} = \frac{w_{\text{in}}}{f_{\text{liq}}} = \frac{8682}{0.208} = \mathbf{41{,}740\,kJ/kg}$$

(d) The total work input for the cycle per unit mass of liquefied hydrogen is

$$w_{\text{in,total}} = \frac{w_{\text{in}} + w_{\text{in,nitrogen}}}{f_{\text{liq}}} = \frac{8682 + 15{,}200}{0.208} = 114{,}800 \text{ kJ/kg}$$

(e) The second-law efficiency is determined from

$$\eta_{II} = \frac{w_{\min}}{w_{\text{in,total}}} = \frac{11{,}963}{114{,}800} = 0.104 = \mathbf{10.4\%}$$

5.4.3 *Claude Cycle*

Claude cycle may be used to liquefy various gases including hydrogen (Figure 5.20). In this cycle, an expander (turbine) makes work production during the expansion process possible. Feed gas is compressed to approximately 40 bar pressure. About 75% of the gas after the primary heat exchanger is expanded in a turbine before mixing with the cold returning gas. An expansion valve is used to obtain liquid. An energy balance on the entire cycle with no heat leak into the cycle gives

$$(\dot{m} - \dot{m}_f)h_1 + \dot{m}_f h_f + \dot{m}_e h_e - \dot{m}h_2 - \dot{m}_e h_3 = 0 \tag{5.24}$$

The fraction of mass flowing through the expander is defined as

$$x = \dot{m}_e / \dot{m}_f \tag{5.25}$$

Then the fraction of mass liquefied becomes

$$y = \frac{\dot{m}_f}{\dot{m}} = \frac{h_1 - h_2}{h_1 - h_f} + x\frac{h_3 - h_e}{h_1 - h_f} \tag{5.26}$$

The last term is greater than zero and thus Claude cycle is a definite improvement over the Linde–Hampson cycle. The work produced in the expander is given by

$$W_e = \dot{m}_e(h_3 - h_e) \tag{5.27}$$

The total work input is the difference between the work consumed in the compressor and the work produced in the expander:

$$w = w_{\text{comp}} - w_e = [T_1(s_1 - s_2) - (h_1 - h_2)] - x(h_3 - h_e) \tag{5.28}$$

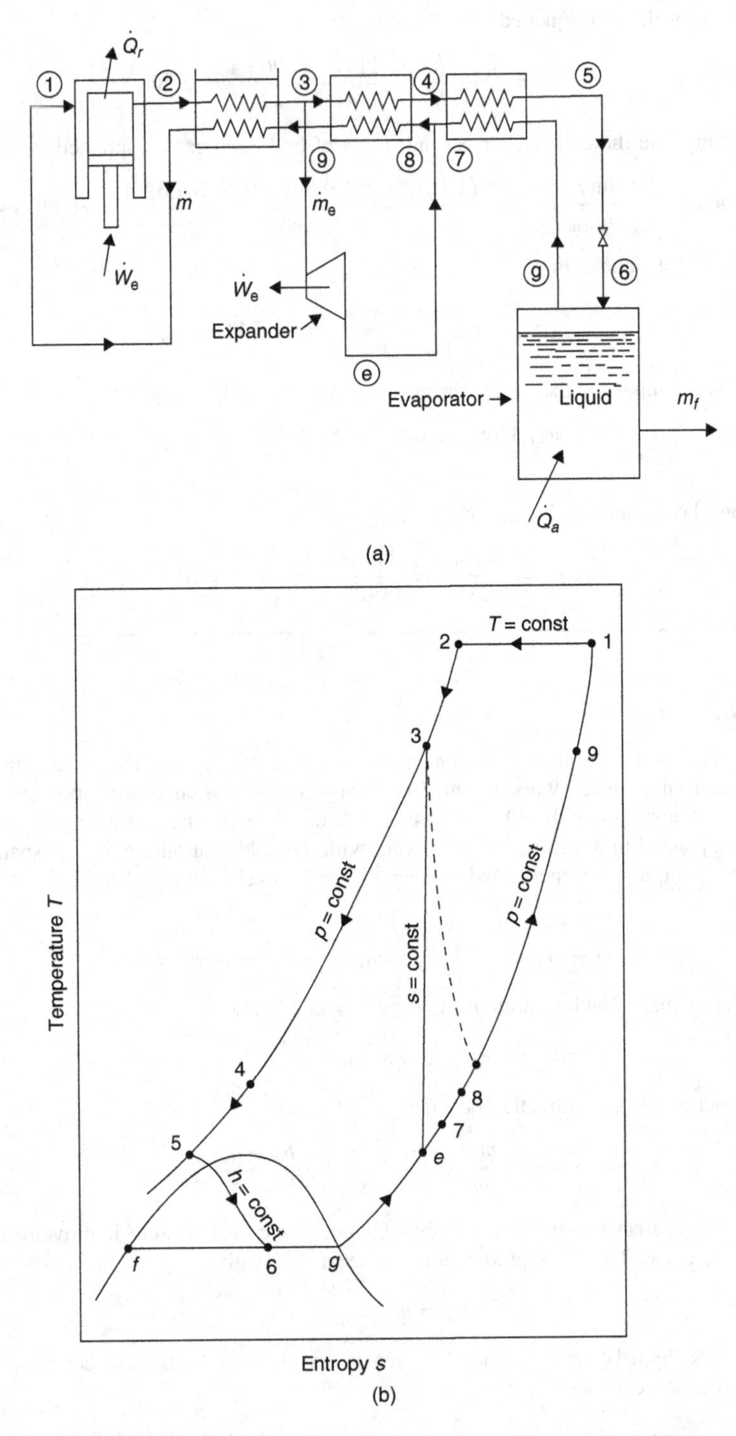

Figure 5.20 A Claude low-pressure process cycle using an (a) expansion machine and (b) its $T - s$ diagram (Adapted from Barron, 1985).

5.4.4 Multistage Cascade Refrigeration Cycle Used for Natural Gas Liquefaction

Importance. Natural gas is a mixture of components consisting mainly of methane (60–98%) with small amounts of other hydrocarbon fuel components. It also contains various amounts of nitrogen, carbon dioxide, helium, and traces of other gases. It is stored as compressed natural gas (CNG) at pressures of 16–25 MPa and around room temperature or as a liquefied natural gas (LNG) at pressures of 70–500 kPa and around −150 °C or lower. When transportation of natural gas in pipelines is not feasible for economic and other reasons, it is first liquefied using nonconventional refrigeration cycles and then it is usually transported by marine ships in specially made insulated tanks. It is regasified in receiving stations before being given off the pipeline for end-use. In fact, different refrigeration cycles with different refrigerants can be used for natural gas liquefaction. The first cycle used (and still commonly used) for natural gas liquefaction was the multistage cascade refrigeration cycle that uses three different refrigerants, namely propane, ethane (or ethylene), and methane in their individual refrigeration cycles. A great amount of work is consumed to obtain LNG at about −150 °C that enters the cycle at about atmospheric temperature in the gas phase. Minimizing the work consumed in the cycle is the most effective measure to reduce the cost of LNG. In this regard, exergy appears to be a potential tool for the design, optimization, and performance evaluation of such systems. Note that identifying the main sites of exergy destruction shows the direction for potential improvements. An important object of exergy analysis for systems that consume work such as liquefaction of gases and distillation of water is finding the minimum work required for a certain desired result.

Description of the cycle. Figure 5.21 shows a schematic of the cascade refrigeration cycle and its components. The cycle consists of three subcycles and each one uses a different refrigerant. In the first cycle, propane leaves the compressor at a high temperature and pressure and enters the condenser where the cooling water or air is used as the coolant. The condensed propane then enters the expansion valve where its pressure is decreased to the evaporator pressure. As the propane evaporates, the heat of evaporation comes from the condensing ethane, cooling methane, and cooling natural gas. Propane leaves the evaporator and enters the compressor, thus

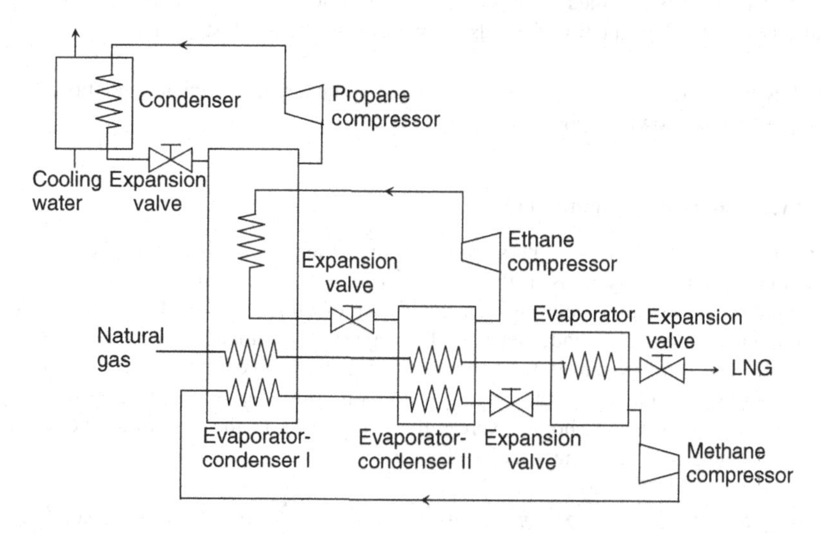

Figure 5.21 Schematic of the cascade refrigeration cycle (showing only one stage for each refrigerant cycle for simplicity) (Adapted from Kanoglu, 2002).

completing the cycle. The condensed ethane expands in the expansion valve and evaporates as methane condenses and natural gas is further cooled and liquefied. Finally, methane expands and then evaporates as natural gas is liquefied and subcooled. As methane enters the compressor to complete the cycle, the pressure of LNG is dropped in an expansion valve to the storage pressure. The three refrigerant cycles have multistage compression and expansion with usually three stages and consequently three evaporation temperature levels for each refrigerant. The mass flows in each stage are usually different. Natural gas from the pipeline goes through a process during which the acid gases are removed and its pressure is increased to an average value of 40 bar before entering the cycle.

Exergy analysis. The flow exergy of any fluid in a control volume can be written as follows (with negligible changes in kinetic and potential energies):

$$\dot{E}x = \dot{m}ex = \dot{m}\,[(h - h_0) - T_0(s - s_0)] \tag{5.29}$$

where T_0 is the dead state temperature, h and s are the enthalpy and entropy of the fluid at the specified state, and h_0 and s_0 are the corresponding properties at the dead (reference) state.

The specific exergy change between two states (e.g., inlet and outlet) is

$$\Delta ex = ex_1 - ex_2 = (h_1 - h_2) - T_0(s_1 - s_2) \tag{5.30}$$

As mentioned earlier, some part of the specific exergy change is lost during the process due to entropy generation; referring to $T_0\Delta s$ for the above equation, $i = T_0\Delta s = T_0 s_{gen}$ known as *specific irreversibility*. Here s_{gen} is the entropy generation. Two main causes for entropy generation are friction and heat transfer across a finite temperature difference. Heat transfer is always accompanied by exergy transfer, which is given by

$$ex_q = \int \delta q \left(1 - \frac{T_0}{T}\right) \tag{5.31}$$

where δq is differential heat transfer and T is the source temperature where heat transfer takes place. Heat transfer is assumed to occur with the surroundings at T_0. If this heat transfer shows an undesired heat loss, Equation 5.31 also expresses the exergy lost by heat.

The following sections give the exergy destruction and exergetic efficiency relations for various cycle components as shown in Figure 5.21.

5.4.4.1 Evaporators and Condensers

The evaporators and condensers in the system are treated as heat exchangers. There are a total of four evaporator–condenser systems in the cycle. The first system, named evaporator–condenser-I, is the evaporator of propane cycle and the condenser of ethane and methane cycles. Similarly, the system named evaporator–condenser-II is the evaporator of ethane cycle and the condenser of methane cycle. The third system is the evaporator of methane cycle and the fourth system is the condenser of propane cycle where the cooling water is used as coolant. An exergy balance written on the evaporator–condenser-I should express the exergy loss in the system as the difference of exergies of incoming and outgoing streams. That is,

$$\begin{aligned} \dot{I} = \dot{E}x_{in} - \dot{E}x_{out} &= \left[\sum (\dot{m}_p ex_p) + \sum (\dot{m}_e ex_e) + \sum (\dot{m}_m ex_m) + (\dot{m}_d ex_d)\right]_{in} \\ &\quad - \left[\sum (\dot{m}_p ex_p) + \sum (\dot{m}_e ex_e) + \sum (\dot{m}_m ex_m) + (\dot{m}_n ex_n)\right]_{out} \end{aligned} \tag{5.32}$$

where the subscripts in, out, p, e, m, and n stand for inlet, outlet, propane, ethane, methane, and natural gas, respectively. The summation signs are due to the fact that there are three stages in each refrigerant cycle with different pressures, evaporation temperatures, and mass flow rates.

The exergetic efficiency of a heat exchanger can be defined as the ratio of total outgoing stream exergies to total incoming stream exergies as follows:

$$\varepsilon = \frac{\sum(\dot{m}_p ex_p)_{out} + \sum(\dot{m}_e ex_e)_{out} + \sum(\dot{m}_m ex_m)_{out} + (\dot{m}_n ex_n)_{out}}{\sum(\dot{m}_p ex_p)_{in} + \sum(\dot{m}_e ex_e)_{in} + \sum(\dot{m}_m ex_m)_{in} + (\dot{m}_n ex_n)_{in}} \tag{5.33}$$

The second definition for the exergy efficiency of heat exchangers can be the ratio of the increase in the exergy of the cold fluid to the decrease in the exergy of the hot. In the system, the only fluid with an exergy increase is propane while the exergies of ethane, methane, and natural gas decrease. Therefore, the equation becomes

$$\varepsilon = \frac{\sum(\dot{m}_p ex_p)_{out} - \sum(\dot{m}_p ex_p)_{in}}{\sum(\dot{m}_e ex_e)_{in} - \sum(\dot{m}_e ex_e)_{out} + \sum(\dot{m}_m ex_m)_{in} - \sum(\dot{m}_m ex_m)_{out} + (\dot{m}_n ex_n)_{in} - (\dot{m}_n ex_n)_{out}} \tag{5.34}$$

The above two methods used to determine the exergetic efficiency of a heat exchanger are sometimes called the *scientific approach* and the *engineering approach*, respectively. The efficiencies calculated using these two approaches are usually very close to each other. Here in this example, the engineering approach will be used in the following relations. The relations for exergy destruction and exergetic efficiency for evaporator–condenser-II are determined as

$$\dot{I} = \dot{E}x_{in} - \dot{E}x_{out} = \left[\sum(\dot{m}_e ex_e) + \sum(\dot{m}_m ex_m) + (\dot{m}_n ex_n)\right]_{in} - \left[\sum(\dot{m}_e ex_e) + \sum(\dot{m}_m ex_m) + (\dot{m}_n ex_n)\right]_{out} \tag{5.35}$$

$$\varepsilon = \frac{\sum(\dot{m}_e ex_e)_{out} - \sum(\dot{m}_e ex_e)_{in}}{\sum(\dot{m}_m ex_m)_{in} - \sum(\dot{m}_m ex_m)_{out} + (\dot{m}_n ex_n)_{in} - (\dot{m}_n x_n)_{out}} \tag{5.36}$$

From the exergy balance on the evaporator of the methane cycle, the following exergy destruction and exergetic efficiency expressions can be written:

$$\dot{I} = \dot{E}x_{in} - \dot{E}x_{out} = \left[\sum(\dot{m}_m ex_m) + (\dot{m}_n ex_n)\right]_{in} - \left[\sum(\dot{m}_m ex_m) + (\dot{m}_n ex_n)\right]_{out} \tag{5.37}$$

$$\varepsilon = \frac{\sum(\dot{m}_m ex_m)_{out} - \sum(\dot{m}_m ex_m)_{in}}{(\dot{m}_n ex_n)_{in} - (\dot{m}_n ex_n)_{out}} \tag{5.38}$$

Finally, for the condenser of the propane cycle the following can be obtained:

$$\dot{I} = \dot{E}x_{in} - \dot{E}x_{out} = \left[\sum(\dot{m}_p ex_p) + (\dot{m}_w ex_w)\right]_{in} - \left[\sum(\dot{m}_p ex_p) + (\dot{m}_w ex_w)\right]_{out} \tag{5.39}$$

$$\varepsilon = \frac{(\dot{m}_w ex_w)_{out} - (\dot{m}_w ex_w)_{in}}{\sum(\dot{m}_p ex_p)_{in} - \sum(\dot{m}_p ex_p)_{out}} \tag{5.40}$$

where the subscript w stands for water.

5.4.4.2 Compressors

There is one multistage compressor in the cycle for each refrigerant. The total work consumed in the cycle is the sum of work inputs to the compressors. There is no exergy destruction in a compressor if irreversibilities can be totally eliminated. This results in a minimum work input for the compressor. In reality, there are irreversibilities due to friction, heat loss, and other dissipative effects. The exergy destruction in propane, ethane, and methane compressors can be expressed, respectively, as

$$\dot{I}_p = \dot{E}x_{in} - \dot{E}x_{out} = \sum (\dot{m}_p ex_p)_{in} + \dot{W}_{p,in} - \sum (\dot{m}_p ex_p)_{out} \tag{5.41}$$

$$\dot{I}_e = \dot{E}x_{in} - \dot{E}x_{out} = \sum (\dot{m}_e ex_e)_{in} + \dot{W}_{e,in} - \sum (\dot{m}_e ex_e)_{out} \tag{5.42}$$

$$\dot{I}_m = \dot{E}x_{in} - \dot{E}x_{out} = \sum (\dot{m}_m ex_m)_{in} + \dot{W}_{m,in} - \sum (\dot{m}_m ex_m)_{out} \tag{5.43}$$

where $\dot{W}_{p,in}$, $\dot{W}_{e,in}$, and $\dot{W}_{m,in}$ are the actual power inputs to the propane, ethane, and methane compressors, respectively. They are part of the exergy inputs to the compressors. The exergetic efficiency of the compressor can be defined as the ratio of the minimum work input to the actual work input. The minimum work is simply the exergy difference between the actual inlet and exit states. Applying this definition to propane, ethane, and methane compressors, respectively, the exergy efficiency equations become

$$\varepsilon_p = \frac{\sum (\dot{m}_p ex_p)_{out} - \sum (\dot{m}_p ex_p)_{in}}{\dot{W}_{p,in}} \tag{5.44}$$

$$\varepsilon_e = \frac{\sum (\dot{m}_e ex_e)_{out} - \sum (\dot{m}_e ex_e)_{in}}{\dot{W}_{e,in}} \tag{5.45}$$

$$\varepsilon_m = \frac{\sum (\dot{m}_m ex_m)_{out} - \sum (\dot{m}_m ex_m)_{in}}{\dot{W}_{m,in}} \tag{5.46}$$

5.4.4.3 Expansion Valves

Beside the expansion valves in the refrigeration cycles, one is used to dropping the pressure of LNG to the storage pressure. Expansion valves are considered essentially isenthalpic devices with no work interaction and negligible heat transfer with the surroundings. From an exergy balance, the exergy destruction equations for propane, ethane, methane, and LNG expansion valves can be written as

$$\dot{I}_p = \dot{E}x_{in} - \dot{E}x_{out} = \sum (\dot{m}_p ex_p)_{in} - \sum (\dot{m}_p ex_p)_{out} \tag{5.47}$$

$$\dot{I}_e = \dot{E}x_{in} - \dot{E}x_{out} = \sum (\dot{m}_e ex_e)_{in} - \sum (\dot{m}_e ex_e)_{out} \tag{5.48}$$

$$\dot{I}_m = \dot{E}x_{in} - \dot{E}x_{out} = \sum (\dot{m}_m ex_m)_{in} - \sum (\dot{m}_m ex_m)_{out} \tag{5.49}$$

$$\dot{I}_n = \dot{E}x_{in} - \dot{E}x_{out} = \sum (\dot{m}_n ex_n)_{in} - \sum (\dot{m}_n ex_n)_{out} \tag{5.50}$$

The exergetic efficiency of expansion valves can be defined as the ratio of the total exergy output to the total exergy input. Therefore, the exergy efficiencies for all expansion valves become

$$\varepsilon_p = \frac{\sum (\dot{m}_p ex_p)_{out}}{\sum (\dot{m}_p ex_p)_{in}};\ \varepsilon_e = \frac{\sum (\dot{m}_e ex_e)_{out}}{\sum (\dot{m}_e ex_e)_{in}};\ \varepsilon_m = \frac{\sum (\dot{m}_m ex_m)_{out}}{\sum (\dot{m}_m ex_m)_{in}};\ \varepsilon_n = \frac{\sum (\dot{m}_n ex_n)_{out}}{\sum (\dot{m}_n ex_n)_{in}} \tag{5.51–5.54}$$

5.4.4.4 Cycle

The total exergy destruction in the cycle is simply the sum of exergy destructions in condensers, evaporators, compressors, and expansion valves. This total can be obtained by adding the exergy destruction terms in the above equations. Then, the overall exergy efficiency of the cycle can be defined as

$$\varepsilon = \frac{\dot{E}x_{\text{out}} - \dot{E}x_{\text{in}}}{\dot{W}_{\text{actual}}} = \frac{\dot{W}_{\text{actual}} - \dot{I}_{\text{total}}}{\dot{W}_{\text{actual}}} \tag{5.55}$$

where given in the numerator is the exergy difference or the actual work input to the cycle $\dot{W}_{\text{actual}}$ minus the total exergy destruction $\dot{I}$. The actual work input to the cycle is the sum of the work inputs to the propane, ethane, and methane compressors, which are as follows:

$$\dot{W}_{\text{actual}} = \dot{W}_{\text{p,in}} + \dot{W}_{\text{e,in}} + \dot{W}_{\text{m,in}} \tag{5.56}$$

In this regard, the exergetic efficiency of the cycle can also be expressed as

$$\varepsilon = \frac{\dot{W}_{\text{min}}}{\dot{W}_{\text{actual}}} \tag{5.57}$$

where $\dot{W}_{\text{min}}$ is the minimum work input to the cycle. Here, a process is proposed to determine the minimum work input to the cycle, or in other words, the minimum work for liquefaction process.

The exergetic efficiency of the natural gas liquefaction process can be defined as the ratio of the minimum work required to produce a certain amount of LNG to the actual work input. An exergy analysis needs to be performed on the cycle to determine the minimum work input. The liquefaction process is essentially the removal of heat from the natural gas. Therefore, the minimum work can be determined by utilizing a reversible or Carnot refrigerator. The minimum work input for the liquefaction process is simply the work input required for the operation of Carnot refrigerator for a given heat removal. It can be expressed as

$$w_{\text{min}} = \int \delta q \left(1 - \frac{T_0}{T}\right) \tag{5.58}$$

where δq is the differential heat transfer and T is the instantaneous temperature at the boundary where the heat transfer takes place. Note that T is smaller than T_0 for liquefaction process and to get a positive work input we have to take the sign of heat transfer to be negative since it is a heat output. The evaluation of Equation 5.52 requires knowledge of the functional relationship between the heat-transfer δq and the boundary temperature T, which is usually not available.

As seen in Figure 5.21, natural gas flows through three evaporator–condenser systems in the multistage refrigeration cycle before it is fully liquefied. Thermodynamically, this three-stage heat removal from natural gas can be accomplished using three Carnot refrigerator as seen in Figure 5.22. The first Carnot refrigerator receives heat from the natural gas and supplies it to the heat sink at T_0

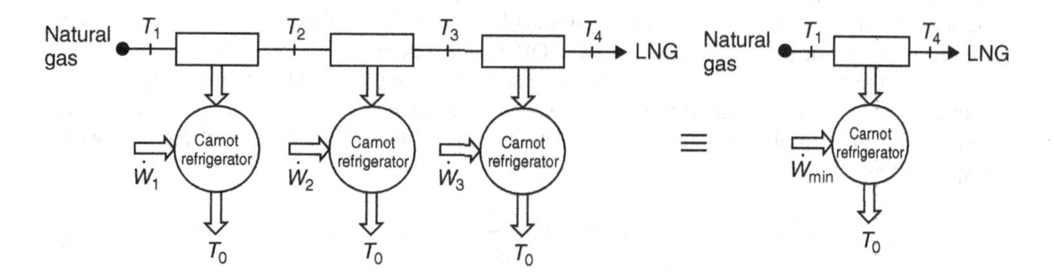

Figure 5.22 Determination of minimum work for the cycle (Kanoglu, 2002).

as the natural gas is cooled from T_1 to T_2. Similarly, the second Carnot refrigerator receives heat from the natural gas and supplies it to the heat sink at T_0 as the natural gas is cooled from T_2 to T_3. Finally, the third Carnot refrigerator receives heat from the natural gas and supplies it to the heat sink at T_0 as the natural gas is further cooled from T_3 to T_4, where it exists as LNG. The amount of power that needs to be supplied to each of the Carnot refrigerator can be determined from

$$\dot{W}_{min} = \dot{W}_1 + \dot{W}_2 + \dot{W}_3 = \dot{m}_n(ex_1 - ex_4) = \dot{m}_n\,[h_1 - h_4 - T_0(s_1 - s_4)] \quad (5.59)$$

where $\dot{W}_1$, $\dot{W}_2$, and $\dot{W}_3$ are the power inputs to the first, second, and third Carnot refrigerators, respectively.

$$\dot{W}_1 = \dot{m}_n(ex_1 - ex_2) = \dot{m}_n\,[h_1 - h_2 - T_0(s_1 - s_2)] \quad (5.60)$$

$$\dot{W}_2 = \dot{m}_n(ex_2 - ex_3) = \dot{m}_n\,[h_2 - h_3 - T_0(s_2 - s_3)] \quad (5.61)$$

$$\dot{W}_3 = \dot{m}_n(ex_3 - ex_4) = \dot{m}_n\,[h_3 - h_4 - T_0(s_3 - s_4)] \quad (5.62)$$

This is the expression for the minimum power input for the liquefaction process. This minimum power can be obtained by using a single Carnot refrigerator that receives heat from the natural gas and supplies it to the heat sink at T_0 as the natural gas is cooled from T_1 to T_4. That is, this Carnot refrigerator is equivalent to the combination of three Carnot refrigerators as shown in Figure 5.22. The minimum work required for liquefaction process depends only on the properties of the incoming and outgoing natural gas and the ambient temperature T_0.

Example 5.4

In this illustrative example, we use numerical values to study multistage cascade refrigeration cycle used for natural gas liquefaction. A numerical value of the minimum work can be calculated using typical values of incoming and outgoing natural gas properties. The pressure of natural gas is around 40 bar when entering the cycle. The temperature of natural gas at the cycle inlet can be taken to be the same as the ambient temperature $T_1 = T_0 = 25\,°C$. Natural gas leaves the cycle liquefied at about 4 bar pressure and at $-150\,°C$. Since the natural gas in the cycle usually consists of more than 95% methane, thermodynamic properties of methane can be used for natural gas. Using these inlet and exit states, the minimum work input to produce a unit mass of LNG can be determined from Equation 5.30 to be 456.8 kJ/kg. The heat removed from the natural gas during the liquefaction process is determined from

$$\dot{Q} = \dot{m}_n(h_1 - h_4) \quad (5.63)$$

For the inlet and exit states of natural gas described above, the heat removed from the natural gas can be determined from Equation 5.63 to be 823.0 kJ/kg. That is, for the removal of 823.0 kJ/kg heat from the natural gas, a minimum of 456.8 kJ/kg work is required. Since the ratio of heat removed to the work input is defined as the COP of a refrigerator, this corresponds to a COP of 1.8. That is, the COP of the Carnot refrigerator used for natural gas liquefaction is only 1.8. This is expected because of the high difference between the temperature T and T_0 in Equation 5.58 An average value of T can be obtained from the definition of the COP for a Carnot refrigerator, which is expressed as

$$\text{COP}_{rev} = \frac{1}{T_0/T - 1} \quad (5.64)$$

Using this equation, for COP = 1.8 and $T_0 = 25\,°C$ we determine $T = -81.3\,°C$. This is the temperature a heat reservoir would have if a Carnot refrigerator with a COP of 1.8 operated between this reservoir and another reservoir at 25 °C. Note that the same result could be obtained by writing Equation 5.52 in the form

$$w_{\min} = q\left(1 - \frac{T_0}{T}\right) \tag{5.65}$$

where $q = 823.0$ kJ/kg, $w_{\min} = 456.8$ kJ/kg, and $T_0 = 25\,°\text{C}$.

As part of the analysis we now investigate how the minimum work changes with the natural gas liquefaction temperature. We take the inlet pressure of natural gas to be 40 bar, inlet temperature to be $T_1 = T_0 = 25\,°\text{C}$, and exit state to be the saturated liquid at the specified temperature. The properties of methane are obtained from thermodynamic tables. Using the minimum work relation in Equation 5.65, the plot shown in Figure 5.23 is obtained. Using Equation 5.64, the variation of COP of the Carnot refrigerator with the natural gas liquefaction temperature is also obtained and shown in Figure 5.24.

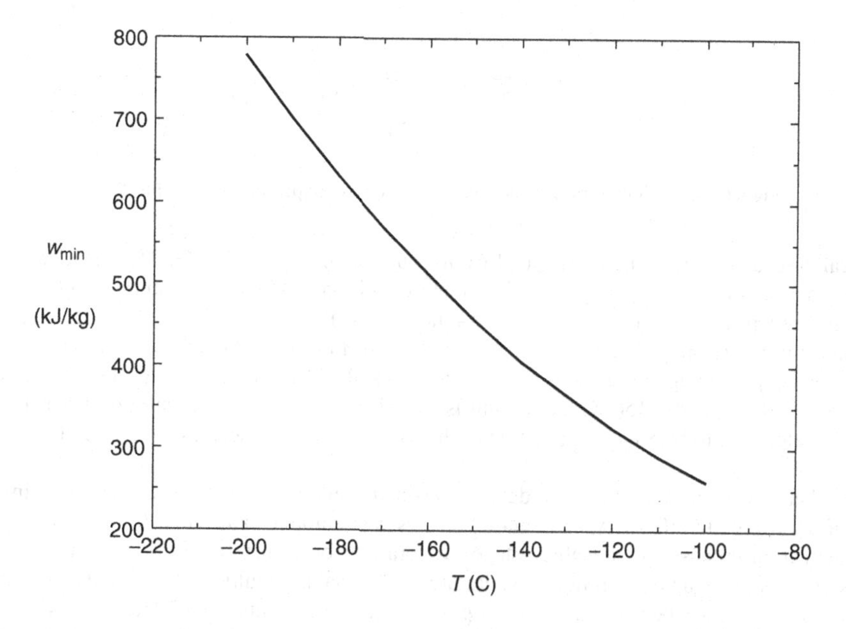

Figure 5.23 Minimum work ($w_{\min}$) versus natural gas liquefaction temperature (Kanoglu, 2002).

As shown in the figure, the minimum work required to liquefy a unit mass of natural gas increases almost linearly with the decreasing liquefaction temperature. Obtaining LNG at −200 °C requires exactly three times the minimum work required to obtain LNG at −100 °C. Similarly, obtaining LNG at −150 °C requires exactly 1.76 times the minimum work required to obtain LNG at −100 °C. The COP of the Carnot refrigerator decreases almost linearly with the decreasing liquefaction temperature as shown in Figure 5.24. The COP decreases almost by half when the liquefaction temperature decreases from −100 to −200 °C. These figures show that the maximum possible liquefaction temperature should be used to minimize the work input. In other words, the LNG should not be liquefied to lower temperatures than needed.

For a typical natural gas inlet and exit states specified in the previous section, the minimum work is determined to be 456.8 kJ/kg of LNG. A typical actual value of work input for a cascade cycle used for natural gas liquefaction may be 1188 kJ/kg of LNG. Then the exergetic efficiency of

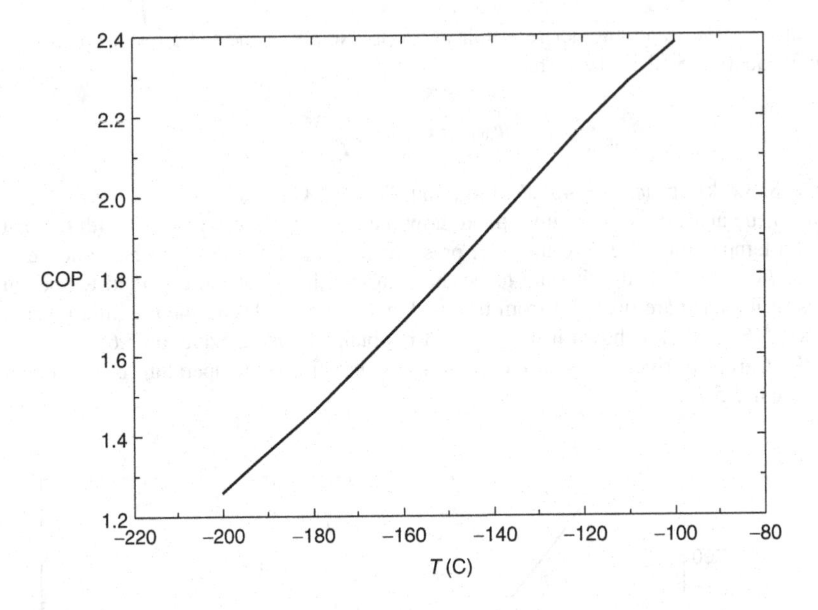

Figure 5.24 COP versus natural gas liquefaction temperature (Kanoglu, 2002).

a typical cascade cycle can be determined from Equation 3.47 to be 38.5%. The actual work input required depends mainly on the feed and ambient conditions and on the compressor efficiency.

It has been possible to replace the Joule–Thomson (JT) valve of the cycle with a cryogenic hydraulic turbine (Kanoglu, 2001). The same pressure drop as in JT valve is achieved with the turbine while producing power. Using the same typical values as before, the cryogenic turbine inlet state is 40 bar and −150 °C. Assuming isentropic expansion to a pressure of 4 bar, the work output is calculated to be 8.88 kJ/kg of LNG. This corresponds to a decrease of 2% in the minimum work input.

Note that the main site of exergy destruction in the cycle is the compressors. Any improvement in the exergetic efficiency of the compressors will automatically yield lower work input for the liquefaction process. Having three-stage evaporation for each refrigerant in the cascade cycle results in a total of nine evaporation temperatures. Also, having multiple stages makes the average temperature difference between the natural gas and the refrigerants small. This results in smaller exergy destruction in the evaporators since the greater the temperature difference the greater the exergy destruction. As the number of evaporation stages increases the exergy destruction decreases. However, adding more stages means additional equipment cost, and more than three stages for each refrigerant are not justified.

Example 5.5

Natural gas at 77 °F and 1 atm (14.7 psia) at a rate of 2500 lbm/h is to be liquefied in a natural gas liquefaction plant. Natural gas leaves the plant at 1 atm as a saturated liquid. Using methane properties for natural gas, determine (a) the temperature of natural gas after the liquefaction process and the rate of heat rejection from the natural gas during this process, (b) the minimum power input, and (c) the reversible COP. (d) If the liquefaction is done by a Carnot refrigerator between temperature limits of $T_H = 77\,°\text{F}$ and T_L with the same reversible COP, determine the temperature T_L (see Figure 5.25). Various properties of methane before and after liquefaction process are given as follows:

$h_1 = -0.4254$ Btu/lbm

$h_2 = -391.62$ Btu/lbm

$s_1 = -0.0006128$ Btu/lbm · R

$s_2 = -1.5946$ Btu/lbm · R

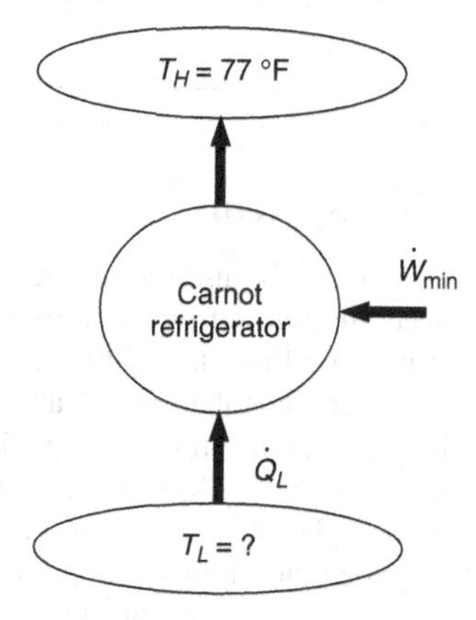

Figure 5.25 A Carnot refrigerator operating between T_L and T_H as considered in Example 5.4.

Solution

(a) The state of natural gas after the liquefaction is 14.7 psia and is a saturated liquid. The temperature at this state is determined from methane tables to be

$$T_2 = \mathbf{-259\,°F}$$

The rate of heat rejection from the natural gas during the liquefaction process is

$$\dot{Q}_L = \dot{m}(h_1 - h_2) = (2500\text{ lbm/h})\,[(-0.4254) - (-391.62)]\text{ Btu/lbm} = \mathbf{978{,}000\,Btu/h}$$

(b) Taking the dead state temperature to be $T_0 = T_1 = 77\,°C = 536$ R, the minimum work input is determined from

$$\begin{aligned}\dot{W}_{min} &= \dot{m}\,[h_2 - h_1 - T_0(s_2 - s_1)]\\ &= (2500\text{ lbm/h})\,[(-391.62) - (-0.4254)]\text{ Btu/lbm} - (537\text{ R})\\ &\quad [(-1.5941 - (-0.0006128)\text{ Btu/lbm}\cdot\text{R}]\\ &= 1.162 \times 10^6\text{Btu/h} = \mathbf{340.5\,kW}\end{aligned}$$

(c) The reversible COP is

$$\text{COP}_{rev} = \frac{\dot{Q}_L}{\dot{W}_{min}} = \frac{9.78 \times 10^5\text{ Btu/h}}{1.162 \times 10^6\text{ Btu/h}} = \mathbf{0.842}$$

(d) The temperature T_L is determined from

$$\mathrm{COP}_{\mathrm{rev}} = \frac{1}{T_H/T_L - 1} \longrightarrow 0.842 = \frac{1}{(537\ R)/T_L - 1} \longrightarrow T_L = \mathbf{245\,R}$$

It may also be determined from

$$\dot{W}_{\min} = -\dot{Q}_L\left(1 - \frac{T_0}{T_L}\right) \longrightarrow 1.162 \times 10^6\ \mathrm{Btu/h}$$

$$= -(978{,}000\ \mathrm{Btu/h})\left(1 - \frac{537\ \mathrm{R}}{T_L}\right) \longrightarrow T_L = 245\ \mathrm{R}$$

5.5 Steam Jet Refrigeration Systems

In steam jet refrigeration systems, water can be used as the refrigerant. Like air, it is perfectly safe. These systems were applied successfully to refrigeration in the early years of this century. At low temperatures the saturation pressures are low (0.008129 bar at 4 °C) and the specific volumes are high (157.3 m^3/kg at 4 °C). The temperatures that can be attained using water as a refrigerant are not low enough for most refrigeration applications but are in the range which may satisfy air-conditioning, cooling, or chilling requirements. Also, these systems are used in some chemical industries for several processes, for example, the removal of paraffin wax from lubricating oils. Note that steam jet refrigeration systems are not used when temperatures below 5 °C are required. The main advantages of this system are the utilization of mostly low-grade energy and relatively small amounts of shaft work.

Steam jet refrigeration systems use steam ejectors to reduce the pressure in a tank containing the return water from a chilled water system. The steam jet ejector utilizes the energy of a fast-moving jet of steam to capture the flash tank vapor and to compress it. Flashing a portion of the water in the tank reduces the liquid temperature. Figure 5.26 presents a schematic arrangement of a steam jet refrigeration system for water cooling. In the system shown, high-pressure steam expands while

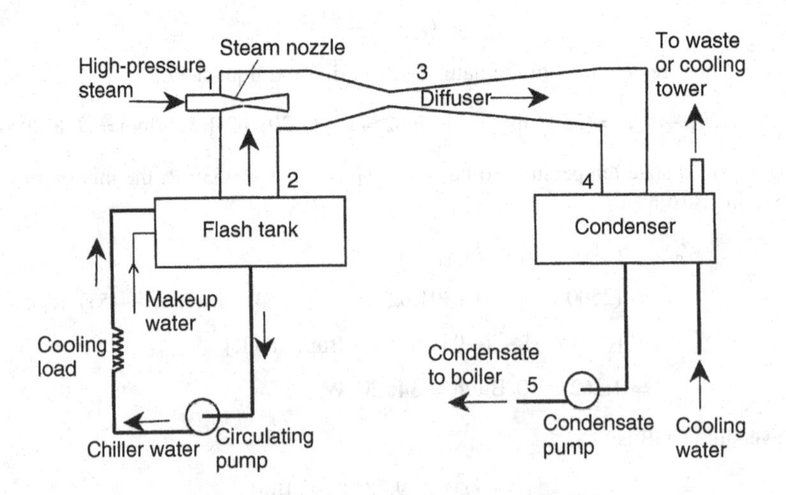

Figure 5.26 A steam jet refrigeration system.

flowing through the nozzle 1. The expansion causes a drop in pressure and an enormous increase in velocity. Owing to the high velocity, flash vapor from the tank 2 is drawn into the swiftly moving steam and the mixture enters the diffuser 3. The velocity is gradually reduced in the diffuser but the pressure of the steam at the condenser 4 is increased 5–10 times more than that at the entrance of the diffuser (e.g., from 0.01 to 0.07 bar).

This pressure value corresponds to the condensing temperature of 40 °C. This means that the mixture of high-pressure steam and the flash vapor may be liquefied in the condenser. The latent heat of condensation is transferred to the condenser water, which may be at 25 °C. The condensate 5 is pumped back to the boiler, from which it may again be vaporized at a high pressure. The evaporation of a relatively small amount of water in the flash tank (or flash cooler) reduces the temperature of the main body of water. The cooled water is then pumped as the refrigeration carrier to the cooling-load heat exchanger.

An ejector was invented by Sir Charles Parsons around 1901 for removing air from steam engine condensers. In about 1910, the ejector was used by Maurice Leblanc in the steam ejector refrigeration system. It experienced a wave of popularity during the early 1930s for air conditioning large buildings. Steam ejector refrigeration cycles were later supplanted by systems using mechanical compressors. Since that time, development and refinement of ejector refrigeration systems have been almost at a standstill as most efforts have been concentrated on improving vapor compression cycles (Aphornratana *et al.*, 2001).

Furthermore, another typical gas-driven ejector is shown schematically in Figure 5.27a. High-pressure primary fluid (P) enters the primary nozzle, through which it expands to produce a low-pressure region at the exit plane (1). The high-velocity primary stream draws and entrains the secondary fluid (S) into the mixing chamber. The combined streams are assumed to be completely mixed at the end of the mixing chamber (2) and the flow speed is supersonic. A normal shock wave is then produced within the mixing chamber's throat (3), creating a compression effect, and the flow speed is reduced to a subsonic value. Further compression of the fluid is achieved as the mixed stream flows through the subsonic diffuser section (b).

Figure 5.27b shows a schematic diagram of an ejector refrigeration cycle. It can be seen that a boiler, an ejector, and a pump are used to replace the mechanical compressor of a conventional system. High-pressure and high-temperature refrigerant vapor is evolved in a boiler to produce the primary fluid for the ejector. The ejector draws vapor refrigerant from the evaporator as its secondary fluid. This causes the refrigerant to evaporate at low pressure and to produce useful refrigeration. The ejector exhausts the refrigerant vapor to the condenser where it is liquefied. The liquid refrigerant accumulated in the condenser is returned to the boiler via a pump while the

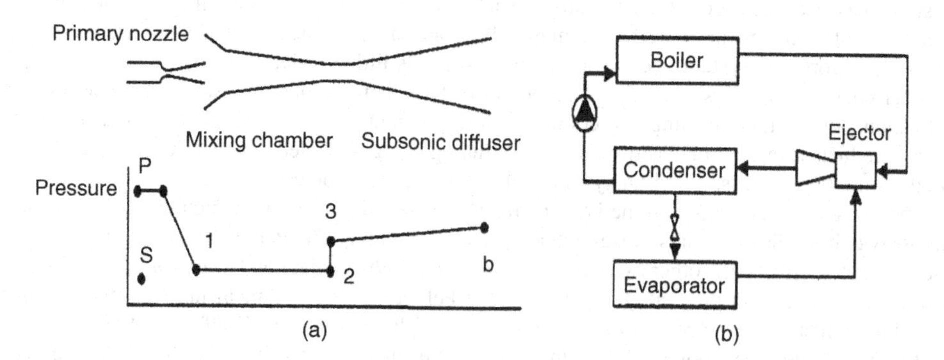

Figure 5.27 Schematic of (a) a jet ejector and (b) a simple jet ejector refrigeration system (Aphornratana *et al.*, 2001).

remainder is expanded through a throttling valve to the evaporator, thus completing the cycle. As the working input required to circulate the fluid is typically less than 1% of the heat supplied to the boiler, the COP may be defined as the ratio of evaporator refrigeration load to heat input to the boiler as follows:

$$\mathrm{COP} = \frac{\dot{Q}_L}{\dot{Q}_\mathrm{B}} \tag{5.66}$$

where $\dot{Q}_L$ is evaporator refrigeration load in kW and $\dot{Q}_\mathrm{B}$ is heat input to the boiler in kW.

In the past, Aphornratana *et al.* (2001) have developed a new jet ejector refrigeration system using R-11 as the refrigerant as shown in Figure 5.28. All vessels in the systems were constructed from galvanized steel. The boiler was designed to be electrically heated, with two 4 kW electric heaters being located at the lower end. At its upper end, three baffle plates were welded to the vessel to prevent liquid droplets being carried over with the refrigerant vapor. The evaporator design was similar to that of the boiler. A single 3 kW electric heater was used to simulate a cooling load. A water-cooled plate type heat exchanger was used as a condenser. Cooling water was supplied at 32 °C. The boiler was covered with 40 mm thickness of glass wool with aluminum foil backing. The evaporator was covered with 30 mm thickness of neoprene foam rubber. A diaphragm pump was used to circulate liquid refrigerant from the receiver tank to the boiler and the evaporator. The pump was driven by a variable-speed 1/4 hp motor. One drawback of using the diaphragm pump is cavitation of the liquid refrigerant in the suction line due to pressure drop through an inlet check valve. Therefore, a small chiller was used to subcool the liquid R-11 before entering the pump. Figure 5.28c shows a detailed schematic diagram of the experimental ejector. The nozzle was mounted on a threaded shaft, which allowed the position of the nozzle to be adjusted. A mixing chamber with throat diameter of 8 mm was used: in the inlet section of the mixing chamber, the mixing section is a constant are a duct while in the exit section, the mixing section is a convergent duct.

Aphornratana *et al.* experiments showed that an ejector refrigeration system using R-11 proved to be practical and could provide reasonably acceptable performance. It can provide a cooling temperature as low as −5 °C. The cooling capacity ranged from 500 to 1700 W with COP ranging from 0.1 to 0.25.

5.6 Thermoelectric Refrigeration

This type of system is used to move heat from one area to another by the use of electrical energy. The electrical energy, rather than the refrigerant, serves as a "carrier." The essential use of thermoelectric systems has been in portable refrigerators, watér coolers, cooling of scientific apparatus used in space exploration, and in aircraft. The main advantage of this system is that there are no moving parts. Therefore, the system is compact, quiet, and needs little service.

Thermoelectric coolers are solid state equipment used in applications where temperature stabilization, temperature cycling, or cooling below ambient temperature are required. There are many products using thermoelectric coolers, including charge-coupled device (CCD) cameras, laser diodes, microprocessors, blood analyzers, and portable picnic coolers.

Thermoelectrics are based on the Peltier Effect, discovered in 1834, by which DC current applied across two dissimilar materials causes a temperature differential. The Peltier effect is one of the three thermoelectric effects, the other two being known as the *Seebeck effect* and *Thomson effect*. Whereas the last two effects act on a single conductor, the Peltier effect is a typical junction phenomenon. The three effects are connected to each other by a simple relationship (Godfrey, 1996).

The typical thermoelectric module is manufactured using two thin ceramic wafers with a series of P- and N-doped bismuth telluride semiconductor materials sandwiched between them. The ceramic material on both sides of the thermoelectric adds rigidity and the necessary electrical insulation.

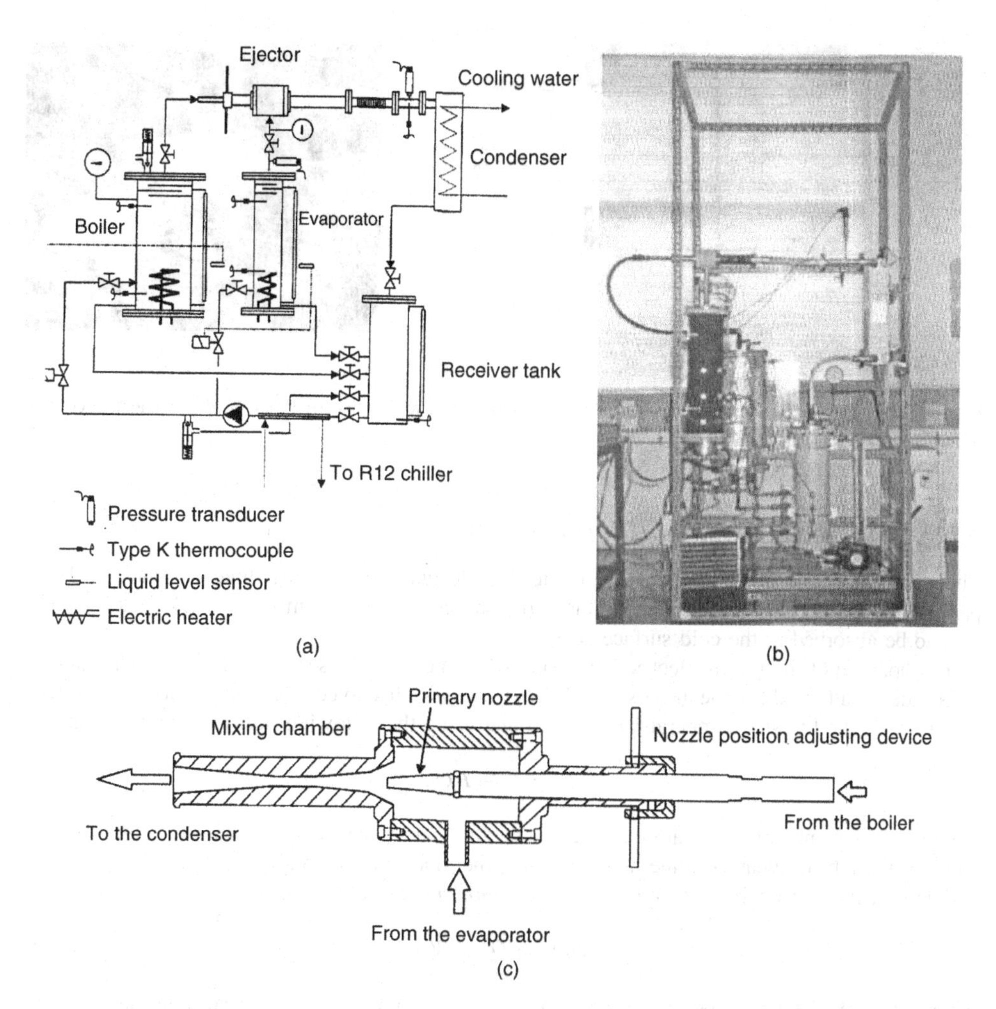

Figure 5.28 (a) Schematic of the experimental ejector refrigerator. (b) Photograph of the experimental refrigerator. (c) The ejector used in the experimental set-up (Aphornratana *et al.*, 2001).

The N type material has an excess of electrons, while the P type material has a deficit of electrons. One P and one N make up a couple, as shown in Figure 5.29. The thermoelectric couples are electrically in series and thermally in parallel. A thermoelectric module can contain one to several hundred couples. As the electrons move from the P type material to the N type material through an electrical connector, the electrons jump to a higher energy state absorbing thermal energy (cold side). Continuing through the lattice of material, the electrons flow from the N type material to the P type material through an electrical connector, dropping to a lower energy state and releasing energy as heat to the heat sink (hot side).

Thermoelectrics can be used to heat and to cool, depending on the direction of the current. In an application requiring both heating and cooling, the design should focus on the cooling mode. Using a thermoelectric in the heating mode is very efficient because all the internal heating (Joulian heat) and the load from the cold side is pumped to the hot side. This reduces the power needed to achieve the desired heating.

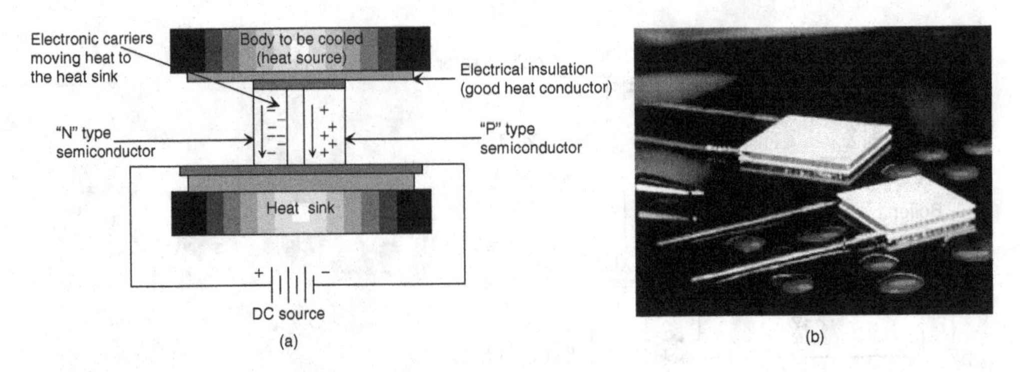

Figure 5.29 (a) Cross-sectional view of a typical thermoelectric cooler. (b) Practical thermoelectric coolers (*Courtesy of Melcor Corporation*).

5.6.1 Significant Thermal Parameters

The appropriate thermoelectric for an application depends on at least three parameters. These parameters are the hot surface temperature (T_h), the cold surface temperature (T_c), and the heat load to be absorbed at the cold surface ($\dot{Q}_c$).

The hot side of the thermoelectric is the side where heat is released when DC power is applied. This side is attached to the heat sink. When using an air-cooled heat sink (natural or forced convection), the hot side temperature can be found using the following heat-transfer equation:

$$T_h = T_a + R\dot{Q}_h \tag{5.67}$$

where T_a is the ambient temperature in °C, R is the thermal resistance of the heat exchanger (°C/W), and $\dot{Q}_h$ is the heat released to the hot side of the thermoelectric in W.

The heat-transfer balance equation for the thermoelectric cooler becomes

$$\dot{Q}_h = \dot{Q}_c + \dot{W} \tag{5.68}$$

where $\dot{Q}_c$ is the heat absorbed from the cold side in W and $\dot{W}$ is the electrical input power to the thermoelectric cooler in W. The COP of a thermoelectric refrigerator is defined as

$$\text{COP} = \frac{\dot{Q}_c}{\dot{W}} \tag{5.69}$$

Note that the thermal resistance of the heat sink causes a temperature rise above ambient temperature. If the thermal resistance of the heat sink is unknown, then estimates of acceptable temperature rise above ambient temperature are as follows (Godfrey, 1996):

- Natural convection: 20 to 40 °C
- Forced convection: 10 to 15 °C
- Liquid cooling: 2 to 5 °C (rise above the liquid coolant temperature).

The heat sink is a key component in the assembly. A heat sink that is too small means that the desired cold side temperature may not be obtained. The cold side of the thermoelectric is the side that gets cold when DC power is applied. This side may need to be colder than the desired temperature of the cooled object. This is especially true when the cold side is not in direct contact with the object, such as when cooling an enclosure.

The temperature difference across the thermoelectric relates to T_h and T_c as follows:

$$\Delta T = T_h - T_c \tag{5.70}$$

In a recent paper, Godfrey (1996) studied the thermoelectric performance curves and the relationship between the temperatures and the other parameters, as well as other parameters that are required to calculate the thermal loads for the design. The thermal loads can be classified as follows:

- **Active loads.** I^2R heat load from the electronic devices and any load generated by a chemical reaction
- **Passive loads.** Radiation (heat loss between two close objects with different temperatures), convection (heat loss through the air where the air has a different temperature than the object), insulation losses, conduction losses (heat loss through leads, screws, etc.), and transient load (time required to change the temperature of an object).

It is also important that all thermoelectrics are rated for I_{max}, V_{max}, Q_{max}, and T_{max}, at a specific value of T_h. Operating at or near the maximum power is relatively inefficient because of internal heating (Joulian heat) at high power. Therefore, thermoelectrics generally operate within 25 to 80% of the maximum current. The input power to the thermoelectric determines the hot side temperature and cooling capability at a given load. As the thermoelectric operates, the current flowing through it has two effects: (i) the Peltier effect (cooling) and (ii) the Joulian effect (heating). The Joulian effect is proportional to the square of the current. Therefore, as the current increases, the Joulian heating dominates the Peltier cooling and causes a loss in net cooling. This cut-off defines I_{max} for the thermoelectric. In fact, for each device, Q_{max} is the maximum heat load that can be absorbed by the cold side of the thermoelectric. This maximum occurs at I_{max} and V_{max}, and with $\Delta T = 0\,°C$. The T_{max} value is the maximum temperature difference across the thermoelectric. This maximum occurs at I_{max} and V_{max} and with no load ($Q_c = 0$ W). These values of Q_{max} and T_{max} are well treated by Godfrey (1996).

Example 5.6

Suppose we have a thermoelectric application with a forced convection type heat sink with a thermal resistance of 0.15 °C/W, an ambient temperature of 25 °C, and an object that needs to be cooled to 5 °C. The cold side of the thermoelectric will be in direct contact with the object. The hot side temperature is 35 °C and the electric current and voltage are 3.6 A and 10 V, respectively. Determine the temperature difference across the thermoelectric ΔT and the heat absorbed from the cold side $\dot{Q}_c$. Also, determine the COP of the system.

Solution

The temperature difference across the thermoelectric is

$$\Delta T = T_h - T_c = 35 - 5 = \mathbf{30\,°C}$$

The heat released to the hot side of the thermoelectric is

$$T_h = T_a + R\dot{Q}_h \Rightarrow \dot{Q}_h = \frac{T_h - T_a}{R} = \frac{(35-25)\ °C}{0.15\ °C/W} = 66.7\ W$$

Then the heat absorbed from the cold side becomes

$$\dot{Q}_c = \dot{Q}_h - \dot{W} \Rightarrow \dot{Q}_c = \dot{Q}_h - IV = 66.7\ W - (3.6\ A)(10\ V)\left(\frac{1\ W}{1\ AV}\right) = \mathbf{30.7\ W}$$

The COP of the system is

$$\mathrm{COP} = \frac{\dot{Q}_\mathrm{c}}{\dot{W}} = \frac{30.7\ \mathrm{W}}{36\ \mathrm{W}} = \mathbf{0.853}$$

5.7 Thermoacoustic Refrigeration

Garrett and Hofler (1992) pointed out that two recent events are responsible for the *new era* in refrigeration before the beginning of the twenty-first century. The most significant of these is the international agreement (signing of the Montreal Protocol) on the production and consumption of chlorofluorocarbons (CFCs), which were found to be causing the depletion of the stratospheric ozone layer. The second event was the discovery of "high-temperature" superconductors and the development of high-speed and high-density electronic circuits, which require active cooling and hence a new approach to refrigeration, or thermoacoustic refrigeration, which was first discovered by Wheatley *et al.* (1993) in August 1983. The simplicity of the hardware involved in thermoacoustic machines is best appreciated by examining a concrete example. In the mid-1990s, S.L. Garrett and his colleagues at the Naval Postgraduate School in Monterey, California, developed two thermoacoustic refrigerators for the Space Shuttle. The first was designed to cool electronic components and the second was intended to replace the refrigerator-freezer unit used to preserve blood and urine samples from astronauts engaged in biomedical experiments (Garrett and Backhaus, 2000).

Thermoacoustic refrigeration is considered a new technology, attaining cooling without the need for refrigerants. The basic mechanism is very simple and efficient. A loudspeaker creates sound in a hollow tube which is filled with an ordinary gas. In fact, thermoacoustic refrigeration utilizes high-density sound waves to transfer heat due to the thermoacoustic effect (i.e., acoustic energy). Therefore, the working fluid in this system is acoustically driven gas. The process itself utilizes standing acoustic waves in an enclosed cavity to generate the mechanical compression and expansion of the working fluid (gas in this case) needed for the cooling cycle. The technique has the potential for high-efficiency operation without the need for cooling liquids or mechanical moving parts. These factors make the concept amenable to miniaturization to chip-scale dimensions for thermal management of electronic components.

The interaction between acoustics and thermodynamics has been known ever since the dispute between Newton and Laplace over whether the speed of sound was determined by the adiabatic or isothermal compressibility of air. At the present time, the efficiency of thermoacoustic refrigerators is 20–30% lower than their vapor-compression refrigerators. Part of that lower efficiency is due to the intrinsic irreversibilities of the thermoacoustic heat transport process. These intrinsic irreversibilities are also the favorable aspects of the cycle, since they make for mechanical simplicity, with few or no moving parts. A greater part of the inefficiency of current thermoacoustic refrigerators is simply due to technical immaturity. With time, improvements in heat exchangers and other subsystems should narrow the gap. It is also likely that the efficiency in many applications will improve only because of the fact that thermoacoustic refrigerators are well suited to proportional control. One can easily and continuously control the cooling capacity of a thermoacoustic refrigerator so that its output can be adjusted accurately for varying load conditions. This could lead to higher efficiencies than for conventional vapor-compression chillers which have constant displacement compressors and are therefore only capable of binary (on/off) control. Proportional control avoids losses due to the start-up surges in conventional compressors and reduces the inefficiencies in the heat exchangers, since such systems can operate over smaller temperature gaps between the coolant fluid and the heat load.

The research focus of the Thermoacoustics Laboratory in ARL at Pennsylvania State University in cooperation with Los Alamos Research Laboratory is the study of acoustically driven heat transport. Their goals include an improved understanding of fundamental thermoacoustic processes and the development of new thermoacoustic refrigerators and heat engines with increased power

(a)

(b)

Figure 5.30 (a) A thermoacoustic refrigerator and (b) its application to a refrigerator (*Courtesy of Pennsylvania State University Applied Research Laboratory*).

density, temperature span, and efficiency, and the commercialization of those devices. The laboratory provides the infrastructure to support research on the basic processes required to understand this emerging, environmentally friendly refrigeration technology. This facility also supports the fabrication and testing required to produce complete, full-scale operational prototype refrigeration systems for military and commercial applications such as food refrigerators/freezers and air conditioners. Their prototypes have been flown on the Space Shuttle and have been used to cool radar electronics onboard a US Navy warship. Thermoacoustic refrigerators with cooling powers ranging from a few watts to chillers with cooling capacities in excess of 10 kW are currently in operation or under construction. Figure 5.30a shows a thermoacoustic refrigerator developed by this laboratory and it is operational for running a small refrigerator such as in Figure 5.30b.

Although thermoacoustic refrigerators have not been commercialized yet and are considered to be still at a developmental stage, it is known that they can be used for any kind of cooling. Conventional, single-stage, electrically operated thermoacoustic refrigerators can reach cold side temperatures that are two-thirds to three-quarters of ambient temperature, so they are not well suited to cryogenic applications below −40 °C. However, thermoacoustically driven pulse-tube style refrigerators can reach the cryogenic temperatures required to liquefy air or natural gas. In their early commercial stages, they will probably be limited to niche applications such as in military systems which are required to operate in closed environments and food merchandizing where toxicity is an important issue. As global environmental mandates and legislations/amendments become essential, one can expect the scope of thermoacoustic applications to expand both domestically and in emerging markets.

5.8 Metal Hydride Refrigeration Systems

For the first time, a group of Japanese companies (JNT, 1996) have succeeded in making operational an innovative, CFC-free, metal hydride (MH) refrigeration system using hydrogen absorbing alloys (MH alloys) for cold storage at low temperatures. Their state-of-the-art MH refrigeration system can keep the temperature in a cold storage area below −30 °C. This was a real landmark in the field. The joint R&D group in 1995 demonstrated an MH refrigeration system under test conditions by cooling a 100 m^3 cold storehouse. They succeeded in continuously operating the system with a store room temperature below −30 °C. The MH system can be made as compact in size as a conventional vapor-compression refrigeration system. The system can be incorporated easily, therefore, into automatic vending machines and display cabinets for frozen foods. At the end of the year, the group also completed a trial model of an automatic vending machine equipped with an MH refrigeration system which can be used for commercial operation once the system size has been reduced. In addition, the system is safe since hydrogen is absorbed and stored in MH alloys. Ammonia absorption refrigerating machines have also been proposed as an alternative CFC-free system for cold storage. However, it is not possible to make ammonia systems as small in size as MH systems, and the strong toxicity and highly irritating odor of ammonia are serious obstacles to their widespread use.

JNT (1996) stated that the MH refrigeration system is a very safe as well as clean and environmentally friendly, CFC-free refrigeration system. Hydrogen is sealed in gas-tight cylinders, and, being far lighter than air, rapidly diffuses into the atmosphere in the accidental event of its leakage. Thus, the danger of explosion caused by hydrogen is slight. In addition, their system has additional advantages as follows:

- By not using CFCs or ammonia, the system is applicable to wide-ranging uses.
- The system needs a heating source to generate energy for refrigeration, but can save energy consumption for this purpose by the use of waste heat or by operating in combination with a cogeneration system.
- It has no moving elements except for pumps circulating water and brine. In particular, its hydrogen system, driven solely by heat input, does not need the manipulation of valves. Thus, rarely suffering breakdowns, the MH system is easy and simple to operate and maintain.
- Driven by heat, the system consumes 20% less electric power than conventional, electrically powered compression refrigeration machines.
- Having no sliding and vibrating components such as compressors, the system operates with low noise levels.
- Because the MH refrigeration system is safe and simple to control, it is possible to design refrigerating units using this technology with wide varieties of cooling capacities ranging from 10–10,000 kW.

5.8.1 Operational Principles

When MH alloys (e.g., TiZrCrFe series) come into contact with hydrogen, the alloys absorb hydrogen by an exothermic reaction and store it as MHs. In reverse, the alloys easily dissociate and discharge hydrogen by an endothermic reaction. Employing this endothermic reaction when MH alloys discharge hydrogen, MH refrigeration systems implement a refrigeration cycle by a combination of two types of alloys. One type works at a higher temperature and the other at a lower temperature, both under their own equilibrium hydrogen pressure. The working principles of MH refrigeration systems are illustrated in Figure 5.31.

MH alloys absorb or discharge hydrogen at certain constant equilibrium hydrogen pressure levels, determined by temperature. MH refrigeration systems use an MH alloy (MH-A) driving hydrogen to carry out the regeneration process on the high-temperature side of the system and another MH alloy (MH-B) refrigerating brine on the low-temperature side. Each of these alloys has the relation between temperatures and equilibrium hydrogen pressures as shown in Figure 5.31a.

Regeneration Process. (1) To raise the hydrogen pressure of the MH-A by raising its temperature, the alloy is heated (Q2-1). Hydrogen is discharged from the MH-A and moves to the MH-B with lower hydrogen pressure. (2) The MH-B absorbs hydrogen, thus generating heat. However,

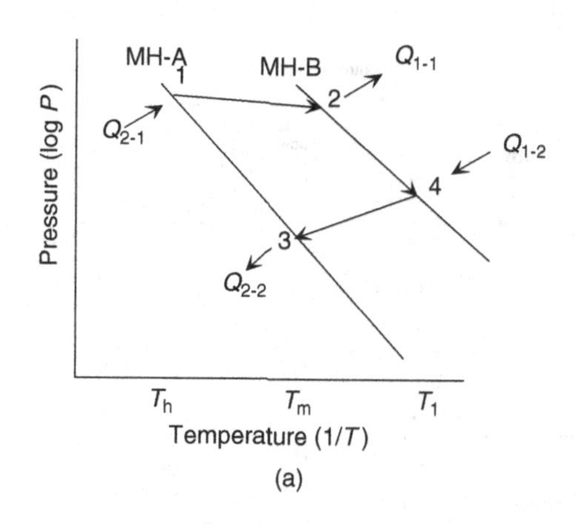

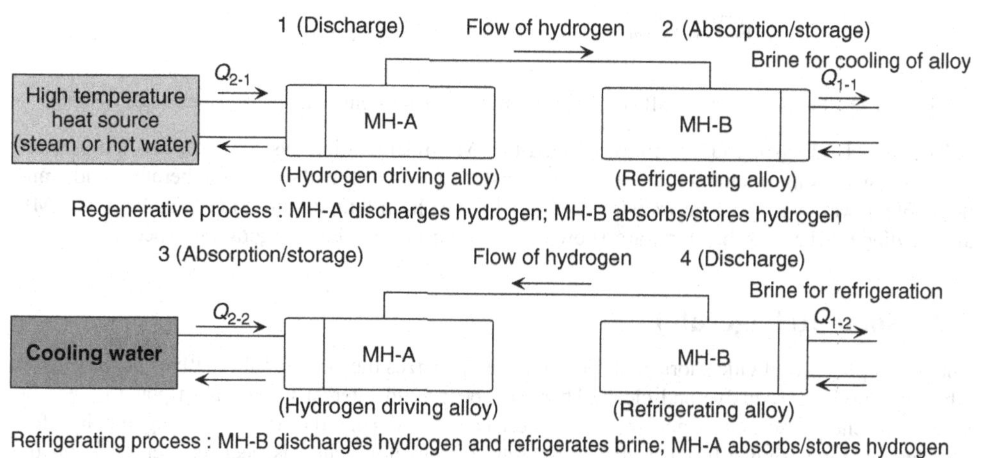

Figure 5.31 (a) Working principle of MH refrigeration system and (b) illustration of MH refrigeration system principle (JNT, 1996).

the circulation of cooling brine (Q1-1) suppresses the rise in the MH-B temperature, thereby preventing the pressure of the MH-B from increasing. The MH-B continues to absorb and store hydrogen in this way.

Refrigeration Process. (3) When all the hydrogen of the MH-A is transferred to the MH-B, the former alloy is cooled by cooling water (Q2-2), thereby lowering its hydrogen pressure. (4) Hydrogen is discharged from the MH-B. This has a lower hydrogen pressure than that of the MH-A and moves to the latter alloy. Reducing its temperature by the hydrogen discharge, the MH-B refrigerates brine (Q1-2).

To continuously cool brine for refrigeration, an MH refrigeration system has two sets of high- and low-temperature MH alloy pairs. While one set of the alloys operates in the regeneration process, the other set works in the refrigeration process. Each of these MH alloys is packed in a heat exchanger cylinder. This enables the exchange of thermal energy with a heating medium (steam, hot water, etc.), cooling water, or brine. Shell and tube type heat exchangers are used.

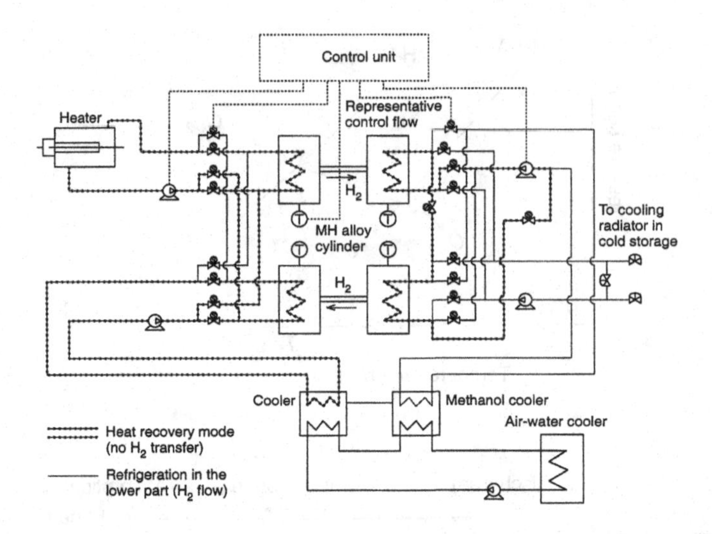

Figure 5.32 Flow chart of the MH refrigeration system for low-temperature cold storage (JNT, 1996).

The joint R&D group (JNT, 1996) applied this MH refrigeration system for a low-temperature cold storage, using methanol as the heat-transfer medium for the low-temperature side and hot/cooling water as the heat-transfer medium for the high-temperature side (Figure 5.32). MH alloy cylinders (heat exchangers) are shown as forming part of the refrigerating process.

5.9 Solar Refrigeration

The developing worldwide shortage of petroleum emphasizes the need for alternative energy sources which are both inexpensive and clean. There has been high interest in, and high potential use of, renewable energy sources since the energy crisis faced during the 1970s. During the last few decades, an increasing effort based on research and development has been concentrated on the utilization of renewable energy sources, for example, solar energy, wind energy, tidal waves, biogas, geothermal energy, hydropower, and hydrogen energy. Among these sources, solar energy for refrigeration applications is very popular because it is direct and easy to use, renewable, and continuous, maintains the same quality, is safe and free, and is environmentally friendly.

The continuous supply of solar energy to the earth's surface is equal to a power of about 100,000 TW. Approximately one-third of the radiation impinging on land area and accumulated over less than 2 hours should suffice to satisfy the entire primary energy demand by humans for the period of 1 year (Dincer, 1997; 2003). More than 25% of the total energy in the world is consumed for heating and cooling of buildings and providing hot water. Therefore, the diversion of this particular energy demand to an alternative source would result in a substantial reduction in the world's dependence on fossil fuels. The annual incidence of solar energy on buildings in the United States is several times the amount required to heat these buildings; approximately 10^{15} kW·h of solar energy is received on earth annually. It has been projected that by the year 2020 from 25 to 50% of the thermal energy for buildings could be provided from the sun (Dincer *et al.*, 1996). Consequently, solar energy is an available energy source for many applications ranging from electricity generation to food cooling.

5.9.1 Solar Refrigeration Systems

Many food products (e.g., fruits, vegetables, meats, dairy products) are stored in cooling units for periods of the order of weeks at temperatures between 0 and 4 °C in order to prevent spoilage and maintain freshness and quality. Food freezing systems are required for longer term storage at −18 to −35 °C. Food storage and transport take place in chambers covering a wide range of sizes from cold stores to household refrigerators. Solar cooling is of great interest especially in developing countries, where food preservation is often as difficult a problem as food production.

From an energy-saving view, a solar cooling system has the capability of saving electrical energy in the range of 25–40% when compared to an equivalent cooling capacity of a conventional water-cooled refrigeration system. Therefore, the use of solar cooling systems will save energy, especially during the summer season. The contribution of these systems to the food processing sector and consequently to the economy will be high (Dincer, 1997).

Solar-powered mechanical cooling, of whatever type, is presently in the developmental phase. The technology is ready, but cost factors stand in the way of vigorous marketing programs. At present, active solar cooling is not in a reasonably competitive position with respect to conventional cooling systems (energized by electricity or fossil fuel). During the last decade, the situation has changed quickly because of increasing interest in renewable energy sources, especially solar energy, for reducing the use of fossil fuels and electricity.

Solar energy can be used in different systems available for cooling applications. These systems are (Dincer and Dost, 1996) as follows:

- Rankine cycle vapor-compression system,
- absorption cycle system,
- adsorption system,
- jet ejector system,
- Rankine cycle-inverse Brayton cycle system,
- nocturnal radiation system.

Among these systems, the solar-powered absorption cooling cycle is the most popular system for solar cooling applications because of the following advantages (Dincer and Dost, 1996):

- quiet operation,
- high reliability,
- long service life,
- effective and economic use of low-grade energy sources (e.g., solar energy, waste energy, geothermal energy, natural gas),
- easy implementation and capacity control,
- no cycling losses during on–off operations, and
- meeting the variable cooling load easily and efficiently.

5.9.2 Solar-Powered Absorption Refrigeration Systems (ARSs)

Solar energy is a renewable and ozone-friendly energy source. Solar cooling is the most attractive subject for many engineers and scientists who work on solar energy applications. Most of the research and development efforts have been carried out using an absorption cooling system. This system is usually a preferable alternative, since it uses thermal energy collected from the sun without the need to convert this energy into mechanical energy as required by the vapor-compression system. Besides, the absorption cooling system utilizes thermal energy at a lower temperature (i.e., in the range 80–110 °C) than that used by the vapor-compression system.

Research and development studies on solar absorption refrigeration systems (ARSs) using different combinations of refrigerants and absorbents as working fluids have been done. These ARSs have good potential where solar energy is available as low-grade thermal energy at a temperature level of 100 °C and above.

The principle of operation of a solar-powered absorption cooling system is the same as that of the absorption cooling system shown in Figure 4.33, except for the heat source to the generator. In Figure 4.33, we presented a solar absorption cooling system using an R-22 (refrigerant)-DMETEG (absorbent) combination as a working fluid (Dincer *et al.*, 1996). Its operation can be briefly explained as follows. In the absorber, the DMETEG absorbs the R22 at the low pressure and absorber temperature supplied by circulating water, and hence a strong solution occurs (2). This strong solution from the absorber enters a solution pump, which raises its pressure and delivers the solution into the generator through the heat exchanger (3–6). The generator, which is heated by a solar hot water system, raises the temperature of the strong solution, causing the R-22 to separate from it. The remaining weak solution flows down to the expansion valve through the heat exchanger and is throttled into the absorber for further cooling as it picks up a new charge of the R22 vapor, becoming a strong solution (6–2) again. The hot R-22 vapor from the generator passes to the condenser and is released to the liquid phase (8–9). The liquid R-22 enters the second heat exchanger and loses some heat to the cool R-22 vapor. The pressure of the liquid R-22 drops significantly in the throttling valve before it enters the evaporator. The cycle is completed when the desired cooling load is achieved in the evaporator (10–12). Cool R-22 vapor obtained from the evaporator enters the absorber while the weak solution comes to the absorber continuously. The R-22 vapor is absorbed here (12–1). This absorption activity lowers the pressure in the absorber, causing the vapor to be taken off from the evaporator. When the vapor goes into liquid solution, it releases both its latent heat and a heat of dilution. This energy release has to be continuously dissipated by the cooling water.

Solar-operated ARSs have so far achieved limited commercial viability because of their high cost–benefit ratios. The main factor which is responsible for this drawback is the low COP associated with these systems, which generally operate on conventional thermodynamic cycles with common working fluids. It is essential to investigate the possibility of using alternative working fluids operating in new thermodynamic cycles. Also, development of more efficient, less expensive solar collectors will be a continuing need for solar energy to reach its full potential.

5.10 Magnetic Refrigeration

Magnetic refrigeration is a cooling technology based on the magnetocaloric effect. This technique can be used to attain extremely low temperatures (well below 1 K) as well as the ranges used in common refrigerators, depending on the design of the system. The magnetocaloric effect is a magneto-thermodynamic phenomenon in which a reversible change in temperature of a suitable material is caused by exposing the material to a changing magnetic field. One of the most notable examples of the magnetocaloric effect is in the chemical element gadolinium and some of its alloys. Gadolinium's temperature is observed to increase when it enters certain magnetic fields. When it leaves the magnetic field, the temperature returns to normal.

In the magnetic refrigeration cycle, depicted in Figure 5.33, initially randomly oriented magnetic moments are aligned by a magnetic field, resulting in heating of the magnetic material. This heat is removed from the material to the ambient temperature by heat transfer. On removing the field, the magnetic moments randomize, which leads to cooling of the material below ambient temperature. Heat from the system to be cooled can then be extracted using a heat-transfer medium. Depending on the operating temperature, the heat-transfer medium may be water (with antifreeze) or air, and for very low temperatures, helium.

Magnetic refrigeration is an environmentally friendly cooling technology. It does not use ozone-depleting chemicals (CFCs), hazardous chemicals (NH3), or greenhouse gases (hydrochlorofluorocarbons [HCFCs] and hydrofluorocarbons [HFCs]). Another key difference between vapor cycle refrigerators and magnetic refrigerators is the amount of energy loss incurred during the refrigeration cycle. The cooling efficiency in magnetic refrigerators working with gadolinium has been

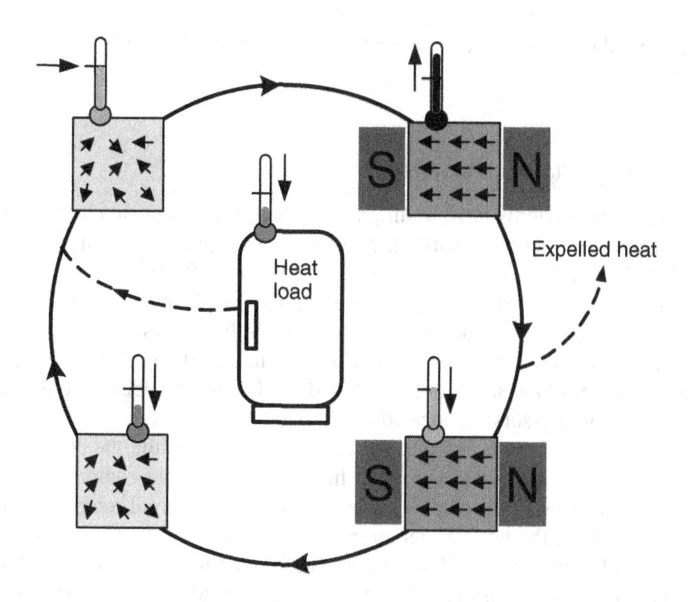

Figure 5.33 Schematic representation of a magnetic refrigeration cycle that transports heat from the heat load to the ambient. Left and right depict material in low and high magnetic field, respectively (Adapted from Bruck, 2005).

shown (Zimm *et al.*, 1988) to reach 60% of the theoretical limit, compared with only about 40% in the best gas compression refrigerators. This higher energy efficiency will also result in a reduced CO_2 release (Bruck, 2005).

Magnetic refrigeration has three prominent advantages compared with compressor-based refrigeration. First, there are no harmful gases involved; second, it may be built more compactly as the working material is a solid; and third, magnetic refrigerators generate much less noise. Recently, a new class of magnetic refrigerant materials for room temperature applications was discovered (Bruck, 2005).

This technology is still at the research stage for mainstream HVAC&R applications. Target future applications include residential central air conditioners and heat pumps, ductless unitary applications, and small chillers. In the first two of these applications, the small-scale, variable capacity chilled water system of the magnetic refrigeration cycle would replace the direct expansion refrigerant system. Ultimately, to achieve commercial success in target applications, magnetic refrigeration will need to achieve similar or superior efficiency at similar or lower cost than conventional vapor compression equipment. Consequently, much recent and ongoing research efforts have focused on enhancing the performance of prototype systems using the magnetocaloric materials and permanent magnet materials currently available, and developing magnetocaloric materials that yield greater temperature changes at lower magnetic field strengths (Dieckmann *et al.*, 2007).

5.11 Supermarket Refrigeration

An important application of refrigeration is supermarket refrigeration. Nearly all supermarkets today use ozone-depleting HCFC refrigerant, usually R-22, or a blend consisting entirely or primarily of HFCs. HCFCs and the HFCs are also greenhouse gases. Most supermarkets use direct expansion refrigeration systems. Two of the most common advanced refrigeration technologies for supermarkets are distributed system and secondary loop system. Below, we present the operation of each system briefly.

5.11.1 Direct Expansion System

Supermarket refrigeration systems have traditionally been direct expansion systems (used in about 70% of the supermarket refrigeration market). These systems typically use refrigerants R-22, R-502 (a blend of R-22 and CFC-115), R-404A (a blend of HFCs), or R-507A (a blend of HFCs).

The average emission rate of direct expansion systems is believed to be between 15 and 30%. Most of the emissions are due to leaks in the system, including leaks in the valves and compressors. In a direct expansion system, the compressors are mounted together and share suction and discharge refrigeration lines that run throughout the store, feeding refrigerant to the cases and coolers (Figure 5.34). The compressors are located in a separate machine room, either at the back of the store or on its roof, to reduce noise and prevent customer access, while the condensers are usually air-cooled and hence are placed outside to reject heat. These multiple compressor racks operate at various suction pressures to support display cases operating at different temperatures (IEA, 2003).

As shown in Figure 5.34, the hot refrigerant gas from the compressors is cooled and condensed as it flows into the condenser. The liquid refrigerant is collected in the receiver and distributed to the cases and coolers by the liquid manifold. The refrigerant is expanded turning a fraction of liquid into vapor before flowing into the evaporator. After cycling through the cases, the refrigerant returns

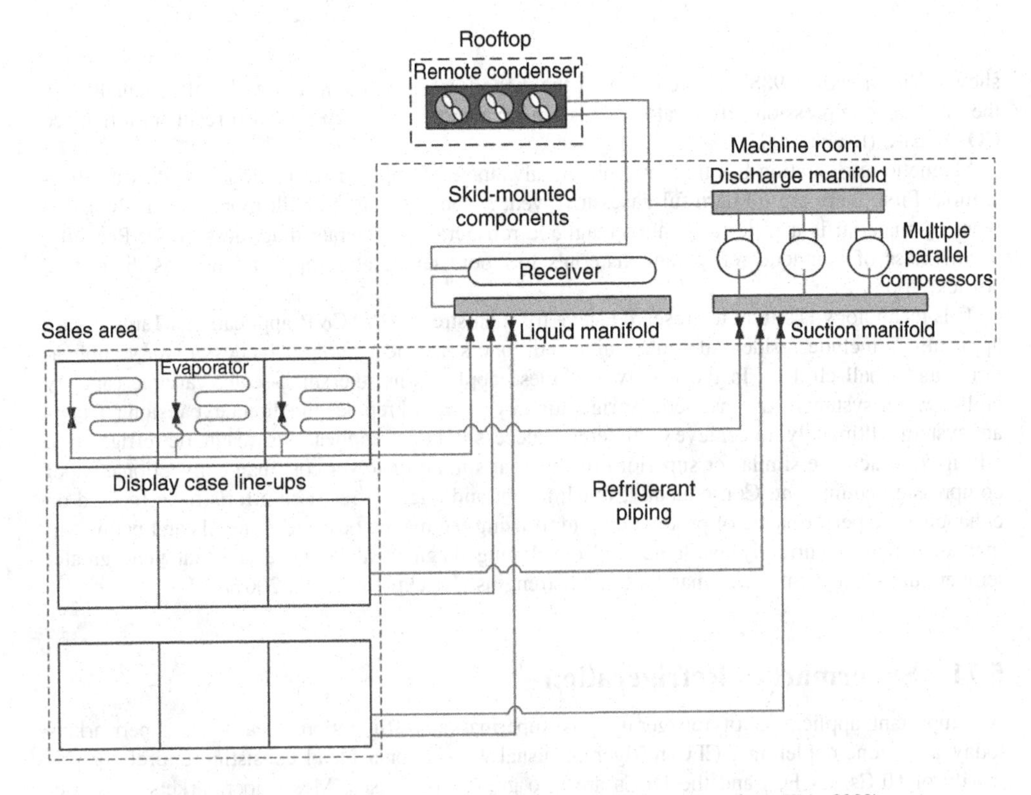

Figure 5.34 The schematic of direct expansion system (Adapted from IEA, 2003).

to the suction manifold and the compressors. Supermarkets tend to have one direct expansion system for "low-temperature" refrigeration (e.g., ice cream, frozen foods, etc.) and one or two direct expansion systems for "medium-temperature" refrigeration (e.g., meat, prepared foods, dairy, refrigerated drinks, etc.).

5.11.2 Distributed System

Unlike traditional direct expansion refrigeration systems, which have a central refrigeration room containing multiple compressor racks, distributed systems use multiple smaller rooftop units that connect to cases and coolers, using considerably less piping (Figure 5.35). The compressors in a distributed system are located near the display cases they serve, for instance, on the roof above the cases, behind a nearby wall, or even on top of or next to the case in the sales area. Thus, distributed systems typically use a smaller refrigerant charge than direct expansion systems and hence have decreased total emissions (IEA, 2003).

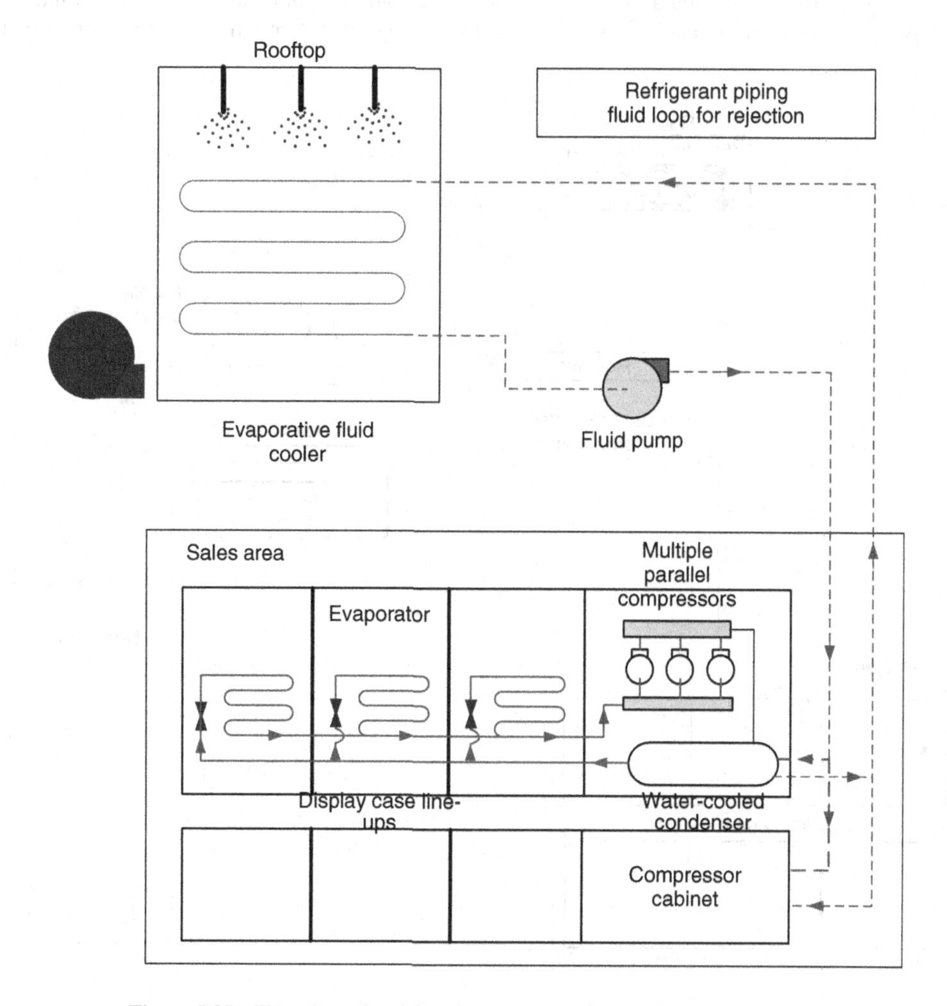

Figure 5.35 The schematic of distributed system (Adapted from IEA, 2003).

As shown in Figure 5.35, the refrigerant is compressed in multiple parallel compressors and the superheated refrigerant gas is cooled and condensed in a water-cooled condenser. The refrigerant is then expanded before entering the evaporator. It absorbs heat from the cooled products before returning to the compressors as a vapor. The water that is heated by the condensing refrigerant in the condenser is sent to an evaporative cooler. It is cooled and pumped back to the condenser to repeat the cycle.

5.11.3 Secondary Loop System

Secondary loop systems have recently seen increased introduction into retail food equipment, and now make up about 4% of the market (Figure 5.36). These systems generally use R-404A or R-507A, although some earlier systems used R-22. Their average leak rate is between 2 and 15%.

Secondary loop systems use a much smaller refrigerant charge than traditional direct expansion refrigeration systems, and hence have significantly decreased total refrigerant emissions. In secondary loop systems, two liquids are used. The first is a cold fluid, often a brine solution, which is pumped throughout the store to remove heat from the display equipment. The second is a refrigerant

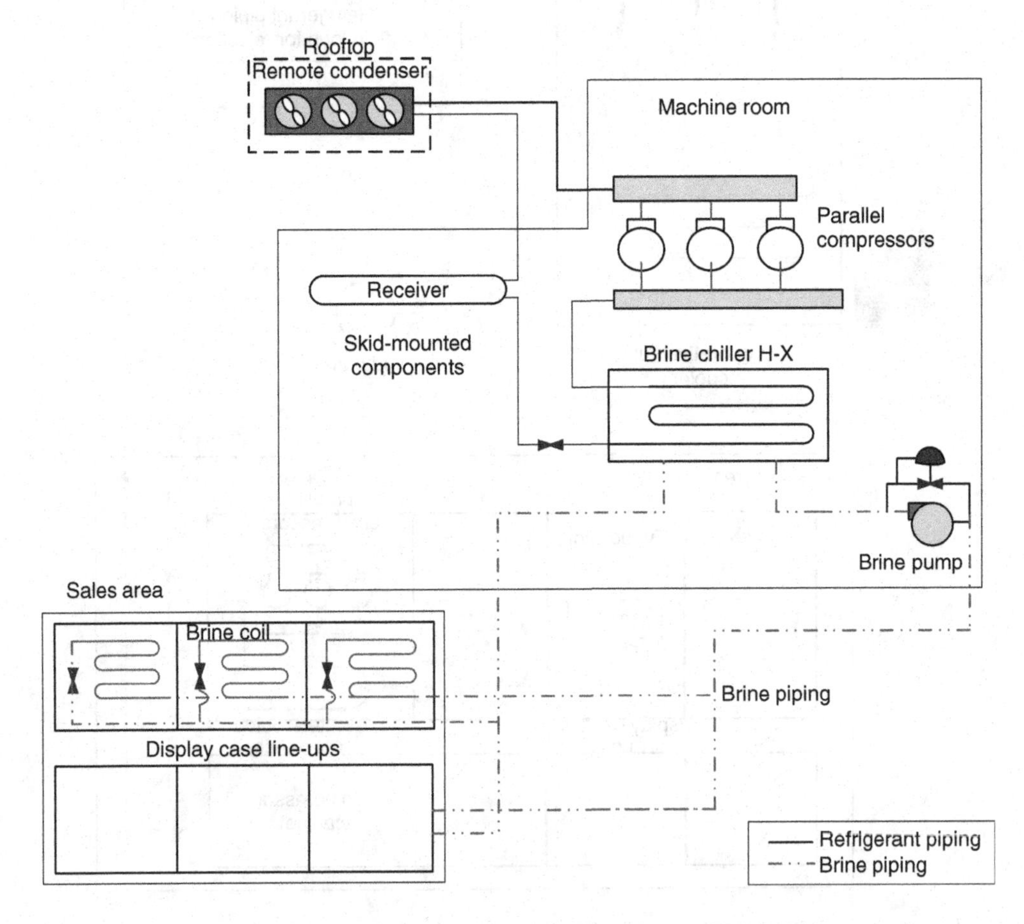

Figure 5.36 The schematic of secondary loop system (SCEFM, 2004).

used to cool the cold fluid that travels around the equipment. Secondary loop systems can operate with two to four separate loops and chiller systems depending on the temperatures needed for the display cases (SCEFM, 2004).

As shown in Figure 5.36, the refrigerant is compressed in parallel compressors and the superheated refrigerant gas is cooled and condensed in a remote condenser. The liquid refrigerant is then collected in the receiver, expanded in a throttling device, and evaporated by absorbing heat from a cold fluid (i.e., brine). The cooled brine is distributed in the sales area (refrigerated area) absorbing heat from the products before returning to the evaporator to repeat the process.

5.12 Concluding Remarks

This chapter has dealt with advanced refrigeration cycles and systems and their energy and exergy analyses. New refrigeration systems and their technical and operational details and applications are provided. Illustrative examples and some practical cases are also provided to better understand the technical details of the advanced refrigeration systems.

Nomenclature

COP coefficient of performance
c_p constant-pressure specific heat, kJ/kg·K
ex specific exergy, kJ/kg
$\dot{E}x$ exergy rate, kW
h enthalpy, kJ/kg
$\dot{m}$ mass flow rate, kg/s
P pressure, kPa
q specific heat, kJ/kg
$\dot{Q}$ heat load; power, kW
s entropy, kJ/kg
$\dot{S}_{gen}$ entropy generation rate, kW/K
T temperature, °C or K
v specific volume, m^3/kg
$\dot{V}$ volumetric flow rate, m^3/s
w specific work, kJ/kg
$\dot{W}$ work input to compressor or pump, kW
X concentration of refrigerant in solution, kg/kg

Greek Letters

η efficiency

Study Problems

Multistage and Cascade Refrigeration Cycles

5.1 Consider a two-stage cascade refrigeration system operating between the pressure limits of 1.4 MPa and 120 kPa with refrigerant-134a as the working fluid. Heat rejection from the lower cycle to the upper cycle takes place in an adiabatic counter-flow heat exchanger where the pressure is maintained at 0.4 MPa. Both cycles operate on the ideal vapor-compression refrigeration cycle. If the mass flow rate of the refrigerant through the upper cycle is 0.12 kg/s, determine (a) the mass flow rate of the refrigerant through the lower cycle, (b) the rate of heat removal from the refrigerated space, and (c) the COP of this refrigerator.

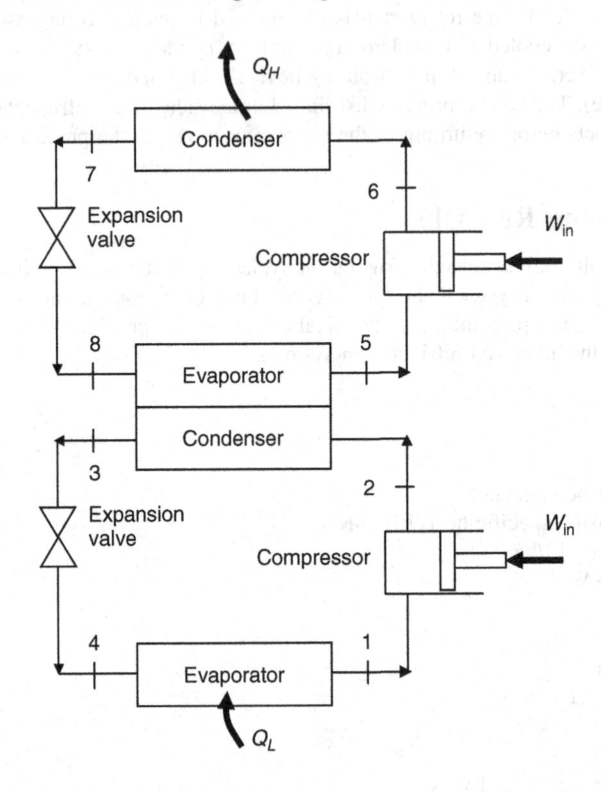

5.2 A two-stage cascade refrigeration system operates between the pressure limits of 1.6 and 0.18 MPa with refrigerant-134a as the working fluid. Heat rejection from the lower cycle to the upper cycle takes place in an adiabatic counter-flow heat exchanger where the pressure is maintained at 0.6 MPa. Both cycles operate on the ideal vapor-compression refrigeration cycle. If the refrigeration load is 11 tons, determine (a) the power input to the cycle, (b) the COP, and (c) the power input and the COP if this refrigerator operated on a single ideal vapor-compression cycle between the same pressure limits and the same refrigeration load.

5.3 Consider a two-stage cascade refrigeration system operating between the pressure limits of 1.2 MPa and 200 kPa with refrigerant-134a as the working fluid. Heat rejection from the lower cycle to the upper cycle takes place in an adiabatic counter-flow heat exchanger where the pressure in the upper and lower cycles are 0.4 and 0.5 MPa, respectively. In both cycles, the refrigerant is a saturated liquid at the condenser exit and a saturated vapor at the compressor inlet, and the isentropic efficiency of the compressor is 82%. If the mass flow rate of the refrigerant through the lower cycle is 0.06 kg/s, determine (a) the mass flow rate of the refrigerant through the upper cycle, (b) the rate of heat removal from the refrigerated space, and (c) the COP of this refrigerator.

5.4 Consider a two-stage cascade refrigeration system operating between the pressure limits of 1.2 MPa and 200 kPa with refrigerant-134a as the working fluid. The refrigerant leaves the condenser as a saturated liquid and is throttled to a flash chamber operating at 0.45 MPa. Part of the refrigerant evaporates during this flashing process and this vapor is mixed with the refrigerant leaving the low-pressure compressor. The mixture is then compressed to the condenser pressure by the high-pressure compressor. The liquid in the flash chamber is throttled to the evaporator pressure and cools the refrigerated space as it vaporizes in the evaporator. The mass flow rate of the refrigerant through the low-pressure compressor is 0.06 kg/s. Assuming the refrigerant leaves the evaporator as a saturated vapor and the isentropic efficiency is 82% for both compressors, determine (a) the mass flow rate of the refrigerant through the high-pressure compressor, (b) the rate of heat removal from the refrigerated space, and (c) the COP of this refrigerator. Also, determine (d) the rate of heat removal and the COP if this refrigerator operated on a single-stage cycle between the same pressure limits with the same compressor efficiency and the same flow rate as in part (a).

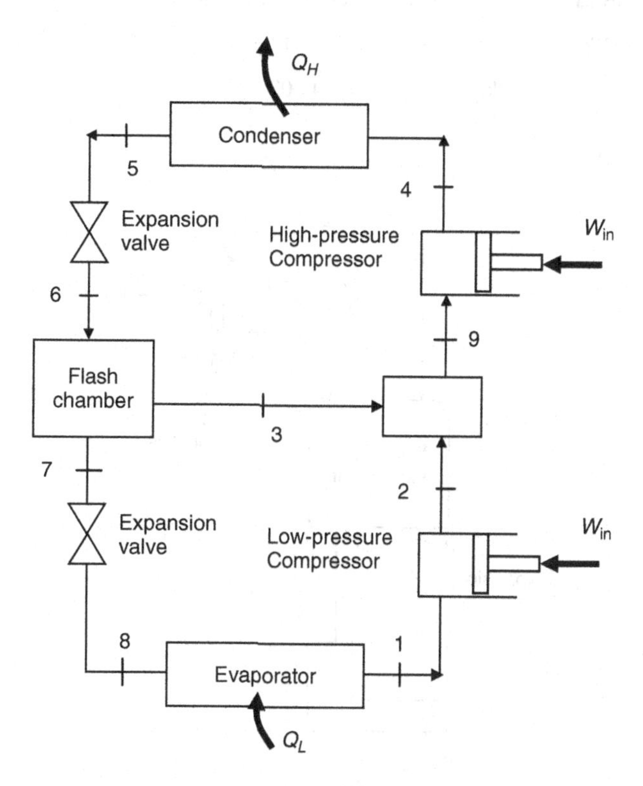

5.5 Consider a two-stage cascade refrigeration system operating between the pressure limits of 1.6 MPa and 100 kPa with refrigerant-134a as the working fluid. The refrigerant absorbs heat from a space at 0 °C and rejects heat to ambient air at 25 °C. Heat rejection from the lower cycle to the upper cycle takes place in an adiabatic counter-flow heat exchanger where the pressures in the lower and upper parts are 0.40 and 0.32 MPa, respectively. Both cycles operate on the ideal vapor-compression refrigeration cycle. If the mass flow rate of the refrigerant through the upper cycle is 0.15 kg/s, determine (a) the mass flow rate of the refrigerant through the lower cycle, (b) the rate of heat removal from the refrigerated space and the COP of this refrigerator, (c) the exergy destruction in the heat exchanger, and (d) the second-law efficiency of the cycle and the total exergy destruction in the cycle. Take $T_0 = 25\,°C$.

Liquefaction of Gases

5.6 Consider the simple Linde–Hampson cycle shown in the figure. The gas is nitrogen and it is at 25 °C and 1 atm (101.325 kPa) at the compressor inlet, and the pressure of the gas is 20 MPa at the compressor outlet. The compressor is reversible and isothermal, the heat exchanger has an effectiveness of 100% (i.e., the gas leaving the liquid reservoir is heated in the heat exchanger to the temperature of the gas leaving the compressor), the expansion valve is isenthalpic, and there is no heat leak to the cycle. Determine (a) the refrigeration load per unit mass of the gas flowing through the compressor, (b) the COP, (c) the minimum work input per unit mass of liquefaction, and (d) the exergy efficiency. Various properties of nitrogen in the cycle are given as follows:

$h_1 = -0.036$ kJ/kg $\quad s_1 = -0.000059$ kJ/kg · K

$h_2 = -32.67$ kJ/kg $\quad s_2 = -1.6806$ kJ/kg · K

$h_5 = h_g = -232.70$ kJ/kg $\quad s_6 = s_f = -4.0025$ kJ/kg · K

$h_6 = h_f = -431.54$ kJ/kg

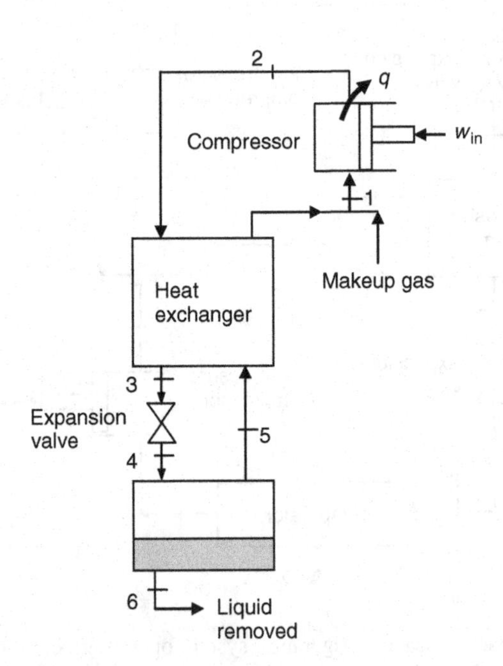

5.7 Repeat Problem 5.6 for argon gas. Various properties of argon in the cycle are given as follows:

$h_1 = -0.1933$ kJ/kg $\quad s_1 = -0.000516$ kJ/kg · K

$h_2 = -33.34$ kJ/kg $\quad s_2 = -1.1930$ kJ/kg · K

$h_5 = h_g = -111.51$ kJ/kg $\quad s_6 = s_f = -2.4985$ kJ/kg · K

$h_6 = h_f = -272.65$ kJ/kg

5.8 Consider the simple Linde–Hampson cycle shown in Problem 5.6. Air enters the compressor at 100 kPa and 15 °C at a rate of 0.36 kg/s and leaves at 15 MPa. The compressor is reversible and isothermal, the heat exchanger has an effectiveness of 100%, the expansion valve is isenthalpic, and there is no heat leak to the cycle. Determine (a) the mass flow rate of liquid withdrawn from the cycle, (b) the rate of heat removed and the power input to the cycle per unit mass of liquefaction, (c) the actual and reversible COPs, and (d) the exergy efficiency of the cycle.

Various properties of air in the cycle are given as follows:

$h_1 = 288.39$ kJ/kg $\quad s_1 = 6.8298$ kJ/kg · K

$h_2 = 257.47$ kJ/kg $\quad s_2 = 5.2933$ kJ/kg · K

$h_5 = h_g = 78.68$ kJ/kg $\quad s_6 = s_f = 2.9761$ kJ/kg · K

$h_6 = h_f = -126.31$ kJ/kg

5.9 Natural gas at 25 °C and 1 atm (101.325 kPa) is to be liquefied in a natural gas liquefaction plant. Natural gas leaves the plant at 1 atm as a saturated liquid. Using methane properties for natural gas determine (a) the heat rejection from the natural gas, (b) the minimum work input, and (c) the reversible COP. (d) If the liquefaction is done by a Carnot refrigerator between temperature limits of $T_H = 25$ °C and T_L with the same reversible COP, determine the temperature T_L.

Various properties of methane before and after liquefaction process are given as follows:

$h_1 = -0.9891$ kJ/kg

$h_2 = -910.92$ kJ/kg

$s_1 = -0.002422$ kJ/kg · K

$s_2 = -6.6767$ kJ/kg · K

5.10 Natural gas at the ambient temperature of 35 °C and 4 bar at a rate of 8500 kg/h is to be liquefied in a natural gas liquefaction plant. Natural gas leaves the plant at 4 bar and −150 °C as a liquid. Using methane properties for natural gas determine (a) the rate of heat removed from the natural gas, and (b) the minimum power input, and (c) the reversible COP. (d) If the actual power input during this liquefaction process is 13,500 kW, determine the exergy efficiency and the actual COP of this plant.

Various properties of methane before and after liquefaction process are given as follows:

$h_1 = 18.69$ kJ/kg

$h_2 = -870.02$ kJ/kg

$s_1 = -0.6466$ kJ/kg · K

$s_2 = -6.3343$ kJ/kg · K

5.11 Hydrogen gas is to be liquefied by a Carnot refrigerator as shown in the figure. Hydrogen enters at 100 kPa and 25 °C and leaves at 100 kPa as a saturated liquid. Determine (a) the heat removed from hydrogen in kJ/kg, (b) the minimum work input in kWh/kg, and (c) the reversible COP of this Carnot refrigerator. (d) If the same liquefaction process is done by an actual refrigeration system and the COP of this system is 0.035, determine the exergy efficiency and the actual work input in kWh/kg.

Various properties of hydrogen before and after liquefaction process are given as follows:

$h_1 = 4199.7$ kJ/kg

$h_2 = 270.67$ kJ/kg

$s_1 = 70.470$ kJ/kg · K

$s_2 = 17.066$ kJ/kg · K

$T_2 = -252.8$ °C

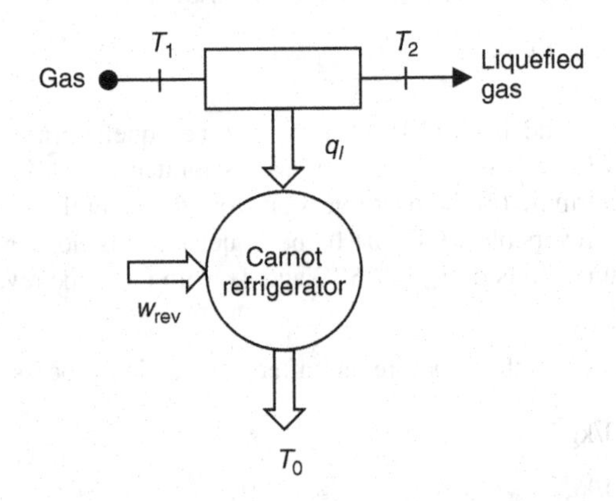

5.12 In a natural gas liquefaction plant, the LNG enters a cryogenic turbine at 4000 kPa and −160 °C at a rate of 28 kg/s and leaves at 400 kPa. The actual power produced by the turbine is measured to be 185 kW. If the density of LNG at the turbine inlet is 423.8 kg/m^3 and LNG remains as a liquid at the turbine outlet, determine the efficiency of the cryogenic turbine.

Thermoelectric Refrigeration

5.13 Consider a thermoelectric application with a forced convection type heat sink with a thermal resistance of 0.42 °C/W, an ambient temperature of 18 °C, and an object that needs to be cooled to 3 °C. The cold side of the thermoelectric will be in direct contact with the object. The hot side temperature is 40 °C and the electric current and voltage are 5.7 A and 8 V, respectively. Determine the temperature difference across the thermoelectric ΔT and the heat absorbed from the cold side $\dot{Q}_c$.

5.14 Consider a thermoelectric application with a forced convection type heat sink with a thermal resistance of 0.14 °C/W, an ambient temperature of 15 °C, and an object that needs to be cooled to 7 °C. The cold side of the thermoelectric will be in direct contact with the object. The hot side temperature is 62 °C and the rate of cooling provided is 45 W. Determine (a) the temperature difference across the thermoelectric, (b) the power input, and (c) the COP of this thermoelectric refrigerator.

References

Aphornratana, S., Chungpaibulpatana, S. and Srikhirin, P. (2001) Experimental investigation of an ejector refrigerator: effect of mixing chamber geometry on system performance. *International Journal of Energy Research*, **25**, 397–411.

ASHRAE (American Society of Heating, Refrigerating and Air-Conditioning Engineers) (2006) *Handbook of Refrigeration*, ASHRAE Inc., Atlanta.

Barron, R. (1985) *Cryogenic Systems*, Oxford University Press, New York.

Bruck, E. (2005) Developments in magnetocaloric refrigeration. *Journal of Physics D: Applied Physics*, **38**, R381–R391.

Cengel, Y.A. and Boles, M.A. (2008) *Thermodynamics: An Engineering Approach*, 6th edn, McGraw Hill, New York.

Dieckmann, J., Roth, K. and Brodrick, J. (2007) Magnetic refrigeration. *ASHRAE Journal*, 74–76.

Dincer, I. (1997) *Heat Transfer in Food Cooling Applications*, Taylor & Francis, Washington, DC.

Dincer, I. (2003) *Refrigeration Systems and Applications*, Wiley, 1st ed, London.

Dincer, I. and Dost, S. (1996) A simple model for heat and mass transfer in absorption cooling systems (ACSs). *International Journal of Energy Research*, **20**, 237–243.

Dincer, I., Edin, M. and Ture, I.E. (1996) Investigation of thermal performance of a solar powered absorption refrigeration system. *Energy Conversion and Management*, **37**, 51–58.

Garrett, S.L. and Backhaus, S. (2000) The power of sound. *American Scientist*, November–December issue, available at: http://www.sigmaxi.org/amsci/articles/00articles/garrettbib.html.

Garrett, S.L. and Hofler, T.J. (1992) Thermoacoustic refrigeration. *ASHRAE Journal*, **34**, 28–36.

Godfrey, S. (1996) An introduction to thermoelectric coolers. *Electronics Cooling*, September issue 4, 1–6.

Haywood, R.W. (1980) *Analysis of Engineering Cycles*, 3rd ed, Pergamon, Oxford.

International Energy Agency (IEA) (2003) *IEA Annex 26: Advanced Supermarket Refrigeration/Heat Recovery Systems, Final Report Volume 1-Executive Summary*, Compiled by Van D. Baxter, Oak Ridge National Laboratory, April.

JNT (1996) Safe, CFC-free, refrigeration system using hydrogen absorbing alloys. *CADDET Energy Efficiency Newsletter*, **4**, 4–7.

Kanoglu, M. (2001) Cryogenic turbine efficiencies. *Exergy, an International Journal*, **1/3**, 202–208.

Kanoglu, M. (2002) Exergy analysis of multistage cascade refrigeration cycle used for natural gas liquefaction. *International Journal of Energy Research*, **26**, 763–774.

Kanoglu, M., Dincer, I. and Rosen, M.A. (2008) Performance analysis of gas liquefaction cycles. *International Journal of Energy Research*, **32** (1), 35–43.

Klein, S.A. (2006) Engineering Equation Solver (EES), F-Chart Software.

Southern California Edison and Foster-Miller, Inc. (2004) Investigation of Secondary Loop Supermarket Refrigeration Systems. Report prepared for California Energy Commission, Public Interest Energy Research Program, March.

Timmerhaus, K.D. and Flynn, T.M. (1989) *Cryogenic Process Engineering*, The International Cryogenic Monographs Series, Plenum Press, New York.

Walker, G. (1983) *Cryocoolers*, Plenum Press, New York.

Wheatley, J.C., Hofler, T., Swift, G.W., and Migliori, A. (1993) Acoustical Heat Pumping Engine, U.S. Patent 4,398,398.

Zimm, C., Jastrab, A., Sternberg, A. (1988) The Magnetocaloric Effect. *Advances in Cryogenic Engineering*, **43**, 1759–1766.

6

Heat Pumps

6.1 Introduction

The heat pump principle was discovered before the turn of the century as the basis of all refrigeration. The principle of using a heat engine in a reverse mode as a heat pump was proposed by Lord Kelvin in the nineteenth century, but it was only in the twentieth century that practical machines came into common use, mainly for refrigeration. Beginning in the 1970s, air source heat pumps came into common use. They have the advantage of being combustion free, and thus there is no possibility of generating indoor pollutants like carbon monoxide. Heat pumps provide central air conditioning as well as heating as a matter of course. Also, they are installation cost competitive, with a central combustion furnace/central air conditioner combination.

It has become common practice now to call a heat pump any device that extracts heat from a source at low temperature and gives off this heat at a higher temperature, which can be useful. If the purpose is to extract heat from a low-temperature source, the device is called a *refrigeration system*. Therefore, the basic objective of heat pumping is exactly the same as the objective of refrigeration: the heat is removed at a low temperature and rejected at a higher temperature. Most heat pumps in use today operate on a vapor-compression cycle. In this respect, the components of a vapor-compression heat pump system are exactly the same as those of a vapor-compression refrigeration system, namely, compressor, condenser, evaporator, and expansion device. The difference between these two systems is that a refrigeration system generally transfers heat from a low temperature to the ambient, whereas a heat pump transfers heat from the ambient to a higher temperature, for example, from a low-temperature heat source (e.g., air, water, or ground) to a higher temperature heat sink (e.g., air, water, or ground). For this reason, heat pump systems are identified as *reverse-cycle refrigeration systems*. Many chlorofluorocarbons and hydrochlorofluorocarbons (HCFCs) have been universally used as refrigerants in air conditioning and refrigeration systems for more than five decades and as working fluids in heat pumps in the past few decades.

Heat pump technology is a promising means of promoting efficient use of thermal energy and thus achieving energy saving. It is capable of recovering low-temperature waste heat, which would otherwise be discharged to the air or water, in an efficient way and converting it to high-temperature heat energy that can be utilized for various purposes. In this regard, the heat pump is considered an optimal heat recovery technology. Commercial development of absorption heat pumps (AHPs) has been carried out during the past three decades. Today, heat pumps are widely used for air conditioning, cooling, and heating; producing hot water; and preheating boiler feed water in various types of facilities including office buildings, computer centers, public buildings, restaurants, hotels, district heating and cooling systems, industrial plants, and so on.

Refrigeration Systems and Applications İbrahim Dinçer and Mehmet Kanoğlu

In the utilization of heat pump systems, there are three main features, namely, environmental impact, economy, and technology. Preservation of a viable environment can be achieved by replacing the combustion of fossil fuels with heat pumps (with the exception of hydro, wind, and solar electricity). In terms of the economy, oil import/transport costs and heating costs (with a concomitant increase in private purchasing power) are reduced, and possible investments in new plants stimulate the economy. The simple technology of heat pumps enables every installer to apply standardized system designs with relatively low initial costs for new construction and retrofit of existing systems. Also, reliability and efficiency are additional benefits of heat pumps. Thus, the heat pump is shown to be an important instrument in the fight for energy conservation. This also explains the sudden breakthrough of heat pumps in several areas since the advent of the energy crisis.

One of the most common heat sources for a heat pump is air, although water is also used. Recently, there has been increasing interest in using soil (ground) as heat source for heating and cooling applications. Ground-source heat pumps (GSHPs) therefore have a good market share, along with low-temperature geothermal sources. It is important to highlight that by utilizing low-temperature resources exergetic efficiencies are dramatically increased. This issue is more important than thermal efficiency [and/or the coefficient of performance (COP)], as will be discussed later in detail.

The primary objective of this chapter is to introduce heat pump cycles, systems, and applications, to discuss their technical, operational, energetic, thermodynamic, and environmental aspects in detail, and to highlight the importance of their utilization with some illustrative examples.

6.2 Heat Pumps

Heat pumps have enormous potential for saving energy, particularly in industrial processes. They are the only heat recovery systems which enable the temperature of waste heat to be raised to more useful levels. Although the principle of the heat pump has been known since the middle of the nineteenth century, there was little incentive to develop them in a time of cheap and abundant energy.

Recent research and development has indicated that heat pump performance is likely to improve over the coming years. Improvements in component design and in the use of waste heat sources will raise heat pump performance. With regard to technical aspects, the many years of experience that have brought about important findings for the planning and design of heat pump systems can be used. Moreover, new ideas and equipment appearing in the last decade have simplified the construction of the heat pump heating and cooling systems.

Heat pumps look and operate very much like air conditioners (only for forced-air systems) with the notable exception that they provide both heating and cooling. While heat pumps and air conditioners require the use of some different components, they both operate on the same basic principles.

Heat flows naturally from a higher to a lower temperature. Heat pumps, however, are able to force the heat flow in the other direction, using a relatively small amount of high-quality drive energy (electricity, fuel, or high-temperature waste heat). Thus, heat pumps can transfer heat from natural heat sources in the surroundings, such as the air, ground or water, or from man-made heat sources such as industrial or domestic waste, to a building or an industrial application. Heat pumps can also be used for cooling. Heat is then transferred in the opposite direction, from the application that is cooled, to surroundings at a higher temperature. Sometimes the excess heat from cooling is used to meet a simultaneous heat demand.

Almost all heat pumps currently in operation are based either on a vapor compression or on an absorption cycle. Theoretically, heat pumping can be achieved by many more thermodynamic cycles and processes, including Stirling and Vuilleumier cycles, single-phase cycles (e.g., with air, CO_2 or noble gases), solid–vapor sorption systems, hybrid systems (notably combining the vapor-compression and absorption cycle), thermoelectric cycle and electromagnetic and acoustic processes. Some of these are entering the market or have reached technical maturity, and are expected to become significant in the future.

A heat pump is essentially a heat engine operating in reverse and can be defined as a device that moves heat from a region of low temperature to a region of higher temperature. The residential air-to-air heat pump, the type most commonly in use, extracts heat from low temperature outside air and delivers this heat indoors. To accomplish this and in accordance with the second law of thermodynamics work is done on the working fluid (i.e., a refrigerant) of the heat pump.

In order to transport heat from a heat source to a heat sink, external energy is needed to drive the heat pump. Theoretically, the total heat delivered by the heat pump is equal to the heat extracted from the heat source, plus the amount of drive energy supplied. Electrically driven heat pumps for heating buildings typically supply 100 kWh of heat with just 20–40 kWh of electricity. Many industrial heat pumps can achieve even higher performance and supply the same amount of heat with only 3–10 kWh of electricity.

For large-scale applications, heat pumps using a combustion furnace for supplemental heat and/or temperature peaking have become popular due to

- their applicability to the retrofit market as add-on units to existing oil or gas furnaces and boilers, and
- the improved performance of the combined system compared with electric resistance heat-supplemented heat pumps.

In this regard, heat pumps operating with supplementary heat are often said to be operating in a *bivalent* mode. A heat pump operating without electric resistance heating or without other backup is said to be operating in a *monovalent* mode. With the exception of certain control components designed to regulate compressor and furnace operation, essentially standard heat pump components are used. The system is operated in the heat pump mode down to a predetermined temperature called the *balance point* and the furnace is switched on when supplementary heat is required or, in the case of air distribution systems, during heat pump defrosting. Some systems switch the compressor off completely below the balance point while others allow parallel heat pump and furnace operation down to $-10\,^{\circ}\mathrm{C}$ for an air source heat pump. The heat pump technology is of special interest in colder climates where the traditional means of heating existing buildings is gas or oil and a requirement for some air conditioning as an add-on arises. The systems can also be used for heating alone in conjunction with conventional furnaces. Even in the coldest climates there are a sufficient number of heating days above the balance point of an existing heat pump to make this combination worthy of consideration.

6.2.1 Heat Pump Efficiencies

There are four different criteria used to describe heat pump efficiency. In all of these criteria, the higher the number the higher the efficiency of the system. Heat pump efficiency is determined by comparing the amount of energy delivered by the heat pump to the amount of energy it consumes. It is important to highlight that efficiency measurements are based on laboratory tests and do not necessarily measure how the heat pump performs in actual use (Dincer, 2003).

6.2.2 Coefficient of Performance (COP)

The COP is the most common measurement used to rate heat pump efficiency. COP is the ratio of the heat pump's heat output to the electrical energy input, as given below:

$$\mathrm{COP} = \text{heat output/electrical energy input} \tag{6.1}$$

For example, air source heat pumps generally have COPs of 2–4; they deliver 2–4 times more energy than they consume. Water and GSHPs normally have COPs of 3–5. The COP of air source heat pumps decreases as the outside temperature drops. Therefore, two COP ratings are usually given for a system: one at 8.3 °C (47 °F) and the other at −9.4 °C (17 °F). When comparing COPs, make sure ratings are based on the same outside air temperature. COPs for ground- and water-source heat pumps do not vary as much because ground and water temperatures are more constant than air temperatures.

While comparing COPs is helpful, it does not provide the whole picture. When the outside temperature drops below 6.4 °C (40 °F), the outdoor coils of a heat pump must be defrosted periodically. It is actually possible for the outdoor coil temperature to be below freezing when a heat pump is in the heating cycle. Under these conditions, any moisture in the air will freeze on the surface of the cold coil. Eventually, the frost could build up enough to keep air from passing over the coil and the coil would then lose efficiency. When the coil efficiency is reduced enough to appreciably affect system capacity, the frost must be eliminated. To defrost the coils, the heat pump reverses its cycle and moves heat from the house to the outdoor coil to melt the ice. This reduces the average COP significantly.

In fact, some heat pump units have an energy-saving feature that will allow the unit to defrost only when necessary. Others will go into a defrost cycle at set intervals whenever the unit is in the heating mode.

Another factor which lowers the overall efficiency of air-to-air heat pumps is their inability to provide enough heat on the coldest days of the winter. This means a back-up heating system is required. This backup is often electric resistance heat, which has a COP of 1 only. Whenever the temperature drops into the −3.8 to −1.1 °C range, or whatever its balance point is, and this electric resistance heat kicks in, overall system efficiency drops.

6.2.3 Primary Energy Ratio (PER)

Heat pumps may be activated either electrically or by engines (like internal combustion engines or gas motors). Unless electricity comes from an alternative source (e.g., hydro, wind, and solar), heat pumps also utilize primary energy sources upstream like a thermoelectric plant or on-spot like a natural gas motor. When comparing heat pump systems driven by different energy sources it is more appropriate to use the primary energy ratio (PER), as defined by Holland, Watson, and Devotta (1982), as the ratio of useful heat delivered to primary energy input. So this can be related to the COP by the following equation:

$$\mathrm{PER} = \eta\,\mathrm{COP} \tag{6.2}$$

where η is the efficiency with which the primary energy input is converted into work up to the shaft of the compressor.

However, due to high COP, the PER, as given below, becomes high relative to conventional fossil-fuel-fired systems.

In the case of an electrically driven compressor where the electricity is generated from a coal burning power plant, the efficiency η may be as low as 0.25 or 25%. The above equation indicates that gas-engine-driven heat pumps are very attractive from a PER point of view since values for η of 0.75 or better can be obtained. However, heat recovery systems tend to be judged on their potential money savings rather than on their potential energy savings.

6.2.4 Energy Efficiency Ratio (EER)

The energy efficiency ratio (EER) is used for evaluating a heat pump's efficiency in the cooling cycle. The same rating system is used for air conditioners, making it easy to compare different units. EER is the ratio of cooling capacity in British thermal units per hour provided to electricity consumed in watts as follows:

$$\text{EER} = \text{cooling capacity (Btu/h)/electrical energy input (W)} \tag{6.3}$$

Since 1 W = 3.412 Btu/h, the relationship between the COP and EER is

$$\text{EER} = 3.412\,\text{COP} \tag{6.4}$$

In practice, EER ratings higher than 10 are the most desirable.

6.2.5 *Heating Season Performance Factor (HSPF)*

A heat pump's performance varies depending on the weather and how much supplementary heat is required. Therefore, a more realistic measurement, especially for air-to-air heat pumps, is calculated on a seasonal basis. These measurements are referred to as the heating season performance factor (HSPF) for the heating cycle. The industry standard test for overall heating efficiency provides a rating known as HSPF. Such a laboratory test attempts to take into account the reductions in efficiency caused by defrosting, temperature fluctuations, supplemental heat, fans, and on/off cycling. HSPF is the estimated seasonal heating output in British thermal units, Btu divided by the seasonal power consumption in watt-hours, Wh, as follows:

$$\text{HSPF} = \text{total seasonal heating output (Btu)/total electrical energy input (Wh)} \tag{6.5}$$

It can be thought of as the "average COP" for the entire heating system. An HSPF of 6.8 corresponds to an average COP of 2. HSPFs of 5–7 are considered good. The higher the HSPF, the more efficient the heat pump. To estimate the average COP, divide the HSPF by 3.412 since 1 Wh = 3.412 Btu.

Most utility-sponsored heat pump programs require that heat pumps have an HSPF of at least 6.8. Many heat pumps meet this requirement. Some heat pumps have HSPF ratings above 9. In general, more efficient heat pumps are more expensive. Compare the energy savings to the added cost.

6.2.6 *Seasonal Energy Efficiency Ratio (SEER)*

As explained above, a heat pump's performance varies depending on the weather and the amount of supplementary heat required. Thus, a more realistic measurement, particularly for air-to-air heat pumps, is calculated on a seasonal basis. These measurements are referred to as the seasonal energy efficiency ratio (SEER) for the cooling cycle. Therefore SEER is rating the seasonal cooling performance of the heat pump. The SEER is the ratio of the total cooling of the heat pump in British thermal units, Btu to the total electrical energy input in watt-hours, Wh during the same period.

$$\text{SEER} = \text{total seasonal cooling output (Btu)/total electrical energy input (Wh)} \tag{6.6}$$

Naturally, the SEER for a unit will vary depending on where in the country it is located. SEERs of 8–10 are considered good. The higher the SEER the more efficiently the heat pump cools. The SEER is the ratio of heat energy removed from the house compared to the energy used to operate the heat pump, including fans. The SEER is usually noticeably higher than the HSPF since defrosting is not needed and there is no need for expensive supplemental heat during air-conditioning weather.

6.3 Sectoral Heat Pump Utilization

As mentioned earlier, a heat pump is a device that gets heat at a certain temperature and releases this heat at a higher temperature. When operated to provide heat (e.g., for space heating or water heating), the heat pump is said to operate in the heating mode; when operated to remove heat (e.g., for air conditioning), it is said to operate in the cooling mode. In both cases, additional energy has to be provided to drive the pump. Overall, this operation becomes energetically attractive if the total energy output is greater than the energy used to drive the heat pump and economically attractive if the total life-cycle cost (including installation, maintenance and operating costs) is lower than that for a competing device.

The common heat source for a heat pump is air, although water is also used in many applications. During the past decade ground or geothermal resources have received increasing attention to be used as a heat source, particularly in residential applications. From the sectoral utilization point of view, air is considered the most common distribution medium where the heat pump provides both heating and cooling. For heating only, air is also a common medium, except in those regions where many water distribution systems are installed in the residential sector. The energy needed to drive a heat pump is normally provided by electricity or fossil fuels, such as oil or gas.

The general characteristics of some typical commercially available heat pump systems are listed in Table 6.1 for the residential, commercial, and industrial sectors. For the commercial sector, all the basic characteristics are similar to those in the residential sector except for the fuel drive. In the former sector, a greater variety of fuels can be used because of the larger-scale operation which suits fossil engine systems. In industry, large-scale uses also result in greater fuel flexibility and the heat source is usually waste hot water, steam, or humid air. The type of heat sink will depend on the particular industrial process.

The heating and cooling of single and multifamily houses has become the most successful application of heat pumps thus far. A large variety of systems exists, depending upon whether they are intended for both heating and cooling or only heating, the nature of the low temperature source, and the medium distributing the heat (cold) to the building (air, water, etc.).

The heating-only heat pump is applicable to the residential sector in many countries where there is no air-conditioning load. Units can be installed separately or as add-on devices. While performance tends to be higher than for existing systems, the major difficulty is that the higher first cost of the unit can be recovered only over the heating season, in contrast to heating and cooling units which operate throughout the year. As indicated earlier, the electric add-on heat pump is a system that can be used in conjunction with fossil-fuel-fired furnaces or with central electric warm air furnaces.

For the residential sector, output requirements from a heat pump vary according to the use to which the output is applied, as indicated in Table 6.2. The requirements of a single-family residence will range from 4 to 30 kW, depending on the size, type, and degree of insulation of the building. Multifamily building needs a range from 20 kW for a two-family residence to 400 kW for an apartment block, although noncentral installations involve smaller sized units. Depending on the size of the grid, district heating schemes can range from 400 kW for a localized application to 10 MW for a large-scale system. The output needs of the commercial sector range from 20 kW for

Table 6.1 Typical heat pump characteristics and applications.

	Residential[a]	Commercial[b]	Industry
Fuel	Electricity	Electricity, gas, or oil	Electricity, gas, or oil
Heat source	Air, soil, or water	Air, soil, or water	Warm water, air, or steam
Heat sink	Air or water	Air or water	Process and/or waste heat
Utilization	Heating and/or cooling	Heating and/or cooling	Heating and/or cooling

[a] 1–2 family houses.
[b] Multifamily houses, industrial space heating, the commercial sector, and so on.

Table 6.2 Output capacities in heat pump applications.

Application	Output Capacity (kW)
Residential hot water heater	1–3
Residential single room	1–4
Residential single-family residence	4–30
Residential multifamily residence	
• (noncentralized units)	1–20
• (centralized units)	20–400
District heating	400–10,000
Commercial	20–1,000
Industry	100–30,000

Source: Berghmans (1983a).

Table 6.3 Delivery temperatures in various applications.

Application	Delivery Temperature (°C)
Space cooling (chilled water)	5–8
Space cooling (cooled air)	10–15
District heating (cool water)	10–30
Warm air heating	30–50
Warm water floor heating	30–50
Warm water radiators (low temperature)	45–55
Warm water radiators (forced convection)	55–70
Warm water radiators (free convection)	60–90
District heating (warm water)	80–100
District heating (hot water/steam)	100–180
Industrial process, water	60–110
Industrial process, steam	100–200

Source: Berghmans (1983a).

shops and small offices to 1 MW for large commercial centers. A greater range, from 100 kW to 30 MW, is found in the industrial sector. The delivery temperature also varies with the requirements of a particular application. Table 6.3 summarizes the temperature requirements for a number of uses.

Heat pumps for residential heating and cooling can be classified into four main categories depending on their operational function:

- Heating-only heat pumps for space heating and/or water heating applications.
- Heating and cooling heat pumps for both space heating and cooling applications.

- Integrated heat pump systems for space heating, cooling, water heating, and sometimes exhaust air heat recovery.
- Heat pump water heaters for water heating.

In residential applications, room heat pumps can be reversible air-to-air heat pumps (ductless packaged or split type units). The heat pump can also be integrated in a forced-air duct system or a hydronic heat distribution system with floor heating or radiators (central system).

They often use air from the immediate surroundings as heat source, but can also be exhaust-air heat pumps, or desuperheaters on air-to-air and water-to-air heat pumps. Heat pumps can be both monovalent and bivalent, where monovalent heat pumps meet the annual heating and cooling demand alone, while bivalent heat pumps are sized for 20–60% of the maximum heat load and meet around 50−95% of the annual heating demand. The peak load is met by an auxiliary heating system, often a gas or oil boiler. In larger buildings the heat pump may be used in tandem with a cogeneration system.

In commercial/institutional buildings the heat pump system can be a central installation connected to an air duct or hydronic system, or a multizone system where multiple heat pump units are placed in different zones of the building to provide individual space conditioning. Efficient in large buildings is the water-loop heat pump system, which involves a closed water loop with multiple heat pumps linked to the loop to provide heating and cooling, with a cooling tower and auxiliary heat source as backup.

In residential, commercial, and institutional buildings, recently there is an increasing interest in room-type controlled heat pumps (Figure 6.1). In addition to some benefits such as greater comfort, reduced noise, and reduced energy use, some features of this type of system are

- preventing operation when connection is made to the wrong supply voltage or if the wiring is incorrect,
- preventing overheating of the compressor, fan motor, and power transistor,
- detecting refrigerant undercharge and evaporator freeze-up, and
- maintaining the pressure balance by controlling the on/off switching cycle of the compressor.

6.3.1 Large Heat Pumps for District Heating and Cooling

Many large electrically driven heat pumps are in operation worldwide today and even more are planned for the future. Large heat pumps are defined as equipment with an output of about 500 kW or more. These are particularly used for district heating and cooling applications.

The point has often been made that the heat pump is competitive and is well established in markets where cooling is required, too. These markets are

- simultaneous cooling and heating (double utilization) such as in more recent HVAC applications or in the classical commercial cases of skating rink plus swimming pool, or refrigeration plus hot tap water production and
- consecutive production of cold and heat in HVAC plants with the same equipment, known as the heating/cooling heat pump (air cooling and dehumidifying in the summer season and heating and possibly humidifying during the winter season).

The large heat pump for district heating and cooling use proves to be well-suited:

- for base load coverage in systems without combined heat and power (CHP) generation;
- for low load, low-temperature summer operation for domestic hot water production;
- where supply and/or return temperatures are low;

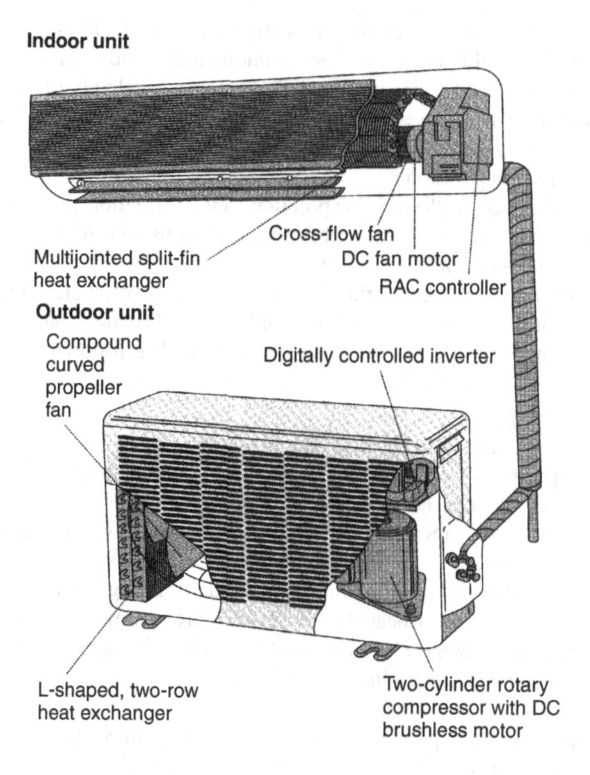

Figure 6.1 A controlled room heat pump (Itoh, 1995).

- where water is available as a heat source, for instance, cleared sewage water, industrial waste water, lake, or sea water. (There are also plants with a heat capacity up to 2.5 MW using ambient air as a heat source.)

6.4 Heat Pump Applications in Industry

The potential for industrial heat pumps has become promising. It is estimated that in highly industrialized countries, up to 40% of industrial primary energy demand can be saved by the application of heat pumps. In several countries strong efforts are being made to assist the introduction of industrial heat pumps.

In terms of the economy, effectiveness, savings, recovery of waste, and the environment, the industrial heat pumps compete with several alternative technologies, for example, boilers, heat pipes, and regenerators. Lehmann (1983) pointed out that the following requirements must be met for the industrial heat pump to prevail against these alternatives:

- manufacture of industrial heat pumps with lower initial cost,
- development of heat pumps with output temperatures between 150 and 300 °C,
- intensive and detailed analyses of different industrial heat pumps for specific industries,
- better adaptation of process technologies to heat pump applications,
- increased cooperation among heat pump users, practitioners, and engineers, and
- increased information collection about existing plants and operating experience.

Process heat pump opportunities in the food industry include both closed and open cycle designs. For closed-cycle heat pumps, this industry offers a unique combination of using large quantities of moderately warm (60 °C) water for processing and cleanup while having a ready source of waste heat that is currently being thrown away by the refrigeration plant condensers. Reclamation of some of this heat makes increasing economic sense to plant operators. For open-cycle designs, a number of food processes involve evaporation.

When heat pumps are used in drying, evaporation, and distillation processes, heat is recycled within the process. For space heating, heating of process streams and steam production, heat pumps utilize waste heat sources between 20 and 100 °C.

Industrial applications show a great variation in the type of drive energy, heat pump size, operating conditions, heat sources, and the type of application. The heat pump units are generally designed for a specific application, and are therefore unique. The following are some major types of industrial heat pumps (IEA-HPC, 2001).

- Mechanical vapor recompression (MVR) systems, classified as open or semi-open heat pumps. In open systems, vapor from an industrial process is compressed to a higher pressure and thus a higher temperature, and condensed in the same process, giving off heat. In semi-open systems, heat from the recompressed vapor is transferred to the process via a heat exchanger. Because one or two heat exchangers are eliminated (evaporator and/or condenser) and the temperature lift is generally small, the performance of MVR systems is high, with typical COPs of 3–9. Present MVR systems work with heat-source temperatures from 70 to 80 °C and deliver heat between 110 and 150 °C, in some cases up to 200 °C. Water is the most common working fluid (i.e., recompressed process vapor), although other process vapors are also used, notably in the petrochemical industry.
- Closed-cycle compression heat pumps are described in detail in Section 6.10. Currently applied working fluids limit the maximum output temperature to 120 °C.
- AHPs are not widely used in industrial applications. Present systems with water–lithium bromide as the working pair achieve an output temperature of 100 °C and a temperature lift of 65 °C. The COP typically ranges from 1.2 to 1.6. The new generation of advanced AHP systems are expected to have higher output temperatures up to 260 °C and higher temperature lifts.
- Heat transformers have the same main components and working principle as AHPs. With a heat transformer waste heat can be upgraded, virtually without the use of external drive energy. Waste heat of a medium temperature (i.e., between the demand level and the environmental level) is supplied to the evaporator and generator. Useful heat of a higher temperature is given off in the absorber. All present systems use water and lithium bromide as the working pair. These heat transformers can achieve delivery temperatures up to 150 °C, typically with a lift of 50 °C. COPs under these conditions range from 0.45 to 0.48.
- Reverse Brayton cycle heat pumps recover solvents from gases in many processes. Solvent loaded air is compressed, and then expanded. The air cools through the expansion, and the solvents condense and are recovered. Further expansion (with the associated additional cooling, condensation, and solvent recovery) takes place in a turbine, which drives the compressor.

Heat pumps are available for many industrial processes ranging from the pulp and paper industry to the food industry for a large number of applications, including:

- space heating,
- process water heating and cooling,
- steam production,
- drying and dehumidification processes, and
- evaporation and distillation and concentration processes.

The most common waste heat streams in industry are cooling water, effluent, condensate, moisture, and condenser heat from refrigeration plants. Because of the fluctuation in waste heat supply, it may be necessary to use large storage tanks for accumulation to ensure stable operation of the heat pump. Some common applications can be summarized as follows (IEA-HPC, 2001):

- **Space heating.** Heat pumps can utilize conventional heat sources for heating of greenhouses and industrial buildings, or they can recover industrial waste heat that could not be used directly, and provide a low to medium temperature heat that can be utilized internally or externally for space heating. Mainly electric closed-cycle compression heat pumps are used.
- **Process water heating and cooling.** In many industries, warm process water in the temperature range from 40 to 90 °C is needed, particularly for washing, sanitation, and cleaning purposes. Heat pumps offer good potential for such applications and may be a part of an integrated system that provides both cooling and heating. Although electric closed-cycle compression heat pumps are mainly installed, some AHPs and heat transformers may find application.
- **Steam production.** In industrial processes, vast amounts of low-, medium-, and high-pressure steam in the temperature range from 100 to 200 °C are consumed. In the market, at present, the high-temperature heat pumps can produce steam up to 300 °C. In this regard, open and semi-open MVR systems, closed-cycle compression heat pumps, cascade systems, and some heat transformers are employed.
- **Drying and dehumidification process.** Heat pumps are used extensively in industrial dehumidification and drying processes at low and moderate temperatures (maximum 100 °C). The main applications are drying of pulp and paper, various food products, wood, and lumber. Drying of temperature-sensitive products is also interesting. Heat pump dryers generally have high COP (with COP of 5–7), and often improve the quality of the dried products as compared with traditional drying methods. Because the drying is executed in a closed system, odors from the drying of food products, and so on, are reduced. Both closed-cycle compression heat pumps and MVR systems are used.
- **Evaporation, distillation, and concentration processes.** Evaporation, distillation, and concentration are energy-intensive processes, and most heat pumps are installed in these processes in the chemical and food industries. In evaporation processes the residue is the main product, while the vapor (distillate) is the main product in distillation processes. Most systems are open or semi-open MVRs, but closed-cycle compression heat pumps are also applied. Small temperature lifts result in high performance, with COPs ranging from 6 to 30.

In addition to the above-mentioned processes, there are some significant applications of heat pumps, covering a very wide range, from connected loads of a few watts for the thermoelectric heating/cooling units in the food industry to loads of several megawatts for large vapor-compression plants in industry, including

- Small, mass-produced, hot water heaters, sometimes combined with refrigerators and with connected loads between 200 and 800 W.
- Heating heat pumps for individual rooms, single-family houses, smaller office buildings, restaurants, and similar projects. Package heat pumps (in closed casings) are also available as split units with an indoor and an outdoor section for installation in the open. Mass-produced, sometimes on a large scale, these have a heat output often with supplementary heating (electric, liquefied gas, warm water) up to about 120 kW, and connected loads from 2 to 30 kW.
- Heat pumps for heating and heat recovery for large air-conditioning plants in office blocks, department stores, and similar projects. Appropriately adapted mass-produced chilled water units as well as systems individually assembled from the usual components for large refrigeration plants are used. Heat output is more than 1200 kW and the connected load is between 20 and

400 kW. If the heat pump is also used for cooling in summer, it is often better to use it to recover heat from the extract air in winter than to use an additional recuperative heat exchanger.
- Heating–cooling heat pumps for cooling and heating of rooms, objects of mass flows. The main task of these plants, also determining the control, is usually for either cooling or heating, not both, since the other effect is an additional gain which is not available during nonoperational periods of the system and can only be supplied by a store, for example, a hot water boiler.
- Waste heat utilization heat pumps for utilizing or reusing discharged heat which cannot be reused immediately because of its low temperature. This is so, for example, in drying processes in which the waste heat contained in the extracted water vapor is used for heating the drying air, or in laundries where practically all the applied heat energy is discharged with the waste water and can be recovered by a heat pump. The plants are controlled by heat demand, often combined with the storing of waste heat.

In the case of the industrial heat pump usually, a payback period of less than 4 years, sometimes even of 2 years, is required. Applications are therefore limited to cases where temperature levels and therefore COPs are favorable and utilization factors are high. Four basic heat pump systems are in use, available, or being developed for industrial low- and high-temperature applications:

- the closed-cycle compression heat pump, up to about 115 °C;
- the open-cycle heat pump (steam compressor) for much higher temperatures;
- the AHP (Type I), up to about 110 °C; and
- the "Type II" AHP (heat transformer), up to about 180 °C.

Various types of heat pumps are widely used in industry. However, as environmental regulations become stricter, it will become more important to use industrial heat pumps that reduce emissions, improve efficiency, and limit the use of groundwater for cooling.

To ensure the sound application of heat pumps in industry, processes should be optimized and integrated. Through process integration improved energy efficiency is achieved by thermodynamically optimizing total industrial processes. An important instrument for process integration is pinch analysis, a technology to characterize process heat streams and identify possibilities for heat recovery. Such possibilities may include improved heat exchanger networks, cogeneration, and heat pumps. Pinch analysis is especially powerful for large, complex processes with multiple operations, and is an excellent instrument to identify sound heat pump opportunities.

6.5 Heat Sources

To understand the basic principle of the heat pump, one must realize that heat is a form of energy, the quantity of which is quite independent of the temperature which happens to exist at the time. In air, soil, and water, in air extracted from buildings, and in waste water of any kind, there are enormous quantities of heat which are useless only because the temperature is too low. From all these sources, heat can be extracted, and with a small expenditure of additional, high-grade energy a heat pump can upgrade the waste heat to a temperature suitable for room heating.

The primary heat sources include air, water, and soil. In practice, air is the most common source for heat pumps while water- and soil-source systems are less commonly applicable. In general, air, soil, and groundwater are considered practicable as heat sources for small heat pump systems while surface water, sea water, and geothermal systems are more suited to larger heat pump systems. As far as low-temperature sources are concerned, ground or surface water, air, and soil are most commonly used.

The technical and economic performance of a heat pump is closely related to the characteristics of the heat source. An ideal heat source for heat pumps in buildings has a high and stable

Table 6.4 Commonly used heat sources.

Heat Source	Temperature Range (°C)
Ambient air	–10–15
Exhaust air	15–25
Groundwater	4–10
Lake water	0–10
River water	0–10
Sea water	3–8
Rock	0–5
Ground	0–10
Waste water and effluent	>10

Source: IEA-HPC (2001).

temperature during the heating season, is abundantly available, is not corrosive or polluted, has favorable thermophysical properties, and requires low investment and operational costs. In most cases, however, the availability of the heat source is the key factor determining its use. Table 6.4 presents commonly used heat sources. Ambient and exhaust air, soil, and groundwater are practical heat sources for small heat pump systems, while sea/lake/river water, rock (geothermal), and waste water are used for large heat pump systems.

Several heat pump configurations can be visualized utilizing a seemingly inexhaustible number of energy sources. Some of these energy sources are outside air, sensible heat from stream or well water, latent heat diffusion from water (ice formation), warm discharge effluents from industry, fireplace waste heat, and heat generation in sewage. Most of these energy sources are not widely available to the general public. Four types of heat pump systems are in common use in practice:

- single-package heat pumps using an air source,
- split-system heat pumps using an air source,
- single-package heat pumps using a water source, and
- split-system heat pumps using a water source.

Single-package heat pumps have all the essential components contained within a single unit while split-system heat pumps house the essential components in two separate units (i.e., one unit outdoors and one unit indoors).

Here, we present the most common heat sources for heat pumps.

6.5.1 Air

While ambient air is free and widely available, there are a number of problems associated with its use as a heat source. In the cooler and more humid climates, some residual frost tends to accumulate on the outdoor heat-transfer coil as the temperature falls below the 2–5 °C range, leading to a reduction in the capacity of the heat pump. Coil defrosting can be achieved by reversing the heat pump cycle or by other less energy-efficient means. This results in a small energy penalty because during the defrost cycles cool air is circulated in the building. Provided the defrost cycle is of short duration, this is not significant. In addition, for thermodynamic reasons the capacity and performance of the heat pump fall in any case with decreasing temperature. As the heating load is greatest at this time, a supplementary heating source is required. This device could be an existing oil, gas, or electric furnace or electric resistance heating; the latter is usually part of the heat pump system. The alternative to the provision of a supplementary heating device is to ensure that the capacity of the heat pump is adequate to cope with the most extreme weather conditions. This can result in over-sizing of the unit at a high additional capital cost and is not cost-effective compared with the cost of supplementary heating devices.

Exhaust (ventilation) air is a common heat source for heat pumps in residential and commercial buildings. The heat pump recovers heat from the ventilation air and provides water and/or space heating. Continuous operation of the ventilation system is required during the heating season or throughout the year. Some units are also designed to utilize both exhaust air and ambient air. For large buildings exhaust air heat pumps are often used in combination with air-to-air heat recovery units.

Outside ambient air is the most interesting heat source as far as availability is concerned. Unfortunately when the space heating load is the highest, the air temperature is the lowest. However, temperatures are not stable. The COP of vapor-compression heat pumps decreases with decreasing cold source temperature. In addition at evaporator temperatures below 5 °C air humidity is deposited on the evaporator surface in the form of ice. This does not improve the heat transfer and leads to lower working fluid temperatures and therefore lower COP values, depending upon the temperature of the air flowing over the evaporator. If ice formation occurs periodic de-icing of the evaporator surface has to be applied. This invariably leads to decreased values of the overall system COP (5–10%).

6.5.2 Water

Water-source units are common in applied or built-up installations where internal heat sources or heat or cold reclamation is possible. In addition, solar or off-peak thermal storage systems can be used. These sources have a more stable temperature, compared to air. The combination of a high first cost solar device with a heat pump is not generally an attractive economic proposition on either a first cost or a life-cycle cost basis.

Groundwater is available with stable temperatures between 4 and 10 °C in many regions. Open or closed systems are used to tap into this heat source. In open systems the groundwater is pumped up, cooled, and then reinjected in a separate well or returned to surface water. Open systems should be carefully designed to avoid problems such as freezing, corrosion, and fouling. Closed systems can be either direct expansion systems, with the working fluid evaporating in underground heat exchanger pipes or brine loop systems. Owing to the extra internal temperature difference, heat pump brine systems generally have a lower performance, but are easier to maintain. A major disadvantage of ground-water heat pumps is the cost of installing the heat source. Additionally, local regulations may impose severe constraints regarding interference with the water table and the possibility of soil pollution.

Most groundwater at depths more than 10 m is available throughout the year at temperatures high enough (e.g., 10 °C) to be used as low temperature source for heat pumps. Its temperature remains practically constant over the year and makes it possible to achieve high seasonal heating COPs (3 and more). The pump energy necessary to pump up this water has a considerable effect upon COP (10% reduction per 20 m pumping height). It is necessary to pump the evaporator water back into the ground to avoid depletion of groundwater layers.

The groundwater has to be of a purity almost up to the level of drinking water to be usable directly in the evaporator. The rather large consumption of water of high purity limits the number of heat pump systems which can make use of this source. Also, surface waters constitute a heat source which can be used only for a limited number of applications.

Groundwater at considerable depth (aquifers) may offer interesting possibilities for direct heating or for heating with heat pump systems. The drilling and operating costs involved require large-scale applications of this heat source. The quality of these waters often presents serious limitations to their use (corrosive salt content).

Groundwater (i.e., water at depths of up to 80 m) is available in most areas with temperatures generally in the 5–18 °C range. One of the main difficulties with these sources is that often the water has a high dissolved solids content, producing fouling or corrosion problems with heat exchangers. In addition, the flow rate required for a single-family house is high, and ground-water systems are difficult to be used widely in densely populated areas. Inclusion of the cost of providing the heat source has a significant impact on the economic attractiveness of these systems. A rule of thumb seems to be that such systems are economic if both the supply and the reinjection sources are available, marginally economic if one is available, and not cost-competitive if neither source is available. In addition, if a well has to be sunk, the necessity for drilling teams to act in coordination with heating and ventilation contractors can pose problems. Also, many local legislatures impose severe constraints when it comes to interfering with the water table and this can pose difficulties for reinjection wells.

Surface water such as rivers and lakes is in principle a very good heat source, but suffers from the major disadvantage that either the source freezes in winter or the temperature can be very close to 0 °C (typically 2–4 °C). As a result, great care is needed to avoid freezing on the evaporator. Where the water is thermally polluted by industry or by power stations, the situation is somewhat improved.

Sea water appears to be an excellent heat source under certain conditions and is mainly used for medium-sized and large heat pump installations. At a depth of 25–50 m, the sea temperature is constant (5–8 °C), and ice formation is generally not a problem (freezing point −1 to −2 °C). Both direct expansion systems and brine systems can be used. It is important to use corrosion-resistant heat exchangers and pumps and to minimize organic fouling in sea water pipelines, heat exchangers, evaporators, and so on. Where salinity is low, however, the freezing point may be near 0 °C, and the situation can be similar to that for rivers and lakes in regard to freezing.

Waste water and effluent are characterized by a relatively high and constant temperature throughout the year. Examples of possible heat sources in this category are effluent from public sewers (treated and untreated sewage water) in a temperature range of 10–20 °C throughout the year, industrial effluent, cooling water from industrial processes or electricity generation, and condenser heat from refrigeration plants. Condenser cooling water for electricity generation or industrial effluent could also be used as heat sources. The major constraints for use in residential and commercial buildings are, in general, the distance to the user and the variable availability of the waste heat flow. However, waste water and effluent serve as an ideal heat source for industrial heat pumps to achieve energy savings in industry.

Apart from surface water systems which may be prone to freezing, water source systems generally do not suffer from the low-temperature problems of air source heat pumps because of the higher year-round average temperature. This ensures that the temperature difference between the source and sink is smaller and results in an improvement of the performance of the heat pump. The evaporator must, however, be cleaned regularly. The heat transfer at the evaporator can drop by as much as 75% within approximately 5 months if it is not kept properly clean. The costs of cleaning become relatively low for larger projects so that the use of this source may become economic.

6.5.3 Soil and Geothermal

Soil or sub-soil (ground source) systems are used for residential and commercial applications and have similar advantages to water source systems, because of the relatively high and constant annual temperatures resulting in high performance. Generally the heat can be extracted from pipes laid horizontally or sunk vertically in the soil. The latter system appears to be suitable for larger heat pump systems. In the former case, adequate spacing between the coils is necessary, and the availability of suitably large areas (about double the area to be heated) may restrict the number of applications. For the vertical systems, variable or unknown geological structures and soil thermal properties can cause considerable difficulties. Owing to the removal of heat from the soil, the soil temperature may fall during the heating season. Depending on the depth of the coils, recharging may be necessary during the warm months to raise the ground temperature to its normal levels. This can be achieved by passive (e.g., solar irradiation) or active means. In the later case, this can increase the overall cost of the system. Leakage from the coils may also pose problems. Both the horizontal and vertical systems tend to be expensive to design and install and, moreover, involve different types of experts (one for heating and cooling and the other for laying the pipe work).

Rock (geothermal heat) can be used in regions with no or negligible occurrence of groundwater. Typical bore hole depth ranges from 100 to 200 m. When large thermal capacity is needed the drilled holes are inclined to reach a large rock volume. This type of heat pump is always connected to a brine system with welded plastic pipes extracting heat from the rock. Some rock-coupled systems in commercial buildings use the rock for heat and cold storage. Because of the relatively high cost of the drilling operation, rock is seldom economically attractive for domestic use.

The ground constitutes a suitable heat source for a heat pump in many countries. At small depths, temperatures remain above freezing. Furthermore, the seasonal temperature fluctuations are much smaller than those of the air. Heat is extracted from the soil by means of a glycol solution flowing through tubing embedded in the ground. If a horizontal grid of tubing is utilized, several hundred square meters of surface area are needed to heat a single family building. In urban areas such space is rarely available. In addition, considerable costs are involved. For these reasons vertical ground heat exchangers are more preferred presently.

Geothermal heat sources for heat pumps are currently utilized in various countries, particularly in the United States, Canada, and France. These resources are generally localized and do not usually coincide with areas of high-density population. In addition, the water often has a high salt component which leads to difficulties with the heat exchangers. Owing to the high and constant temperatures of these resources, the performance is generally high.

6.5.4 Solar

Solar energy, as either direct or diffuse radiation, is similar to air in its characteristics. A solar-source heat pump or a combined solar/heat pump heating system has all the disadvantages of the air source heat pump, such as low performance and extreme variability, with the additional disadvantage of high capital cost, particularly because in all cases a heat-store or back-up system is required. In areas with high daily irradiation, this may not be the case.

Each of the above-mentioned heat sources for heat pumps presents some drawbacks. Presently, considerable research is devoted to the technical problems involved and alternative heat sources. Also, solar energy may provide a suitable heat source. Unfortunately, solar systems presently are very costly. Furthermore, the intermittent character of solar energy requires the use of large and costly storage volumes.

6.6 Classification of Heat Pumps

A systematic classification of the different types of heat pumps is difficult because the classification can be made from numerous points of view, for example, purpose of application, output, type of heat source, and type of heat pump process. If the heat is distributed via a mass flow, for example, warm air or warm water, this mass flow is called the *heat carrier*.

Customarily in the United States heat pumps are classified for the heating of buildings according to the type of heat source (first place) and type of heat carrier (second place). A distinction can be made between the terms:

- heat pump, covering only the refrigeration machine aspect, and
- heat pump plant, which, besides the heat pump itself, also contains the heat source.

This differentiation is due to heat from the heat source being transferred to the cold side of the heat pump by an intermediate circuit, the cold carrier.

Another usual classification differentiates between

- primary heat pumps which utilize a natural heat source present in the environment, such as external air, soil, groundwater, and surface water,
- secondary heat pumps which reuse waste heat as heat source, that is, already used heat, such as extract air, waste water, waste heat from rooms to be cooled, and
- tertiary heat pumps which are in series with a primary or secondary heat pump in order to raise the achieved, but still relatively low, temperature further, for example, for hot water preparation.

Furthermore, heat pumps are generally classified by their respective heat sources and sinks. Depending on cooling requirements, various heat source and heat sink arrangements are possible in practical applications. The six basic types of heat pump are as follows:

- water-to-water,
- water-to-air,
- air-to-air,
- air-to-water,
- ground-to-water, and
- ground-to-air.

In each of these types the first term represents a heat source for heating or a heat sink for cooling applications. Schematics of the common types of heat pumps are also shown in Figure 6.2.

6.6.1 Water-to-Water Heat Pumps

In these heat pump systems, the heat source and the heat sink are water. The heat pump system takes heat from a water source (by coil A) while simultaneously rejecting it to a water heat sink (by coil B) and either heats or cools a space or a process. In practice, there are many sources of water, for example, waste water, single or double well, lake, pool, and cooling tower. These heat pumps use less electricity than other heat pumps when they are properly maintained. However, without proper maintenance the operating costs increase dramatically.

Table 6.5 shows typical COPs for a water-to-water heat pump operating in various heat distribution systems. The temperature of the heat source is 5 °C, and the heat pump's Carnot efficiency is 50%.

6.6.2 Water-to-Air Heat Pumps

Some heat pumps have been designed to operate utilizing a water source instead of an air source simply by designing the outdoor heat exchanger to operate between the heat pump working fluid

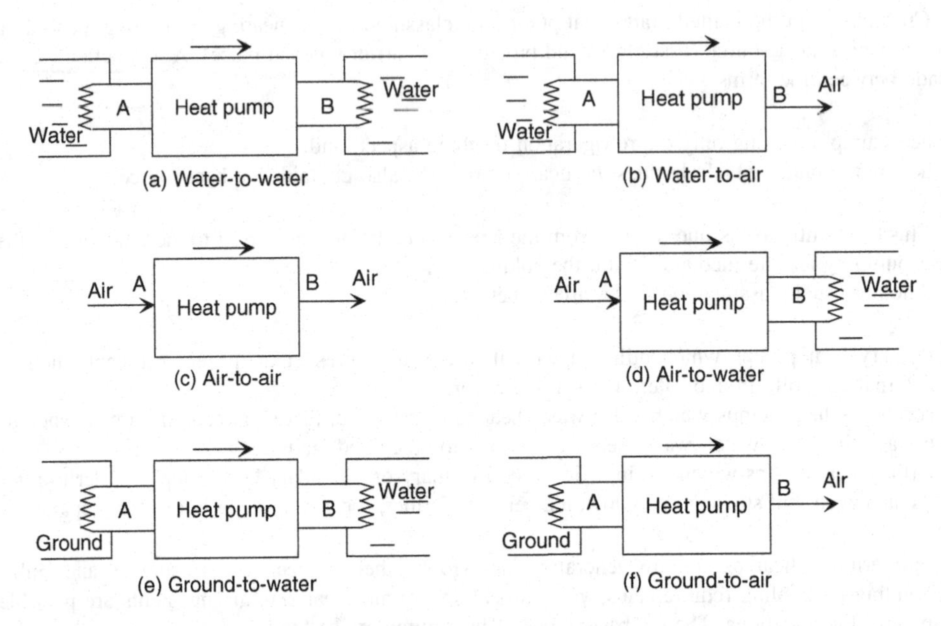

Figure 6.2 Types of heat pumps (here, the first part refers to the heat source for the outdoor coil during the heating process and the second part indicates the medium treated by the refrigerant in the indoor coil).

Table 6.5 Example of how the COP of a water-to-water heat pump varies with the distribution/return temperature.

Heat distribution system (supply/return temperature)	COP
Conventional radiators (60/50 °C)	2.5
Floor heating (35/30 °C)	6.0
Modern radiators (45/35 °C)	3.5

Source: IEA-HPC (2001).

and water instead of between the working fluid and air. These so-called water-to-air heat pumps have advantages over the air-to-air type if a relatively warm source of water is available which does not require an excessively large amount of pumping power. In particular, industrial waste heat might be used.

In this case, the difference from the water-to-water heat pump is in the method of treating the air. This system provides heating and cooling of air with water as the heat sink or source. The same sources of water can be used in these systems. These systems are less efficient than the water-to-water systems because of the much lower heat-transfer coefficient of air. These systems are commonly used in large buildings and sometimes in industrial applications to provide hot or cold water.

The water-to-air heat pump removes heat from water and converts it to hot air in exactly the same way that a cold water drinking fountain removes the heat from the water and discharges the heat from the side or back of the drinking fountain.

6.6.3 Air-to-Air Heat Pumps

These systems use air on both sides (on coils) and provide heating or cooling. In the cooling mode, heat is removed from the air in the space and discharged to the outside air. In the heating mode, heat is removed from the outside air and discharged to air in the space. In these units, it is necessary to provide defrost controls and periods to maintain maximum efficiency. These are the most popular systems for residential and commercial applications because of easy economical installation and lower maintenance cost.

Depending on climate, air source heat pumps (including their supplementary resistance heat) are about 1.5–3 times more efficient than resistance heating alone. Operating efficiency has improved since the 1970s, making their operating cost generally competitive with combustion-based systems, depending on local fuel prices. With their outdoor unit subject to weathering, some maintenance should be expected.

The most popular heat pump is the air source type (air-to-air) which operates in two basic modes:

- As an air conditioner, a heat pump's indoor coil (heat exchanger) extracts heat from the interior of a structure and pumps it to the coil in the unit outside where it is discharged to the air outside (hence the term air-to-air heat pump).
- As a heating device, the heat pump's outdoor coil (heat exchanger) extracts heat from the air outside and pumps it indoors where it is discharged to the air inside.

6.6.4 Air-to-Water Heat Pumps

In Figure 6.2, these systems work in reverse of the water-to-air heat pumps: they extract heat from ambient or exhaust air to heat or preheat water used for space or process heating. The system is simply reversed. Heat is extracted from the air inside the home and transferred to water and put back in the ground. All the householders select the temperature that makes their homes as cool as they wish.

6.6.5 Ground-to-Water and Ground-to-Air Heat Pumps

In these systems, coil A in Figure 6.2 is buried underground and heat is extracted from the ground. These heat pump systems have limited use. Practical applications are limited to space heating where the total heating or cooling effect is small, and the ground coil size is equally small. This system requires the burial of several meters of pipe per ton of refrigeration, thus requiring a large amount of land.

6.6.6 Basic Heat Pump Designs

The four basic heat pump designs for space heating and cooling employ

- air as the heat source/sink and air as the heating and cooling medium,
- air as the heat source/sink and water as the heating and cooling medium,
- water as the heat source/sink and air as the heating and cooling medium, or
- water as heat source/sink and water as the heating and cooling medium.

Each of these basic designs can supply the required heating and cooling effect by changing the direction of the refrigerant flow, or by maintaining a fixed refrigerant circuit and changing the direction of the heat source/sink media. A third alternative is to incorporate an intermediate transfer fluid in the design. In this case the direction of the fluid is changed to obtain heating or

Table 6.6 Typical delivery temperatures for various heat and cold distribution systems.

Application	Supply Temperature Range (°C)
Air distribution	
• Air heating	30–50
• Floor heating (low-temperature)	30–45
Hydronic systems	
• Radiators	45–55
• High-temperature radiators	60–90
• District heating (hot water)	70–100
District heating	
• District heating (hot water/steam)	100–180
• Cooled air	10–15
Space cooling	
• Chilled water	5–15
• District cooling	5–8

Source: IEA-HPC (2001).

cooling and both the refrigerant and heat source/sink circuits are fixed. The fixed refrigerant circuit designs, generally referred to as the indirect type of application, are becoming increasingly popular, particularly in the larger capacities.

6.6.7 Heat and Cold-Air Distribution Systems

Air is one of the most widely used distribution media in the mature heat pump markets, especially in the United States. The air is either delivered directly to a room by the space-conditioning unit or distributed through a forced-air ducted system. The output temperature of an air distribution system is usually in the range of 30–50 °C in heating applications.

Water distribution systems (so-called *hydronic systems*) are predominantly used in many countries, particularly in Europe, Canada, and the northeastern part of the United States. Conventional radiator systems require high distribution temperatures, typically 60–90 °C. Today's low-temperature radiators and convectors are designed for a maximum operating temperature of 45–55 °C, while 30–45 °C is typical for floor-heating systems. Table 6.6 summarizes typical supply temperature ranges for various heat and cold distribution systems.

Because a heat pump operates most effectively when the temperature difference between the heat source and heat sink (distribution system) is small, the heat distribution temperature for space heating heat pumps should be kept as low as possible during the heating season.

6.7 Solar Heat Pumps

During the past two decades, there has been increasing interest in solar-assisted heat pumps. In this system, solar energy is used as the heat source for a heat pump. The main advantage of this system

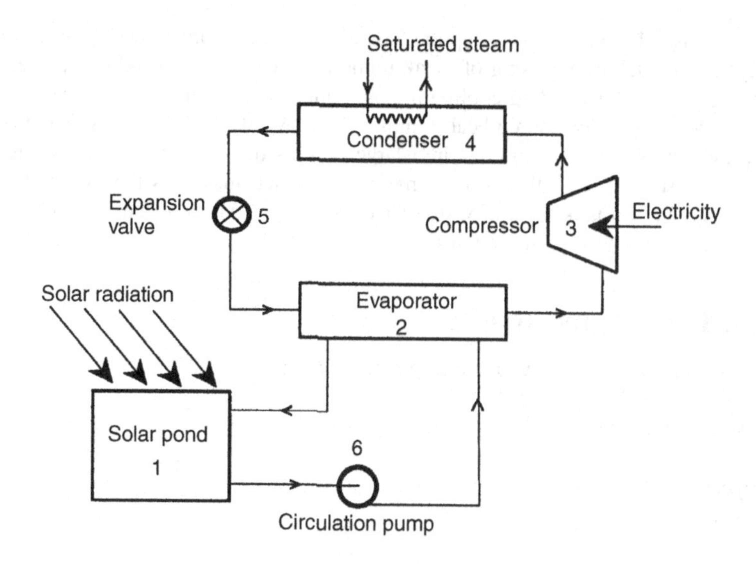

Figure 6.3 A solar heat pump (Tu, 1987).

is that solar energy supplies heat at a higher temperature level than other sources and, therefore, provides an increase in the COP. As compared to a solar heating system without a heat pump, collector efficiency and capacity are materially increased because of the lower collector temperature required. Direct utilization of solar energy depends on sufficient solar radiation. However, there is still an extremely large amount of solar energy stored in the atmosphere and the ground at temperatures between 0 and 20 °C, and this source is also available during the winter months, which can be used by everyone. Research and development efforts for using solar energy for heat pump operation have been focused on direct and indirect systems. In a direct system, refrigerant evaporator tubes are incorporated with a solar collector system, usually flat-plate collectors, and their surface extracts heat from the outdoor air. The same surface may be employed as a condenser using outdoor air as a heat sink for cooling. An indirect system utilizes either water or air circulated through the solar collector.

Figure 6.3 shows a schematic diagram of a simple solar-assisted heat pump system that was employed by Tu (1987). In the operation, heated water (~80 °C) from the solar pond (1) and the heat pump working fluid (water) exchange heat in the evaporator (2). Water vapor from the evaporator (2) is compressed by a compressor (3) and then condensed in the condenser (4), giving up its heat to the user to produce steam at 120 °C from saturated water at 120 °C.

Thermal storage coupled heat pumps are primarily associated with heat reclaim units for domestic water heating. Sometimes they are used for heat storage as part of large commercial cool storage installations. Thermal energy storage works well in such applications. Use of thermal storage has recently become common in residential applications.

6.8 Ice Source Heat Pumps

Ice-making systems can be configured to provide heating alone, or heating and cooling, for a building or process. The conversion of water to ice occurs at a relatively high evaporator temperature and COP compared to air source heat pumps operating at low ambient temperatures. These systems can provide the necessary heating, with the resulting ice disposed of by melting with low-grade heat, for example, solar energy. In addition, they can be used for cooling through diurnal and seasonal storage applications. The energy consumption savings resulting from the COP of a conventional heat pump system are achieved. Using off-peak night and weekend rates reduces power costs, and the ice produced is used for building cooling requirements. As a result, a system can be developed that consumes less energy. Ice source heat pumps follow two basic approaches. The first involves using the ice builder principle, with coils in a large tank as the evaporator components of a heat pump system. The second one utilizes a fragmentary-type ice maker as the evaporator of a heat pump, with ice stored in a tank (as a mixture of ice and water). Many variations and combinations may be developed from these basic systems.

6.9 Main Heat Pump Systems

Numerous heat pump cycles are available in practice, including

- vapor-compression cycle,
- Stirling cycle,
- Brayton cycle,
- Rankine cycle,
- absorption cycle,
- compression cycle,
- open cycle,
- recompression cycle, and
- compression/resorption cycle.

In conjunction with the above classification, in the following sections, we will focus on the following heat pump systems:

- vapor-compression heat pump systems,
- MVR heat pump systems,
- cascaded heat pump systems,
- Rankine powered heat pump systems,
- AHP systems (including heat transformers, resorption heat pumps, diffusion heat pumps),
- thermoelectric heat pump systems,
- vapor jet heat pump systems,
- quasi-open-cycle heat pump systems,
- chemical heat pump systems, and
- metal hydride heat pump systems.

6.10 Vapor-Compression Heat Pump Systems

A great majority of heat pumps on the market are simple four-component vapor-compression cycle systems. The term *vapor compression* refers to the use of a mechanical compressor. During the last three decades, many forms of these heat pump systems have been developed. Figure 6.4 shows the most common four-component vapor-compression heat pump system, consisting of four main components:

- an evaporator (where heat is absorbed into a boiling refrigerant),
- a compressor (which raises the pressure, and hence temperature, of the refrigerant),
- a condenser (where the absorbed energy and compressor power are released), and
- an expansion device (where the refrigerant liquid changes from a high temperature liquid to a low-pressure and low-temperature mixture of liquid and vapor).

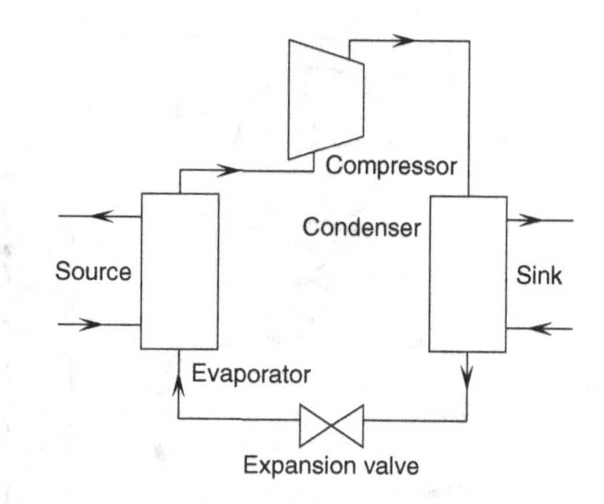

Figure 6.4 A single-stage vapor-compression heat pump.

It is important to recognize that any vapor-compression heat pump system has three distinct fluid circuits:

- A source from which heat is removed.
- A sink to which heat is delivered.
- A refrigerant circuit through which the energy is transferred.

Furthermore, the components of a vapor-compression heat pump are connected to form a closed circuit, as shown in Figure 6.5. A volatile liquid, known as the working fluid or refrigerant, circulates through the four components. In the evaporator the temperature of the liquid working fluid is kept lower than the temperature of the heat source, causing heat to flow from the heat source to the

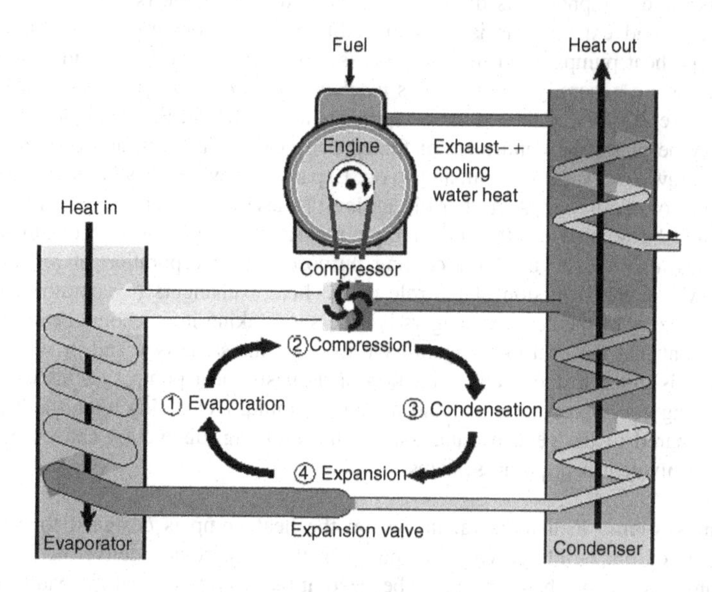

Figure 6.5 Closed cycle, engine-driven vapor-compression heat pump (*Courtesy of IEA-HPC*).

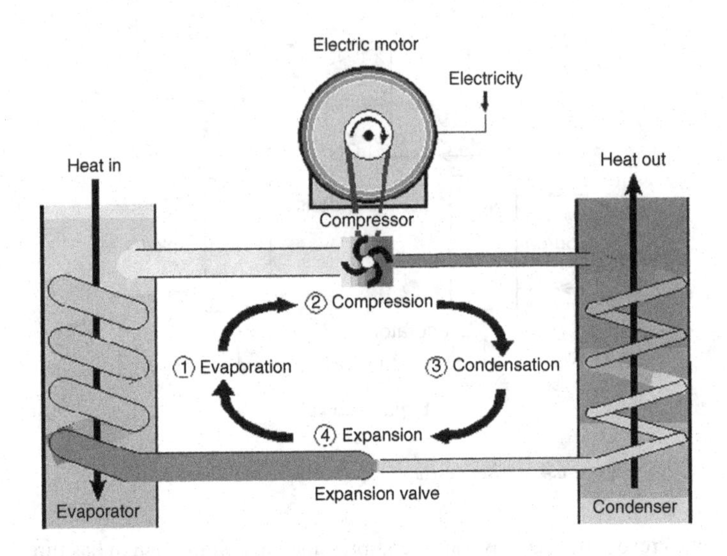

Figure 6.6 Closed-cycle, electric-motor-driven vapor-compression heat pump (*Courtesy of IEA-HPC*).

liquid, and the working fluid evaporates. Vapor from the evaporator is compressed to a higher pressure and temperature. The hot vapor then enters the condenser, where it condenses and gives off useful heat. Finally, the high-pressure working fluid is expanded to the evaporator pressure and temperature in the expansion valve. The working fluid is returned to its original state and once again enters the evaporator. The compressor is usually driven by an electric motor and sometimes by a combustion engine.

An electric motor drives the compressor (see Figure 6.6) with minimal energy losses. The overall energy efficiency of the heat pump strongly depends on the efficiency by which the electricity is generated. When the compressor is driven by a gas or diesel engine (see Figure 6.5), heat from the cooling water and exhaust gas is used in addition to the condenser heat. Industrial vapor-compression type heat pumps often use the process fluid itself as working fluid in an open cycle. These heat pumps are generally referred to as *mechanical vapor recompressors*, or MVRs.

In practice, several types of these four components are available. For example, evaporators and condensers may be plate, shell-and-tube, or finned coil heat exchangers, and compressors may be reciprocating, screw, rotary, centrifugal, and so on. Expansion devices may be float and thermostatic expansion valves or capillaries. A further discussion of these components is given in Chapter 3.

First, it should be pointed out that in principle the vapor-compression heat pumps utilized for heating purposes also can be used for cooling purposes by incorporation of a four-way valve. The latter allows for the inversion of the role of the heat exchangers (evaporator–condenser). In practice, proper sizing of the heat exchangers is necessary, taking into account heating and cooling loads. It is evident that the heat exchangers have to be of the same type (air or water). Hereafter, special attention is given to the heating function of domestic heat pump systems while the above remark concerning cooling operation should be kept in mind. Depending upon the capacity of the heat pump compared to the peak heating load of the building, distinction can be made between monovalent and bivalent heat pump systems:

- **Monovalent systems.** In a monovalent system the heat pump is designed to cover the peak heating losses of the building. The heat pump is the only device delivering heat during the whole heating season and therefore has to be sized at the maximum heating load of the building. This leads to considerable overcapacity of the heat pump, which leads to more intermittent use and therefore less efficient heat pump operation and lower average COP values. Therefore, monovalent systems are applied only when heat sources like groundwater and ground heat are available. Such systems are characterized by large investments. On the other hand, the seasonal COP values usually are higher than those which can be obtained with air systems.

- **Bivalent systems.** Bivalent systems are used whenever the performance of the heat pump deteriorates because of low source temperatures. Air heat pumps are of this type. In these systems heat pumps of smaller capacity are utilized. They can provide the heat to the building down to the outside temperature (e.g., design temperature). If it goes below this temperature, an auxiliary system has to be called into operation. Depending upon the operation of the system one may distinguish between parallel bivalent systems and alternative bivalent systems.

As a measure of the energetic performance of space heating heat pumps, one may take the average COP over a heating season. The value of this COP will depend upon the characteristics of the heating season. Table 6.7 lists typical seasonal COP for a number of systems (lowest average temperature being $-10\,°C$). Table 6.7 also lists the type and capacity of the auxiliary system and the percentage of the heat load provided by the heat pump. Of course only bivalent heat pump systems require auxiliary heat, the parallel systems requiring less of this heat than the alternative ones. The seasonal COP values listed in Table 6.7 take into account the electricity consumed by fans, pumps, and so on. The water–water monovalent heat pump shows the highest COP values because of the high temperature of the heat source available throughout the whole heating season. The lower temperatures of ground sources considerably reduce the COP values. For the bivalent systems the heat pump COP loses a lot of its meaning since it does not take into account auxiliary heat. Finally, Table 6.7 also lists the seasonal primary energy efficiency which is the ratio of heat produced to the primary energy consumed to produce this heat. Here one has taken into account the fact that all the heat pumps are electrically driven and that the efficiency of electricity production

Table 6.7 Heat pump systems and their performance comparisons.

Heat Pump System	Heat Source	Source Temperature (°C)	Auxiliary Heating Type	Auxiliary Heating Capacity (%)	Heat Pump Delivered Heat (%)	Seasonal COP	Primary Energy Efficiency
Monovalent soil/water	Soil	–2 to 10	–	–	100	2.0 to 3.0	66 to 100
Monovalent water/water	Ground or surface water	10	–	–	100	3.0 to 4.0	100 to 130
Bivalent parallel air/water	Outside air	–10 to 15	Oil or gas boiler	40	90	2.5 to 3.0	70
Bivalent alternative air/water	Outside air	–10 to 15	Oil or gas boiler	100	60 to 70	1.6 to 1.8	75
Bivalent extracted air/air or water	Extracted air	18 to 20	Direct electric heating	75 to 85	50 to 60	1.6	50

Source: Berghmans (1983b).

is about 33%. The last column shows that only the monovalent heat pump systems using the soil or ground may give rise to primary energy savings loads compared with conventional oil or natural gas heaters.

The heat pump runs on an evaporation–condensation cycle, just like traditional air conditioners and refrigerators. A typical vapor-compression heat pump has several vital components as mentioned earlier: a refrigerant that circulates continuously in a closed-cycle, a motor-driven compressor, a pair of heat exchangers which can alternate in the roles or condenser and evaporator, and some form of expansion valve that can be used to control a pressure drop and hence the temperature change of the working fluid. A heat pump is essentially an air conditioner with a few additions. A heat pump has a reversing valve, two metering devices and two bypass valves. This allows the unit to provide both cooling and heating. Here, we discuss the vapor-compression heat pump for both cooling and heating.

6.10.1 The Cooling Mode

In the cooling mode (e.g., cooling operation in summer), the outdoor coil becomes the condenser and the indoor coil the evaporator. By condensing water vapor out of the circulated interior air, the heat pump can also dehumidify like a traditional air conditioner. Figure 6.7 shows a vapor-compression heat pump in cool mode. The cycle operates as follows: the compressor (1) pumps the refrigerant to the reversing valve (2). The reversing valve directs the flow to the outside coil (condenser) where the fan (3) cools and condenses the refrigerant to liquid. The air flowing across the coil removes heat (4) from the refrigerant. The liquid refrigerant bypasses the first metering device and flows to the second metering device (6) at the inside coil (evaporator) where it is metered. Here it picks up heat energy from the air blowing (3) across the inside coil (evaporator) and the air comes out cooler (7). This is the air that blows into the home. The refrigerant vapor (8) then travels back to the reversing valve (9) to be directed to the compressor to start the cycle all over again (1).

6.10.2 The Heating Mode

To provide heat from this same unit the evaporator and condenser must essentially switch places. That is, heat must be moved from the outside air to the indoor coil for discharge. This is

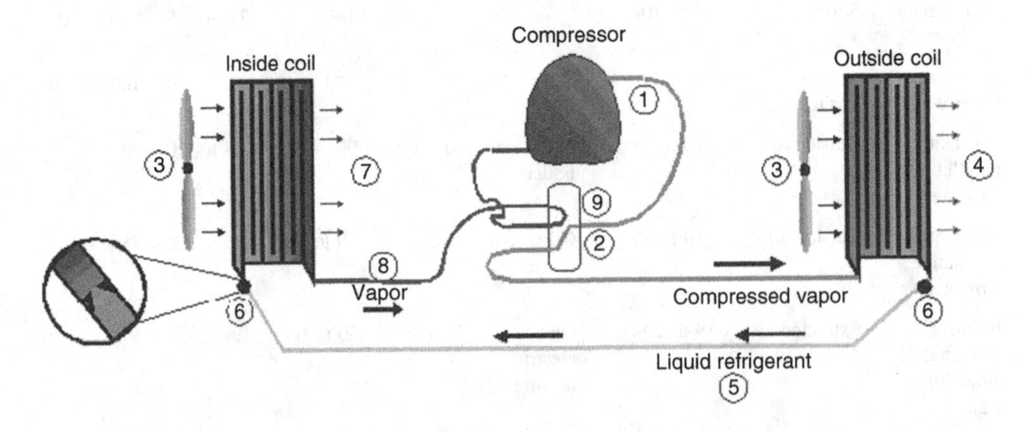

Figure 6.7 A vapor-compression heat pump running for cooling (*Courtesy of DHClimate Control*).

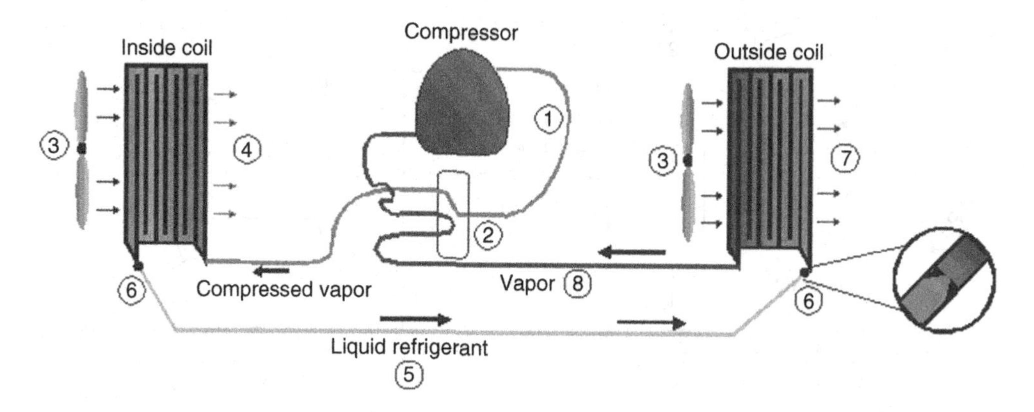

Figure 6.8 A vapor-compression heat pump running for heating (*Courtesy of D&H Climate Control*).

accomplished by reversing the flow of refrigerant through a device found in heat pumps knows as a "reversing valve". This valve is automatically controlled through the thermostat when switched to heat.

Figure 6.8 shows the heat pump in the heating mode of operation (e.g., in winter the heat exchanger located outside the house functions as an evaporator, absorbing low-temperature heat from the environment). Switching the heat pump from the cooling mode to the heating mode is achieved simply by switching the direction of the refrigerant flow. The difference between Figures 6.7 and 6.8 is that the reversing valve (2) directs the compressed refrigerant to the inside coil first. This makes the inside coil the condenser and releases the heat energy (3-4). This heated air is ducted to the home. The outside coil is used to collect the heat energy (3-7). This now becomes the evaporator. Both heating and A/C modes do exactly the same thing. They pump heat from one location to another. In these examples, the heat in the air is moved out of or into the home.

There is usable heat in outdoor air at temperatures as low as −8.5 °C. As the temperature of the outdoor air decreases, however, the heating capacity of the heat pump diminishes proportionately, resulting in lower discharge air temperatures at the air registers and gradual cooling of the house. To supplement the heating capacity of the heat pump, electric resistance heating elements are used, which automatically engage via the thermostat when this condition occurs.

6.10.3 Single-Stage Vapor-Compression Heat Pump with Subcooler

A simple modification can be made to the four-component cycle to make it, in some situations, considerably more efficient. This is the addition of a refrigerant subcooler, as shown in Figure 6.9a. This extra heat exchanger extracts heat from the hot liquid refrigerant before it goes through the expansion valve. This leads to less flash gas formation through the expansion valve. Hence, the same compressor is doing more useful cooling and heating, with no extra power consumption (note that the compressor is "unaware" of whether its suction vapors are formed across the expansion valve or through boiling in the evaporator). Figure 6.9b shows the effect of subcooling on a Mollier chart. In both cases illustrated, the primary heating is carried out between 70 and 75 °C. In the case of a subcooler the hot refrigerant (80 °C when leaving the condenser) is cooled to 30 °C by a stream of air. This extra free heat can successfully be used by integrating an air space heating unit with the main hot water system. The financial advantage is considerable.

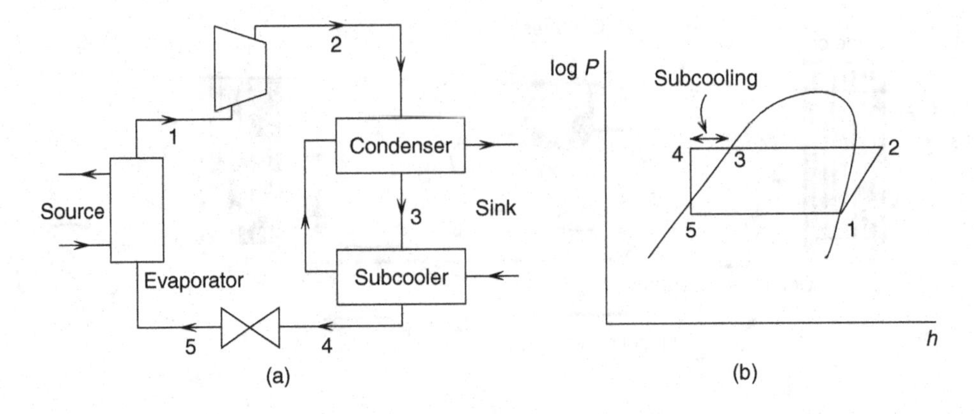

Figure 6.9 (a) A single-stage vapor-compression heat pump with subcooler and (b) its log P–h diagram.

Table 6.8 ARI standard rating conditions for variable capacity compressors and compressor units used in heat pumps.

Rating Test Point	Intended Use	Suction Dew Point Temperature (°C)	Discharge Dew Point Temperature (°C)	Return Gas Temperature (°C)	Capacity Setting[a]
A	Air source (cooling)	7.2	56.4	18.3	Max.
B	Air source (cooling)	7.2	46.1	18.3	Max.
C	Air source (cooling and heating)	7.2	37.8	18.3	Min.
D	Air source (heating)	–1.1	43.3	10.0	Max.
E	Air source (heating)	–15.0	35.0	–3.9	Max.
F	Air source (cooling)	7.2	26.7	18.3	Min.
G	Air source (heating)	1.7	32.2	12.8	Min.
H	Water source (cooling and heating)	7.2	48.9	18.3	Max. and Min.

Ratings based on 35 °C temperature surrounding compressor. If air flow across the compressor is used to determine ratings, it shall be specified by the compressor manufacturer.
[a]The maximum and minimum capacity setting is the highest and lowest displacement capacity obtainable by the compressor or compressor unit.
Source: ARI (2000).

6.10.4 Standard Rating Conditions for Compressors

The standard rating(s) of a compressor or compressor unit used in a heat pump is its compressor rating(s) based on the tests performed at standard rating conditions at the test points from Table 6.8.

6.10.5 ARI/ISO Standard 13256-1

This standard covers those heating and cooling systems usually referred to as *water source heat pumps*. These electrically driven vapor-compression systems consist of one or more factory-made matched assemblies which normally include an indoor conditioning coil with air moving means, a compressor, and a refrigerant-to-liquid (water or brine) heat exchanger. A system may provide cooling-only, heating-only, or both functions and is typically designed for use within one or more of the following liquid heat source/sink applications: (i) water-loop heat pump using temperature-controlled water circulating in a common piping loop, (ii) groundwater heat pump using water pumped from a well, lake, or stream, and (iii) ground-loop heat pump using brine circulating through a subsurface piping loop.

These three applications were previously separately covered by ARI Standards 320, 325, and 330. Rating and performance test conditions for ARI/ISO 13256-1 (Ellis, 2001), as compared to the previous ARI standards, are summarized in Tables 6.9 and 6.10. As can be seen, the differences in rating test temperatures are relatively minor, and consist mainly of rounding to the Celsius scale and

Table 6.9 Comparison of ARI and ISO rating test conditions.

Rating Tests	Water-Loop Heat Pumps		Ground-Water Heat Pumps			Ground-Loop Heat Pumps	
	ARI/ISO	ARI 320	ARI/ISO	ARI 325 Hi	ARI 325 Lo	ARI/ISO	ARI 330
Standard cooling:							
Air dry bulb, °C	27	26.7	27	26.7	26.7	27	26.7
Air wet bulb, °C	19	19.4	19	19.4	19.4	19	19.4
Air flow rate, l/s	per mfr[a]	per mfr	per mfr	per mfr	per mfr	per mfr	per mfr
Liquid full load, °C	30	29.4	15	21.1	10.0	25	25.0
Liquid part load, °C	30	23.9	15	21.1	10.0	20	21.1
Liquid flow rate, l/s	per mfr	5.6 °C rise	per mfr	per mfr	per mfr	per mfr	per mfr
Standard heating:							
Air dry bulb, °C	20	21.1	20	21.1	21.1	20	21.1
Air wet bulb, °C	15	15.6	15	15.6	15.6	15	15.6
Air flow rate, l/s	per mfr	std clg[b]	per mfr	std clg	std clg	per mfr	std clg
Liquid full load, °C	20	21.1	10	21.1	10.0	0	0.0
Liquid part load, °C	20	23.9	10	21.1	10.0	5	5.0
Liquid flow rate, l/s	per mfr	std clg	per mfr	per mfr	per mfr	per mfr	per mfr
External static:							
Air, Pa	0	25.0–75.0	0	25.0–75.0	25.0–75.0	0	25.0–75.0
Liquid, kPa	0	NA	0	150.0	150.0	0	50.0

[a]per mfr: per manufacturer.
[b]std clg: standard catalog.
Source: Ellis (2001).

Table 6.10 Comparison of ARI and ISO performance test conditions.

Rating Tests	Water-Loop Heat Pumps ARI/ISO	Water-Loop Heat Pumps ARI 320	Ground-Water Heat Pumps ARI/ISO	Ground-Water Heat Pumps ARI 325 Hi	Ground-Loop Heat Pumps ARI/ISO	Ground-Loop Heat Pumps ARI 330
Maximum cooling:						
Air dry bulb, °C	32	35.0	32	35.0	32	35.0
Air wet bulb, °C	23	21.7	23	21.7	23	21.7
Liquid, °C	40	35.0	25	23.9	40	35.0
Maximum heating:						
Air dry bulb, °C	27	26.7	27	26.7	27	26.7
Liquid, °C	30	32.3	25	23.9	25	23.9
Minimum cooling:						
Air dry bulb, °C	21	19.4	21	NA	21	26.7
Air wet bulb, °C	15	13.9	15	NA	15	19.4
Liquid, °C	20	18.3	10	NA	10	0.0
Minimum cooling:						
Air dry bulb, °C	15	NA	15	15.6	15	15.6
Liquid, °C	15	NA	5	7.2	−5	−3.9
Enclosure sweat:						
Air dry bulb, °C	27	26.7	27	26.7	27	26.7
Air wet bulb, °C	24	23.9	24	23.9	24	23.9
Liquid, °C	20	26.7	10	10.0	10	10.0

Source: Ellis (2001).

eliminating the dual rating points for ARI 325. Performance test temperatures vary more, but these tests are concerned only with verification of proper equipment operation under extreme conditions, and results are not published as ratings.

The ARI/ISO standard is not design prescriptive and provides a means for manufacturers to specify unique air and liquid flow rates for both heating and cooling, and for each step of capacity, in each chosen application. Additionally, the ARI/ISO standard introduces the concept of "effective power input" to the heat pump, which includes the power input of the compressor and controls as well as the proportional power input of fans and pumps, whether internal or external, and whether provided by the manufacturer or not. The power input of fans and pumps is proportional in that it only includes that power required to transport air and liquid through the heat pump, and again avoiding design prescription, does not include arbitrary external static conditions for each application. Unlike the previous ARI standards, the power input is calculated in a consistent manner, inclusive of fan and pump power, across all applications.

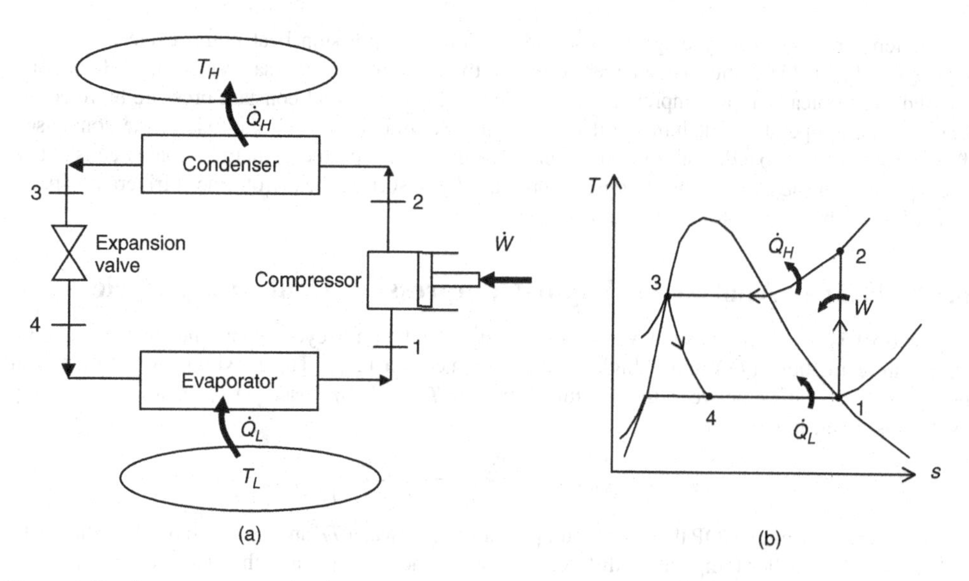

Figure 6.10 A vapor-compression heat pump system for analysis and its temperature–entropy diagram for the ideal case.

6.11 Energy Analysis of Vapor-Compression Heat Pump Cycle

Energy analysis of a vapor-compression heat pump cycle is very similar to the energy analysis of vapor-compression refrigeration cycle as given in Chapter 4. Applying conservation of energy principle to each of the processes of the cycle as shown in Figure 6.10 for steady-flow operation with negligible kinetic and potential energy changes gives

Compressor:

$$\dot{W} = \dot{m}(h_2 - h_1) \tag{6.7}$$

Condenser:

$$\dot{Q}_H = \dot{m}(h_2 - h_3) \tag{6.8}$$

Expansion valve:

$$h_3 = h_4 \tag{6.9}$$

Evaporator:

$$\dot{Q}_L = \dot{m}(h_1 - h_4) \tag{6.10}$$

An energy balance on the entire system gives

$$\dot{W} + \dot{Q}_L = \dot{Q}_H \tag{6.11}$$

A heat pump is used to supply heat to the high-temperature space. Therefore, the COP of the heat pump cycle is defined as

$$\text{COP} = \frac{\dot{Q}_H}{\dot{W}} \tag{6.12}$$

The temperature–entropy diagram of an ideal vapor-compression heat pump cycle is given in Figure 6.10b. In this cycle, the refrigerant enters the compressor as a saturated vapor. It is compressed isentropically in a compressor; it is cooled and condensed at constant pressure by rejecting heat to high-temperature medium until it exists as a saturated vapor at the exit of the condenser. The refrigerant is expanded in an expansion valve during which the enthalpy remains constant: it is evaporated in the evaporator at constant pressure by absorbing heat from the refrigerated space, and it leaves the evaporator as a saturated vapor.

6.12 Exergy Analysis of Vapor-Compression Heat Pump Cycle

Figure 6.10 is a schematic of a vapor-compression heat pump cycle operating between a low-temperature medium (T_L) and a high-temperature medium (T_H). The maximum COP of a heat pump cycle operating between temperature limits of T_L and T_H based on the Carnot heat pump cycle was given in Chapter 1 as

$$COP_{Carnot} = \frac{T_H}{T_H - T_L} = \frac{1}{1 - T_L/T_H} \tag{6.13}$$

This is the maximum COP that a heat pump operating between T_L and T_H can have. Equation 6.13 indicates that a smaller temperature difference between the heat sink and the heat source ($T_H - T_L$) provides greater heat pump COP.

The aim in an exergy analysis is usually to determine the exergy destructions in each component of the system and to determine exergy efficiencies. The components with greater exergy destructions are also those with more potential for improvements. Exergy destruction in a component can be determined from an exergy balance on the component. It can also be determined by first calculating the entropy generation and using

$$\dot{E}x_{dest} = T_0 \dot{S}_{gen} \tag{6.14}$$

where T_0 is the dead-state temperature or environment temperature. In a heat pump, T_0 is usually equal to the temperature of the low-temperature medium T_L. Exergy destructions and exergy efficiencies for major components of the cycle are as follows:

Compressor:

$$\dot{E}x_{dest,1-2} = \dot{W} + \dot{E}x_1 - \dot{E}x_2 = \dot{W} - \Delta\dot{E}x_{12} = \dot{W} - \dot{m}\left[h_2 - h_1 - T_0(s_2 - s_1)\right] = \dot{W} - \dot{W}_{rev} \tag{6.15}$$

or

$$\dot{E}x_{dest,1-2} = T_0 \dot{S}_{gen,1-2} = \dot{m} T_0 (s_2 - s_1) \tag{6.16}$$

$$\eta_{ex,Comp} = \frac{\dot{W}_{rev}}{\dot{W}} = 1 - \frac{\dot{E}x_{dest,1-2}}{\dot{W}} \tag{6.17}$$

Condenser:

$$\dot{E}x_{dest,2-3} = \dot{E}x_2 - \dot{E}x_3 - \dot{E}x_{\dot{Q}_H} = \dot{m}\left[h_2 - h_3 - T_0(s_2 - s_3)\right] - \dot{Q}_H\left(1 - \frac{T_0}{T_H}\right) \tag{6.18}$$

or

$$\dot{E}x_{dest,2-3} = T_0 \dot{S}_{gen,2-3} = \dot{m} T_0 \left(s_3 - s_2 + \frac{q_H}{T_H}\right) \tag{6.19}$$

$$\eta_{ex,Cond} = \frac{\dot{E}x_{\dot{Q}_H}}{\dot{E}x_2 - \dot{E}x_3} = \frac{\dot{Q}_H\left(1 - \frac{T_0}{T_H}\right)}{\dot{m}\left[h_2 - h_3 - T_0(s_2 - s_3)\right]} = 1 - \frac{\dot{E}x_{dest}}{\dot{E}x_2 - \dot{E}x_3} \tag{6.20}$$

Expansion valve:

$$\dot{E}x_{\text{dest},3-4} = \dot{E}x_3 - \dot{E}x_4 = \dot{m}\,[h_3 - h_4 - T_0(s_3 - s_{43})] \tag{6.21}$$

or

$$\dot{E}x_{\text{dest},3-4} = T_0 \dot{S}_{\text{gen},3-4} = \dot{m} T_0 (s_4 - s_3) \tag{6.22}$$

$$\eta_{\text{ex,ExpValve}} = \frac{0}{\dot{E}x_3 - \dot{E}x_4} = 1 - \frac{\dot{E}x_{\text{dest},3-4}}{\dot{E}x_3 - \dot{E}x_4} = 1 - \frac{\dot{E}x_3 - \dot{E}x_4}{\dot{E}x_3 - \dot{E}x_4} \tag{6.23}$$

Evaporator:

$$\begin{aligned}\dot{E}x_{\text{dest},4-1} &= (\dot{E}x_4 - \dot{E}x_1) - \dot{E}x_{\dot{Q}_L} \\ &= \dot{m}\,[h_4 - h_1 - T_0(s_4 - s_1)] - \left[-\dot{Q}_L\left(1 - \frac{T_0}{T_L}\right)\right]\end{aligned} \tag{6.24}$$

or

$$\dot{E}x_{\text{dest},4-1} = T_0 \dot{S}_{\text{gen},4-1} = \dot{m} T_0 \left(s_1 - s_4 - \frac{q_L}{T_L}\right) \tag{6.25}$$

$$\eta_{\text{ex,Evap}} = \frac{\dot{E}x_{\dot{Q}_L}}{\dot{E}x_1 - \dot{E}x_4} = \frac{-\dot{Q}_L\left(1 - \frac{T_0}{T_L}\right)}{\dot{m}\,[h_1 - h_4 - T_0(s_1 - s_4)]} = 1 - \frac{\dot{E}x_{\text{dest},4-1}}{\dot{E}x_1 - \dot{E}x_4} \tag{6.26}$$

The total exergy destruction in the cycle can be determined by adding exergy destructions in each component:

$$\dot{E}x_{\text{dest,total}} = \dot{E}x_{\text{dest},1-2} + \dot{E}x_{\text{dest},2-3} + \dot{E}x_{\text{dest},3-4} + \dot{E}x_{\text{dest},4-1} \tag{6.27}$$

It can be shown that the total exergy destruction in the cycle can also be expressed as the difference between the exergy supplied (power input) and the exergy recovered (the exergy of the heat transferred to the high-temperature medium):

$$\dot{E}x_{\text{dest,total}} = \dot{W} - \dot{E}x_{\dot{Q}_H} \tag{6.28}$$

where the exergy of the heat transferred to the high-temperature medium is given by

$$\dot{E}x_{\dot{Q}_H} = \dot{Q}_H \left(1 - \frac{T_0}{T_H}\right) \tag{6.29}$$

This is in fact the minimum power input to accomplish the required heating load $\dot{Q}_H$:

$$\dot{W}_{\min} = \dot{E}x_{\dot{Q}_H} \tag{6.30}$$

The second-law efficiency (or exergy efficiency) of the cycle is defined as

$$\eta_{\text{II}} = \frac{\dot{E}x_{\dot{Q}_H}}{\dot{W}} = \frac{\dot{W}_{\min}}{\dot{W}} = 1 - \frac{\dot{E}x_{\text{dest,total}}}{\dot{W}} \tag{6.31}$$

Substituting $\dot{W} = \dot{Q}_H/\text{COP}$ and $\dot{E}x_{\dot{Q}_H} = \dot{Q}_H\,(1 - T_0/T_H)$ into the second-law efficiency equation,

$$\eta_{\text{II}} = \frac{\dot{E}x_{\dot{Q}_H}}{\dot{W}} = \frac{\dot{Q}_H\,(1 - T_0/T_H)}{\dot{Q}_H/\text{COP}} = \dot{Q}_H \left(1 - \frac{T_0}{T_H}\right)\frac{\text{COP}}{\dot{Q}_H} = \frac{\text{COP}}{T_H/(T_H - T_L)} = \frac{\text{COP}}{\text{COP}_{\text{Carnot}}} \tag{6.32}$$

since $T_0 = T_L$. Thus, the second-law efficiency is also equal to the ratio of actual and maximum COPs for the cycle. This second-law efficiency definition accounts for irreversibilities within the heat pump since heat transfers with the high- and low-temperature reservoirs are assumed reversible.

Example 6.1

A heat pump is used to keep a room at 25 °C by rejecting heat to an environment at 5 °C. The total heat loss from the room to the environment is estimated to be 45,000 kJ/h and the power input to the compressor is 4.5 kW. Determine (a) the rate of heat absorbed from the environment in kJ/h, (b) the COP of the heat pump, (c) the maximum rate of heat supply to the room for the given power input, and (d) the second-law efficiency of the cycle. (e) Also, determine the minimum power input for the same heating load and the exergy destruction of the cycle.

Solution

(a) The rate of heat absorbed from the environment in kJ/h is

$$\dot{Q}_L = \dot{Q}_H - \dot{W} = 45,000\,\text{kJ/h} - (4.5\,\text{kW})\left(\frac{3600\,\text{kJ/h}}{1\,\text{kW}}\right) = \mathbf{28{,}800\,kJ/h}$$

(b) The COP of the heat pump is

$$\text{COP} = \frac{\dot{Q}_H}{\dot{W}} = \frac{(45,000/3600)\,\text{kW}}{4.5\,\text{kW}} = \mathbf{2.78}$$

(c) The COP of the Carnot cycle operating between the same temperature limits and the maximum rate of heat supply to the room for the given power input are

$$\text{COP}_{\text{Carnot}} = \frac{T_H}{T_H - T_L} = \frac{298}{298 - 278} = 14.9$$

$$\dot{Q}_{H,\text{max}} = \dot{W}\text{COP}_{\text{Carnot}} = (4.5\,\text{kW})\left(\frac{3600\,\text{kJ/h}}{1\,\text{kW}}\right)(14.9) = \mathbf{241{,}380\,kJ/h}$$

(d) The second-law efficiency of the cycle is

$$\eta_{\text{II}} = \frac{\text{COP}}{\text{COP}_{\text{Carnot}}} = \frac{2.78}{14.9} = 0.186 = \mathbf{18.6\%}$$

(e) The minimum power input for the same heating load and the exergy destruction of the cycle are

$$\dot{W}_{\text{min}} = \dot{E}x_{\dot{Q}_H} = \dot{Q}_H\left(1 - \frac{T_0}{T_H}\right) = (45,000\,\text{kJ/h})\left(1 - \frac{278}{298}\right) = \mathbf{3020\,kJ/h}$$

$$\dot{E}x_{\text{dest}} = \dot{W} - \dot{W}_{\text{min}} = (4.5 \times 3600)\,\text{kJ/h} - 3020\,\text{kJ/h} = \mathbf{13{,}180\,kJ/h}$$

The second-law efficiency may alternatively be determined from

$$\eta_{\text{II}} = \frac{\dot{W}_{\text{min}}}{\dot{W}} = \frac{3020\,\text{kJ/h}}{(4.5 \times 3600)\,\text{kJ/h}} = 0.186 = 18.6\%$$

The result is the same as expected.

Example 6.2

A heat pump operates on the ideal vapor-compression refrigeration cycle with refrigerant-134a as the working fluid. The refrigerant evaporates at −20 °C and condenses at 1200 kPa. The refrigerant absorbs heat from ambient air at 4 °C and transfers it to a space at 24 °C. Determine (a) the work input and the COP, (b) the exergy destruction in each component of the cycle and the total exergy destruction in the cycle, (c) the minimum work input and the second-law efficiency of the cycle. (d) Determine the COP, the minimum power input, the total exergy destruction, and the exergy efficiency of the cycle if a ground-source heat pump is used with a ground temperature of 18 °C. The evaporating temperature in this case is −6 °C. Take everything else the same.

Solution

(a) Temperature–entropy diagram of the cycle is given in Figure 6.11.

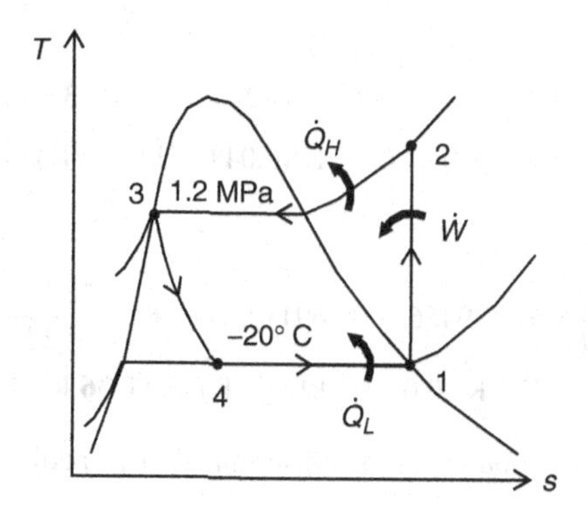

Figure 6.11 Temperature–entropy diagram of the cycle considered in Example 6.2.

From the refrigerant-134a tables (Tables B.3 through B.5)

$$\left.\begin{array}{l} T_1 = -20\,°\text{C} \\ x_1 = 1 \end{array}\right\} \begin{array}{l} h_1 = 238.41\,\text{kJ/kg} \\ s_1 = 0.9456\,\text{kJ/kg}\cdot\text{K} \end{array}$$

$$\left.\begin{array}{l} P_2 = 1200\,\text{kPa} \\ s_2 = s_1 \end{array}\right\} h_2 = 284.43\,\text{kJ/kg}$$

$$\left.\begin{array}{l} P_3 = 1200\,\text{kPa} \\ x_3 = 0 \end{array}\right\} \begin{array}{l} h_3 = 117.77\,\text{kJ/kg} \\ s_3 = 0.4244\,\text{kJ/kg}\cdot\text{K} \end{array}$$

$$h_4 = h_3 = 117.77\,\text{kJ/kg}$$

$$\left.\begin{array}{l} T_4 = -20\,°\text{C} \\ h_4 = 117.77\,\text{kJ/kg} \end{array}\right\} s_4 = 0.4691\,\text{kJ/kg}\cdot\text{K}$$

$$q_L = h_1 - h_4 = 238.41 - 117.77 = 120.6\,\text{kJ/kg}$$

$$q_H = h_2 - h_3 = 284.43 - 117.77 = 166.7\,\text{kJ/kg}$$

$$w = h_2 - h_1 = 284.43 - 238.41 = \mathbf{46.0\,kJ/kg}$$

The COP of the cycle is

$$\text{COP} = \frac{q_H}{w} = \frac{166.7 \text{ kJ/kg}}{46.0 \text{ kJ/kg}} = \mathbf{3.62}$$

(b) The exergy destruction in each component of the cycle is determined as follows:
Compressor:

$$s_{\text{gen},1-2} = s_2 - s_1 = 0$$

$$ex_{\text{dest},1-2} = T_0 s_{\text{gen},1-2} = \mathbf{0}$$

Condenser:

$$s_{\text{gen},2-3} = s_3 - s_2 + \frac{q_H}{T_H} = (0.4244 - 0.9456) \text{ kJ/kg} \cdot \text{K} + \frac{166.7 \text{ kJ/kg}}{297 \text{ K}} = 0.03991 \text{ kJ/kg} \cdot \text{K}$$

$$ex_{\text{dest},2-3} = T_0 s_{\text{gen},2-3} = (277 \text{ K})(0.03991 \text{ kJ/kg} \cdot \text{K}) = \mathbf{11.06 \text{ kJ/kg}}$$

Expansion valve:

$$s_{\text{gen},3-4} = s_4 - s_3 = 0.4691 - 0.4244 = 0.04473 \text{ kJ/kg} \cdot \text{K}$$

$$ex_{\text{dest},3-4} = T_0 s_{\text{gen},3-4} = (277 \text{ K})(0.04473 \text{ kJ/kg} \cdot \text{K}) = \mathbf{12.39 \text{ kJ/kg}}$$

Evaporator:

$$s_{\text{gen},4-1} = s_1 - s_4 - \frac{q_L}{T_L} = (0.9456 - 0.4691) \text{ kJ/kg} \cdot \text{K} - \frac{120.6 \text{ kJ/kg}}{277 \text{ K}} = 0.04100 \text{ kJ/kg} \cdot \text{K}$$

$$ex_{\text{dest},4-1} = T_0 s_{\text{gen},4-1} = (277 \text{ K})(0.04100 \text{ kJ/kg} \cdot \text{K}) = \mathbf{11.36 \text{ kJ/kg}}$$

The total exergy destruction can be determined by adding exergy destructions in each component:

$$ex_{\text{dest,total}} = ex_{\text{dest},1-2} + ex_{\text{dest},2-3} + ex_{\text{dest},3-4} + ex_{\text{dest},4-1}$$
$$= 0 + 11.06 + 12.39 + 11.36 = \mathbf{34.8 \text{ kJ/kg}}$$

(c) The exergy of the heat transferred to the high-temperature medium is

$$ex_{q_H} = q_H \left(1 - \frac{T_0}{T_H}\right) = (166.7 \text{ kJ/kg}) \left(1 - \frac{277}{297}\right) = 11.22 \text{ kJ/kg}$$

Thus, the minimum work input is

$$w_{\min} = ex_{q_H} = \mathbf{11.22 \text{ kJ/kg}}$$

The second-law efficiency of the cycle is

$$\eta_{\text{II}} = \frac{ex_{q_H}}{w} = \frac{11.22}{46.0} = 0.244 = \mathbf{24.4\%}$$

The total exergy destruction may also be determined from

$$ex_{\text{dest,total}} = w - ex_{q_H} = 46.0 - 11.22 = 34.8 \text{ kJ/kg}$$

The result is identical as expected.

(d) Repeating calculations for $T_L = 18\,°C$ and $T_1 = -6\,°C$, we obtain

$$\text{COP} = \frac{q_H}{w} = \frac{163.2\ \text{kJ/kg}}{34.0\ \text{kJ/kg}} = \mathbf{4.80}$$

$$w_{\min} = ex_{q_H} = \mathbf{3.30\ kJ/kg}$$

$$\eta_{\text{II}} = \frac{ex_{q_H}}{w} = \frac{3.30}{34.0} = 0.097 = \mathbf{9.7\%}$$

$$ex_{\text{dest,total}} = w - ex_{q_H} = 34.0 - 3.30 = \mathbf{30.7\,kJ/kg}$$

The COP in the ground-source heat pump case is 32.6% higher than that in the air source cycle. The second-law efficiency of the cycle decreases and the total exergy destruction decreases.

6.13 Mechanical Vapor-Recompression (MVR) Heat Pump Systems

The open-cycle vapor recompression evaporator provides a very efficient means of concentrating dilute solutions using the solvent removed as the operating fluid. The latent heat of vaporization is recovered when the evaporated vapor is condensed following compression and the excess solvent is then available for recovery if required. Alternatively, the process may be used to obtain a purer solvent, as, for example, in desalination of sea water. The ejection of gas from a nozzle into an expander can be used to increase the pressure in a secondary circuit in which the same gas is used as a refrigerant. This method has been applied using steam as the working fluid, primarily to obtain cooling using a conventional steam boiler, but the efficiency of all such systems is low. Heat pumping applications may exist where there is spare steam, possibly in large-scale total energy schemes.

Although the heat pump's operational method of transferring heat from a colder to a hotter reservoir remains the same under various conditions, the means of realizing this method differ according to temperature level, range of temperature rise, variety of heat sources, applicable processes, and so on. These differences characterize the system's design and the choice of hardware for an industrial heat pump. Compared to various other heat pump systems, the MVR heat pump type for latent heat recovery with a high thermal efficiency was thought to be better. It completely replaces conventional systems by means of newly developed component technology, improved performance, operation and service characteristics and reduced installation costs. This tendency has been particularly important since the first oil crisis. This system, which recovers and reuses latent heat, is remarkably effective for processes such as concentration, distillation, and rectification, where the COP is often higher than 20. The compressor used for this MVR system is a specially designed steam compressor. The goal of this system is to achieve a high COP (3–6) by restricting the temperature rise to the minimum required for a refrigerant vapor which has a high latent heat (539 kcal/kg, 1 atmospheric pressure, 100 °C in the case of water) thus also minimizing the power required. The essentials of this technology are as follows (Kuroda, 1986): to efficiently compress water vapor (which has a high specific volume and a comparatively low compression ratio) and raise its temperature; to efficiently exchange this latent heat with a smaller difference in temperature; and to combine both technologies. The combination of these technologies was successful (Figure 6.12). The application fields of these technologies, however, were limited with regard to the availability of process streams containing liquid water and similar materials. On the other hand, the extension of application fields of such a highly efficient system was considered to be important for various industrial processes. If the compression ratio of water vapor could be increased, resulting in higher temperatures, the application fields could be further extended.

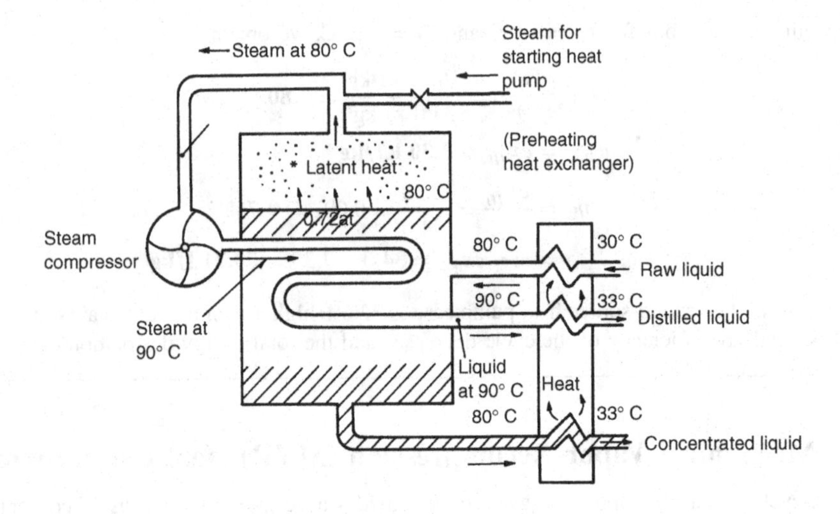

Figure 6.12 A MVR heat pump system (Kuroda, 1986).

The recently developed high-pressure, two-stage, centrifugal compressor with a high compression ratio has permitted improvement of the technology and extension of the application fields. As a basis for the development of heat pump technology, the above-mentioned technology is available not only for working with water vapor but also for working with fluids such as ammonia, Freon, hydrocarbons, and so on. These technologies have been developed to provide higher temperatures and higher efficiencies. Although water is considered to be most suitable for heat pumps used for high-temperature ranges, it is desirable to produce optimum combinations for various conditions and systems.

6.14 Cascaded Heat Pump Systems

In the case of cascaded systems, it is possible to use different refrigerants in the two cycles, each refrigerant being selected as the optimum for the particular temperature range of operation. Figure 6.13 illustrates a schematic of a cascaded system and its log $P-h$ diagram. Another way of increasing the COP is to use cascaded heat pumps. The combination of heat pumps will have a higher COP_H than the COP_H (heat delivered/power absorbed) of a single heat pump performing the same duty. A cascade is probably applicable only to very large systems and will mostly be applied for process heating in industry. However, the use of cascade systems in geothermal and district heating schemes looks extremely promising.

6.15 Rankine-Powered Heat Pump Systems

The Rankine-cycle heat pump converts low-temperature heat (100–200 °C) to usable process steam. In this system, the resource temperature, normally industrial waste heat, is raised through the addition of mechanical work. An organic Rankine cycle turbine extracts energy for this mechanical work from the heat source to drive the heat pump compressor. Figure 6.14 illustrates the operation of a Rankine-cycle heat pump system. The heat stream first passes through the water evaporator to generate low-pressure steam, which is then compressed by a centrifugal compressor to the desired pressure. Steam is discharged at superheated conditions and may be desuperheated at the

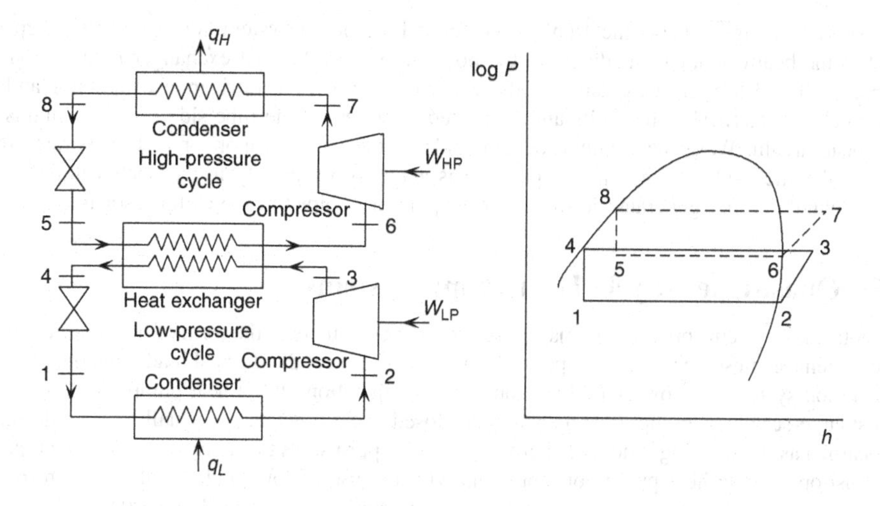

Figure 6.13 Horizontal cascade heat pump cycle and its log $P-h$ diagram.

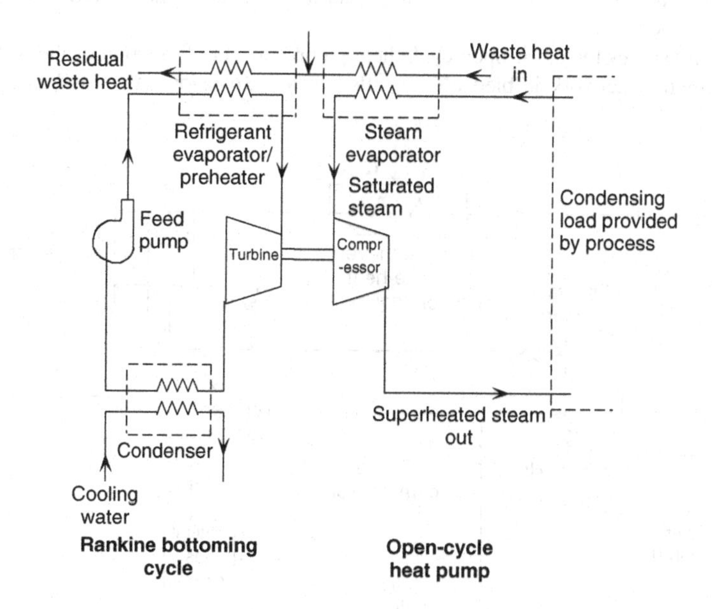

Figure 6.14 A Rankine-powered heat pump system (Adapted from Koebbeman, 1982).

user's option. The highest temperature is utilized in the water evaporators, minimizing the steam compressor pressure ratio and also the power required by the Rankine drive. After the water evaporator, the waste heat passes through the refrigerant evaporator and into the refrigerant preheater. Therefore, the Rankine cycle working fluid (R-113) is heated from condenser conditions to saturated liquid at evaporator conditions. The refrigerant evaporator vaporizes the working fluid, which is then expanded through the turbine to produce the required compressor power. The turbine exhausts into the condenser, where the refrigerant vapors are condensed to liquid and returned to the hot well to repeat the cycle.

As shown in Figure 6.14, the Rankine-powered heat pump system consists of two separate modules: the heat exchanger module and the power module. The heat exchanger module consists of three shell and tube heat exchangers: the water evaporator, the refrigerant evaporator and the refrigerant preheater. All three units are connected in series on the tube side, which contains the waste heat stream. Waste heat enters the tube side of the water evaporator and evaporates fresh water on the shell side. Then, vapor separation is achieved by gravity in the open volume above the tube bundle. Vapor generated in the water evaporator is piped to the compressor inlet.

6.16 Quasi-Open-Cycle Heat Pump Systems

District heating systems provide thermal energy to their customers in the form of hot water or steam. These systems can use one or more types of heat sources to meet the thermal load, including boilers, cogeneration systems, or low-grade heat sources in conjunction with a heat pump.

Most large-scale heat pumps operate using the closed-cycle concept, and usually use a chlorinated fluorocarbon as the working fluid. An alternative to this approach is the quasi-open-cycle heat pump. The quasi-open-cycle heat pump concept deals with the use of low-grade local energy resources in district heating systems that use hot water as the transport medium. The particular low-grade energy resources of interest in this concept are water resources, such as flooded mines and sewage ponds, and waste heat resources, such as low-pressure waste steam, hot water, and exhaust gases from industrial processes.

Figure 6.15 illustrates the quasi-open-cycle heat pump concept. A fraction of the water returning from the distribution network is bled off, throttled in an expansion valve, and injected into the

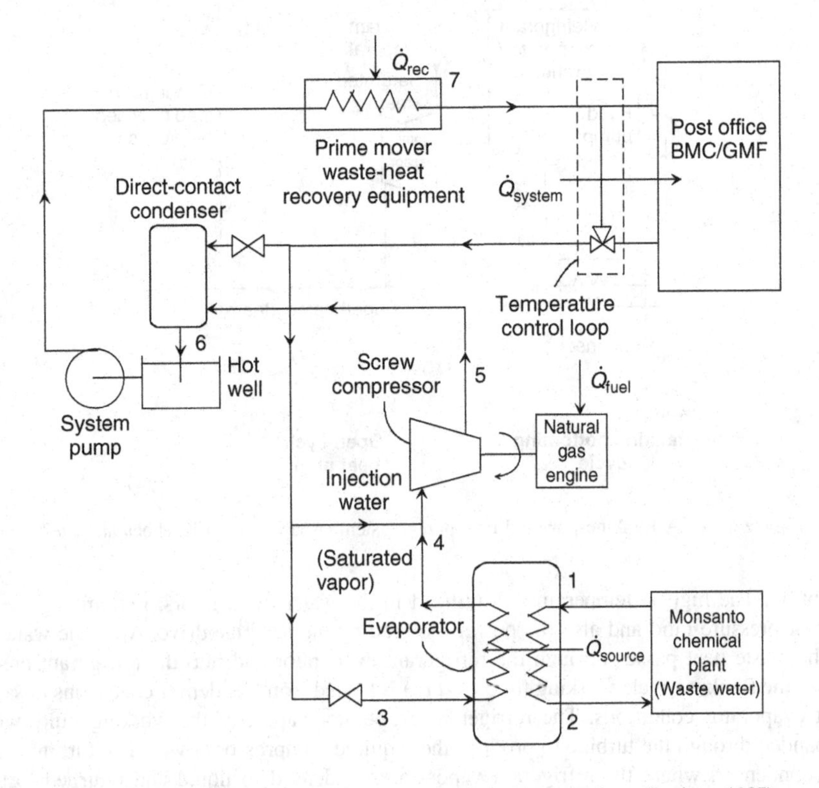

Figure 6.15 A quasi-open-cycle heat pump system (Adapted from Kunjeer, 1987).

evaporative heat exchanger. The heat exchanger is connected to the low-grade energy resource which transfers heat to the low-pressure water and produces saturated steam. The steam produced in the heat exchanger is then compressed in the compressor. The superheated steam that exits the compressor then enters a direct-contact condenser where the remaining fraction of the district heating system return water is added. The water that exits the direct-contact condenser is then pressurized by the district heating system pump where it enters the prime mover's waste heat recovery equipment and, if applicable, is distributed throughout the district heating system where it transfers its thermal energy to various users (Kunjeer, 1987).

The quasi-open-cycle heat pump is "open" in the sense that the compressor working fluid is the same as the district heating hot water transport media. The system, however, resembles the closed cycle in that two heat exchangers, a direct-contact condenser, and a surface-area type evaporator, are used. The quasi-open-cycle heat pump has several advantages over the closed cycle, including the following:

- The working fluid is water, which is nontoxic and has excellent thermal properties.
- Since the high-temperature heat exchanger is a direct-contact condenser as opposed to a primary surface heat exchanger, the capital cost of this heat exchanger is much lower.

The quasi-open-cycle heat pump is found to be best suited for the higher temperature heat resources such as those found in the waste streams of industrial processes. This is due mainly to the thermodynamic properties of steam. At low temperatures, the vapor-specific volume is quite large and, because of the pressure–temperature relationship, a reasonable temperature rise is obtained when the compressor operates with a large pressure ratio.

6.17 Vapor Jet Heat Pump Systems

In the vapor jet heat pump, the kinetic energy of a vapor jet, produced by heat input, is utilized for compressing the refrigerant vapor. In principle, this is a compression process which is, however, operated without input of mechanical energy. The operation of a vapor jet compressor was explained earlier in Chapter 5. In the injection nozzle, the drive vapor at pressure P_i is expanded, and a vapor jet with a velocity several times the velocity of sound is produced. This carries forward the expansion vapor at pressure P_0 and accelerates it. Because of the decreased pressure on the suction side, evaporation takes place and the vapor is cooled by extracting the evaporation enthalpy. The pressure of the vapor mixture is increased in the diffuser to the condensing pressure P at which condensing can take place in the condenser. The definition of a COP for vapor jet heat pumps leads to difficulties. For industrial purposes, the so-called specific vapor consumption, that is, the ratio of drive vapor quantity to suction vapor quantity, is given; relevant tables are available from the manufacturers. In thermodynamic terms, the definition of a heat ratio analogous to the AHP would be logical using the enthalpy of the vapor mixture and the enthalpy of the drive vapor. Because the evaporation enthalpy is very high for water at about 2000 kJ/kg, the process is mostly carried out with water vapor. But, for technical process reasons other media are also used.

6.18 Chemical Heat Pump Systems

A number of other methods of heat pumping have been proposed, which have not, to date, been tested in practical devices. These include vortex tubes (so-called *chemical heat pumps*). The vortex tube heat pump makes use of an effect known as the Ranque effect. If a high-pressure gas is injected tangentially into a tube, a vortex is formed and the gas at the center of the tube is at a lower temperature and pressure than the gas near the tube wall. The gas can be extracted separately from these two regions, yielding heated or cooled gas as required. Although, in principle, air can be used to provide heat by this means, a sufficiently efficient device worth developing has never been demonstrated.

Chemical heat pump systems have been proposed based on the mixing and dissolution of two components. Thermodynamic analysis and consideration of material properties have been carried out theoretically, but no practical machine has emerged, the most common problem being irreversibility. A sorption and desorption system for hydrogen on and from lanthanum penta-nickel has been proposed. Such systems are unlikely to be used other than in extremely specialized applications.

The basis of a chemical heat pump can provide temperature upgrading at high temperatures. This type of temperature upgrading has been difficult to accomplish with conventional technologies. Basically, the chemical heat pump uses a high- and low-temperature heat source, needs only a small amount of mechanical energy input, and, depending on which component element reactions are selected, can deliver useful heat at a desired temperature. A wide variety of combinations of working reactants are conceivable for chemical heat pumps. The most common type is the reversible thermochemical reaction of $CaO/Ca(OH)_2$ that is used along with the evaporation and condensation of water to complete a heat pump cycle (see Hasatani *et al.*, 1988).

The operation of this chemical heat pump consists of the reactions given in the following equations:

$$CaO(s) + H_2O(g) \Rightarrow Ca(OH)_2(s) + 1.858 \times 10^3 \text{ kJ/kg} \tag{6.33}$$

and

$$H_2O(g) \Rightarrow H_2O(l) + 2.316 \times 10^3 \text{ kJ/kg (average value of 293–641 K)} \tag{6.34}$$

Figure 6.16 shows the relationship between the reaction equilibrium pressure P_e and temperature for the reaction in Equation 6.33 which is given by line 2–4 in the figure. Also shown is the relationship between the saturated steam pressure P_s and temperature for Equation 6.34, which is given by line 1–3.

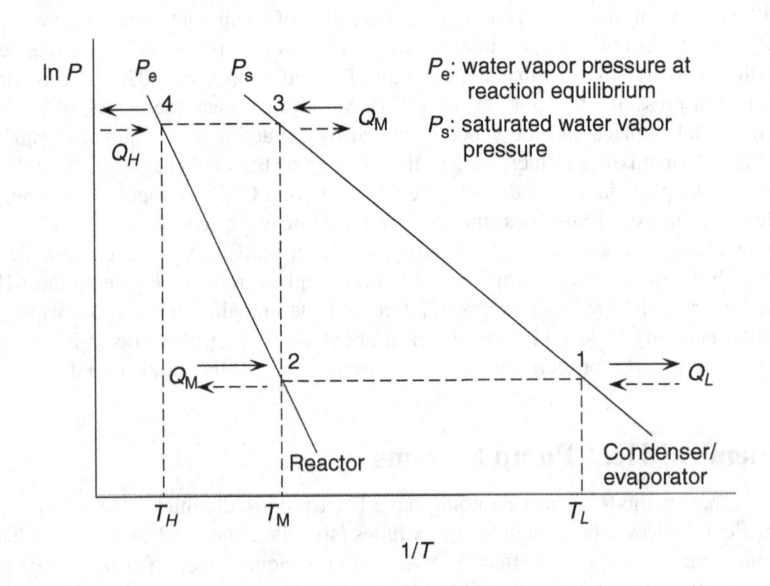

Figure 6.16 ln $P-1/T$ diagram for a chemical heat pump (Adapted from Hasatani *et al.*, 1988).

For the heat release mode, consider a hermetically sealed reaction system having a reactor and an evaporator filled with CaO and water, respectively. If heat (Q_M) is added to the evaporator from a medium temperature (T_M) heat source, the water in the evaporator becomes pressurized steam (path 1–3 in Figure 6.16). Owing to the pressure difference between P_s and P_e, this steam enters the reactor to undergo an exothermic hydration reaction with CaO. This causes the temperature in the reactor to rise (2–4) to temperature T_H at which point high-temperature heat (Q_H) becomes available.

In the heat storage (regeneration mode), heat Q_M from a medium-temperature (T_M) source is added to the reactor which contains $Ca(OH)_2$ formed as described above. At the same time, the condenser is cooled to a temperature T_L. Under these conditions, the $Ca(OH)_2$ undergoes an endothermic dehydration reaction to release steam. The steam shifts from the reactor to the condenser (path 2–1 in Figure 6.16) due to the pressure difference between the two chambers. There it condenses by releasing its latent heat of condensation to the low-temperature heat sink (T_L).

An experimental unit developed by Hasatani *et al.* (1988) is shown schematically in Figure 6.17. The evaporator/condenser (1) and the reactor (2) are made of stainless steel and are cylindrical in shape, having both an inside diameter and a height of 150 mm. Both containers are equipped with a cooling coil (4), thermocouple insertion tube (5), and an electric heater (6). Each of these items is arranged symmetrically about the center of the container. Both containers are also equipped with a pressure gauge (7). In addition, the evaporator/condenser has a water level gauge (8) and the reactor has an auxiliary external heater (9). The auxiliary heater consists of nichrome wire wound around the reactor's outer surface. The two containers are connected to each other by stainless steel piping via valve V2. A vacuum pump (3) is used to obtain the proper pressure level in the reactor. The equipment is insulated with an adiabatic material to reduce heat loss. For the reaction system employed, the temperature T_M shown in Figure 6.16 is in principle about 640 K. At this temperature, the pressure of the saturated steam in the evaporator is 27.5 MPa. The present experimental apparatus was not designed for such high pressures. For this reason the evaporator temperature was kept below about 430 K for this experiment.

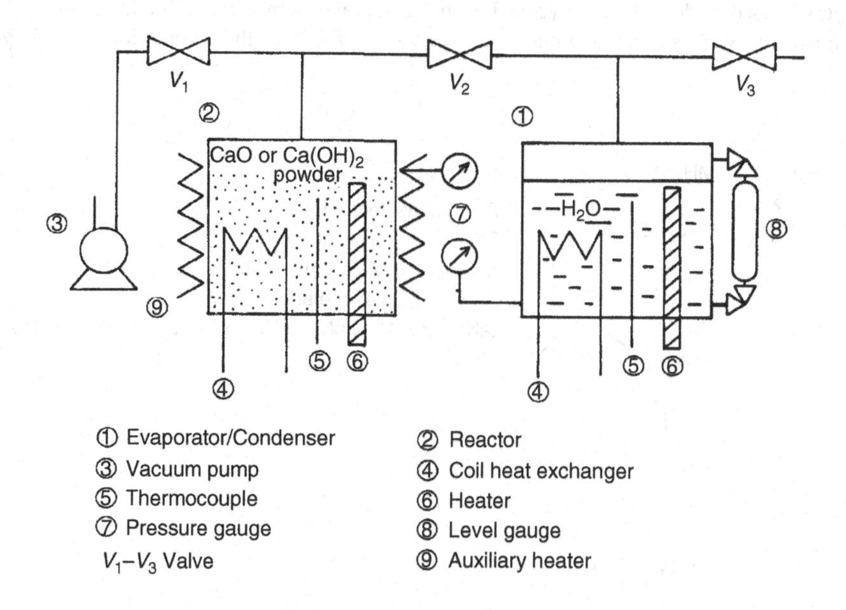

Figure 6.17 Schematic diagram of the experimental unit (Hasatani *et al.*, 1988).

6.19 Metal Hydride Heat Pump Systems

A large amount of thermal energy is involved in the dehydriding and hydriding reactions of hydrogen with hydride-forming materials. Heat involved in each reaction can be used to extract and supply heat for thermal energy storage and conversion, air conditioning, heating, drying, and humidity control devices. Such hydride-forming material is called a *metal hydride* and has been considered as the material extensively applicable to energy storage and conversion. A heat pump is a typical application for the dehydriding and hydriding characteristics of metal hydrides, particularly as a cooling and temperature upgrading device. In the principle of these heat pumps, the reaction between hydrogen and hydride-forming materials is reversible. It is characterized by rapid kinetics and large amounts of heat in an endothermic desorption reaction and exothermic absorption reaction in the following equation:

$$MH_{x+y} = MH_y + x/2H_2 \pm \Delta H \tag{6.35}$$

The amount of heat (ΔH) in each reaction has an average value of 6.4–9.2 kcal/mole of H_2 or 160–230 kJ/kg of alloy. In the case of Mg-based alloys, much higher values are obtained. In the former case, approximately 1000 kg of the alloy are necessary for a heat pump with a capacity of 100 kW.

To run a heat pump cycle, a set of two paired metal hydride heat exchangers are necessary. With this system, a continuous countercurrent flow of hydrogen can be maintained even though the hydriding and dehydriding reactions are executed in a batch-wise manner. Figure 6.18 shows a metal hydride heat pump cycle providing a low temperature (T_l) sink which can be used for air conditioning and refrigeration purposes. The points on 1–4 indicate the final states that the metal hydride and metal achieve during execution of the cycle. Lines labeled MH_l and MH_h are lines of constant concentration for the low vapor pressure (low concentration of H_2) and high vapor pressure (high concentration of H_2) metal hydrides. These equilibrium temperature and pressure relations are known as van Hoff plots.

The cycle is executed as follows (for clarity, a single pair of high and low pressure hydride heat exchangers is used to describe the cycle). Let metal M_h be at temperature T_l and pressure P_3, and let metal hydride MH_l be at temperature T_m and pressure P_6. With these conditions, metal hydride

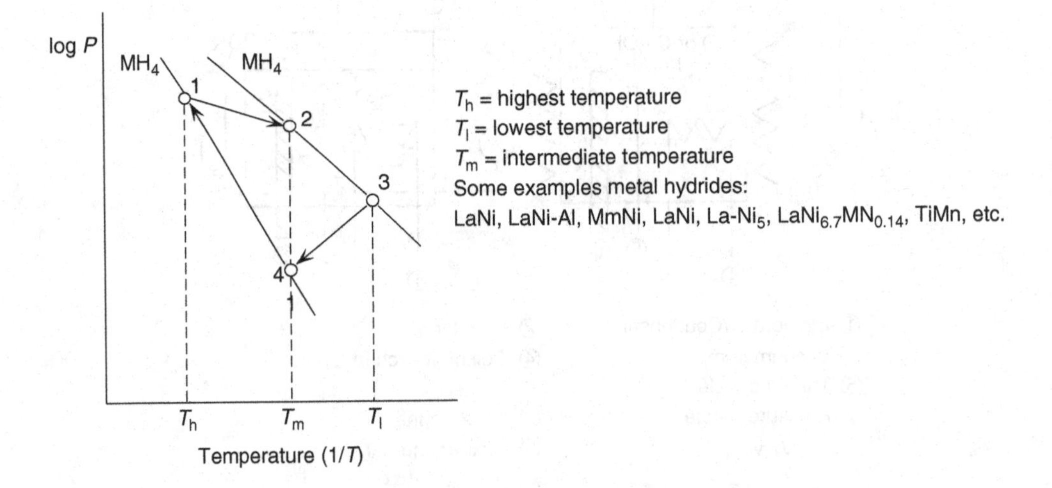

Figure 6.18 Log P and $1/T$ diagram of the heat hydride heat pump (Adapted from Suda, 1987).

MH_l is heated at point 1 to raise its temperature from T_m to T_h. At the same time, it is used to supply hydrogen to metal M_h. Metal M_h, reacting with hydrogen from MH_l, releases energy causing its temperature to rise from T_l to T_m. At point 2, metal M_h has absorbed all the hydrogen to become MH_h at T_m and P_2. Metal M_l is now cooled to T_m by rejecting heat to the atmosphere. MH_h with its vapor pressure $P_2 > P_4$ is then used to supply H_2 to metal M_l. Since the release of H_2 is an endothermic reaction, the temperature of MH_h drops. Once it reaches temperature T_l, heat is added to the metal hydride, MH_h, to maintain this temperature until all the hydrogen has been driven off. The hydrogen is absorbed by the metal M_l until the metal hydride MH_l is formed. The energy released during this exothermic absorption of H_2 is rejected at point 4, thereby maintaining the low-pressure metal hydride at T_{n-m} and P_6. The cycle can now be repeated. As summarized in Figure 6.18, heat is rejected to the atmosphere (T_m) at points 2 and 6. At point 1, a high temperature source (T_h) adds heat to the cycle; and at point 3, heat is absorbed from a low-temperature source (T_l).

6.20 Thermoelectric Heat Pump Systems

Thermoelectrics are based on the *Peltier effect*, discovered by Peltier in 1834, that when a direct electric current passes round a circuit incorporating two different metals, one contact area is heated and the other cooled, depending on the direction of current flow. The Peltier effect provides a means for pumping heat without using moving parts. In a circuit containing two junctions between dissimilar conductors, heat may be transferred from one junction to the other by applying a DC voltage (Figure 6.19). To be effective, the conductors must provide high thermoelectric power and low thermal conductivity, combined with adequate electrical conductivity. Such a combination of properties is not to be found in metallic conductors, and this principle could not be applied to heat pumping until the advent of semiconductor materials, typically bismuth, antimony, selenium, and tellurium alloys. The effectiveness of materials for such applications is measured by the "figure of merit," Z, defined by Heap (1979):

$$Z = \frac{a^2}{k\rho} \tag{6.36}$$

where a is the Seebeck coefficient, k is the thermal conductivity, and ρ is the electrical resistivity of the material. Using presently available materials for which $Z = 0.003\,\mathrm{K}^{-1}$, thermoelectric heat pump performances about half as good as those of typical vapor-compression machines can be predicted. Practical devices only achieve half these predicted values.

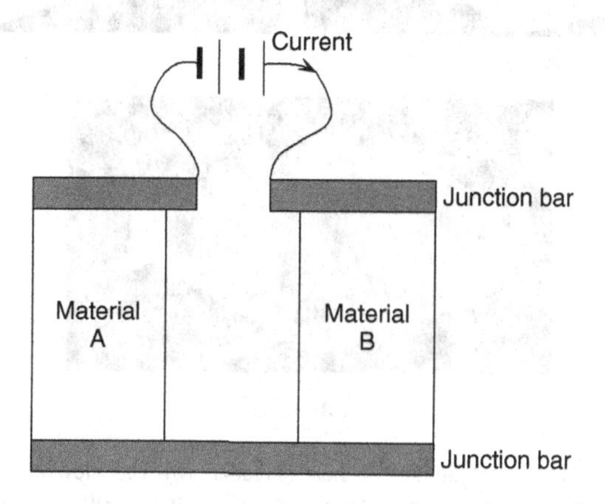

Figure 6.19 Schematic representation of the Peltier effect.

The limitations of known materials and the scope for possible future developments of new materials have been considered by various researchers. If semiconductors were to be developed with $Z = 0.006\,K^{-1}$ or greater there could be a considerable widening of thermoelectric heat pumping applications, and a COP comparable with those obtainable with vapor-compression machines might be achieved. At present, thermoelectrical devices are not competitive with vapor-compression heat pumps, but they may find increasing use in specialized cooling applications where power requirements are low or where silent operation is necessary. Close temperature control of electronic components and cold stores in nuclear submarines are examples of these.

To operate semiconductor Peltier cells, a high direct current at a low voltage is required and consequently a large number of cells are connected together in series. Heat exchangers are also required at hot and/or cold junctions to transfer heat as needed. During the past two decades, custom Peltier (thermoelectric) modules and Peltier coolers have been designed, with heat pump capacities ranging from a few watts to thousands of watts. Currently, some companies are conducting research and development to optimize all Peltier performance parameters independently for maximum heat pump performance.

By using IsoFilm heat spreaders (Figure 6.20a) to effectively transfer heat to and from the Peltier module (Figure 6.20b), higher power density Peltier coolers without sacrificing their maximum temperature differentials are designed and manufactured. Since the manufacturing cost of a Peltier module is highly dependent on its size and not its heat pumping capacity, higher power density designs have a higher performance versus cost ratio. A very high power density ($50\,W/cm^2$) Peltier technology using deposited thin film bismuth–telluride (Bi_2Te_3) with LIGA-processed copper junctions and a proprietary insulation scheme is under development. This entire device is manufactured using wafer fabrication techniques. This twenty-first century wafer-scale processing is a paradigm shift from the mechanical processing currently employed by today's Peltier module manufacturers.

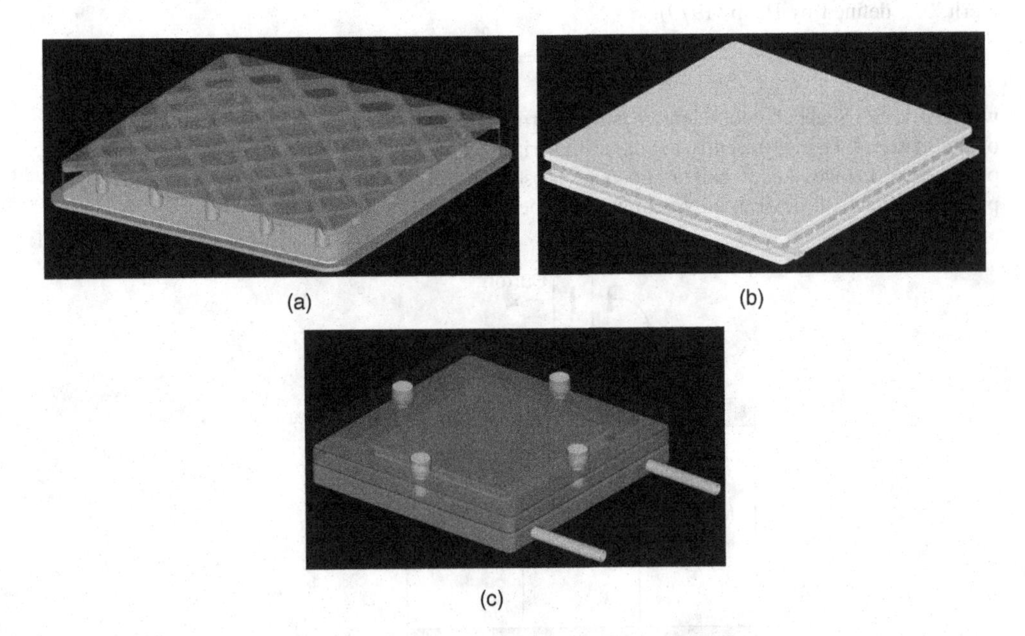

Figure 6.20 (a) Translucent view of an IsoDie heat spreader. (b) The view of a custom Peltier module. (c) The view of a custom-made liquid cooled Peltier (thermoelectric) cooling assembly (*Courtesy of Novel Concepts, Inc.*).

Figure 6.20a shows a translucent view of an IsoDie heat spreader used to planarize the uneven junction temperatures caused by the nonuniform power distribution of an integrated circuit. This IsoDie has five parts; starting at the bottom is the evaporator sidewall, followed by the evaporator sidewall microstructure region (actual microstructures not shown for simplicity), planar capillary form, condenser sidewall microstructure region, and finally, the condenser sidewall. This IsoDie measures 15 mm^2 and is 2 mm thick. It is constructed entirely from oxygen-free high conductivity (OFHC) copper and uses water as its working fluid. Thermal analysis suggests that it will handle over 100 W from a 10 to 100 mm^2 nonuniform power source, including power densities as high as 3.0 W/mm^2, with a thermal resistance of less than 0.16 °C/W (three times better than solid copper).

Figure 6.20b shows a custom-made Peltier module (40 × 40 × 3.3 mm) which has a maximum heat pump capacity of 160 W (10 W/cm^2), and a maximum temperature differential of 67 °C (zero load), under the following conditions: 16.2 V, 17.6 A, and a hot-side temperature of 50 °C.

Figure 6.20c shows a custom-made liquid cooled Peltier (thermoelectric) cooling assembly which measures 60 mm^2 and is 14 mm high (including copper cold plate), and has a maximum heat pump capacity of 100 W (2.8 W/cm^2), with a maximum temperature differential of 63 °C (zero load), under the following conditions: 12.0 V, 11.1 A, and a hot-side liquid input temperature of 50 °C. This Peltier heat pump assembly uses a liquid-cooled copper heat sink, which has a thermal resistance of 0.031 °C/W, with a flow rate of 0.5 L/min, and a pressure drop of 20 kPa. The volumetric thermal efficiency equals 1.984 W/°C/cm^3. Total weight is 281 g.

6.21 Resorption Heat Pump Systems

Renewed interest in ammonia has been evident in the wake of the Montreal protocol and mixtures of ammonia and water appear to be particularly suitable as the working fluid in high-temperature heat pump applications. The advantages over a single component working fluid are related to the possibility of matching the temperature glide of the working fluid to that of the source/sink and the variation of the circulation composition to enable better matching of the source/sink conditions and load. Furthermore, it is possible to configure the system and working fluid to allow heat rejection at temperatures in the range of 80–120 °C. This temperature range would be typical of a number of industrial process heating applications, with the heat pump utilizing what might otherwise be waste heat at temperatures between 40 and 80 °C (Mongey *et al.*, 2001). The practical application of ammonia–water mixtures cannot be achieved using a conventional vapor-compression cycle. The temperature glide associated with complete phase change is of the order of 100 °C, so it could be expected that only partial phase change will be achieved in any specific application. Wet compression does not appear to be a realistic proposition, because of the large liquid fraction remaining after any typical heat exchange process.

A more practical alternative is to separate the phases after the working fluid has come into thermal contact with the heat source. The vapor passes through the compressor while a solution pump is used to transfer the liquid to the high side before recombining with the vapor. This approach is referred to as *a* resorption cycle, with the desorber and resorber performing the same functions as the evaporator and condenser in the vapor-compression cycle. The resorption cycle is shown schematically in Figure 6.21. Changes in the circulating composition can be achieved by varying the flow ratios of the vapor and liquid phases. Modulation of the flow velocity through the solution pump is thought to be the most practical means of achieving this end. Because of this, a receiver is required to store a charge of working fluid that has a much greater volume than that necessary for circulation purposes. In order to achieve circulating compositions that differ significantly from the original bulk charge, a considerable proportion of the charge must be removed from circulation. This excess fluid can be stored at the point where phase separation occurs, since equilibrium liquid and vapor phases differ markedly.

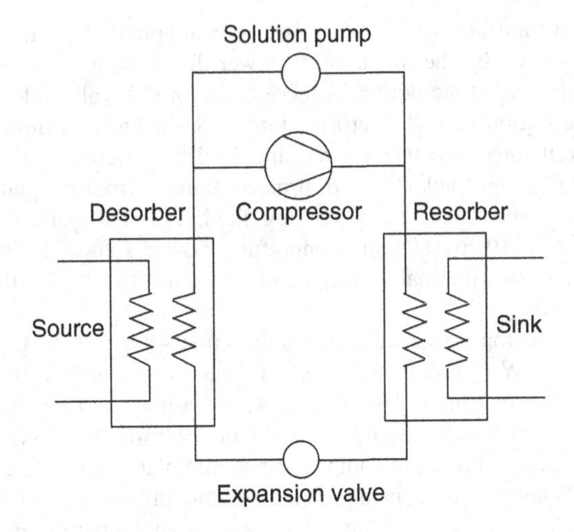

Figure 6.21 Schematics of a resorption cycle (Adapted from Mongey *et al.*, 2001).

The temperature of cooling water in the condenser of cold vapor machines (in this category we have compression and AHPs of conventional type and style) is usually about 5–10 °C lower than the constant condensation temperature, and in the evaporator the temperature of the incoming heat source is higher than the constant temperature of the boiling refrigerant. This results in irreversibilities, because the absorber has to take in the cold vapor from a temperature which is lower than the heat source temperature. Similarly, the refrigerant in the expeller is expelled at a higher temperature than the corresponding heat sink temperature. These losses can be avoided in part if, according to a recommendation of Altenkirch, the condensation and the evaporation do not proceed at a constant temperature, but within a given temperature range. This can be obtained by substituting the branch condenser, refrigerant injection valve, and evaporator by a second working fluid circuit with resorber (instead of condenser) and desorber (instead of evaporator) in an AHP. In the resorber a suitable weak solution absorbs the refrigerant set free in the expeller. The solution temperature in the resorber varies according to the concentration of the solution. The resorption of the refrigerant vapor does not take place at a constant temperature as in the condenser, but within a temperature range. The same applies to the desorber which substitutes the evaporator. The advantages of resorption heat pumps compared with common AHPs are

- no rectification unit (e.g., ammonia/water),
- lower pressure difference,
- reduced losses and increased heat ratio.

However, a high premium has to be paid for the above-mentioned advantages by the higher first cost (e.g., an additional working fluid pump).

The single-stage resorption heat transformer should be mentioned as the fundamental configuration (Figure 6.22) from which a large number of variations can be developed. The condenser is replaced by the resorber and the evaporator by the desorber; like the resorption heat pump this requires the addition of a second solution circuit. This design permits heat transfer at varying temperature differences. Thereby, lower exergy losses and reduced consumption of cooling water are attained. Higher heat ratios in comparison with the Type I AHP, and lower electric power consumption are possible for the resorption heat transformer.

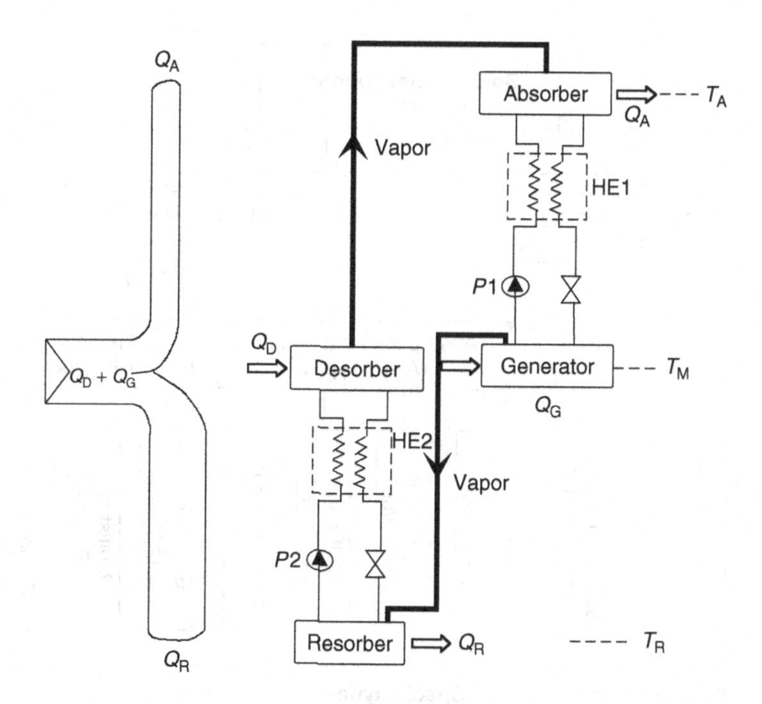

Figure 6.22 Principal flow sheet of the resorption heat transformer (Adapted from Podesser, 1984).

6.22 Absorption Heat Pump (AHP) Systems

The beginning of the absorption cooling technology dates back to the middle of the nineteenth century. At that time, the brothers Ferdinand and Edmund Carre were building a periodic and, some years later, a continuously working absorption cooling machine for ice production. As a result, names such as E. Altenkirch, R. Planck, F. Merkel, F. Bosnjakovit, W. Niebergall, and some others are inseparably connected with the research and development of the absorption cooling machine. Altenkirch, in particular, made various suggestions for different methods and measures for the reduction of irreversible losses, and, as early as 1911, suggested a central heating system using AHP. In 1959, Niebergall gave a comprehensive presentation of absorption technology.

The heat pump is one of the two classical processes to generate large amounts of low- and medium-temperature energy by means of relatively small amounts of exergy (available energy), the other process being the cogeneration of heat and power. The implementation of the heat pump process requires mechanical (or thermal) equipment, increasing first cost but decreasing energy cost. It is not surprising that increased cost of conventional fuel and decreasing first cost of heat pumps (by mass production) has greatly improved the economy of the heat pump.

The main topics of research and development on heat pump units are technical improvements of the units on the one hand, and, on the other hand, reduction of their production cost by large-scale, partly or fully automated production and by a design geared to it. Generally, technical improvements will cause higher production cost, but the actual cost increase may be small if modern production techniques are applied. Some of the many ways to improve the heat pump units are increased motor efficiency, lower heat losses, speed control by pole changing and by thyristors, new improved compressor types, double cycle, economizers, nonazeotropic refrigerant mixtures, improved heat exchangers and expansion valves, and use of waste heat from the compression side

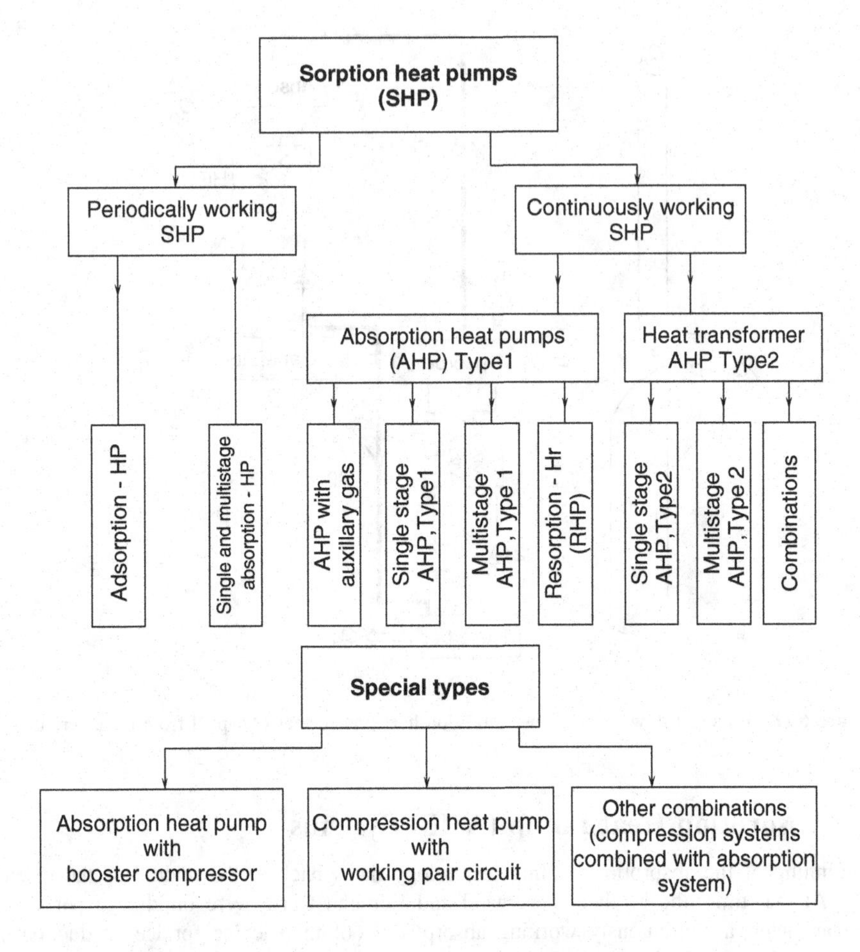

Figure 6.23 Classification of diffusion heat pumps (Adapted from Podesser, 1984).

for superheating suction gas. The technical development potential is large, even if it cannot be fully utilized for economic reasons.

The sorption heat pump process differs fundamentally from the compression heat pump process in that the mechanical compressor is replaced by a thermal compressor. The thermal compressor consists of expeller (generator), where the drive energy (heat) is supplied, solution heat exchanger, absorber, working fluid pump, and expansion valve. It is powered by heat. The entire group of possible heat pumps with thermal compressors is, according to Niebergall, called *sorption heat pumps*, because the procedure of "sucking in" can occur both through absorption into liquid or nonliquid substances and through adsorption from nonliquid substances. Figure 6.23 shows a possible categorization of the better known types of sorption machines.

So far, only the periodically working absorption systems and, especially, the periodic AHP systems with a liquid working pair, show significance. Basically, the single-stage periodic AHP consists of two parts, one of which acts as both evaporator and condenser, and the other as both expeller and absorber, but in a shift of time. The single-stage periodically working absorption system has been applied many times to cooling machines, but as an AHP has not become very significant, till date.

The advantage of the multistage type in comparison to the single-stage type lies in the attainment of low evaporation temperatures or better heat ratio (the ratio between available heat and driving heat).

The majority of plants built to date fall into this group of AHPs. In most cases the experience from absorption refrigeration technology could be used in their design and construction. In general, heat requirements of from at least 15 kW up to several megawatts are required for the application of AHPs. In this connection, a COP as high as possible for the capacity of the AHP should be aimed at. For this reason, AHPs with an auxiliary gas and no fluid pumps are scarcely used because of the low heat ratio of approximately 1.25, all in spite of the captivating simplicity of the system.

The continuously working, single-stage AHP has become the most famous type of AHP through applications in experimental plants of small capacities (up to 50 kW) for heating houses or large capacities up to several megawatts. All the experience from absorption refrigeration technology can be adapted directly to the design of such AHPs up to output temperatures of about 60 °C. Mostly used working pairs are $LiBr/H_2O$, $LiBr/CH_3OH$, and H_2O/NH_3. The limitations to the application of the working pairs mentioned are determined by the danger of recrystalization, of chemical decomposition, and of increased susceptibility to corrosion at high working temperatures.

Absorption refrigeration machine, AHP (Type I), and heat transformer (Type II) are shown. In the above-mentioned summer and winter operations a $LiBr/H_2O$ AHP and a $LiBr/CH_3OH$ AHP work together, the only connection being by means of the external water circuits. In summer this system is used for ice (−13 °C) production and for supplying the air-conditioning system (+6 °C) while in winter it works as a heat transformer with temperatures of +90 °C for space heating. Waste heat with a temperature of 60 °C is used to drive the system. As an AHP (Type I), this plant achieves a heat ratio of 1.67, and as heat transformer (Type II), the heat ratio is 0.51. All main elements of these machines are taken directly from absorption refrigeration technology. The COP ranges from 1.45 to 1.55 for single-stage machines at design conditions, because of the irreversible losses caused by dephlegmation and rectification (increase of concentration of the expelled working vapors).

The conventional application of the absorption refrigeration machine has been clearly abandoned. It is worth mentioning that the working pair NH_3/H_2O in the expeller is operated nearly at the limit temperature of about 180 °C, where the pressure reaches almost 40 bar.

AHPs are thermally driven, which means that heat rather than mechanical energy is supplied to drive the cycle. AHPs for space conditioning are often gas fired, while industrial installations are usually driven by high-pressure steam or waste heat. Most attention presently is going to the AHPs in industry. The phenomenon of vapor absorption in a liquid has been applied successfully in refrigeration equipment. However, only very few AHPs are presently available on the market. The transition from cooling device to heating device, in contrast with vapor-compression systems, has not been achieved thus far.

As with vapor-compression cycle heat pumps, there are many varieties of AHPs, using different working fluids or different cycles. The AHPs use a combination of a refrigerant and an absorbent, called the *working fluid*.

Absorption systems utilize the ability of liquids or salts to absorb the vapor of the working fluid. The most common working pairs for absorption systems are

- water (working fluid) and lithium bromide (absorbent) and
- ammonia (working fluid) and water (absorbent).

A profound understanding of the operation of an AHP requires acquaintance with the thermodynamics of binary mixtures. Based on the properties of these mixtures, thermodynamic absorption cycles are developed and discussed. Distinction here is made with heat transformer cycles, the latter being characterized by the utilization of low-temperature heat as driving heat source. Special attention is devoted to the requirements to be filled by the working pairs suited for AHPs and transformers.

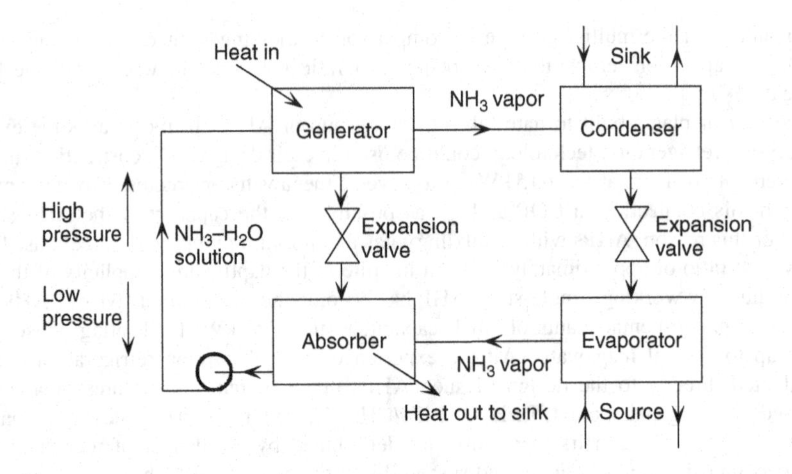

Figure 6.24 An ammonia–water absorption cycle heat pump.

AHPs, in addition to condensers and evaporators, require components such as: generators, rectifiers, and absorbers. The principles underlying the operation of each of these components are discussed in detail. It is shown that design and optimization of these items are very complex and require extensive knowledge of the thermodynamic and thermal properties of the working pairs. Such knowledge is often lacking. Furthermore, it will become apparent as to how a number of limitations originating from the working pair influence the performance of the heat pump. It is felt that more suitable working pairs have to be developed.

A schematic diagram of a basic ammonia–water AHP is shown in Figure 6.24. It can be seen in Figure 6.24 that the absorption cycle is similar to the vapor-compression cycle. In the vapor-compression cycle, there is a vapor-compression part (including the compressor). In the absorption cycle there is an absorption part (including the absorber, solution pump, and generator). Therefore, the difference is that the compressor is replaced with an absorber, a solution pump, and a generator. As can be seen in Figure 6.24, the ammonia refrigerant absorbs heat in the evaporator and rejects heat in the condenser in the same way as in vapor-compression heat pump systems. In the absorber, the water, which is the absorbent fluid, absorbs the cool ammonia vapor, creating a strong solution. This solution enters a solution pump and is pumped to the high-pressure side, to the generator. From the generator, hot ammonia vapor is then condensed, expanded, and returned to the evaporator in the normal way.

In absorption systems, compression of the working fluid is achieved thermally in a solution circuit which consists of an absorber, a solution pump, a generator, and an expansion valve as shown in Figure 6.25. Low-pressure vapor from the evaporator is absorbed in the absorbent. This process generates heat. The solution is pumped to high pressure and then enters the generator, where the working fluid is boiled off with an external heat supply at a high temperature. The working fluid (vapor) is condensed in the condenser while the absorbent is returned to the absorber via the expansion valve. Heat is extracted from the heat source in the evaporator. Useful heat is given off at medium temperature in the condenser and in the absorber. In the generator high-temperature heat is supplied to run the process. A small amount of electricity may be needed to operate the solution pump.

For heat transformers, which, through the same absorption processes, can upgrade waste heat without requiring an external heat source. The details of the heat transformers are given in Section 6.21.

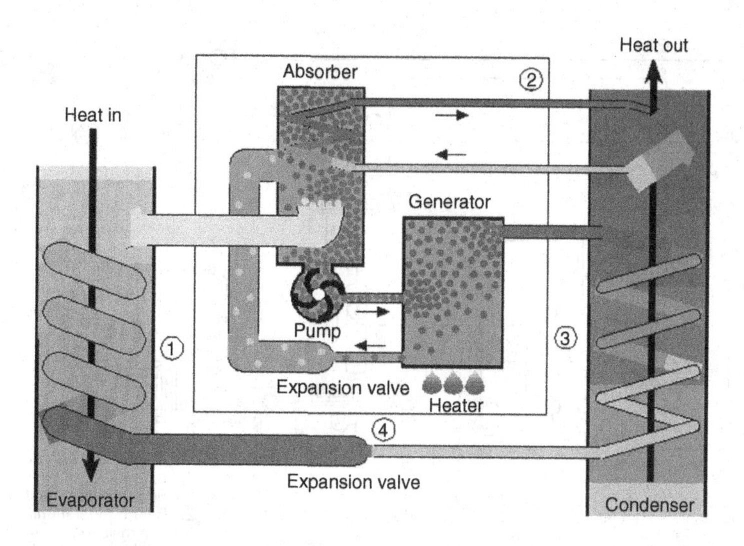

Figure 6.25 An absorption heat pump (*Courtesy of IEA-HPC*).

Example 6.3

Figure 6.26 shows a schematic diagram of a single-stage, ammonia–water AHP which started operating at the Technische Werke der Stadt Stuttgart AG. The general and technical features and details of this system are as follows (Lehmann, 1986):

- Heat use: space heating, ventilating system, local heaters
- Refrigerant: ammonia
- Absorbent: water
- Fuel: natural gas
- Heat source: ambient air
- Operation mode: parallel
- Heat output: 310 kW
- Fuel consumption in boiler: 235 kW
- Heat ratio (COP): 1.32
- Cooling capacity: 88 kW
- Heating water temperature supply and return: 50 and 41.5 °C
- Electric power demand for pumps and fan: 12 kW

In Figure 6.26, the operating principle of the AHP is described: in two outdoor air heat exchangers ammonia is evaporated at a low pressure extracting heat from outdoor air (16). This vapor is dissolved in the poor water–ammonia solution in the absorber (8) releasing heat to the heating system. The resulting rich solution is pumped (15) through a heat exchanger (7) and the rectifier (5) into the generator (14). The generator is indirectly heated via an intermediate heating circuit by a gas-fired boiler (1). Thermal oil is used as heat transport fluid in this intermediate circuit. In the generator the working fluid ammonia is evaporated from the ammonia–water solution. This vapor still containing some water is purified in a rectifier (5) and a reflux heat exchanger (6) before being condensed in the condenser (12).

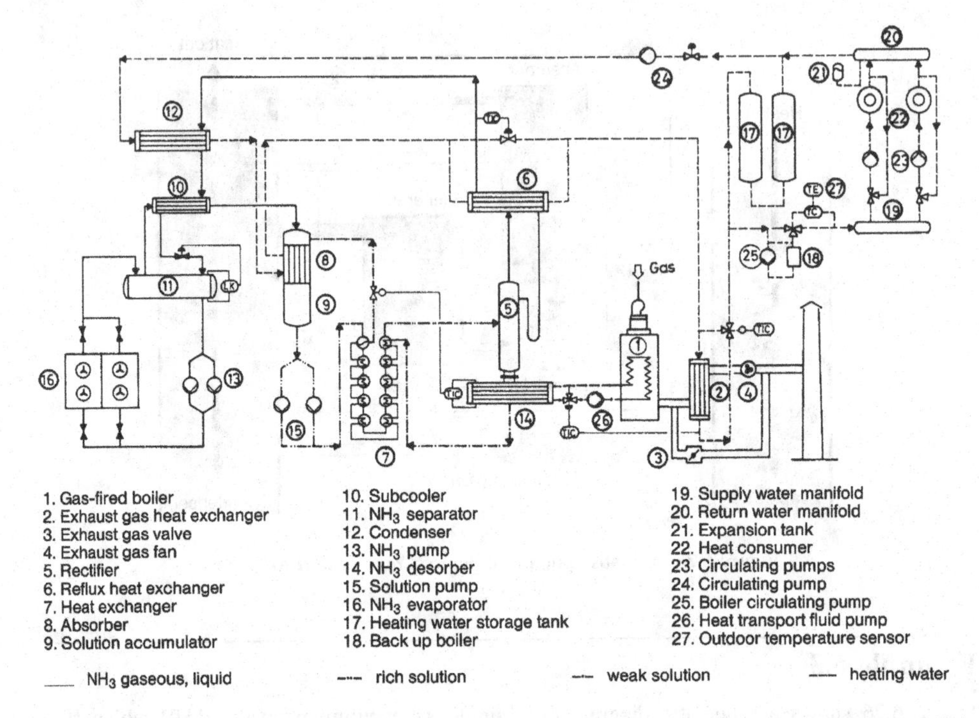

Figure 6.26 Schematic diagram of a single-stage, ammonia–water absorption heat pump (*Courtesy of IEA-HPC*).

6.22.1 Diffusion Absorption Heat Pumps

In the past, attempts made to solve the problems of AHPs have almost all been with absorption refrigeration machines using mechanical solution pumps. These pumps are used successfully on large absorption cooling machines of 50–1000 kW and more. Miniaturizing such machines presents difficulties which involve the small inefficient solution pump. This led to the evaluation of the diffusion-type absorption cycle. Absorption cooling units working with ammonia–water and hydrogen as auxiliary gas are well known. Millions have been built and are mainly used in domestic, camping, and caravan refrigerators. They are powered electrically, by LPG, natural gas, or kerosene. The cooling performance of such a cooling unit is in the order of 20–50 W, in contract to 1000 W required for a 3-kW heat pump. Contrary to the miniaturizing problems mentioned above, one now faces enlargement problems of the diffusion absorption units.

6.22.2 Special-Type Absorption Heat Pumps

In this group, we find different systems combining vapor compression with AHP elements. Special types of AHP-like combinations of elements originating from the compression and absorption cycle promise competitive solutions. Test plants of systems with a compressor solution circuit show that at high output temperatures comparatively low process pressures can be achieved compared with one-substance compression heat pumps.

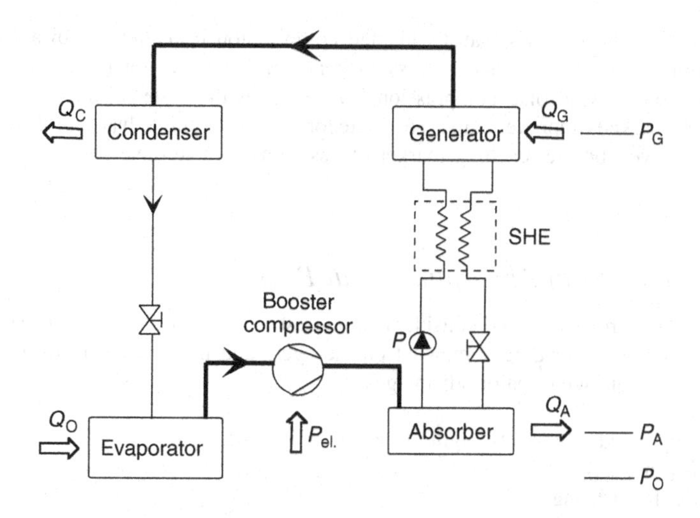

Figure 6.27 An AHP with booster compressor (Adapted from Podesser, 1984).

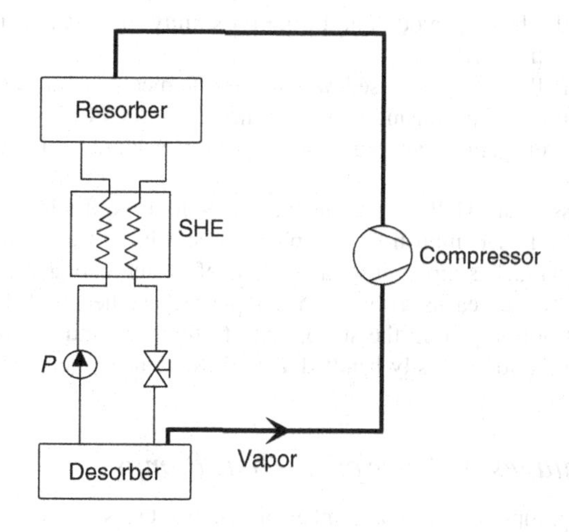

Figure 6.28 Compression heat pump with solution circuit (Adapted from Podesser, 1984).

In refrigeration technology, designs of absorption refrigeration machines using a booster compressor are known. These are used to achieve the features of dual and multistage absorption plants by means of single-stage, continuously working absorption machines. A booster compressor as shown in Figure 6.27 is connected either between the expeller and the condenser or between the evaporator and the absorber. Good performance figures can be attained by a compression heat pump that includes a sorption circuit (Figure 6.28). This can be achieved by designing the resorber and desorber in such a way that the irreversible losses can be kept low by operation according to the Lorenz process (large temperature differences between the outlet and inlet of the mass flow and comparatively small temperature differences between both mass flows). A further advantage is the lower pressure difference compared to compression heat pumps, resulting in a reduced required drive power.

A comparison of COPs shows that an adapted compression heat pump with a solution circuit brings slight improvements over a compression heat pump with a nonazeotropic working fluid mixture. When used instead of a compression heat pump with a one-substance refrigerant (e.g., R-22), it brings marked improvements. This category of heat pump has interesting application possibilities. However, before reaching market fitness, considerable development work has still to be carried out.

6.22.3 *Advantages of Absorption Heat Pumps*

In recent years, AHPs have received considerable interest because of a number of features. Absorption cycle heat pumps, depending somewhat on the cycle and the particular design configuration employed, have the following major advantages:

- Inherently simple and potentially highly reliable equipment.
- No compressor needed.
- Possibility for direct firing.
- High primary energy efficiency.
- Long life expectancy due to the lack of components with moving parts.
- No vibration or noise problems.
- High efficiencies in the heating mode due to the possibility of heat transfer from on-site combustion into the absorption cycle.
- Refrigerant–absorbent fluids can be used which are chemically nonreactive with atmospheric ozone, although some of the promising working pairs include CFCs.
- Cycle reversal or hot refrigerant bypass is practicable for evaporator defrost.

The cost-effectiveness of an AHP is very much a function of the fluids used. The gas-fired AHP benefits from having the fuel combustion take place on-site just as a thermal engine heat pump does, but the former is much simpler insofar as lack of reciprocating or rotating machinery is concerned. The on-site waste heat is available as supplementary heat and, for evaporator defrost, this aspect is important not only from the standpoint of efficiency and reliability but also because refrigerant flow reversal is not as easily handled as with Rankine cycle machines.

6.22.4 *Disadvantages of Absorption Heat Pumps*

Absorption cycle heat pumps also have a number of disadvantages:

- Comparatively large heat-exchanger areas (at attendant high first cost) required for realization of acceptable performance (expensive rectifiers may also be needed).
- Usually lower cooling-mode performance than obtainable with Rankine refrigeration machinery (including electric heat pumps).
- Toxicity of some refrigerants, for example, ammonia, where used, requires that direct refrigerant-conditioned air contact in water/ammonia systems must be avoided (requiring the use of either a water loop or other intermediate means of heat transfer).
- Corrosiveness of some working fluids, for example, water–ammonia systems requires the use of steel rather than aluminum or copper in the manufacture of the heat pump (the organic absorption system uses aluminum).
- Diameter limitation of the absorber requires the use of a tall unit with attendant bulkiness of equipment.

- Unfamiliarity of technical experts with the plumbing of absorption machines.
- Poor operating performance compared with engine-driven systems.
- Dependence on the availability of depleting and polluting fossil fuels (e.g., natural gas), unless waste heat or clean energy sources like solar energy are used.
- Relatively heavier units.
- Reduced performance due to losses resulting from use of on/off control systems.

For these reasons, intense research efforts are being undertaken presently to eliminate some of the above disadvantages.

As described earlier all vapor-compression heat pumps use a mechanically driven refrigeration compressor. This leads to the requirement for large amounts of shaft power and the use of a relatively complex piece of rotating or reciprocating machinery. It is possible, however, to design a cycle that uses a source of heat as its motive power. Such a cycle is called an *absorption refrigeration cycle* and might be familiar in the form of the gas refrigerator. As with vapor-compression systems there are many varieties of absorption systems, using different pairs of fluids or different cycles. Figure 6.29 shows a common type of ammonia–water absorption cycle. To understand an absorption cycle it is easiest to start with the parts of the cycle that are the same as a compression cycle, that is, the evaporator and the condenser. In Figure 6.29 it can be seen that the ammonia refrigerant absorbs heat in the evaporator and rejects heat in the condenser in the same way as vapor-compression systems. The difference is that the compressor is replaced with an absorber and regenerator. The low-pressure suction vapors are absorbed into an ammonia–water solution. The solution is pumped to a high pressure (note that pumping a liquid uses considerably less energy than the compression of a gas across the same pressure ratio). At the higher pressure the solution enters the regenerator in which heat is applied; ammonia vapor is formed. These high-pressure "discharge" vapors are then condensed, expanded, and returned to the evaporator in the normal way.

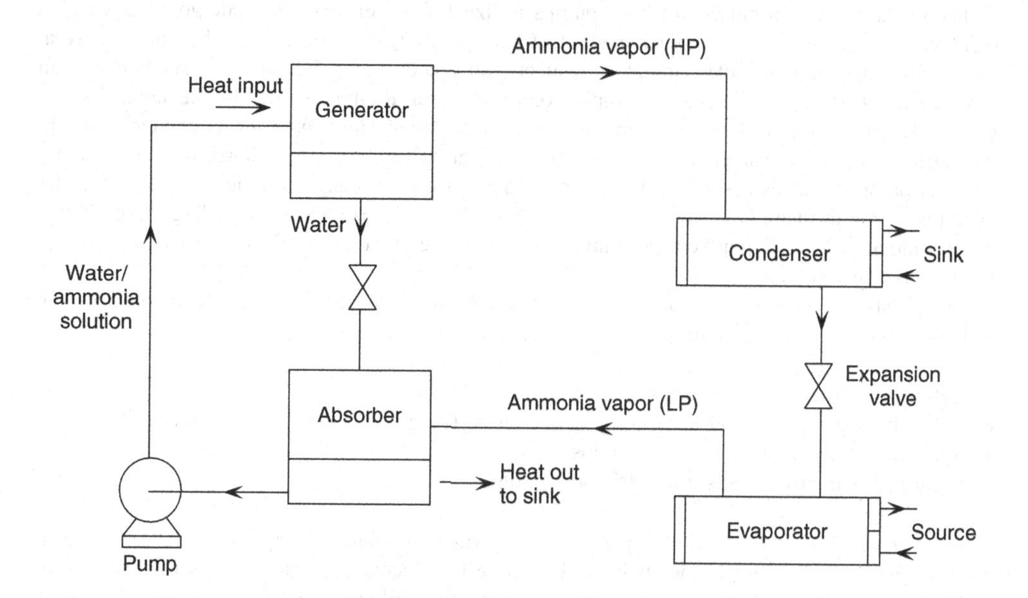

Figure 6.29 An ammonia–water absorption heat pump.

Obviously there must be some intrinsic drawbacks in the absorption cycle or else we would be using these systems in all refrigeration and heat pumping applications. The basic problems are as follows:

- Achievable COPs are low.
- Ammonia–water systems are restricted in top temperature because of high pressure.
- Lithium bromide–water systems require large equipment because of the low density of water vapor.
- Concentrations of the fluids are very critical for stable operation. This can cause practical problems, particularly at part load.
- Lithium bromide–water systems cannot operate below the freezing point of water.
- Practical systems are more complex than the basic arrangement shown in Figure 6.29.

During the past two decades, there has been increasing interest in research and development on new pairs of absorber-refrigerant without some of the disadvantages mentioned above.

AHPs probably offer the best hopes on the basis of design, operation, and initial cost criteria. With a solution pump as the only moving part, they hold promise of a long life, easy maintenance, and high reliability. The COP is fairly low in the first-generation machines, that is, from 1.1 to 1.3. In more advanced designs, it is possible to improve this COP to 1.4–1.5. This is, however, a practical limit which is exceeded only by going to multistage operation, with inherent increases in complexity and cost.

The future of absorption cycle heat pumps will ultimately be dependent on success in the search for an ideal working fluid pair capable of prolonged operation between the large temperature differences encountered in the cycle. In the initial stages of development it is probable that the full performance requirements will not be met at both ends of the operating temperature range. Recently, the development of AHPs has taken place in many countries, for example, the United States, Germany, Japan, the United Kingdom, and Switzerland.

The electric and thermal engine heat pumps utilize fluids (refrigerants) undergoing a Rankine or other refrigeration cycle. Another method of heat pumping can be achieved by one of several absorption refrigeration cycles. In many respects, an analogy may be said to exist between an absorption refrigeration cycle and a vapor-compression or Rankine cycle. In the latter case, a compressor is used to increase the pressure of the refrigerant vapor prior to its condensation; in the former case, the physical affinity between a refrigerant vapor and an absorbent solution and a solution pump produces a similar effect. Absorption air-conditioning machines are commercially available today in many countries, for example, the United States and Japan, in large sizes. Some smaller unitary residential and commercial absorption cycle air conditioners are presently produced by various manufacturers.

On the basis of experience to date, it appears that the main constraints to the development of an AHP which could achieve significant market penetration include

- high initial cost,
- high natural gas prices and uncertainty about future (long-term) supplies of natural gas,
- unfavorable gas-to-electricity price ratios, and
- performance improvements in electric heat pumps.

Another problem is that initial experience with water–ammonia absorption cycle gas air conditioners has not been good, and their market share has decreased steadily. This is attributed to high cost and equipment reliability problems. In addition, while new installations will use primary energy more efficiently, they are also liable to increase consumption of and dependence on fuels such as natural gas, and, so on.

As with thermal engine heat pumps, it is likely that economics and gas availability rather than performance will determine the future of the absorption cycle heat pump. Traditionally, absorption machines have been deficient in cooling-mode performance in comparison to electric air conditioners. It is possible to improve the cooling-mode performance dramatically by using a multieffect machine, but at a higher first cost. Nevertheless, from the standpoint of system simplicity and potentially maintenance-free operation, the AHP might be preferable to the generally more efficient thermal engine heat pumps for residential applications.

6.22.5 Mesoscopic Heat-Actuated Absorption Heat Pump

In this section, a recent project carried out by Pacific Northwest National Laboratory (PNL) is introduced. This is apparently a new topic in the field of AHP technology. The purpose of their Defense Advanced Research Projects Agency (DARPA) project is to demonstrate a mesoscopic absorption cycle heat-actuated heat pump for a range of military microclimate control applications. Although currently available cooling systems can be integrated with protective suits to provide some degree of cooling, these systems are based on a vapor-compression cycle that requires significant amounts of electricity. An AHP is similar to a vapor-compression device except that compression is accomplished in the AHP through the use of a thermochemical compressor. The simple thermochemical compressor consists of an absorber, a solution pump, a heat exchanger, and a desorber (gas generator). While several heat-actuated heat pump cycles have been investigated, current research at Battelle is focused on a single-effect lithium bromide and water ($LiBr–H_2O$) AHP.

While the mesoscopic heat-actuated heat pump can be applied to many military cooling applications, one specific application of the device is being demonstrated: man-portable cooling. Personnel performing labor-intensive tasks in a hot environment are vulnerable to heat stress, especially when using nuclear, biological, and chemical protective clothing. Supplemental cooling will permit the soldier to perform tasks under these conditions in hot climates with enhanced efficiency.

By using a heat-actuated cooling cycle, the Battelle mesoscopic AHP has radically reduced the requirements for electric power by substituting thermal energy for electric energy. When fuel, batteries for fan and pump power, and an air-cooled heat exchanger are added to the mesoscopic heat-actuated heat pump, the complete system weight is projected to be between 3.7 and 5.0 kg for an 8-h mission. This is less than one-half of the weight of competing conventional microclimate control systems.

The preliminary estimates predict that the mesoscopic AHP will have a volume of 420 cm^3, weigh approximately 0.72 kg, and will be capable of providing 350 Wt (referring to a watt of thermal energy) of cooling.

Within the above-mentioned project, their technical achievements in completing the design of the mesoscopic heat-actuated heat pump are the following:

- Computer simulations of the heat pump to find out the impact of the cooling system variables on the heat pump COP and weight by considering four system values: (i) cooling water temperature entering the absorber (temperature of water leaving the system radiator), (ii) chilled water temperature leaving the evaporator (temperature of water entering vest), (iii) thermal capacitance of cooling water loop, and (iv) thermal capacitance of chilled water loop, and to optimize the man-portable cooling system;
- Computer simulations to optimize the man-portable cooling system design;
- Screening the ranges of materials for use as desorber and absorber effectively.

6.23 Heat Transformer Heat Pump Systems

The heat transformer is an AHP; it releases about 50% of the waste heat supplied to it at a higher temperature level and requires electricity only to run a few pumps. The heat transformer operates on the principle that the temperature at which water vapor condenses (is absorbed) in a salt solution is above the temperature at which water evaporates, provided both processes are at the same pressure. At reduced pressure, a salt solution circulates through special heat exchangers. Absorption of water vapor releases heat, and this heat is generally used to generate steam (10 tons/h) for the production process.

A flow diagram of the heat-transfer system is shown in Figure 6.30. Waste heat is used to evaporate the condensate and to evaporate the water in the salt solution at reduced pressure. This latter vapor is condensed using cooling water, the condensate pressurized and evaporated. This water vapor is absorbed in the salt solution, releasing the heat of absorption at an elevated temperature. Waste heat is used in two heat exchangers, the regenerator and the evaporator, where the same quantity of water is evaporated. Only the waste heat used in the evaporator is retained; this is about 50% of the waste heat supplied. The only expensive energy required is electricity to run circulation pumps.

The heat transformer can be used with waste heat temperatures above 60 °C. The temperature level attainable in the absorber is determined by the temperature of the waste heat and the temperature of the condenser.

The heat transformer is also known as the reverse AHP process or AHP Type II. The objective of such an AHP is to convert waste heat of medium temperature (from 60 to 80 °C) to useful heat at high temperature. Figure 6.31 shows a flow diagram of a single-stage unit which is suitable for this purpose. The working pair ammonia–water can be used, up to about 190 °C. NH_3–H_2O

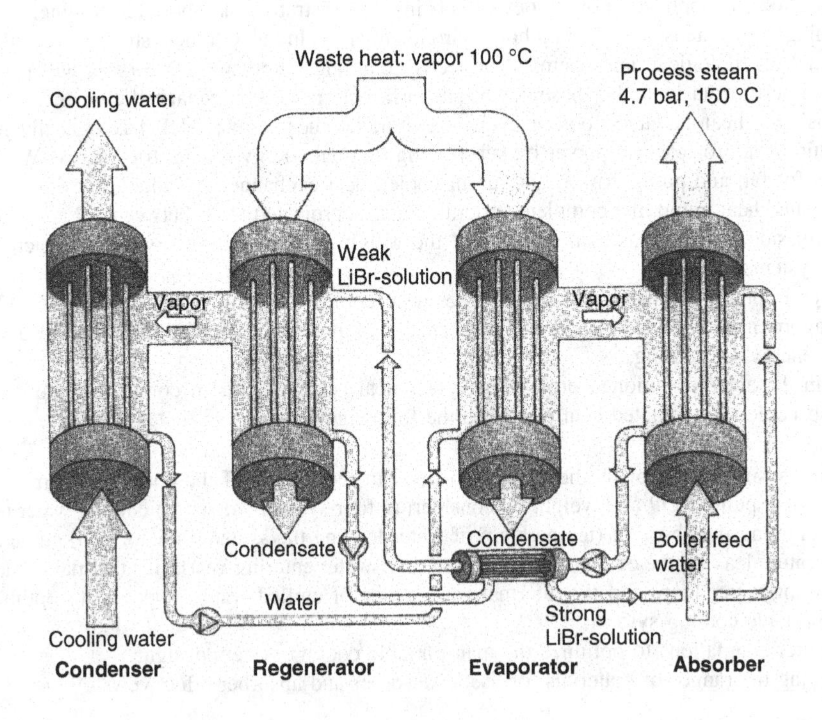

Figure 6.30 An industrial heat transformer (*Courtesy of IEA-HPC*).

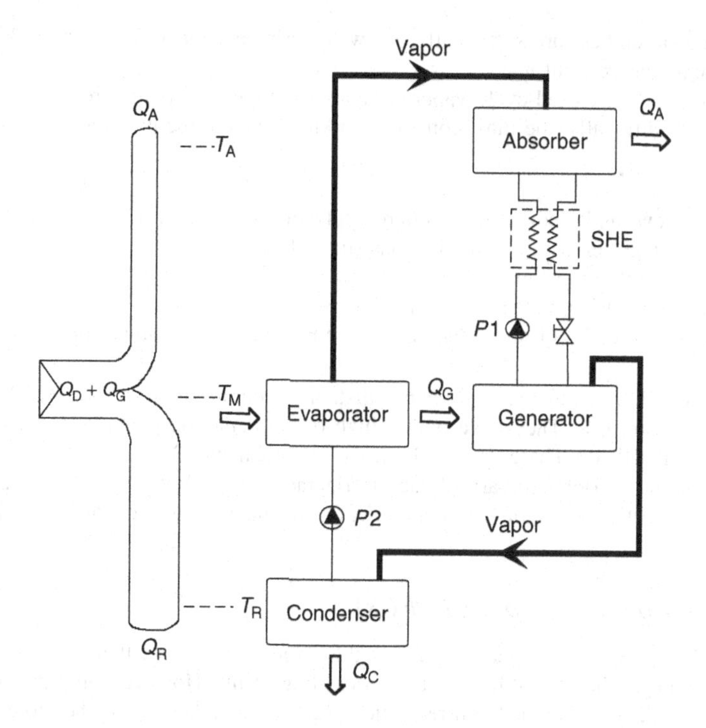

Figure 6.31 A flow sheet of a heat transformer (AHP, Type II) (Adapted from Podesser, 1984).

AHPs have the disadvantage of high pressures, when high output temperatures are necessary, which means that substantially more electric power is required for the working fluid pumps compared with low-pressure working pairs (e.g., LiBr–H_2O or LiBr–CH_3OH).

6.24 Refrigerants and Working Fluids

Within the last decade, several efforts have been undertaken to develop analytical models for research and development of heat pump systems using alternative refrigerants (azeotropic or non-azeotropic mixtures). Moreover, a number of researchers have begun to measure and develop techniques to predict essential thermodynamic and thermophysical data. An impartial evaluation of equations of state for HFC-134a and HFC-123, which are refrigerants friendly to the environment and ozone layer, has resulted in the selection of formulations for these fluids that are expected to become internationally acceptable standards. Consequently, in heat pump technology, further research and development activities have been concentrated on the following topics:

- improvement of system components, leading to system optimization,
- alternative refrigerants and working fluids,
- development of automatic defrosting systems,
- regulation and management of bivalent systems,
- thermal and acoustic optimization of the constituent parts, and
- expansion of heat pump use in industrial applications.

It is important to point out that the critical temperature of the working fluid provides the upper limit at which a condensing vapor heat pump can deliver heat energy. The working fluid should be condensed at a temperature sufficiently below the critical temperature to provide an adequate amount of latent heat per unit mass.

As discussed earlier, closed-cycle vapor-compression type heat pumps require a refrigerant as working fluid. Traditionally, the most common working fluids in the past for heat pumps were the following:

- **CFC-12.** For low- and medium temperature applications (max. 80 °C).
- **CFC-114.** For high-temperature applications (max. 120 °C).
- **R-500.** For medium-temperature applications (max. 80 °C).
- **R-502.** For low-medium temperature applications (max. 55 °C).
- **HCFC-22.** For virtually all reversible and low-temperature heat pumps (max. 55 °C).

Owing to their chlorine content and chemical stability, chlorofluorocarbons (CFCs) are harmful to the global environment. They have both a high ozone depletion potential (ODP) and a global warming potential (GWP). These were fully discussed in Chapter 3.

Here, we provide a brief summary of each refrigerant commodity from the heat pump point of view and a phase-out schedule of CFCs and HCFCs for industrialized countries in Table 6.11.

6.24.1 Chlorofluorocarbons (CFCs)

Owing to their high ODP, the manufacture of these refrigerants, and their use in new plants, is now banned although they are still permitted in existing plants. However, only purified (recycled) refrigerants from decommissioned and retrofitted plants are available. It is therefore expected that these refrigerants will become more and more expensive, and at some point will no longer be available. This group includes the following refrigerants: R-11, R-12, R-13, R-113, R-114, R-115, R-500, R-502, and R-13B1. As a general requirement, heat pumps using alternative working fluids should have at least the same reliability and cost effectiveness as HCFC systems. Moreover, the energy efficiency of the systems should be maintained or be even higher, in order to make heat pumps an interesting energy-saving alternative. In addition to finding new and environmentally acceptable working fluids, it is also important to modify or redesign the heat pumps. Generally speaking, the energy efficiency of a heat pump system depends more on the heat pump and system design than on the working fluid.

Table 6.11 Phase-out schedule for CFCs and HCFCs for developed countries.

Date	Control Measure
1 January 1996	CFCs phased out HCFCs frozen at 1989 levels of HCFC + 2.8% of 1989 consumption of CFCs (base level)
1 January 2004	HCFCs reduced by 35% below base levels
1 January 2010	HCFCs reduced by 65%
1 January 2015	HCFCs reduced by 90%
1 January 2020	HCFCs phased out allowing for a service tail of up to 0.5% until 2030 for existing refrigeration and air-conditioning equipment

Source: IEA-HPC (2001).

6.24.2 Hydrochlorofluorocarbons (HCFCs)

HCFC working fluids also contain chlorine, but they have much lower ODP than CFCs, typically 2–5% of CFC-12, owing to a lower atmospheric chemical stability. The GWP is typically 20% of that of CFC-12. H-CFCs are so-called *transitional refrigerants*. They should only be used for retrofit applications. H-CFCs include R-22, R-401, R-402, R-403, R-408, and R-409. As listed in Table 6.11, industrialized countries agreed on the phase-out schedule of CFCs and HCFCs under the Montreal Protocol and its amendments and adjustments. HCFCs should be phased out for industrialized countries by the year 2020, and should be phased out entirely by 2040. The European Union has adopted an accelerated phase-out schedule for these substances, which requires them to be phased out by January 2015. Some countries in Europe (Sweden, Germany, Denmark, Switzerland, and Austria) also have an accelerated schedule and have been phasing out R-22 for new systems since 1998.

6.24.3 Hydrofluorocarbons (HFCs)

HFCs can be considered long-term alternative refrigerants, meaning that they are chlorine-free refrigerants such as R-134a, R-152a, R-32, R-125, and R-507. Since they do not contribute to ozone depletion, these are long-term alternatives to R-12, R-22, and R-502. However, they do still contribute to global warming. Special attention must be given to the use of lubricants. Mineral oils are nonmiscible with these refrigerants. Normally only ester-based lubricant oils recommended by the refrigerant manufacturer should be used. Mineral oil residues must be completely removed during retrofitting.

HFC-134a is quite similar to CFC-12 in thermophysical properties. The COP of a heat pump with HFC-134a will be practically the same as for CFC-12. At low evaporating temperatures (below $-1\,°C$) and/or high temperature lifts the COP will be slightly lower. HFC-152a has mainly been used as a part of R-500, but it has also been successfully applied in a number of small heat pump systems and domestic refrigerators. HFC-152a is currently applied as a component in blends. Because of its flammability, it should only be used as a pure working fluid in small systems with low working fluid charge. HFC-32 is moderately flammable and has a GWP close to zero. It is considered as a suitable long-term replacement for HCFC-22 in space conditioning, heat pump, and industrial refrigeration applications. Owing to its flammability, HFC-32 is usually applied as a main component in nonflammable mixtures replacing R-502 and HCFC-22. HFC-125 and HFC-143a have properties fairly similar to R-502 and HCFC-22. They are mainly applied as components in ternary mixtures, replacing R-502 and HCFC-22. The GWPs are, however, about three times as high as that of HFC-134a.

6.24.4 Hydrocarbons (HCs)

HCs are well-known flammable working fluids with favorable thermodynamic properties and material compatibility. Presently, propane, propylene, and blends of propane, butane, isobutane, and ethane are regarded as the most promising HC working fluids in heat pumping systems. HCs are widely used in the petroleum industry, sporadically applied in transport refrigeration, domestic refrigerators/freezers, and residential heat pumps (particularly in Europe). Owing to the high flammability, hydrocarbons should only be retrofitted and applied in systems with low working fluid charge. To ensure necessary safety during operation and service, precautions should be taken such as proper placing and/or enclosure of the heat pump, fail-safe ventilation systems, addition of tracer gas to the working fluid, use of gas detectors, and so on.

6.24.5 Blends

Blends or mixtures represent an important possibility for replacement of CFCs, for both retrofit and new applications. A blend consists of two or more pure working fluids, and can be zeotropic, azeotropic, or near-azeotropic. Azeotropic mixtures evaporate and condense at a constant temperature, the others over a certain temperature range. The temperature glide can be utilized to enhance performance, but this requires equipment modification. The advantage of blends is that they can be custom-made to fit particular needs.

Early blends for replacement of CFC-12 and R-502 all contained HCFC-22 and/or other HCFC working fluids, such as HCFC-124 and HCFC-142b, and are therefore considered as transitional or medium-term working fluids. The new generation of blends for replacement of R-502 and HCFC-22 are chlorine-free, and will mainly be made from HFCs (HFC-32, HFC-125, HFC-134a, HFC143a) and HCs (e.g., propane). Two of the most promising alternative working fluids for eventually replacing R-22 in heat-pumping applications are the blends R-410A and R407-C that are discussed below in more detail. The main difference between the two is the chemical composition: R-410A is a mixture of R-32 and R-125 with minimal temperature glide, while R-407C consists of R-32, R-125, and R-134A and has a large temperature glide. Annex 18 of the IEA Heat Pump Programme has performed a detailed study on thermophysical properties of blends (IEA-HPC, 2001).

R-407C is the only refrigerant available for immediate use in existing R-22 plants. Its thermal properties and operating conditions are close to those of R-22. However, because of its temperature glide it is only suitable for certain systems. The use of this refrigerant is increasing, although there are still some engineering difficulties for service companies and manufacturers. Research has shown that the use of R-410A can result in an improved COP compared to R-22. Using R-410A means that overall cost reductions can be achieved, because the system components, particularly the compressor, can be significantly downsized since it has a higher volumetric capacity. The main disadvantage is the higher operating pressure compared to R-22, which indicates that the pressure-proof design of most components should be reviewed. R-410A is very popular, mainly in the Unites States and Japan, for packaged heat pumps and air-conditioning units. Commercial R-410A components for small- and medium-sized refrigeration systems are either already available or are under development.

6.24.6 Natural Working Fluids

Natural working fluids are substances naturally existing in the biosphere. They generally have negligible global environmental drawbacks (zero or near-zero ODP and GWP). They are therefore long-term alternatives to the CFCs. Examples of natural working fluids are ammonia (NH_3), hydrocarbons (e.g., propane), carbon dioxide (CO_2), air, and water. Some of the natural working fluids are flammable or toxic. The safety implications of using such fluids may require specific system design and suitable operating and maintenance routines.

Ammonia is in many countries the leading working fluid in medium- and large refrigeration and cold storage plants. Codes, regulations, and legislation have been developed mainly to deal with the toxic and, to some extent, the flammable characteristics of ammonia. Thermodynamically and economically, ammonia is an excellent alternative to CFCs and HCFC-22 in new heat pump equipment. It has so far only been used in large heat pump systems, and high-pressure compressors have raised the maximum achievable condensing temperature from 58 to 78 °C. Ammonia can also be considered in small systems, the largest part of the heat pump market. In small systems the safety aspects can be handled by using equipment with low working fluid charge and measures such as indirect distribution systems (brine systems), gas-tight rooms or casing, and fail-safe ventilation. Copper is not compatible with ammonia, so that all components must be made of steel. Ammonia is not yet used in high-temperature industrial heat pumps because there are currently no suitable high-pressure compressors available (40 bar maximum). If efficient high-pressure compressors are developed, ammonia will be an excellent high-temperature working fluid.

Water is an excellent working fluid for high-temperature industrial heat pumps because of its favorable thermodynamic properties and the fact that it is neither flammable nor toxic. Water has mainly been used as a working fluid in open and semi-open MVR systems, but there are also a few closed-cycle compression heat pumps with water as working fluid. Typical operating temperatures are in the range from 80 to 150 °C. In a test plant in Japan, 300 °C has been achieved, and there is a growing interest in utilizing water as a working fluid, especially for high-temperature applications. The major disadvantage with water as a working fluid is the low volumetric heat capacity (kJ/m^3) of water. This requires large and expensive compressors, especially at low temperatures.

CO_2 is a potentially strong refrigerant that is attracting growing attention from all over the world. CO_2 is nontoxic, nonflammable, and is compatible to normal lubricants and common construction materials. The volumetric refrigeration capacity is high and the pressure ratio is greatly reduced. However, the theoretical COP of a conventional heat pumping cycle with CO_2 is rather poor, and effective application of this fluid depends on the development of suitable methods to achieve a competitively low power consumption during operation near and above the critical point. CO_2 products are still under development, and research continues to improve systems and components. A prototype heat pump water heater has already been developed in Norway. CO_2 is now used as a secondary refrigerant in cascade systems for commercial refrigeration.

6.25 Technical Aspects of Heat Pumps

In this section, the following technical aspects are discussed briefly.

6.25.1 Performance of Heat Pumps

The performance of any heating and/or cooling device can be measured in two basic ways – either under steady-state conditions, referring to the COP, or under normal operating conditions, referring to seasonal performance factors (SPFs), for example, SEER and HSPF. The latter takes account of the fluctuations and changes in the heating and/or cooling loads as the external temperature changes over a period of time. Because the capacity and efficiency of a heat pump fall with declining temperatures, the operating performance taken over an annual cycle is always lower than the steady-state performance. Owing to the widely varying temperature regimes in different regions, the SPF varies from region to region. In addition, since heating or cooling devices are normally tailored to the most extreme conditions, transient operation will reduce the overall performance of the system. This can be attributed to several factors, including on–off cycling under conditions where heating or cooling capacity exceeds demand, cycle reversal during the course of defrosting and cycling induced by control limits, dirty air filters, and so on. Auxiliary power consumption for pumps, fans, and so on, will also reduce performance. The SPF therefore is taken to reflect the energy actually used over an entire season including the heat pump and all auxiliaries plus any supplemental heat that may be needed. The SPF provides, therefore, a useful indicator of the relative operational performance of different heating and cooling systems under similar climatic conditions. The main determinants of the seasonal performance of a heat pump are the operating temperature of the heat source (air, water, or soil), the temperature difference between the heat source and heat distribution medium (air or water), the effectiveness of the heat pump cycle itself, the sizing of the heat pump in relation to the demand, and the operating regime of the heat pump.

6.25.2 Capacity and Efficiency

The heating and cooling capacity as well as the performance of a heat pump vary as functions of the outdoor temperature, particularly for air source. Heating capacity decreases with a fall in ambient temperature and cooling capacity decreases with an increase in the ambient temperature. This is the reverse of the temperature variation of the heating and cooling demand of most buildings. In addition, the efficiency of the heat pump is least when the heating or cooling demand is at its greatest. Resulting from these factors, the system size must be increased so as to cope with maximum load (which is a costly process), or supplementary heating can be used. The latter, as indicated earlier, can be an electric resistance heating element incorporated at minimal cost in the heat pump itself. This can, however, result in higher fuel costs, particularly if very low temperatures are experienced for long time periods. Alternative supplementary heating devices may be oil or gas furnaces which, while more costly to install than electric resistance heating, are less expensive to operate. An existing furnace may also be used, in which case no additional capital cost apart from the heat pump is incurred until the furnace needs to be replaced.

6.25.3 Cooling, Freezing and Defrost

As indicated earlier, some residual frost tends to accumulate on the outdoor heat pump coil under certain temperature and humidity conditions and the defrost cycle is initiated to remove the ice. The most common method of defrosting is by activating the reversal valve whereby the heat pump is switched from a heating to a cooling cycle while resistance heating elements are used to temper the cold-air flowing into the conditioned space. Defrosting, which usually requires 3–5 min to complete, can be controlled in a number of ways. One system uses a timer to initiate and terminate the defrosting operation at definite time intervals. A second method involves the use of a control that senses the air pressure drop across the outdoor coil. The pressure drop rises as the coil frosts up, restricting the outdoor fan-driven airflow. At a pre-set point, the pump is reversed. Defrost operations are terminated by a signal derived from outdoor refrigerant temperature (pressure). Unless properly carried out, poor results can be obtained from this procedure. A third method of control is by using the temperature differential between two thermostats (one located to measure the outdoor temperature and the other placed in the evaporator coil). As frost accumulates, the temperature difference increases and the derived signal is used to initiate the defrost cycle and to terminate the operation when the temperature inside the outdoor coil rises to a predetermined value. A fourth method is to initiate defrost according to the coil temperature or the evaporator pressure. All defrost schemes will reduce the performance of the system below the already limited low-temperature performance by a further 2–5%.

6.25.4 Controls

Adequate control devices are required for heat pump operation to ensure reliability and optimal performance. Heat pumps employ a number of controls for on/off operation, motor compressor protection, valve control, refrigerant flow regulation, defrost initiation and termination, indoor and outdoor temperature sensing, and user and service diagnostics. Microprocessor technologies are beginning to have an impact and enable the monitoring, analysis, and regulation of a large number of parameters. Other control features include an adjustable balance point control which prevents supplemental heat operation above the balance point. Plug-in contacts also enable interconnection with diagnostic devices to facilitate corrective maintenance.

6.25.5 Fan Efficiency and Power Requirements

A major source of energy loss in a heat pump is outdoor and indoor fan power consumption for air source and air sink units. For example, a slower, larger diameter fan uses less power than a normal fan at the same capacity; many units now incorporate these larger fans.

6.25.6 Compressor Modification

Compressor developments are directed toward improved reliability, better performance under all load conditions, and improved low-temperature heating capacity. The basic mismatch between the variation with ambient temperature of typical building losses (or gains) and the heating (or cooling) capacity of the heat pump remains. This mismatch arises, as indicated earlier, because of the necessity for supplementing the compressor at low outdoor temperatures with electric resistance or other heating devices as well as to on–off cycling under part-load conditions. Capacity modulation tries to overcome these difficulties by achieving a better match between capacity and load.

6.25.7 Capacity Modulation

Capacity modulation can be brought about in a number of ways. A number of two-speed compressor systems are now available involving speed-halving, which is achieved by switching from a two-pole to a four-pole motor configuration. Modulation is also achieved by using multiple compressors and cylinder off-loading as the load varies. Variable speed systems use an electronic "black-box" static inverter unit to achieve power conversion. The high first cost of some of those high-performance systems is a barrier to greater penetration of the market, which could be overcome, in the case of the variable speed systems, by significant cost reductions in the electronic inverter technology. Capacity and speed controls have been applied cost-effectively for large compressors. For smaller units, capacity control can be expensive and part-load operation is usually less efficient than full-load operation.

6.25.8 Heat Exchangers

Heat exchanger effectiveness can be improved by increasing the heat-transfer coefficient across the heat exchanger or by increasing the coil area. The geometry of the heat exchangers greatly influences the performance of heat pumps. To achieve this, many of the systems commercially available have larger coil areas than the older models.

6.25.9 Refrigerants

Refrigerants that may be useful on the basis of suitable boiling points, modest pressures, reasonable performance, and so on, can be classified as HCFCs, HFCs, blends, and natural working fluids (Table 6.12). The gases are much cheaper than the fluorocarbons and are preferred in some industrial/commercial applications. They may, however, be toxic, noxious, poisonous, and flammable and many pose explosion hazards. The most widely used fluorocarbon refrigerants were R-11, R-12, R-22, and R-502, accounting for nearly 92% of the consumption in the world. Concern about the potential impact of fluorocarbons on the atmosphere has been widespread for some time. The

Table 6.12 Some common heat pump refrigerants.

Common Groups	Examples
HCFCs	Phased out as stated above.
HFCs	R-134a, R-152a, R-32, R-125, R-143a
Blends	R-407C, R-410A
Natural working fluids	Ammonia, Water, Carbondioxide, HCs

sources of fluorocarbons vary from aerosols to refrigerators, manufacturing processes, and heat pumps. Normally, alternative working fluids for heat pumps are expected to fulfill the requirements (e.g., reliability, cost effectiveness, environmental friendliness, commercial availability, etc.) as given in Chapter 2. Moreover, the COP of the systems should be maintained or be even higher, in order to make heat pumps a potential energy-saving alternative.

6.26 Operational Aspects of Heat Pumps

As mentioned earlier, the task of the heat pump is to transport heat for either cooling or heating. The heat carrier of the system is the refrigerant. The term heat pump includes the refrigeration part of the total plant, that is, generally the heat exchanger on the cold side, the temperature-raising device with the introduction of drive energy, the heat exchanger on the warm side, and, in most cases, an expansion device for completing the refrigeration cycle. All refrigeration machines are suitable for use as heat pumps. The following have been used (Dincer, 2003)

- the cold-air machine, using air as the working fluid.
- the cold vapor machine, using evaporation and condensation of the working fluid which can be water vapor or a coolant; the energy can be supplied using compression, absorption, or the steam jet principle.
- the thermoelectric principle (Peltier effect). As in refrigeration, the cold vapor heat pump with mechanical compression of the coolant vapor is by far the most important. The highly developed technology of this design can also be fully utilized for heat pumps.

In all the designs mentioned, it is possible to use simultaneously both the cooling and the heating effects to good purpose. Varying utilizations are also possible at different times, for example, cooling in summer and heating in winter, as required, in particular, for air conditioning. There are two methods for this type of operation:

- Changing the flow medium in the heat exchangers; for example, in the heat exchanger on the cold side, using cold water during summer for air conditioning (cooling) and groundwater as the heat source during winter or in the heat exchanger on the warm side, using water for the cooling system in summer and warm water for heating the building in winter.
- Changing the coolant circuit so that, during summer operation, the heat exchanger on the warm side, that is, that which transfers heat to groundwater, becomes the heat exchanger on the cold side in winter by extracting heat from the groundwater. In the same way, the other heat exchanger, which in summer cools the cold water circuit on the cold side for the air conditioning, in

winter becomes the heat exchanger on the warm side which heats the heating water for the air conditioning. The coolant circuit changeover is possible only in thermoelectric systems and in cold vapor systems with piston compressors and appropriate reversing valves, because the difficulty of changeover in other systems is too great.

The following are some guidelines that should be followed to provide efficient, comfortable operation of air source heat pumps (Carrier, 2000).

- Do not set the temperature back at night or when you are at work unless a programmable heat pump thermostat is used. Since heat pumps operate differently than fossil fuel heating systems, setback of a standard heat pump thermostat can actually increase energy consumption. This is due to the use of supplemental heaters to bring the house temperature back to the desired setpoint. Use of supplemental heaters will reduce the efficiency of the heat pump system and result in higher energy costs.
- Keep the temperature setpoint consistent. A standard heat pump thermostat has two controls, one for the heat pump and one for the supplemental heat. If the temperature difference between the room and thermostat setpoint is more than 2 or 3 °C, the supplemental heat will be activated. Manually adjusting the thermostat will result in greater reliance on the supplemental heaters and will reduce the efficiency of the heat pump system and increase operating costs.
- Replace filters regularly. Vacuum dirt and dust from the indoor coil once a year to prevent restricted air flow.
- Adequate air flow through a heat pump system is critical to ensure efficient and comfortable operation.
- Keep supply vents open and free from obstruction. Closing off supply vents will restrict airflow, and reduce system efficiency and the life of the compressor.

A heat pump is one of the most energy-efficient heating and cooling systems available today. Unlike other types of heating systems which convert fuel or electricity directly to heat, a heat pump is designed to move heat from one place to another. Even at temperatures as cold as −18 °C or below, the heat pump is able to extract heat from outside air to use in heating a house.

6.27 Performance Evaluation Aspects of Heat Pumps

The heat delivered by a heat pump is theoretically the sum of the heat extracted from the heat source and the energy needed to drive the cycle. The steady-state performance of an electric compression heat pump at a given set of temperature conditions is referred to as the COP which is defined as the ratio of heat delivered by the heat pump and the electricity supplied to the compressor.

For engine and thermally driven heat pumps the performance is indicated by the PER as given earlier. The energy supplied is then the higher heating value of the fuel supplied. For electrically driven heat pumps, a PER can also be defined by multiplying the COP with the power generation efficiency. The COP or PER of a heat pump is closely related to the temperature lift, that is, the difference between the temperature of the heat source and the output temperature of the heat pump. The COP of an ideal heat pump is determined solely by the condensation temperature and the temperature lift (condensation–evaporation temperature).

Figure 6.32 shows the COP for an ideal heat pump as a function of temperature lift, where the temperature of the heat source is 0 °C. Also shown is the range of actual COPs for various types and sizes of real heat pumps at different temperature lifts. The ratio of the actual COP of a heat pump and the ideal COP is defined as the Carnot efficiency. The Carnot efficiency varies

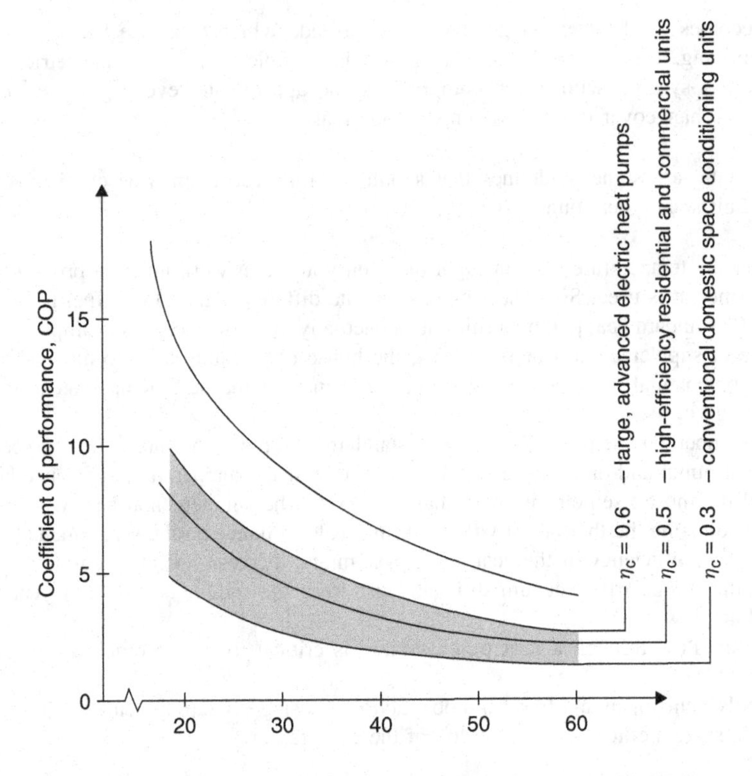

Figure 6.32 Coefficient of performance profiles (*Courtesy of IEA-HPC*).

Table 6.13 Typical COP/PER range for heat pumps with different drive energies.

Heat Pump Type	COP	PER
Electric (compression)	2.5–5.0	–
Engine (compression)	–	0.8–2.0
Thermal (absorption)	–	1.0–1.8

Source: IEA-HPC (2001).

from 0.30 to 0.5 for small electric heat pumps and 0.5 to 0.7 for large, very efficient electric heat pump systems. The COP/PERs for different heat pump types at evaporation 0 °C and condensing temperature 50 °C are given in Table 6.13.

As stated earlier, the operating performance of heat pumps over the season is rated by the seasonal factors, for example, SEER and HSPF. In fact, it takes into account the variable heating and/or cooling demands, the variable heat source and sink temperatures over the year, and includes the energy demand, for example, for defrosting. These seasonal factors can also be used for comparing heat pumps with conventional heating systems (e.g., boilers), with regard to primary energy saving and reduced CO_2 emissions. For evaluating electric heat pumps the power generation mix and the efficiencies of the power stations must be considered.

Table 6.14 Typical COP/PER for heat pumps with different drive energies.

Heat Pump Type	COP	PER
MVR	10.0–30.0	–
Closed cycle, electric	3.0–8.0	–
Closed cycle, engine	–	1.0–2.0
Absorption (Type I)	–	1.1–1.8
Heat transformer (Type II)	–	0.45–0.48

Source: IEA-HPC (2001).

6.27.1 Factors Affecting Heat Pump Performance

The performance of heat pumps is affected by a large number of factors. For heat pumps in buildings these include

- the climate (annual heating and cooling demand and maximum peak loads),
- the temperatures of the heat source and heat distribution system,
- the auxiliary energy consumption (pumps, fans, supplementary heat for bivalent system, etc.),
- the technical standard of the heat pump,
- the sizing of the heat pump in relation to the heat demand and the operating characteristics of the heat pump, and
- the heat pump control system.

Industrial heat pumps often have a higher COP/PER than the heat pumps for buildings. This is mainly due to small temperature lifts and stable operating conditions. Typical COP/PER ranges for industrial heat pumps are given in Table 6.14.

As mentioned earlier, a heat pump may use only one-third as much energy as electric resistance heat (e.g., electric furnace and baseboards) during mild winter weather (outdoor temperature about 7 °C). In the heat pump industry, this is described as a COP of 3. COP is the ratio of heat output to electrical energy input. A number of factors prevent air source heat pumps from maintaining COPs of 3 throughout the heating season:

- **Air temperature.** Heat pumps operate at temperatures colder than 7 °C much of the winter. When the temperature is −6.6 °C, the COP of the heat pump will be closer to 2 than 3.
- **Defrost.** If there is very cold refrigerant flowing through the outdoor heat exchanger, ice can form on the coils, just as it does in freezers. When outdoor temperature is below 6.4 °C, the heat pump may need to defrost periodically. To melt the ice, the heat pump takes heat from the house to heat the outdoor coils, which reduces average heat pump efficiency.
- **Supplemental heat.** As the air gets colder outside, the heat pump fails to provide the necessary heat to keep the house comfortable. At some outdoor temperature it will be too cold for the heat pump to provide all the heat the house needs. To make up the difference, heat pumps have a supplemental heating system, usually electric resistance coils (basically an electric furnace inside the heat pump indoor cabinet). This part of the system is sometimes called "back-up" or "emergency" heat because the same coils can be used to provide some or all of the heat in the event of heat pump failure. Since the supplemental electric heating system does not operate with the same efficiency as the heat pump (the COP of electric resistance heat is 1), the total heat

pump COP will be much lower when the supplemental heat is on. In this regard, gas and oil furnaces provide supplemental heat in some new homes with heat pumps. Existing gas and oil furnaces can also be used as supplemental heat with "add-on" heat pumps that allow a heat pump to be added to an existing system. Controls for these systems are different since the combustion system and the heat pump do not operate at the same time. Special care is required to ensure that there are proper air flows for both the heat pump and the furnace. The economics of purchasing and operating this type of system will depend on local energy costs.

- **Cycling losses.** When heating systems first start up, they need to operate for a while just to get warm enough to heat the house. When they are shut off, there is still heat in the system that does not get into the house. The losses associated with stopping and starting the heat pump are referred to as *cycling losses*.
- **Heating season performance factor (HSPF).** The industry standard test for overall heating efficiency provides this rating known as HSPF. This laboratory test attempts to take into account the reductions in efficiency caused by defrosting, temperature fluctuations, supplemental heat, fans, and on/off cycling. The higher the HSPF, the more efficient the heat pump. A heat pump with an HSPF of 6.8 has an average COP of 2 for the heating season. To estimate the average COP, we divide the HSPF by 3.6. In fact, HSPF is a rough predictor of the actual installed performance. HSPF assumes specific conditions that are unlikely to coincide with the climate. Most utility-sponsored heat pump programs require that heat pumps have an HSPF of at least 6.8. In practice, many heat pumps meet this requirement. Some heat pumps have HSPF ratings above 9. In general, more efficient heat pumps are more expensive.
- **Seasonal energy efficiency ratio (SEER).** Cooling performance is rated using the SEER. The higher the SEER the more efficiently the heat pump cools. The SEER is the ratio of heat energy removed from the house compared to the energy used to operate the heat pump, including fans. The SEER is usually noticeably higher than the HSPF since defrosting is not needed and there is no need for expensive supplemental heat during air-conditioning weather. Except in an area where cooling is more important than heating, the HSPF is a more important measure of efficiency than the SEER.

6.28 Ground-Source Heat Pumps (GSHPs)

Efficient heating performance makes GSHP a good choice for the heating and cooling of commercial and institutional buildings, such as offices, stores, hospitals, hotels, apartment buildings, schools, restaurants, and so on. The three main parts are the heat pump itself which includes the compressor, blower, air, and water coils; the liquid heat exchange medium which is either well water, pond, lake, or river water, or a buried earth loop filled with water and glycol; and the air delivery system (ductwork).

GSHP systems can heat water or heat/cool the interior space by transferring heat from the ground outside, but they can also transfer heat within buildings with a heat-producing central core. GSHP technology can move heat from the core to perimeter zones where it is required, thereby simultaneously cooling the core and heating the perimeter.

GSHP systems are also used as heat recovery devices to recover heat from building exhaust air or from the waste water of an industrial process. The recovered heat is then supplied at a higher temperature at which it can be more readily used for heating air or water.

As with air-to-air heat extraction technology, geothermal (groundwater/ground source) technology utilizes a type of heat pump known as a GSHP. This type of geothermal heat pump device extracts its heat from water rather than from air. While the principles are fundamentally similar, the methodology varies in that water is pumped through a special type of heat exchanger and is either "chilled" by the evaporating refrigerant (in the heating mode) or heated by the condensing refrigerant (in the cooling mode).

A GSHP is essentially an air conditioner that also runs in reverse: in winter, it extracts heat from the ground for heating. These systems have been used more efficiently than most other systems. For example, when you put one unit of energy into a furnace, you get somewhat less than one unit back out as heat. But when we put the same unit of energy into a heat pump, we get more than triple as the return. That is because heat pumps do not use energy to create heat; instead, they move heat that already exists.

1. GSHPs start with a closed loop of buried pipes containing a fluid that can carry heat. The pipes may lie in a shallow long curved trench, or they may jut deep into the ground.
2. For heating (Figure 6.33), the pipe fluid absorbs heat from the earth. The fluid passes through a heat exchanger (acting as an evaporator), where it transfers heat to a refrigerant.
3. The refrigerant, which flows through another closed loop in the heat pump, then boils. The vaporized refrigerant travels to the compressor, where its temperature and pressure are increased.
4. The hot gas continues to two heat exchangers (acting as condensers), one to heat the house's water and the other for space heating. At each, the refrigerant gives up some heat. A fan blows across the space-heat condenser to move the warmed air through the house. The refrigerant, again a liquid, repeats the process.
5. In summer, the cycle reverses to remove heat from the house. Some of the heat is used for hot water; the remainder is dumped into the earth via the ground loop.

GSHP systems use the earth's stored energy and simply transfer it into a home or business for heating, cooling, and generating hot water. These efficient and economical heating and cooling systems draw in well water, extract energy from it, and transfer heat into the home or business.

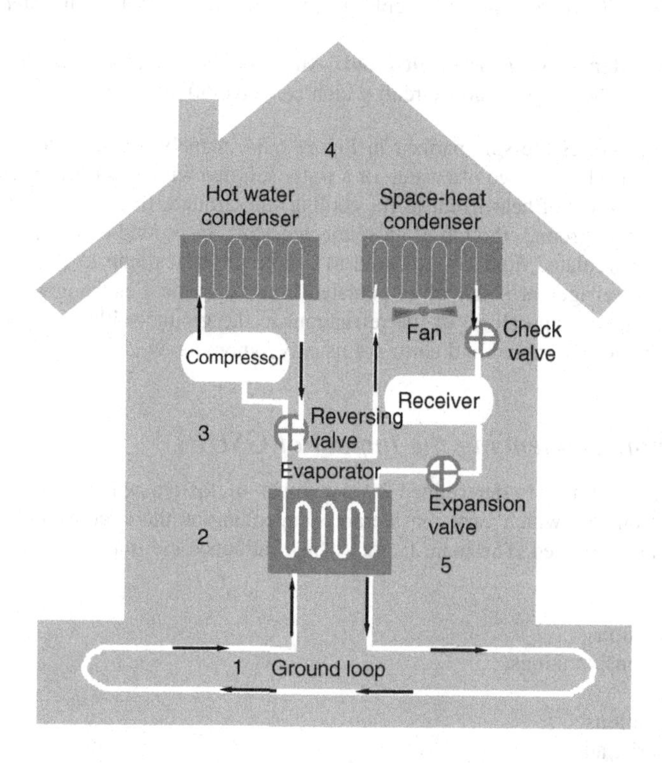

Figure 6.33 Schematic of a ground-source heat pump.

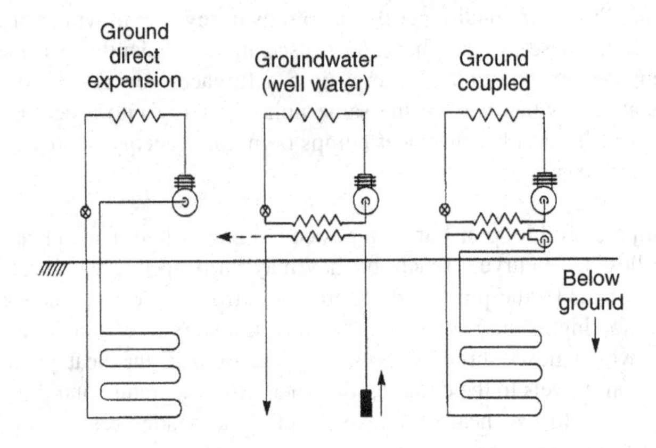

Figure 6.34 Three GSHP categories.

There are several reasons to consider a GSHP for applications, including the following:

- **Efficient heating and cooling.** When measured against other existing systems, the heat pump provides higher COP.
- **Durability and low maintenance.** They require 30–50% less maintenance than a fossil-fuel-based heating and cooling system. Most geothermal systems can run without any repair maintenance for over 20 years. The only replacement required is for the air filter (e.g., every 3 months).
- **Low domestic water heating cost on demand.** All residential and many commercial heat pumps use an optional hot water generator circuit which can save 50–65% on water heating costs.

Three GSHP categories are summarized in Figure 6.34, namely groundwater, ground coupled, and direct expansion. The first two use water, or a water solution with an antifreeze (brine or glycol), for intermediate transport of heat from (to for cooling) the ground. In groundwater systems, water is removed from underground, though often reinjected later. In ground-coupled systems, a closed loop or loops of recirculated fluid is (are) used to couple the heat pump to the ground. For direct expansion systems, refrigerant is actually circulated in a buried heat exchanger; the name derives from direct expansion (evaporation) of the refrigerant in the ground without an intermediate heat transport fluid. The terms ground and earth are usually synonymous, depending on the context.

6.28.1 Factors Influencing the Impact of GSHPs

The impact of heat pumps is determined by a number of factors which are common to many countries and regions but which vary considerably, depending on the circumstances under which a heat pump is being deployed. The main factors which influence the impact of heat pumps include the following:

- climatical conditions,
- energy policy considerations,
- building factors,
- utility considerations,
- economic criteria, and
- competing systems.

6.28.2 *Benefits of GSHPs*

A GSHP is one of the most efficient means available to provide space heating/cooling for homes and offices. It transfers the heat located immediately under the earth's surface (or in a body of water) into a building in winter, using the same principle as a refrigerator that extracts heat from food and rejects into a kitchen. A heat pump takes heat from its source at low temperature and discharges it at a higher temperature, allowing the unit to supply more heat than the equivalent energy supplied to the heat pump. Many people are familiar with air-to-air heat pumps, which use outdoor air as the source of heat. These units are well suited for moderate climates, but they do not operate efficiently when the outdoor temperature drops below $-10\,^{\circ}\mathrm{C}$ and there is little heat left in the air to extract. Here we list some of the key benefits of GSHP systems as follows:
In terms of cost savings,

- lower annual heating and cooling bills – 30–60% on average,
- option to use excess heat for water heating,
- high reliability, low maintenance, and
- less space requirements for heating and cooling equipment-frees space for manufacturing or work facilities.

In terms of enhanced comfort and appearance,

- individual zone control,
- outdoor equipment not necessary,
- safe, noncombustion process,
- quiet operation,
- greater convenience,
- simultaneous heating and cooling of different zones, and
- no more fuel deliveries.

In terms of improved environment,

- renewable, environmentally friendly energy source,
- reduced harmful emissions (no on-site combustion of fossil fuels), and
- self-contained system – less potential for refrigerant leaks.

In addition to heating and cooling the building interior, GSHP units can also provide domestic water. Some major features and technical details of GSHPs are as follows.

6.28.2.1 Efficiency and COP

The major advantage of a GSHP system is that the heat obtained from the ground (via the condenser) is much greater than the electrical energy that is required to drive the various components of the system. The efficiency of a unit is the ratio of heat energy provided versus the electrical energy consumed to obtain that heat, referring to the COP. In some countries, there is a minimum limit for the COP of the GSHP units. For example, in Canada it must exceed 3.0 (i.e., for every kilowatt of electricity needed to operate the system, the GSHP provides 3 kW of heat energy).

6.28.2.2 Cost

With a COP of 3.0, the cost of heating would be one-third (i.e., two-thirds less) of the cost to operate an electric resistance heating system, such as baseboards or electric furnace. With a COP of 6.0, the savings can be as much as three-quarters of the price of electric heating and cooling. As earth energy technology improves and the COP increases above 6.0, the operating savings also increase.

6.28.2.3 Comfort

A GSHP system warms air in smaller increases over a longer period of time, compared to the burst of a combustion oil or gas furnace. As a result, homeowners notice a stable level of heat with no peaks or troughs, less drafts, and so on.

6.28.2.4 Environment

GSHP is preferred by many people and institutions because it is an environmentally benign technology, with no emissions or harmful exhaust. For example, the GSHP industry was the first in Canada to move away from damaging CFCs. Although most GSHP units require electricity to operate the components, a high COP means that GSHP systems provide a significant reduction in the level of CO_2, SO_2, and NO_x emissions (all linked with the issue of greenhouse gas emissions and global warming).

6.28.2.5 Suitability for Public and Commercial Utilization

GSHPs are ideally suited for public and commercial places, due to the following advantages:

- Sports fields and parking lots provide an excellent heat sink for the thermoplastic loops, and are easy and inexpensive to install.
- Decentralized control allows simultaneous heating or cooling of different zones, allowing each teacher to set a different temperature for any room in the school.
- The heating/cooling loads of a school are well-matched to the performance output of a heat pump.
- Locating equipment inside the building reduces possible damage from adverse weather conditions and vandalism, and eliminates the need for roof penetrations (further securing the building envelope and allowing enhanced design with nonhorizontal rooftops).
- The smaller mechanical space requirements for heat pumps and the elimination of large air ducts result in more available space for classrooms and lower costs (the physical dimensions of a school building can drop by 5%).
- Student health and safety is increased by the elimination of combustion heat and the need for fuel storage, as well as the absence of possible carbon monoxide leaks and superior levels of indoor air quality.
- Modularity allows reduced inventory for boilers and chillers.
- GSHP installations are known for quiet operation, high reliability, greater ventilation and dehumidification, and less drafts.

6.28.2.6 Other Possible Applications

GSHP units can be used for the dehumidification of indoor swimming pool areas, where the unit can dehumidify the air and provide condensation control with a minimum of ventilation air. The heat recovered from the condensed moisture is then used for heating domestic/pool water or for space heating.

6.28.3 Types of GSHP Systems

Heat pumps tap the natural geothermal energy underground to heat and cool houses. A fluid is circulated through an underground loop of polyethylene pipe. The fluid absorbs heat from the earth during the winter and dissipates heat from the house during the summer. The heat or cold from the fluid is converted to hot or cool air by circulating it through water-to-refrigerant and refrigerant-to-air heat exchangers (similar to a car radiator). GSHPs are very efficient, utilizing the relatively constant temperature of the ground to obtain heat during the winter and provide cooling during the summer. The loop of polyethylene pipe buried in the ground can be installed vertically (to conserve space) or horizontally around the outside of the house. Depending on location, the pipes can be connected to a well system or coiled in a pond rather than buried underground.

Most loops for residential GSHP systems are installed either horizontally or vertically in the ground, or submersed in water in a pond or lake. In most cases, the fluid runs through the loop in a closed system, but open-loop systems may be used where local codes permit. Each type of loop configuration has its own, unique advantages and disadvantages, as explained below. There are two basic types of geothermal systems, open loop and closed loop.

6.28.3.1 The Open-Loop System

The term *open loop* is commonly used to describe a geothermal heat pump system that uses groundwater from a conventional well as a heat source. The groundwater is pumped into the heat pump unit where heat is extracted and the water is disposed of in an appropriate manner. Since groundwater is relatively constant all year around, it is highly efficient and is an excellent heat source. An open-loop system (Figure 6.35a) uses a conventional well as its heat source. Water is pumped from the well through the heat pump's heat exchanger, where heat is extracted and transferred to a refrigerant system. The heat is then transferred to the air in the home. The water is then returned to a pond, stream, or second well. Local conditions such as quantity and quality of available water can affect the use of this type of system. Local water use and disposal regulations may also limit the use of open loop systems. An open loop is a loop established between a water source and a discharge area in which the water is collected and pumped to a groundwater heat

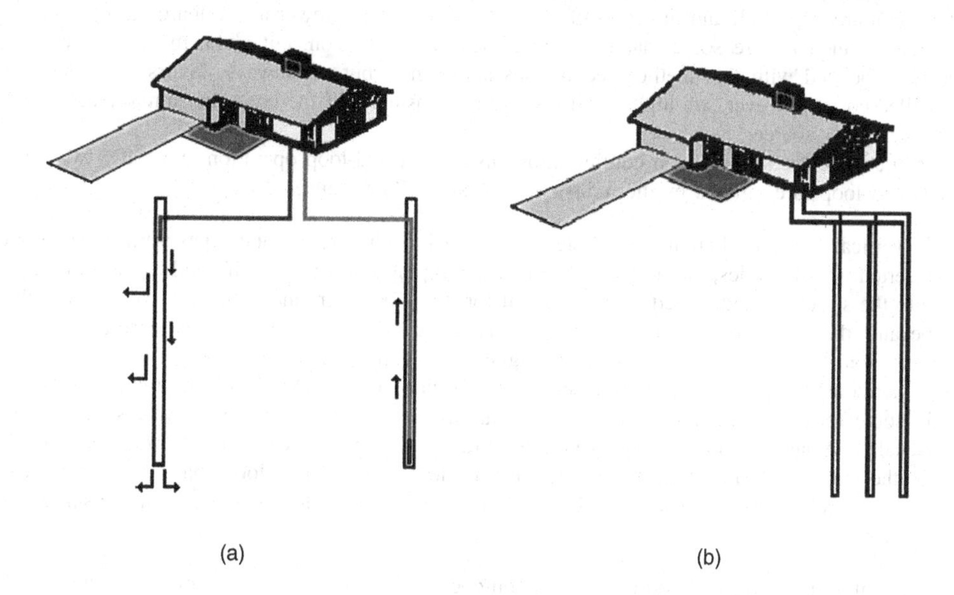

Figure 6.35 (a) An open-loop GSHP system. (b) A closed-loop GSHP system (*Courtesy of Sunteq Geo Distributors*).

pump (GWHP) and then discharged to its original source or to another location. The piping for such configuration is open at both ends and the water is utilized only once. Examples of such loops include systems operating off wells wherein water is pumped from a supply well, through the unit and discharged to a return well and open systems operating from such surface water sources as ponds, lakes, streams, and so on, where the source water is pumped to the unit and returned to the source. Open loops have the advantage of higher equipment performance since the source water is used only once and then discharged, but have two significant disadvantages:

- Water quality needs to be carefully analyzed and treated if such corrosives as sulfur, iron, or manganese are present, if pH is low, or if there are abrasives in it.
- The costs of pumping water through an open loop are usually somewhat higher than those associated with circulating water through a closed loop.

6.28.3.2 The Closed-Loop Systems

Closed-loop systems circulate a heat-transfer fluid (usually a water/antifreeze solution) through a system of buried plastic piping, arranged either horizontally or vertically. Horizontal loop systems draw their heat from loops of piping buried 2–2.5 m deep in trenches or ponds. Vertical loop systems use holes bored 45–60 m deep with U-shaped loops of piping. They work the same as horizontal loop systems, but can be installed in locations where space is limited due to size, landscaping, or other factors.

A closed loop (Figure 6.35b) is one in which both ends of the loop's piping are closed. The water or other fluid is recirculated over and over and no new water is introduced into the loop. The heat is transferred through the walls of the piping to or from the source, which could be ground, groundwater, or surface water. As heat is extracted from the water in the loop the temperature of the loop falls and the heat from the source flows toward the loop. In closed-loop operation water quality is not an issue because corrosives become rapidly spent or used up and corrosion caused by poor water quality is quickly curtailed. The wire-to-water efficiencies of circulators used in closed-loop operation are very high and the costs of pumping the water are lower as compared to open loops. System efficiencies are somewhat lower in closed-loop operation, but given the lower pumping costs associated with this method, economics sometimes, but not always, favors this approach. Installed costs, however, are higher and need to be considered if the consumer already has a well or other water source.

While there are several loop configurations used in closed-loop operation, generally two types of closed loops are utilized by the industry (vertical and horizontal).

- In vertical loop installation, deep holes are bored into the ground and pipes with U-bends are inserted into the holes, the holes are grouted, the piping loops are manifolded together, brought into the structure, and closed. The argument for this type of ground-loop heat exchanger is that because the piping is in the deeper ground (unaffected by surface temperatures) performance will be higher. Generally, installed costs are higher than with a horizontal loop.
- In horizontal loop installation, trenches are dug, usually by a backhoe or other trenching device, in some form of horizontal configuration. Various configurations of piping are installed in the trenches. A larger number of horizontal loop designs have been tried and utilized successfully by the industry. While installed costs have been lower, horizontial loops have been thought to be less efficient than vertical loops because of the effect of air temperatures near the surface of the ground.

Generally the payback period is 1–3 years on open loops, and 5–7 years on closed-loop systems. Payback will also vary depending upon the insulation used and how well the delivery systems (duct work) are designed.

Closed ground-loop systems should be considered when groundwater is unavailable (no well), the groundwater supply is insufficient or of poor quality (sulfur, biological iron), regulations prohibit the use of groundwater, or drilling for the disposal of groundwater is impractical. The open-loop well system, when compared to a closed ground loop, will provide higher operating efficiencies, resulting in lower heating/cooling costs, smaller unit size, lower installation costs, and a payback in less than 3 years. If the homeowner has sizable property, a well, a place to discharge or leach the water, and local codes permit surface discharge, the open loop system is the first choice. An average $20\,m^2$ house needs only about $0.04\,m^3$ for the house and heat pump.

6.28.3.3 The Direct Exchange System

Another type of geothermal heat pump is called a direct exchange (DX) system.This type of system uses a much shorter loop of piping buried below ground, through which the refrigerant itself is circulated; heat transfer directly between the refrigerant and the ground allows heat-transfer fluid used in other geothermal systems to be replaced and the amount of piping to be drastically reduced. This type of system is ideal for situations where the amount of space for the piping loop is very limited.

Another type of geothermal heating and cooling is direct GSHP–DX systems, which utilize copper piping placed underground. As refrigerant is pumped through the loop, heat is transferred directly through the copper to the earth. The length of the loop depends upon a number of factors, including the type of loop configuration used, a house's heating and air-conditioning load, soil conditions, local climate, and landscaping. Larger homes with larger space conditioning requirements generally need larger loops than smaller homes. Houses in climates where temperatures are extreme also generally require larger loops. A heat loss/heat gain analysis should be conducted before the loop is installed.

6.28.4 Types of GSHP Open- and Closed-Loop Designs

In practice, there are the following four basic GSHP loop designs available.

6.28.4.1 Open-Water or Open-Well Loops

If an abundant supply of good quality well water is available, an open-loop system can be installed. A well must have enough capacity to provide adequate water flow for domestic use and for the geothermal unit throughout the year. A good way to discharge the water once it has been used must also be available. Ditches, field tile, streams, and rivers are the most common discharge areas. However, all local codes should be checked out before selecting a discharge method.

If the installation area meets the guidelines requirement, an open-loop system can be used. And, since no closed loop is necessary, this installation usually costs less to install and delivers the same high efficiency.

These systems take water from a well or a drilled well, direct the flow through the GSHP unit where the heat is extracted, and then return the cooled water to the lake or well, in accordance with environmental regulations. If the source of water is a lake, the body of water must be large enough to provide a sufficient heat sink capacity. Rivers can be used as a source of water, but sources with high levels of salt, chlorides, or other minerals are not recommended for most units. If the source of water is a well, most countries have regulations concerning the extraction and reinjection of water to protect natural aquifers. Although a GSHP system only extracts heat from the water in winter, there is a difference of opinion over whether this change in temperature then classifies the discharge water as sewage.

Some advantages of the open-loop systems are

- higher thermal efficiency of heat transfer,
- lower installation cost, and
- lack of need to use a chemical fluid for the heat-transfer medium.

Some disadvantages are

- environmental concerns with disruption of water tables and aquifers, and
- effects of fluctuations in water temperature on system performance.

This type of loop configuration is used less frequently, but may be employed cost-effectively if groundwater is plentiful. Open-loop systems, in fact, are the simplest to install and have been used successfully for decades in areas where local codes permit. In this type of system, groundwater from an aquifer is piped directly from the well to the house, where it transfers its heat to a heat pump. After it leaves the building, the water is pumped back into the same aquifer via a second well (so-called *discharge well*) located at a suitable distance from the first.

Standing-column wells (Figure 6.36a), (so-called *turbulent wells* or *energy wells*), have become an established technology, particularly in some regions of the United States. Standing wells are typically 0.15 m in diameter and may be as deep as 457 m. Temperate water from the bottom of the well is withdrawn, circulated through the heat pump's heat exchanger, and returned to the top of the water column in the same well. Usually, the well also serves to provide potable water. However, groundwater must be plentiful for a standing well system to operate effectively. If the standing well is installed where the water table is too deep, pumping would be prohibitively costly. Under normal circumstances, the water diverted for building (potable) use is replaced by constant-temperature groundwater, which makes the system act like a true open-loop system. If the well-water temperature climbs too high or drops too low, water can be bled from the system to allow groundwater to restore the well-water temperature to the normal operating range. Permitting conditions for discharging the bleed water vary from locality to locality, but are eased by the fact that the quantities are small and the water is never treated with chemicals.

In some places, for example, builders install large community loops, which are shared by all of the buildings in a housing development.

6.28.4.2 Lake or Pond Closed Loops

Since water transfers heat much better than soil, closed loops can also be sunk in lakes or ponds. The coiled pipe can be placed on the bottom of the pond or lake, where it transfers heat to or from the water. A 2500–5000 m^2, 1.8 m-deep pond is acceptable. Pond or lake loops often require less excavation than vertical and horizontal loops. Therefore, they are often less expensive to install.

A closed-loop system is positioned on the floor of a body of water, instead of buried in the ground. The pipe must be weighted properly to remain on the bottom of the lake and to avoid shifting caused by spring ice movement.

Some advantages of the lake loop are that it is

- less expensive to install than trenching into the ground and
- relatively easy to diagnose leaks in the loop.

Some disadvantages are

- potential environmental damage to aquaculture and
- water temperature fluctuation (especially in spring runoff) affecting system performance.

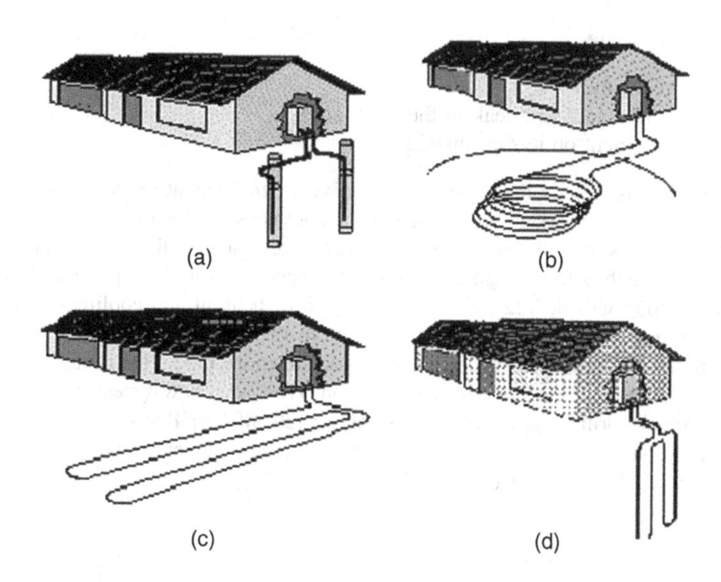

Figure 6.36 (a) A standing-column well system. (b) A pond closed-loop system. (c) A horizontal ground closed-loop system. (d) A vertical ground closed-loop system *(Courtesy of Geothermal Heat Pump Consortium, Inc.)*.

Normally, if a house is near a body of surface water, such as a pond (Figure 6.36b) or lake, this type of loop design may be the most economical. The fluid circulates through polyethylene piping in a closed system, just as it does in the ground loops. Typically, workers run the pipe to the water, then submerge long sections under water. The pipe may be coiled in a slinky shape to fit more of it into a given amount of space. A pond loop is recommended only if the water level never drops below 1.8–2.5 m at its lowest level to assure sufficient heat-transfer capability. Pond loops (Figure 6.36b) used in a closed system result in no adverse impacts on the aquatic system.

6.28.4.3 Horizontal Closed Loops

If adequate land is available, horizontal loops (see Figure 6.36c) can be installed. Loops are placed in trenches 1.2–1.8 m deep. One layer or multiple layers of pipe can be laid in a trench with one foot of soil backfilled between the layers.

These are a very common configuration in North America, particularly in the United States and Canada. A trench is dug on the property, and pipe is buried in a continuous or parallel loop (depending on size of unit). In Canada, the national installation standard (CSA C445) states that the loop must be located at least 0.6 m below ground, but industry guidelines are at least twice that depth. It is possible to lay more than two pipes in each trench, thereby reducing the cost of digging. It is important to backfill the trench properly, to avoid air pockets that can reduce the transfer of heat and to ensure that the pipe is not damaged by large rocks.

Some advantages of the horizontal closed loops are

- relative ease of installion,
- high design flexibility, and
- greater control capability of entering fluid temperature.

Some disadvantages are

- higher cost of trenching,
- greater difficulty in detecting a leak in the loop, and
- landscaping requirement on retrofit installations.

This configuration is usually the most cost-effective when adequate yard space is available and trenches are easy to dig. Workers use trenchers or backhoes to dig the trenches 1–1.8 m below the ground, then lay a series of parallel plastic pipes. They backfill the trench, taking care not to allow sharp rocks or debris to damage the pipes. Fluid runs through the pipe in a closed system. A typical horizontal loop will be 122–183 m long per ton of heating and cooling capacity. The pipe may be curled into a slinky shape in order to fit more of it into shorter trenches, but while this reduces the amount of land space needed it may require more pipe. Horizontal ground loops are easiest to install while a house is under construction. However, new types of digging equipment that allow horizontal boring are making it possible to retrofit GSHP systems into existing homes with minimal disturbance to lawns. Horizontal boring machines can even allow loops to be installed under existing buildings or driveways.

6.28.4.4 Vertical Closed Loops

These are the most expensive but the most efficient configuration, because of the fact that the under-earth level of heat increases and stabilizes with depth. This option is viable when surface property is limited or in difficult terrain, but care must be taken to ensure that the vertical borehole is drilled according to provincial regulations.

Some advantages of the vertical closed loops are

- highest efficiency,
- less property space requirement, and
- high security from accidental post-installation damage.

Some disadvantages are

- usually the highest-cost option and
- potential environmental damage to aquifers if not installed properly.

If the land area is limited, closed loops can be inserted into vertical boreholes. Holes are drilled to a depth of about 38–60 m per ton of unit capacity. U-shaped loops of pipe are inserted into the holes. The holes are then backfilled with a sealing solution.

This type of loop configuration (Figure 6.36d) is also ideal for homes where yard space is insufficient to permit horizontal buildings with large heating and cooling loads, when the Earth is rocky close to the surface, or for retrofit applications where minimum disruption of the landscaping is desired. Contractors bore vertical holes in the ground 45–137 m deep. Each hole contains a single loop of pipe with a U-bend at the bottom. After the pipe is inserted, the hole is backfilled or grouted. Each vertical pipe is then connected to a horizontal pipe, which is also concealed underground. The horizontal pipe then carries fluid in a closed system to and from the GSHP system. Vertical loops are generally more expensive to install, but require less piping than horizontal loops because the Earth deeper down is alternately cooler in summer and warmer in winter.

6.28.5 Operational Principles of GSHPs

Mostly, a GSHP heat pump (Figure 6.37) uses a vapor-compression refrigeration cycle to transfer heat. The temperature at which the refrigerant vaporizes or condenses is controlled by regulating

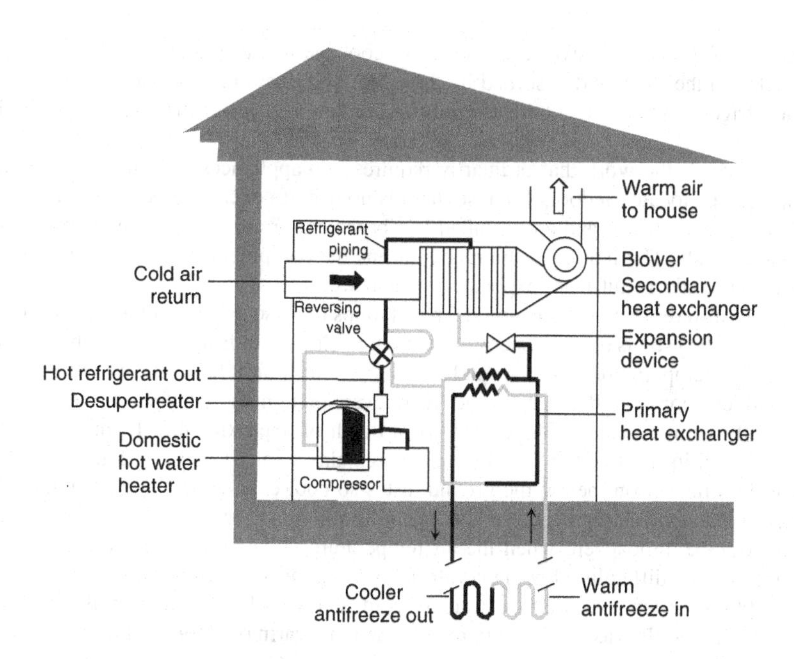

Figure 6.37 A GSHP using a vapor-compression system (*Courtesy of Earth Energy Society of Canada*).

the pressure in different parts of the system. A low pressure is maintained in the evaporator, so the refrigerant can vaporize at low temperatures. Conversely, the condenser is maintained at a high pressure so that the vapor is forced to condense at relatively high temperatures. The compressor compresses the refrigerant vapor and adds heat of compression. The hot vapor (under high pressure) flows from compressor to condenser, where the air or water to be heated passes over and absorbs heat, thereby cooling the refrigerant vapor. As this happens, the vapor condenses to a liquid and gives up its latent heat of condensation. The warm liquid refrigerant (still under high pressure) flows from condenser to a metering valve which controls the flow of liquid refrigerant. The downstream side of this device is under low pressure, being connected through the evaporator to the suction side of the compressor. The liquid passing through the metering device begins to evaporate under low pressure as it enters the evaporator heat exchanger. The temperature of the liquid drops as it releases the latent heat of vaporization to the refrigerant vapor. The cold fluid in the evaporator absorbs heat from the source, and the liquid evaporates. The cool vapor from the evaporator is drawn to the suction side of the compressor where it is compressed and the cycle is repeated. A schematic of such a system is shown in Figure 6.37.

Most heat pumps have a reversing valve which allows them to cool as well as heat the building. This valve changes the flow of the fluid such that the coil in the building becomes the evaporator and the outdoor coil becomes the condenser.

It is known that GSHP systems work on a different principle than an ordinary furnace/air-conditioning system, and they require little maintenance or attention. Furnaces must create heat by burning a fuel (typically natural gas, propane, or fuel oil). With GSHP systems, there is no need to create heat, and hence no need for chemical combustion. Instead, the earth's natural heat is collected in winter through a series of pipes, called a loop, installed below the surface of the ground or submersed in a pond or lake. Fluid circulating in the loop carries this heat to the home. An indoor GSHP system then uses electrically driven compressors and heat exchangers in a vapor-compression cycle to concentrate the earth's energy and release it inside the house at a higher temperature. In typical systems, duct fans distribute the heat to various rooms.

In summer, the process is reversed in order to cool the home. Excess heat is drawn from the home, expelled to the loop, and absorbed by the earth. GSHP systems provide cooling in the same way that a refrigerator keeps its contents cool (by drawing heat from the interior, not by injecting cold air).

GSHP systems do the work that ordinarily requires two appliances, a furnace and an air conditioner. They can be located indoors because there is no need to exchange heat with the outdoor air. Typically, they are compacts and are installed in a basement or attic, and some are small enough to fit atop a closet shelf. The indoor location also means the equipment is protected from mechanical breakdowns that could result from exposure to harsh weather.

GSHP works differently than conventional heat pumps that use the outdoor air as their heat source or heat sink. GSHP systems use less energy because they draw heat from a source whose temperature is moderate. The temperature of the ground or groundwater a few feet beneath the earth's surface remains relatively constant throughout the year, even though the outdoor air temperature may fluctuate greatly with the change of seasons. At a depth of approximately 1.8 m, for example, the temperature of soil in most of the world's regions remains stable between 7.2 and 21.1 °C. This is why well water drawn from below the ground tastes so cool even on the hottest summer days.

In winter, it is much easier to capture heat from the soil at a moderate temperature, for example, 10 °C than from the atmosphere when the air temperature is below zero. This is also why GSHP systems encounter no difficulty blowing comfortably warm air through the ventilation system, even when the outdoor air temperature is extremely cold. Conversely, in summer, the relatively cool ground absorbs a house's waste heat more readily than the warm outdoor air. From the applications, it appears that approximately 70% of the energy used in a GSHP heating and cooling system is renewable energy from the ground. The remainder is clean electrical energy, which is employed to concentrate heat and transport it from one location to another. In winter, the ground soaks up solar energy and provides a barrier to cold air. In summer, the ground heats up more slowly than the outside air.

6.28.6 *Installation and Performance of GSHPs*

There are a number of factors that will have a major influence on the installation and performance of a GSHP system. Therefore, it is important to understand the following issues clearly (EESC, 2001).

6.28.6.1 Heat Loss Calculations

The most important first step in the design of a GSHP installation is to determine how much heat is required to satisfy one's comfort level. For example, in Canada, the national installation standard for residential earth energy units (CSA C445) states that the heat loss must be calculated in accordance with the F280 program. This method needs to know the insulation levels of all walls and windows, the number of occupants, the geographic location in Canada, and soil type, and many other factors, to determine the total annual heat loss in kilowatts (kW). It will also calculate the cooling load for summer (all units will provide sufficient cooling if the unit is large enough to provide sufficient heat) and for hot water heating, if included. With this final heat loss, the installed unit will match the demand.

6.28.6.2 Terminology

Because of the large demand for GSHP as cooling devices, particularly in the United States, the GSHP industry uses the term ton to describe a unit that will provide approximately 13 kJ of cooling capacity. On average, a typical 186 m^2 new residence would require a 4-ton unit for sufficient heat.

6.28.6.3 Sizing

GSHP units do not need to meet 100% of the calculated heat loss of a building, as long as they have an auxiliary electric heating source for backup and for emergencies. Almost 90% of a house's heat load can be met by a GSHP unit that is sized to 70% of the heat loss, with the remaining 10% of load supplied by the auxiliary plenum heater. Oversizing can result in control and operational problems in the cooling mode (especially if the GSHP unit has a single-speed compressor), and the installed cost will increase significantly for little operational savings. Conversely, undersizing will lower the installed cost, but the additional length of time that the GSHP unit will operate for will place excessive demand on many components and may result in unacceptable chill. For example, although the CSA standard for installations says that 60% is the minimum, the industry has moved to a sizing level of 75–80% of heat loss as an optimal design size.

6.28.6.4 Air Flow

GSHP units work efficiently because they provide a small temperature rise, but this means that the air coming through the register on the floor is not as hot as the air from a gas or oil furnace. A GSHP unit must heat more air to supply the same amount of heat to the house, and duct sizes must be larger than those used for combustion furnaces to accommodate the higher volumetric air flow rates.

6.28.6.5 Soil Type

Loose dry soil traps air and is less effective for the heat-transfer required in GSHP technology than moist packed soil. Each manufacturer provides specifications on the relative merits of soil type; low-conductive soil may require as much as 50% more loop than a quality high-conductive soil.

6.28.6.6 Loop Depth

GSHP technology relies on stable underground (or underwater) temperature to function efficiently. In most cases, the deeper the loop is buried, the more efficient it will be. A vertical borehole is the most efficient configuration, but this type of digging can be very expensive.

6.28.6.7 Loop Length

The longer the amount of piping used in a GSHP outdoor loop, the more heat can be extracted from the ground (or water) for transfer to the house. Installing less loop than specified by the manufacturer will result in lower indoor temperature, and more strain on the system as it operates longer to compensate for the demand. However, excessive piping can also create a different set of problems, as well as additional cost. Each manufacturer provides specifications for the amount of pipe required. As a broad rule of thumb, a GSHP system requires 122 m of horizontal loop or 91 m of vertical loop to provide heat for each ton of unit size.

6.28.6.8 Loop Spacing

The greater the distance between buried loops, the higher the efficiency. It is suggested that there should be 3 m between sections of buried loop, in order to allow the pipe to collect heat from the surrounding earth without interference from the neighboring loop. This spacing can be reduced under certain conditions.

6.28.6.9 Type of Loop

GSHP pipe comes in two common diameters: 1.9 cm (0.75 in) and 3.2 cm (1.25 in). Two coiled loops (commonly called *the Svec Spiral and the Slinkey*) require less trenching than conventional straight pipe. As a result, the higher cost of the coiled pipe is offset by the lower trenching costs and the savings in property disruption.

6.28.6.10 Water Quality

Open water systems depend on a source of water that is adequate in temperature, flow rate, and mineral content. GSHP units are rated under the national performance standard (CSA C446) based on their efficiency when the entering water temperature is 10 °C (0 °C for closed-loop units), but this efficiency drops considerably if the temperature of water is lower when it comes from the lake or well. Each GSHP model has a specified flow rate of water that is required, and its efficiency drops if this rate is reduced. The CSA installation standard demands an official water well log to quantify a sustainable water yield. Water for open-loop systems must be free of many contaminants such as chlorides and metals, which can damage the heat exchanger of a GSHP unit.

6.28.6.11 Water Discharge

There are environmental regulations which govern how the water used in an open-loop system can be returned to the ground. A return well is acceptable, as long as the water is returned to the same aquifer or level of water table. A discharge pit is also acceptable, as long as certain conditions are followed.

6.28.6.12 Balance Point

The outdoor temperature at which a GSHP system can fully satisfy the indoor heating requirement is referred to as the *balance point*, and is usually −10 °C in most regions of Canada. At outdoor air temperatures above this balance point, the GSHP cycles on and off to satisfy the demand for heat indoors. At temperatures below this point, the GSHP unit runs almost continuously, and also turns on the auxiliary heater to meet the demand.

6.28.6.13 Auxiliary Heat

When the outdoor air temperature drops below the design balance point, the GSHP unit cannot meet the full heating demand inside the house (for units sized to 100% of heat loss, this is not an issue). The difference in heat demand is provided by the supplementary or auxiliary heat source, usually an electric resistance element positioned in the unit's plenum. Like a baseboard heater, the COP of this auxiliary heater is 1.0, so excessive use of backup heat decreases the overall efficiency of the GSHP system and increases operating costs for the homeowner.

6.28.6.14 Heat Transfer Fluids

Closed-loop GSHP units can circulate any approved fluid inside the pipe, depending on the performance characteristics desired. Each manufacturer must specify which fluids are acceptable to any particular unit, with the most common being denatured ethanol or methanol.

6.28.6.15 Savings

GSHP systems are frequently more expensive to install than gas, oil, or electric heating units, but they are very competitive with any type of combination heating/cooling system. For this reason, heat pumps are most attractive for applications requiring both heating and cooling. For example, in Canada, an open-loop water-source system for an average residence may cost $8000, while a closed-loop ground-source system may cost as much as $15,000. However, annual operating costs would be as low as $850, compared to $2000 or more for conventional heating/cooling systems.

The savings available with a heat pump will reflect the size of the house and its quality of construction (particularly the level of insulation), the building's heat loss and the sizing level of the GSHP unit, the balance point, the COP of the GSHP unit, local climate and energy costs, the occupier's lifestyle habits, the efficiency of alternate heating systems, the configuration of loop, interior temperature setting, ductwork required on a retrofit, site accessibility for equipment, and the options selected.

6.28.7 Hybrid Heat Pump Systems

Heat pump systems can be combined with natural gas, liquefied petroleum, or oil-fired heating systems. These are known as add-on, dual-fuel, or hybrid heat pumps (e.g., Figure 6.38). During the heating season, when the outdoor temperatures drop below the thermal or economic balance point of the heat pump, the heat pump turns "off" and the gas or oil furnace comes "on" to provide

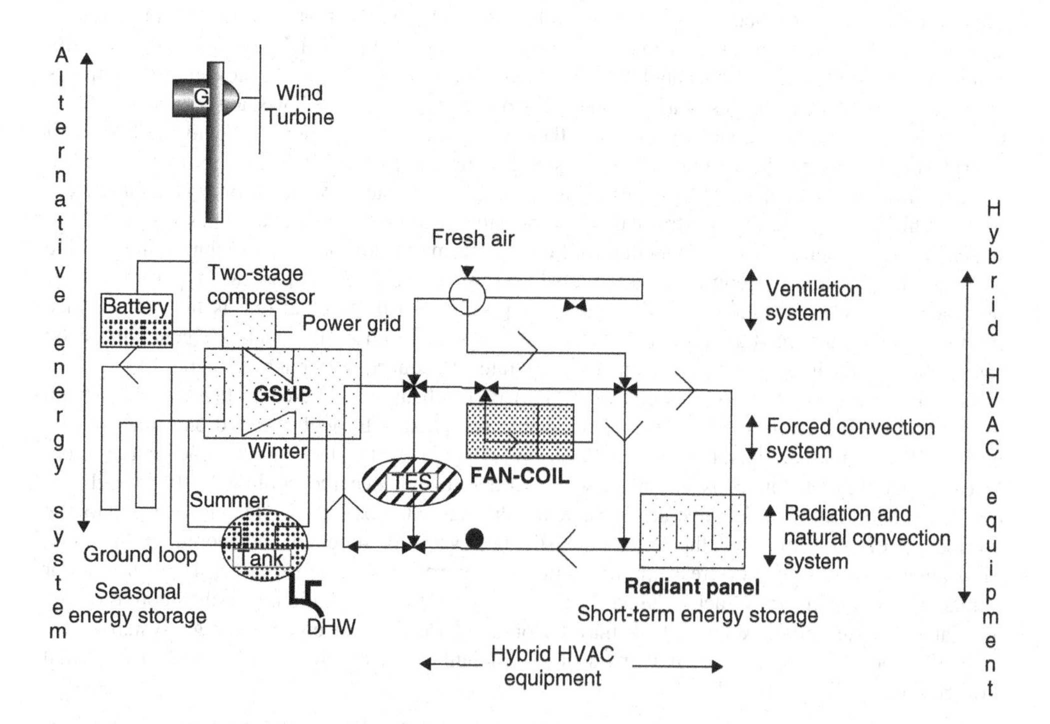

Figure 6.38 A hybrid system coupling HVAC and GSHP systems (Kilkis, 1995).

heating. In other words, the electric resistance auxiliary heat found in conventional air-to-air heat pump systems is replaced by the gas or oil furnace.

If a building's heating and cooling requirements are substantially different, a hybrid system is recommended to reduce costs. A hybrid heat pump combines GSHP with a conventional boiler or cooling tower. In a hybrid system, contractors size the ground loop in order to meet either the building's heating load or its cooling load, whichever is smaller. Depending upon the need, additional heating or cooling requirements beyond the capacity of the GSHP system are provided by a boiler (if the building's heating load is high) or a cooling tower (if the cooling load is high). A hybrid system uses a smaller loop than would otherwise be necessary, thus reducing the initial installation cost.

A hybrid variation is the night evaporative system, in which cooling towers are used at night to expel excess heat that builds up in the loop as a result of heavy daytime use of the GSHP system. This prevents the temperature of the fluid in the loop from losing its efficiency as a heat sink. Such systems may be ideal in climates where the days are unusually hot but the nights are cool, or where time-of-use rates are available from the local utility.

Besides providing air conditioning, add-on heat pumps increase the overall efficiency of the heating system because both the heat pump and the fossil furnaces are operating at their optimal efficiency levels.

Example 6.4

A Hybrid System Coupling HVAC and GSHP Systems

This case study was carried out in Turkey by Kilkis (1995) as the first case study and proposed as a viable solution to balance winter and summer loads on the GSHP. In order to optimally couple the attributes of hybrid panel HVAC systems and wind energy, the key element appears to be a GSHP which enables seasonal energy storage. Figure 6.38 shows a synectic combination of a wind turbine, GSHP, short-term thermal and electrical storage system, and a panel hybrid HVAC system. In design, the word synectic means to combine known elements and/or systems in an unusual manner in order to innovate a new system. Wind turbine generates electricity which drives directly or from the electric storage unit (batteries) compressors of GSHP which are in tandem in order to better follow the thermal loads of the building. Wind turbine and the GSHP form the combined alternative energy system. GSHP has two types of HVAC interface: one is a hydronic interface for radiant panels, and the second one is the air interface for the undersized forced-air (convective) system. In winter fan coils which are generally sized for cooling loads in a two-pipe system may operate at reduced water temperatures like radiant panels. Therefore, heating COP is enhanced. Part of the heat is delivered to the system through the ground circuit. Domestic hot water is available on demand through the hydronic interface of the GSHP. In summer operation, latent loads are satisfied by the fan coil. Whenever the latent loads are negligible, fan coils are bypassed in order to enhance the cooling COP through the use of radiant panels which operate at higher cold water temperatures. Heat is rejected to the ground circuit in summer. Otherwise, domestic hot water is supplied and stored in the tank. Preheating or precooling of fresh air for ventilation purposes is also possible. Thus the indoor human thermal comfort is achieved at three stages, namely ventilation, sensible conditioning, and latent conditioning. A water tank may be optional for thermal energy storage systems both in winter and summer. The thermal mass of the radiant panel provides some degree of thermal storage too.

6.28.8 *Resistance to Heat Transfer*

Two significant factors need to be considered when designing and sizing a ground loop: (i) resistance of the heat source to heat transfer, for example, ground, pond, and lake and (ii) resistance of the pipe to heat transfer. Of the two factors, pipe resistance is the dominant one. But, while little control can be exercised over source resistance, a great deal of influence can be exercised by the designer over the pipe resistance. Thermoplastic pipes are generally poor conductors as compared with metal. Increasing the ratio of pipe surface area to trench length yields significant gains in loop performance.

6.28.9 *Solar Energy Use in GSHPs*

A heat pump is a device that transfers heat efficiently from one place to another; refrigerators and air conditioners are common applications. A GSHP receives the solar energy that has been absorbed in the earth's surface, and moves it into a building where it is used to heat air or water. In summer, the flow is reversed, and the interior of the building is cooled by moving heat out into the earth's crust.

The solar heat is transferred into the building by means of an outdoor loop.

- A closed-loop configuration uses an array of buried plastic pipe to circulate a fluid that takes heat from the surrounding soil. As the warmed fluid passes through a compressor inside the building, the heat is released and the cooled fluid re-enters the pipes to start the cycle again.
- An "open loop" design takes water directly from a lake or underground aquifer, passes it through a compressor to extract the small level of heat, and then returns the cooled water back to the earth with no physical changes (except in temperature).

Regardless of the thermometer reading of the outside air, the earth's surface retains sufficient heat for any demand. The larger the demand for heat, the more pipe must be installed (or more water used) to transfer more solar energy into the building space.

6.28.10 *Heat Pumps with Radiant Panel Heating and Cooling*

Radiant panel heating systems use underfloor or ceiling-mounted water distribution to supply heat or cooling. Underfloor heating reduces heat demand by providing comfortable conditions at a lower temperature. Furthermore, by operating at a much lower water supply temperature than conventional hydronic systems, it allows heat pumps to meet heat demand at lower outside temperatures. This reduces or eliminates the need for a supplementary boiler and improves the COP. Ceiling-mounted radiant panels offer similar advantages for cooling.

Usually, a high-temperature hydronic heating system is designed for 80 °C mean water temperature. In cooling, the chilled water mean temperature is about 10 °C. According to the basic nature of radiant panel systems, their temperature requirements in heating and cooling are very moderate and can be easily suited to the heat pump characteristics. The mean water temperature (T_W) requirement may be just 35 °C for heating, and up to 18 °C for cooling. In order to demonstrate the attributes of radiant panel systems, an air-to-water [air-to-water heat pump (AWHP)] and a ground-source heat pump (GSHP) are analyzed by Kilkis (1993) as explained below.

6.28.10.1 Air-to-Water

Figure 6.39 depicts the heating characteristics of an air-to-water heat pump serving a central heating system with radiators. A boiler supplements the heat pump in either parallel or tandem configuration

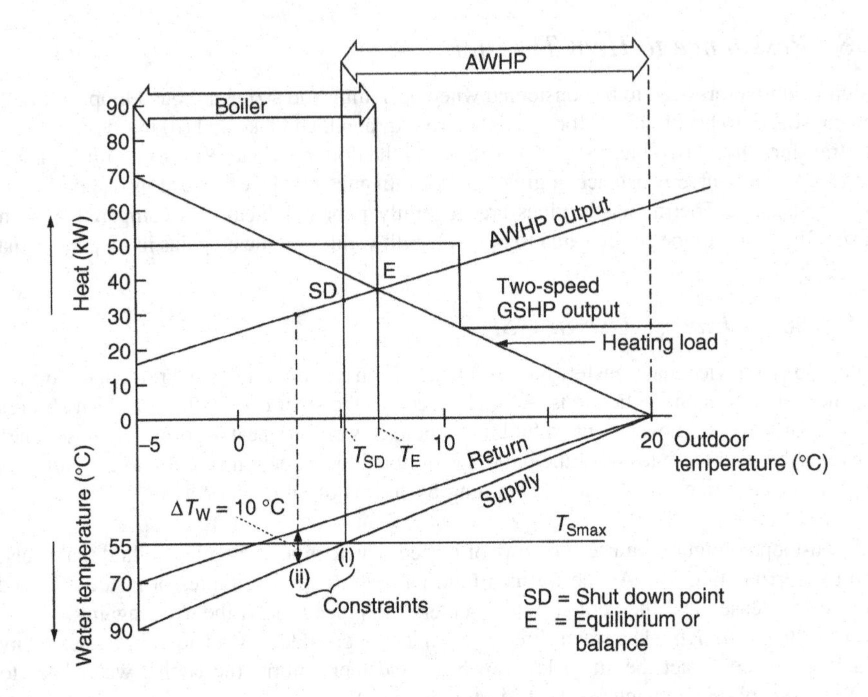

Figure 6.39 Characteristic of an air-to-water heat pump for radiant heating and cooling (Kilkis, 1993).

below the equilibrium temperature (T_E) and takes over at the shutdown temperature (T_{SD}) when the heat pump is forced to switch off. Shutdown occurs when one of the following additional constraints, imposed by the hydronic heating system, occurs: (i) the maximum water temperature that can be delivered by a heat pump (T_{Smax}) is exceeded or (ii) the condenser side water temperature drop (ΔT_W) is exceeded.

T_{Smax} is generally 55 °C, and in parallel mode operation (illustrated in Figure 6.39) corresponds with the supply temperature and occurs at +5 °C outside temperature point (i). In tandem mode, T_{Smax} corresponds with the return water temperature. ΔT_W is generally limited to 10 °C whereas a radiator heating system is normally designed for 20 °C (maximum load), unless the circulation pump is oversized. Constraint (ii) can thus be effective in parallel mode operation. In this illustration, constraint (i) occurs first as the outdoor air temperature reduces, and determines the shutdown point (SD). In parallel mode, T_{SD} is +5 °C. In tandem mode, it is +1 °C. In order to avoid an early shutdown, heat pump sizing must also satisfy the condition $T_E > T_{SD}$.

6.28.10.2 Ground-Source

The heating capacity of a ground-source heat pump is virtually free from the effects of the outdoor temperature. However, a GSHP will also benefit from a radiant panel heating system. In order to follow the heat load line closely, a multiple-speed heat pump may be desirable, as shown in Figure 6.39.

6.28.10.3 Advantages

A radiant panel heating system has several attributes which will improve the performance and cost-effectiveness of a heat pump:

- **Low supply temperature.** The design and operation of a radiant panel ensures that the supply water temperature does not exceed 55 °C by using large radiant panel surfaces to deliver the required heat. This low temperature makes it more practical to use a 10 °C temperature drop at design conditions. These factors will eliminate constraints (i) and (ii), and consequently T_{SD}. Therefore, the $T_E > T_{SD}$ condition will be automatically satisfied, which consequently enables a free choice of any optimum heat pump size. As a result, if other conditions are also satisfied, elimination of the supplementary boiler is possible. The low supply temperature also enhances the heat pump COP.
- **Reduced load.** It is clear that with radiant panels, the indoor air temperature may be 2–3 °C lower than with conventional heating systems, without any sacrifice of the desired human comfort. This moves the lower end of the load line in Figure 6.39 to the left and also reduces heat transmission losses. In addition, a more uniform indoor air temperature and pressure distribution and a slow air movement contribute to reduced natural infiltration heat losses. Heat (or cool) storage in the panel and the exposed walls and partitions allow peak load shaving of the heating or cooling loads. This enables the heat pump to be sized according to the leveled load instead of the peak load. This overall reduction of the heating load shifts the upper end of the load line downward. The net result is that T_E moves toward the origin and a deficit in heat delivered by the heat pump is minimized. Consequently the need for a supplementary boiler is either minimized or eliminated for a given heat pump capacity.

6.28.11 The Hydron Heat Pump

The hydron module GSHPs are available on the market. The following features are standard equipment on all hydron module heat pumps:

- stainless steel cabinet, control panel, condensate pan, framework, nuts and bolts, and a lifetime warranty on the cabinetry;
- corrosion-proof drain pan;
- oversized evaporator and condensing coils;
- cupro-nickel coaxial water to freon heat exchangers;
- water flow switch for freeze protection;
- large whisper flow blower modules;
- high-efficiency scroll compressors.

6.29 Heat Pumps and Energy Savings

Most heat pumps are designed to be used for cooling as well as heating and it is in this form that the heat pump is most efficient. Because of its energy-saving characteristics renewed interest is being shown in the heat pump.

It is clear that heat pumps are very energy efficient, and therefore environmentally benign. Heat pumps offer the most energy-efficient way to provide heating and cooling in many applications, as they can use renewable heat sources in our surroundings. Even at temperatures we consider to be cold, air, ground, and water contain useful heat that is continuously replenished by the sun. By applying a little more energy, a heat pump can raise the temperature of this heat energy to the level needed. Similarly, heat pumps can also use waste heat sources, such as from industrial processes, cooling equipment, or ventilation air extracted from buildings. A typical electrical heat pump will just need 100 kWh of power to turn 200 kWh of freely available environmental or waste heat into 300 kWh of useful heat.

Through this unique ability, heat pumps can radically improve the energy efficiency and environmental value of any heating system that is driven by primary energy resources such as fuel or power. The following six facts should be considered when any heat supply system is designed (IEA-HPC, 2001):

- Direct combustion to generate heat is never the most efficient use of fuel.
- Heat pumps are more efficient because they use renewable energy in the form of low-temperature heat.
- If the fuel used by conventional boilers is redirected to supply power for electric heat pumps, about 35–50% less fuel will be needed, resulting in 35–50% less emissions.
- Around 50% savings are made when electric heat pumps are driven by CHP (or cogeneration) systems.
- Whether fossil fuels, nuclear energy, or renewable power is used to generate electricity, electric heat pumps make far better use of these resources than do resistance heaters.
- The fuel consumption, and consequently the emissions rate, of an absorption or gas engine heat pump is about 35–50% less than that of a conventional boiler.

In the past, most heat pumps were of the air-to-air or air source type. Air source heat pumps rely on outdoor air for their heat source. Although cold outdoor air contains some heat, as temperatures drop, the heat pump must work harder and efficiency decreases. In very cold weather, the air source heat pump alone will not be able to provide enough heat, and supplemental or backup heat must be provided. This can significantly increase heating costs. GSHPs extract heat from the ground or from water below the surface. Because ground and groundwater temperatures are a constant 10–13 °C year round, this type of system is much more efficient.

This varies with the cost of electricity, oil, and propane in your area. Generally, a GSHP can produce heat with average savings of 10–15% over natural gas, 40% savings over fuel oil, and 50% savings over propane; air-conditioning savings average 40–60% over conventional systems (EESC, 2001).

Heat pump water heaters extract heat from surrounding air to heat water in a storage tank and can be fueled by electricity or gas. These heaters have essentially the same performance as electric resistance storage water heaters, except that efficiencies are typically 2–2.5 times higher. The energy factor for heat pump water heaters ranges from 1.8 to 2.5, compared to 0.88–0.96 for electric resistance systems. Heat pump water heaters cool and dehumidify the air surrounding the evaporator coil. This can be an advantage where cooling is desirable, and a disadvantage when cooling is undesirable. Some heat pump water heaters are designed to recover waste heat from whole house ventilation systems.

Heat pump water heaters are commercially available, with payback typically ranging from 2 to 6 years, depending on the hot water use and the efficiency of the water heater system being replaced.

When purchasing a new heat pump, the buyer should check the efficiency rating of the proposed unit. A higher efficiency rating will result in lower operating costs. Heat pump efficiency is designated by the SEER, particularly ranging from 10.0 to over 15.0. For split systems with an outdoor unit and an indoor coil, the efficiency varies with the match between the indoor cooling coil and outdoor condensing unit. The manufacturer should be consulted to determine the combined efficiency. The American Refrigeration Institute publishes an annual directory listing various combinations of outdoor units and indoor coils with their SEER rating. Most major manufacturers' product lines are included in this directory.

Over the past several years, the SEER for the highest efficiency heat pumps has increased from 12.0 to over 15.0 because of the incorporation of the following improvements (JEMC, 2001):

- **Variable speed blowers, compressors, and motors.** This equipment provides variable speeds of operation to optimize performance and efficiency. Heat pumps utilizing multispeed components will typically start in the first stage or at low speed. If comfort levels or control settings cannot be satisfied with the first stage, the second stage or high speed will be activated. Some heat pump systems have more than two stages or speeds of operation.

- **Larger coil surface areas.** Large surface coils provide maximum heat-transfer efficiency.
- **Time delays.** Time delays vary the on and off cycles of compressors, motors, and supplemental heat packages.
- **Expansion valves.** Expansion valves control the flow of refrigerant in proportion to the load on the evaporator. Compared with other types of fixed metering devices, expansion valves are able to exercise control over a much wider range of operating conditions.

In addition to a unit's SEER for its performance, there are some additional energy-saving features to look for when selecting a heat pump for the house, as follows (JEMC, 2001):

- **Dual-fuel backup.** Dual-fuel heat pump systems are supplemented by a fossil fuel furnace or boiler instead of the traditional electric resistance coils. When outdoor temperatures are moderate, the building heat requirements can be satisfied by the heat pump alone. When outdoor temperatures are below the economic balance point, the heat pump is switched off and the furnace or boiler supplies heat at close to its peak efficiency.
- **Programmable setback thermostats.** Programmable thermostats with adaptive recovery or "ramping" are designed specifically for use with heat pumps. They allow the thermostat to be programmed for one or more "setback" periods per day. Their microprocessor unit senses the temperature differential to be overcome when bringing the space temperature back up, and brings the temperature up gradually over a longer period of time. This allows the heat pump alone to provide the temperature increase and minimizes the use of electric resistance auxiliary heat.

The high efficiencies of GSHP systems allow commercial users to save up to 70% in operating costs compared to electric resistance heating, up to 50% over air source heat pumps, and up to 45% over fossil fuel furnaces (GHPC, 2001).

GSHP systems also offer other ways for commercial businesses to save money. For example, GSHP systems are extraordinarily long lasting and reliable. Generally, they require little maintenance other than the need to change the air filter periodically. In the event that one of the GSHP units in a large building does malfunction, the modular nature of the equipment means the problem can be isolated without affecting the entire heating and cooling system. Many commercial GSHP systems installed 30–40 years ago are still operating today (GHPC, 2001).

These combined cost savings associated with GSHP typically provide a significant return on investment for any business or institutional organization. Although the initial installation cost of a GSHP system may be higher, the systems typically pay for themselves in less than 5 years (often in less than 2 years). Another advantage is their extraordinarily low operating costs. Recently developed new trenching and drilling techniques have brought down the costs of installing the loops and hence, the initial costs of GSHP systems.

6.30 Heat Pumps and Environmental Impact

With its applicability in both the building and the industrial sectors, the heat pump looks set to play a major role in meeting heating and cooling needs. Since these applications currently consume a significant proportion of the world's fossil fuel reserves, heat pump technology can make a major contribution to limiting environmental problems such as global warming. The potential environmental effects of heat pumps and their operations are summarized in Figure 6.40. A heat pump is the best choice to help the environment. Table 6.15 presents a comparison of electric heat pumps with fossil fuel systems.

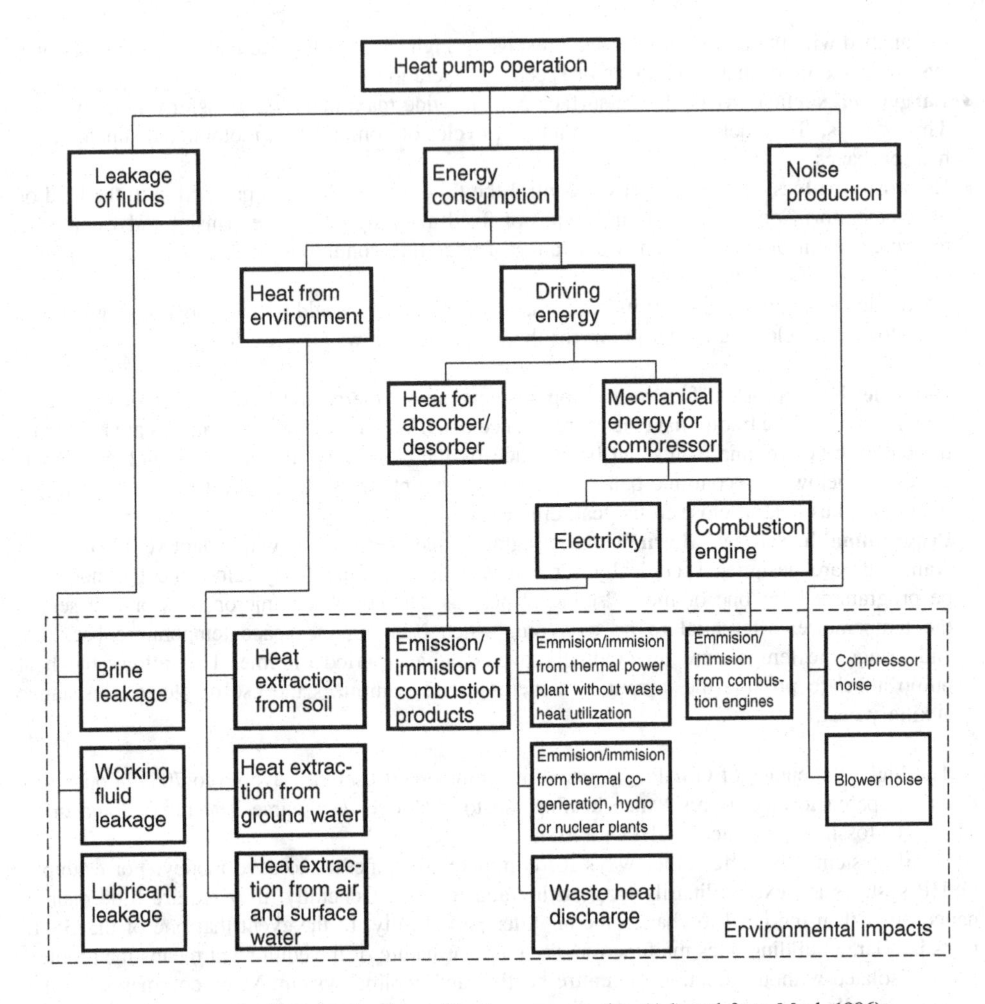

Figure 6.40 Environmental effects of heat pump operation (Adapted from Meal, 1986).

Because heat pumps consume less primary energy than conventional heating systems, they are an important technology for reducing gas emissions that harm the environment, such as carbon dioxide (CO_2), sulfur dioxide (SO_2), and nitrogen oxides (NO_x). However, the overall environmental impact of electric heat pumps depends very much on how the electricity is produced. Heat pumps driven by electricity from, for instance, hydropower or renewable energy reduce emissions more significantly than if the electricity is generated by coal, oil, or gas-fired power plants.

If it is further considered that heat pumps can meet space heating, hot water heating, and cooling needs in all types of buildings, as well as many industrial heating requirements; it is clear that heat pumps have a large and worldwide potential.

Of the global CO_2 emissions that amounted to 22 billion tons in 1997, heating in building causes 30% and industrial activities cause 35%. The potential CO_2 emissions reduction with heat pumps is calculated as follows: 6.6 billion tons CO_2 come from heating buildings (30% of total emissions); 1.0 billion ton can be saved by residential and commercial heat pumps, assuming that they can provide 30% of the heating for buildings, with an emission reduction of 50%; and minimum of 0.2 billion ton can be saved by industrial heat pumps (IEA-HPC, 2001).

Table 6.15 Technical comparison of electric heat pumps with fossil fuel systems.

Features	Electric Heat Pump	Fossil Fuel System
Heats and cools	Yes	Yes
Highest efficiencies	Yes	No
Transfers versus generates heat	Yes	No
Environmentally safer	Yes	No
Higher heating bills	No	Yes
Qualifies for lower electric rate	Yes	No
More even temperature	Yes	No
Open flames	No	Yes
Odor/fumes	No	Yes
Uses free outdoor heat	Yes	No
Requires separate AC unit	No	Yes
Combustion air required	No	Yes

Source: IPALCO (2001).

The total CO_2 reduction potential of 1.2 billion tons is about 6% of the global emissions! This is one of the largest benefits that a single technology can offer, and this technology is already available in the marketplace. And with higher efficiencies in power plants as well as for the heat pump itself, the future global emissions saving potential is even greater, about 16%.

In some regions of the world, heat pumps already play a crucial role in energy systems. But if this technology is to achieve more widespread use, a decisive effort is needed to stimulate heat pump markets and to further optimize the technology. It is encouraging that a number of governments and utilities are strongly supporting heat pumps. In all cases, it is important to ensure that both heat pump applications and policies are based on a careful assessment of the facts, drawn from as wide an experience base as possible.

A heat pump is the best choice to help protect the environment, because it does not burn natural gas or oil, but simply transfers warm air into or out of the house. So the occupier is not adding to air pollution or consuming scarce natural resources, but actually helping the environment. Energy saving has a positive impact on global warming. That is why it is important to continually engineer more energy-efficient products such as the advanced technology heat pump. And it is just as important for consumers to demand products that live up to tougher, more stringent energy standards.

For example, in Canada, space conditioning in schools accounts for 15% of CO_2 emissions, or about 7 Mton each year. Implementation of earth energy heating and cooling will allow schools to help reduce climate change impacts and to provide a role model for their students on environmental issues (EESC, 2001).

In summary, electrical heat pumps, particularly those that replace directly fired heating boilers, offer a significant CO_2 emission reduction potential, even if the electricity to drive the heat pump is generated with oil or gas. A significant potential remains to improve the efficiency of heat pumps where the potential for heating boilers is now virtually exhausted. The economic potential of heat pumps for reducing the total CO_2 emissions in some countries is estimated to be up to 4%, with a technical potential of up to 9%. The report states that the greenhouse effect from the loss of working fluids other than CFCs is negligible. Consequently, it is concluded that the heat pump as an energy-efficient technology which benefits the environment should play an important role in an increasingly environmentally conscious world.

The use of heat pumps is certain to result in environmental improvement and conservation of energy. Assuming the power-generating efficiency at 35%, the boiler heat efficiency at 90%, and the average COP of a heat pump at 3.5, a heat pump requires only 74% of the primary energy needed to produce the same thermal capacity as an oil-fired boiler in homes and office buildings. This means that the use of heat pumps for heating could result in energy conservation of 26% when replacing oil-fired boilers. The amount of energy conserved has been calculated for each application area for all of Japan. The result showed that, in the year 2000, energy equivalent to roughly 1.4 million cubic meters of crude oil could be conserved annually by replacing oil-fired boilers and kerosene stoves with heat pump air conditioners. This amount corresponds to 0.3% of Japan's primary energy supply.

6.31 Concluding Remarks

In this chapter, a large number of topics on heat pump systems and applications are presented, covering all kinds of heat pumps and their basic principles. In particular, various technical, design, installation, operational, evaluation, energetic, exergetic, performance, energy-saving, industrial and sectoral utilization, environmental, and working fluid aspects of heat pumps are discussed from various practical perspectives with many illustrative and practical examples.

Nomenclature

a	Seebeck coefficient
COP	coefficient of performance
c_p	constant-pressure specific heat, kJ/kg · K
ex	specific exergy, J/kg or kJ/kg
$\dot{E}x$	exergy rate, kW
h	enthalpy, J/kg or kJ/kg
k	thermal conductivity, W/m · K or kW/m · K
$\dot{m}$	mass flow rate, kg/s
P	pressure, kPa
$\dot{Q}$	heat load, kW
s	entropy, kJ/kg
$\dot{S}_{gen}$	entropy generation rate, W/K or kW/K
T	temperature, °C or K
v	specific volume, m^3/kg
$\dot{V}$	volumetric flow rate, m^3/s
w	specific work, J/kg or kJ/kg
$\dot{W}$	work input to compressor or pump, W or kW
Z	effectiveness "figure of merit," K^{-1}

Greek Letters

η	efficiency
ρ	density, kg/m^3; electrical resistivity, K/W or °C/W

Study Problems

Heat Pumps – Concept Questions

6.1 What is a heat pump?

6.2 What is the similarity and the difference between a heat pump and a refrigerator?

6.3 What are bivalent and monovalent modes of heat pump operation?

6.4 For which city will a heat pump be more cost-effective during winter season? Chicago or Istanbul? Why?

6.5 Consider a heat pump operating entirely on back-up electric resistance heater because the ambient temperature is very low. What is the COP of the heat pump during this operation?

6.6 Define primary energy ratio (PER) and explain its use.

6.7 What is the relationship between COP and energy efficiency ratio (EER)?

6.8 What is heating season performance factor (HSPF)?

6.9 What is seasonal energy efficiency ratio (SEER)?

6.10 Classify heat pumps for residential heating and cooling depending on their operational function.

6.11 What are the characteristics of absorption heat pumps?

6.12 What are the commonly used heat sources in heat pump applications?

6.13 What are the basic advantages and disadvantages of using groundwater as a heat source?

6.14 Classify heat pumps according to heat source and type of heat carrier.

6.15 Describe the operation of a solar heat pump with a schematic.

6.16 Provide a list of heat pump cycles.

6.17 What are the two basic types of geothermal heat pump systems?

6.18 What are the four basic GSHP open and closed-loop designs?

6.19 What are hybrid heat pump systems? How does it operate?

Heat Pump Efficiencies

6.20 A heat pump is driven by a natural gas-fired internal combustion engine, which has a thermal efficiency of 35%. The COP of the heat pump is 2.5. Determine the primary energy ratio.

6.21 A heat pump is driven by a natural gas-fired internal combustion engine and it supplies heat to a building at a rate of 130,000 kJ/h. The engine has a thermal efficiency of 42% and it consumes 2 kg of natural gas in 1 hour. The heating value of natural gas is 50,050 kJ/kg. Determine the primary energy ratio.

6.22 Consider the following heating systems and determine the primary energy ratio in each case.

(a) Heat pump is driven by an internal combustion engine with $\eta = 0.33$, COP = 2.6.
(b) Heat pump is driven by a combined power plant with $\eta = 0.50$, COP = 3.3.
(c) A natural gas furnace with $\eta = 0.90$.

Energy Analysis of Heat Pumps

6.23 A heat pump is used to keep a room at 20 °C by rejecting heat to an environment at 8 °C. The total heat loss from the room to the environment is estimated to be 24,000 kJ/h and the power input to the compressor is 2.2 kW. Determine (a) the rate of heat absorbed from the environment in kiloJoules per hour, (b) the COP of the heat pump, and (c) the maximum rate of heat supply to the room for the given power input.

6.24 A heat pump system operates between temperature limits of −5 and 18 °C. The heating load of the space is 48,000 Btu/h and the COP of the heat pump is estimated to be 1.7. Determine (a) the power input, (b) the rate of heat absorbed from the cold environment, and (c) the maximum possible COP of this heat pump.

6.25 A room is kept at 25 °C by an 18,000 Btu/h (on heating basis) split air conditioner when the ambient temperature is 6 °C. The air conditioner is running at full load under these conditions. The power input to the compressor is 2.4 kW. Determine (a) the rate of heat removed from the ambient air in British thermal units per hour, (b) the COP of the air conditioner, and (c) the rate of heating supplied in British thermal units per hour if the air conditioner operated as a Carnot heat pump for the same power input.

6.26 Water is continuously heated by a heat pump from 20 to 40 °C. The heat absorbed in the evaporator is 34,000 kJ/h and the power input is 3.8 kW. Determine the rate at which water is heated in liters per minute and the COP of the heat pump. The specific heat of water is 4.18 kJ/kg · °C.

6.27 Water is continuously heated by a heat pump from 70 to 110 °F. The heat absorbed in the evaporator is 32,000 Btu/h and the power input is 4.1 kW. Determine the rate at which water is heated in gallons per minute and the COP of the heat pump. The specific heat of water is 1.0 Btu/lbm · °F.

6.28 Refrigerant-134a enters the compressor of a heat pump system at 160 kPa as a saturated vapor and leaves at 900 kPa and 85 °C. The refrigerant leaves the condenser as a saturated liquid. The rate of heating provided by the system is 2600 W. Determine the mass flow rate of R-134a and the COP of the system.

6.29 Refrigerant-22 enters the condenser of a residential heat pump at 1200 kPa and 55 °C at a rate of 0.024 kg/s and leaves at 1200 kPa as a saturated liquid. If the compressor consumes 1.8 kW of power, determine (a) the COP of the heat pump and (b) the rate of heat absorbed from the outside air. The enthalpies of R-22 at the condenser inlet and exit are 435.62 kJ/kg and 237.07 kJ/kg, respectively.

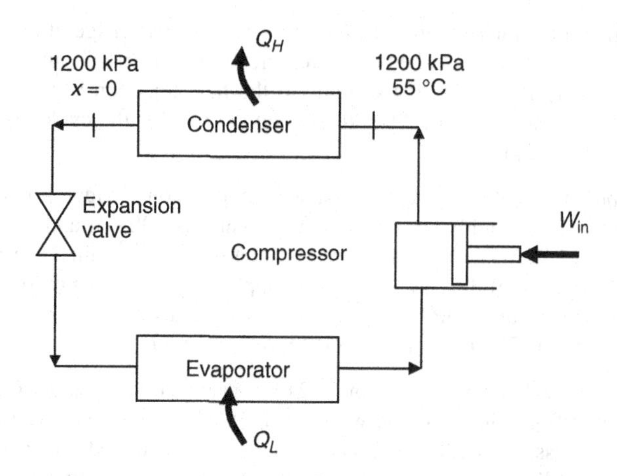

6.30 A heat pump with refrigerant-134a as the working fluid is used to keep a space at 23 °C by absorbing heat from geothermal water that enters the evaporator at 60 °C at a rate of 0.045 kg/s and leaves at 45 °C. The refrigerant enters the evaporator at 20 °C with a quality of 15% and leaves at the same pressure as saturated vapor. If the compressor consumes 1.3 kW of power, determine (a) the evaporator pressure and the mass flow rate of the refrigerant and (b) the heating load and the COP. Take specific heat of water to be 4.18 kJ/kg · °C.

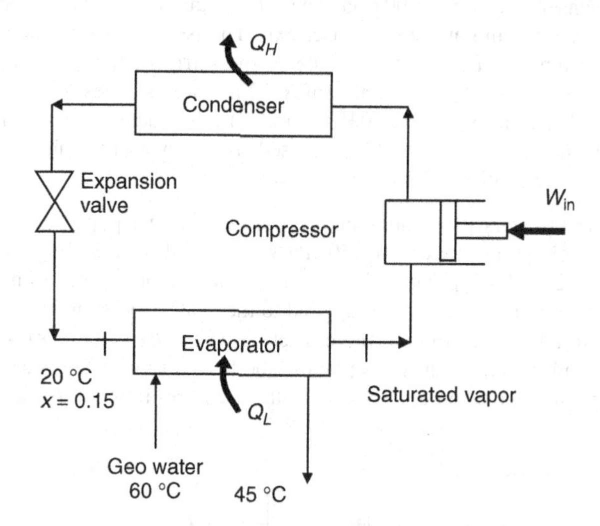

6.31 A heat pump operates on the ideal vapor-compression refrigeration cycle with refrigerant-134a as the working fluid between pressure limits of 200 kPa and 1400 kPa. Determine (a) the heat absorption in the evaporator, (b) the heat rejection in the condenser, (c) the work input, and (d) the COP. (e) Also, determine the COP if the cycle operates as a refrigerator instead of a heat pump.

6.32 A heat pump operates on the ideal vapor-compression refrigeration cycle with refrigerant-134a as the working fluid between pressure limits of 30 psia and 200 psia. Determine (a) the heat absorption in the evaporator, (b) the heat rejection in the condenser (c) the work input, and (d) the COP. (e) Also, determine the COP if the cycle operates as a refrigerator instead of a heat pump.

6.33 A heat pump operates on the ideal vapor-compression refrigeration cycle with refrigerant-22 as the working fluid. The evaporator pressure is 300 kPa and the condenser pressure is 2000 kPa. The flow rate at the compressor inlet is 0.090 m^3/min. Determine (a) the mass flow rate in the cycle, (b) the rate of heating supplied by the heat pump in British thermal units per hour, and (c) the COP. The properties of R-22 at various states are $h_1 = 399.18$ kJ/kg, $v_1 = 0.07656$ m^3/kg, $h_2 = 447.85$ kJ/kg, $h_3 = 265.17$ kJ/kg.

6.34 A heat pump with a heating load of 13.2 kW operates on the vapor-compression refrigeration cycle with refrigerant-22 as the working fluid. The evaporator pressure is 200 kPa and the condenser pressure is 1200 kPa. The refrigerant is saturated vapor at the compressor inlet and saturated liquid at the condenser exit. The isentropic efficiency of the compressor is 84%. Determine (a) the mass flow rate of the refrigerant, (b) the power input, and (c) the COP. The properties of R-22 at various states are $h_1 = 394.67$ kJ/kg, $h_{2s} = 439.97$ kJ/kg, $h_3 = 237.07$ kJ/kg. (d) Also, determine the mass flow rate, the power input and the COP if R-134a is used as the working fluid with the same cycle conditions and the heating load.

6.35 A heat pump with a heating load of 36,000 Btu/h operates on the vapor-compression refrigeration cycle with refrigerant-22 as the working fluid. The evaporator pressure is 40 psia and the condenser pressure is 300 psia. The refrigerant is saturated vapor at the compressor inlet and saturated liquid at the condenser exit. The isentropic efficiency of the compressor is 82%. Determine (a) the mass flow rate of the refrigerant, (b) the power input in kilowatts, and (c) the COP. The properties of R-22 at various states are $h_1 = 171.21$ Btu/lbm, $h_{2s} = 193.53$ Btu/lbm, $h_3 = 114.90$ Btu/lbm. (d) Also, determine the mass flow rate, the power input and the COP if R-134a is used as the working fluid with the same cycle conditions and the heating load.

6.36 Refrigerant-134a enters the condenser of a residential heat pump at 800 kPa and 55 °C at a rate of 0.080 kg/s and leaves at 750 kPa subcooled by 6 °C. The refrigerant enters the compressor at 200 kPa superheated by 10.1 °C. Determine (a) the isentropic efficiency of the compressor, (b) the rate of heat supplied to the heated room, and (c) the COP of the heat pump. The enthalpy of R-134a at the condenser exit is $h_3 = 83.65$ kJ/kg. Also, determine (d) the COP and the rate of heat supplied to the heated room if this heat pump operated on the ideal vapor-compression cycle between the pressure limits of 200 kPa and 800 kPa.

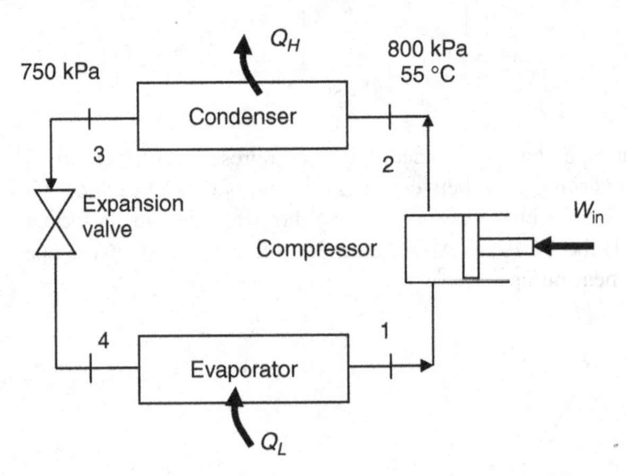

6.37 A heat pump with refrigerant-134a as the working fluid is used to keep a space at 24 °C by absorbing heat from geothermal water that enters the evaporator at 50 °C at a rate of 0.125 kg/s and leaves at 40 °C. The refrigerant enters the evaporator at 20 °C with a quality of 23% and leaves at the inlet pressure as a saturated vapor. The refrigerant leaves the compressor at 1.4 MPa and 65 °C. Determine (a) the isentropic efficiency of the compressor and the degrees of subcooling done on the refrigerant in the condenser, (b) the mass flow rate of the refrigerant, and (c) the heating load and the COP of the heat pump.

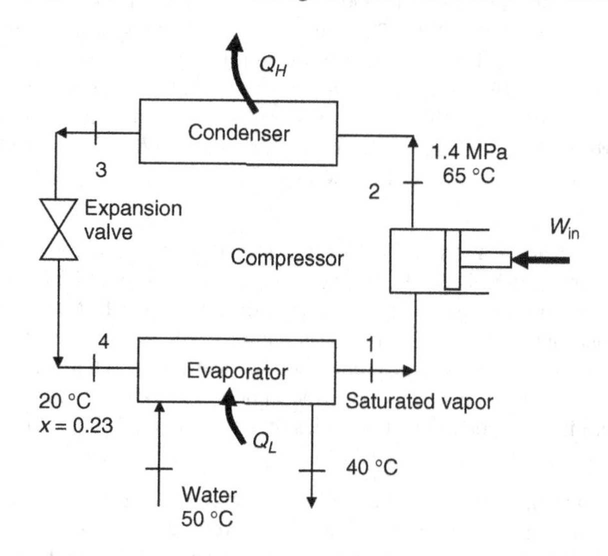

6.38 A ground-source heat pump operates on the vapor-compression refrigeration cycle with refrigerant-134a as the working fluid. The evaporator pressure is 120 kPa and the condenser pressure is 1600 kPa. The refrigerant is saturated vapor at the compressor inlet and saturated liquid at the condenser exit. The isentropic efficiency of the compressor is 82%. The refrigerant absorbs heat from the ground at a rate of 5.7 kW as it flows in the evaporator. Determine (a) the volume flow rate at the inlet and exit of the evaporator in liters per second, (b) the heating load and the COP. (c) Determine the same parameters if the evaporator pressure is 200 kPa. (d) Determine the same parameters if the condenser pressure is 1200 kPa.

6.39 An air source heat pump operates between the ambient air at 0 °C and the indoors at 20 °C. The working fluid R-134a evaporates at −10 °C in the evaporator. The refrigerant is saturated vapor at the compressor inlet and saturated liquid at the condenser exit. The refrigerant condenses at 39.4 °C. Heat is lost from the indoors at a rate of 18,000 kJ/h. The isentropic efficiency of the compressor is 78%. Determine (a) the temperature of the refrigerant at the exit of the compressor and the mass fraction of the refrigerant vapor at the exit of the throttling valve and (b) the power input and the COP. (c) Also, determine the COP if a ground-source heat pump is used with a ground temperature of 15 °C, and the evaporating temperature in this case is 5 °C. Take everything else the same. What is the percentage increase on COP when ground source heat pump is used instead of air source heat pump.

Exergy Analysis of Heat Pumps

6.40 A heat pump is used to keep a room at 23 °C by rejecting heat to an environment at 4 °C. The total heat loss from the room to the environment is estimated to be

24,000 kJ/h and the power input to the compressor is 2.6 kW. Determine (a) the rate of heat absorbed from the environment in kiloJoules per hour, (b) the COP of the heat pump, (c) the maximum rate of heat supply to the room for the given power input, and (d) the second-law efficiency of the cycle. (e) Also, determine the exergy of the heat transferred to the high-temperature medium and the exergy destruction of the cycle.

6.41 A house is kept at 24 °C by a 45,000 Btu/h (on heating basis) split air conditioner when the ambient temperature is 2 °C. The heat pump is operating at 80% load under the given conditions. The power input to the compressor is 4.8 kW. Determine (a) the COP and (b) the second-law efficiency of the heat pump. (c) It is proposed to replace the existing air source heat pump with a ground source heat pump. This heat pump absorbs heat from the ground at 12 °C under the same conditions. What is the COP of this ground source system if the second-law efficiencies of the air source and ground source systems are the same.

6.42 A house is kept at 77 °F by a 45,000 Btu/h (on heating basis) split air conditioner when the ambient temperature is 35 °F. The heat pump is operating at 80% load under the given conditions. The power input to the compressor is 4.3 kW. Determine (a) the COP and (b) the second-law efficiency of the heat pump. (c) It is proposed to replace the existing air source heat pump with a ground source heat pump. This heat pump absorbs heat from the ground at 55 °F under the same conditions. What is the COP of this ground source system if the second-law efficiencies of the air source and ground source systems are the same.

6.43 A heat pump operates on the ideal vapor-compression refrigeration cycle with refrigerant-134a as the working fluid. The refrigerant evaporates at −20 °C and condenses at 1400 kPa. The refrigerant absorbs heat from ambient air at 5 °C and transfers it to a space at 26 °C. Determine (a) the work input and the COP, (b) the exergy destruction in each component of the cycle and the total exergy destruction in the cycle, and (c) the minimum work input and the second-law efficiency of the cycle.

6.44 A heat pump operates between a heated space at 22 °C and the ambient air at −2 °C. Refrigerant-134a enters the compressor of a heat pump system at 160 kPa as a saturated vapor and leaves at 900 kPa and 55 °C. The refrigerant leaves the condenser as a saturated liquid. The rate of heating provided by the system is 2600 W. Determine (a) the mass flow rate of R-134a and the COP of the system, (b) the isentropic and exergetic efficiencies of the compressor, (c) the exergy destruction in each component of the cycle and the total exergy destruction in the cycle, and (d) the second-law efficiency of the cycle.

6.45 A heat pump with refrigerant-134a as the working fluid is used to keep a space at 25 °C by absorbing heat from tap water that enters the evaporator at 15 °C at a rate of 0.125 kg/s and leaves at 5 °C. The refrigerant enters the compressor at 4 °C as a saturated vapor and leaves at 1600 kPa and 75 °C. The refrigerant leaves the condenser at 1600 kPa and 52 °C. Determine (a) the mass flow rate of the refrigerant, the heating load, and the COP of the heat pump, (b) the isentropic and exergetic efficiencies of the compressor, (c) the exergy destructions in the expansion valve and the evaporator, and (d) the exergetic efficiency of the expansion valve and the second-law efficiency of the cycle.

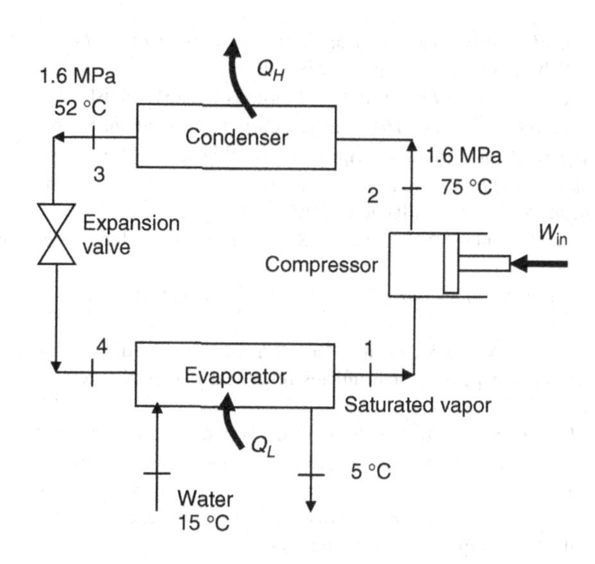

6.46 An air source heat pump operates between the ambient air at −5 °C and the indoors at 25 °C. The working fluid R-134a evaporates at −20 °C in the evaporator. The refrigerant is saturated vapor at the compressor inlet and saturated liquid at the condenser exit. The refrigerant condenses at 46.3 °C. Heat is lost from the indoors at a rate of 45,000 kJ/h. The isentropic efficiency of the compressor is 83%. Determine (a) the evaporator and condenser pressures and the temperature of the refrigerant at the exit of the compressor, (b) the power input and the COP, and (c) the minimum power input, the total exergy destruction, and the exergy efficiency of the cycle. (d) Determine the COP, the minimum power input, the total exergy destruction, and the exergy efficiency of the cycle if a ground source heat pump is used with a ground temperature of 10 °C. The evaporating temperature in this case is −5 °C. Take everything else the same.

6.47 An air source heat pump operates between the ambient air at 24 °F and the indoors at 77 °F. The working fluid R-134a evaporates at −5 °F in the evaporator. The refrigerant is saturated vapor at the compressor inlet and saturated liquid at the condenser exit. The refrigerant condenses at 109.5 °F. Heat is lost from the indoors at a rate of 24,000 Btu/h. The isentropic efficiency of the compressor is 86%. Determine (a) the evaporator and condenser pressures and the temperature of the refrigerant at the exit of the compressor, (b) the power input and the COP, and (c) the minimum power input, the total exergy destruction, and the exergy efficiency of the cycle. (d) Determine the COP, the minimum power input, the total exergy destruction, and the exergy efficiency of the cycle if a ground source heat pump is used with a ground temperature of 50 °F. The evaporating temperature in this case is 25 °F. Take everything else the same.

References

ARI (2000) *Standard for Variable Capacity Positive Displacement Refrigerant Compressors and Compressor Units for Air-Conditioning and Heat Pump Applications*, Standard 500, Air-Conditioning & Refrigeration Institute, Arlington, VA.

Berghmans, J. (1983a) Domestic heat pump applications, in *Heat Pump Fundamentals* (ed. J. Berghmans), Martinus Nijhoff Publishers, The Hague, pp. 279–301.

Berghmans, J. (1983b) *Heat Pump Fundamentals*, Martinus Nijhoff Publishers, The Hague.

Carrier (2000) *Operation of Air Source Heat Pumps*, Carrier Corporation, Farmington, CT.

Dincer, I. (2003) *Refrigeration Systems and Applications*, Wiley, 1st ed, London.

EESC (2001) *Installation and Performance of Geothermal Source Heat Pumps*, Earth Energy Society of Canada, Ottawa (available at: http://www.earthenergy.org).

Ellis, D. (2001) An Overview of ARI/ISO Standard 13256-1 for Water-Source Heat Pumps, ISO TC86/SC6/WG3.

GHPC (2001) *Geothermal Heat Pumps*, Geothermal Heat Pump Consortium, Washington, DC (available at: http://www.ghpc.org).

Hasatani, M., Matsuda, H., Miyazaki, M. and Yanadori, M. (1988) Studies on the basic principles of a high temperature chemical heat pump which utilizes reversible chemical reactions. *Newsletter of the IEA Heat Pump Center*, **6** (2), 29–32.

Heap, R.D. (1979) *Heat Pumps*, E. & F.N. Spon Ltd., a Halsted Press (Wiley), New York, NY.

Holland, F.A., Watson, F.A. and Devotta, S. (1982) *Thermodynamic Design Data for Heat Pump Systems*, Pergamon Press, Oxford.

IEA-HPC (2001) *Heat Pumps in Industry*, International Energy Agency-Heat Pump Center, A.A. Sittard, The Netherlands (available at: http://www.heatpumpcentre.org).

IPALCO (2001) *How Does a Heat Pump Work?* IPALCO Enterprises, Inc., Indianapolis, IN (available at: http://www.heatpumpcentre.org).

Itoh, H. (1995) The world's best selling heat pump. *Newsletter of the IEA Heat Pump Center*, **13** (3), 31–36.

JEMC (2001) *Heat Pump Systems*, Jackson Electric Membership Corporation, Jefferson, GA.

Kavanaugh, S.P. and Rafferty, K. (1997) *Ground Source Heat Pumps: Design of Geothermal Systems for Commercial and Institutional Buildings*, American Society of Heating, Refrigerating and Air Conditioning Engineers, Inc., Atlanta, GA.

Kilkis, I.B. (1993) Advantages of combining heat pumps with radiant panel heating and cooling systems. *Newsletter of the IEA Heat Pump Center*, **11** (4), 28–31.

Kilkis, I.B. (1995) Panel heating and cooling wind with energy, in *Turkish Energy, Electrical, Electronic and Automation Periodicals 85(95-4)*, Kaynak Yayinevi, Istanbul, Turkey, pp. 102–108.

Koebbeman, W.F. (1982) *Industrial Applications of a Rankine Powered Heat Pump for the Generation of Process Steam*. Proceedings of the International Symposium on Industrial Applications of Heat Pumps, March 24–26, Cranfield. pp. 221–230.

Kunjeer, P.B. (1987) Research by the US Department of Energy on application of heat pumps to district heating and cooling systems. *Newsletter of the IEA Heat Pump Center*, **5** (4), 8–10.

Kuroda, S. (1986) Development of a high-efficiency and high-compression-ratio heat pump system with water as working fluid. *Newsletter of the IEA Heat Pump Center*, **4** (4), 22–26.

Lehmann, A. (1983) The industrial heat pump. *Newsletter of the IEA Heat Pump Center*, **4** (2), 1–2.

Lehmann, A. (1986) The Stuttgart air-source absorption heat pump. *Newsletter of the IEA Heat Pump Center*, **4** (3), 7–9.

Meal, M. (1986) Environmental aspects of heat pump applications. *Newsletter of the IEA Heat Pump Center*, **4** (1), 1–5.

Mongey, B., Hewitt, N.J., McMullan, J.T., Henderson, P.C. and Molyneaux, G.A. (2001) Performance trends and heat transfer considerations in an ammonia-water resorption cycle. *International Journal of Energy Research*, **25**, 41–51.

Podesser, E. (1984) The absorption heat pumps – state of the art and prospects. *IEA-HPC Newsletter*, **2** (1/2), 2–6.

Suda, S. (1987) Recent developments of metal hydride heat pumps in Japan. *Newsletter of the IEA Heat Pump Center*, **4** (4), 22–26.

Tu, M. (1987) Thermodynamic and economic evaluation of a solar-assisted heat pump. *International Journal of Energy Research*, **11**, 559–572.

7

Heat Pipes

7.1 Introduction

During the past two decades, heat pipe technology has received a great deal of attention and miniature and conventional heat pipes can be successfully used in many industrial applications, especially in the cooling of electronic components and devices.

The idea of heat pipes was first suggested by Gaugler in 1944. However, it was not until 1962 when G.M. Grover invented it that its remarkable properties were appreciated and serious development began. Since the heat pipe was first patented by Grover in 1963, elementary theories have been advanced and developments in aerospace and terrestrial applications have progressed to the point where the heat pipe is now used commercially. While heat pipe technology has reached a rather high level, its market has not yet met expectations. More recently, increasing environmental problems have attracted a great deal of attention.

Many countries started their research and development programs on heat pipes and their utilization in the 1970s. Since then, both academic institutions and research organizations have been involved in research activities while the industry has gained experience in application. Various laboratories, for example, Los Alamos National Laboratory in the United States, have played a significant role in the research and development of heat pipe technology. Most of them have established the design and manufacturing capability of thermosiphon heat pipes and wicked heat pipes. Such heat pipes have found applications in waste heat recovery and electronic enclosure cooling. Recently, several companies have also developed micro heat pipes as cutting edge technology, particularly for cooling notebook personal computers.

The heat pipe is an efficient heat conductor device for transferring heat from one part to another part. They are often referred to as the *superconductors* of heat as they possess an extraordinary heat-transfer capacity and with almost no heat loss. The heat-transfer rate is thousands of times greater than that possible with a solid heat conductor of the same size (e.g., solid rod and simple fin) because of its thermophysical properties. In this device, one end of the heat pipe (or tube) is filled with a working fluid. Heat is applied to this end, vaporizing the fluid. The vapor comes to the other end, which is cooler, by the capillary effect there and condenses, releasing heat. A simple heat pipe consists essentially of a metal tube (a sealed aluminum or copper container whose inner surfaces have a capillary wicking material) lined with a wick and filled with a fluid. A heat pipe is similar to a thermosiphon. It differs from a thermosiphon by virtue of its ability to transport heat against gravity by an evaporation–condensation cycle with the help of porous capillaries that form the wick. The wick provides the capillary driving force to return the condensate to the evaporator. The quality and type of wick usually determines the performance of the heat pipe, for this is the

Refrigeration Systems and Applications İbrahim Dinçer and Mehmet Kanoğlu

heart of the product. Different types of wicks are used, depending on the application for which the heat pipe is being used.

Some of the well-known working fluids (so-called *heat pipe liquids*) are liquid hydrogen, ammonia, acetone, methanol, water, sodium, potassium, lithium, mercury, and silver. In addition, a number of heat pipe wall and wick materials are recommended, for example, aluminum, carbon steel, nickel, copper, tungsten, molybdenum, and refractory metals and alloys. Much research effort has been carried out to find the most efficient and reliable working fluids and wall/wick materials and the best configurations of the heat pipes for various ranges of temperature, depending on the applications (Dincer, 1997).

It is obvious that elimination of the fluid pump and power supply leads to greater reliability of the heat transport system and reduced weight, in addition to the saving in power consumption.

Heat pipe technology was suggested as a way to transfer solar heat passively and effectively from a solar absorber plate to the inside of a building. However, heat pipes are now of great importance in a variety of industrial applications ranging from mechanical engineering to food process engineering, for example, solar thermal applications using heat pipes as evacuated solar collectors, food cooking, and cooling applications. Further information on heat pipe technology, operating characteristics, heat-transfer limits, and heat pipe design technologies for cooling and heat exchange applications may be found in Peterson (1994) and Faghri (1995).

The main objective of this chapter is to introduce the heat pipes for various thermal applications, highlight the importance of their utilization for various cooling and heating applications, and discuss their technical, design and manufacturing, and operational aspects, as well as their benefits, along with practical examples.

7.2 Heat Pipes

Heat pipes are characterized by their excellent heat-transfer capability, fast heat-transfer rate, uniform temperature distribution, simple construction, compactness, high reliability, high efficiency, small heat losses, low manufacturing costs, environmentally benign nature, and versatile applications. Their most attractive feature is that they do not require external energy.

The concept of a passive two-phase heat-transfer device capable of transferring large quantities of heat with a minimal temperature drop was first introduced by Gaugler in 1942. This device received little attention until 1964 when Grover and his colleagues at Los Alamos National Laboratories published the results of an independent investigation and first applied the term heat pipe. Since that time, heat pipes have been employed in many applications ranging from temperature control of the permafrost layer under the Alaska pipeline to the thermal control of optical surfaces in spacecraft.

A heat pipe is a heat-transfer device with an extremely high effective thermal conductivity. Heat pipes are evacuated vessels, typically circular in cross-section, which are backfilled with a small quantity of a working fluid. They are totally passive and are used to transfer heat from a heat source to a heat sink with minimal temperature gradients, or to isothermalize surfaces.

A heat pipe consists typically of a sealed container with a wicking material. The container is evacuated and filled with just enough liquid to fully saturate the wick. As illustrated in Figure 7.1, a heat pipe consists of three distinct regions: an evaporator or heat addition region of the container, a condenser or heat rejection region, and an adiabatic or isothermal region. When the evaporator region is exposed to a high temperature, heat is added and the working fluid in the wicking structure is heated until it evaporates. The high temperature and the corresponding high pressure in this region cause the vapor to flow to the cooler condenser region where the vapor condenses, giving up its latent heat of vaporization. The capillary forces existing in the wicking structure then pump the liquid back to the evaporator. The wick structure thus ensures that the heat pipe can transfer heat if the heat source is below the cooled end (bottom heat mode) or if it is above the cooled end (top heat mode).

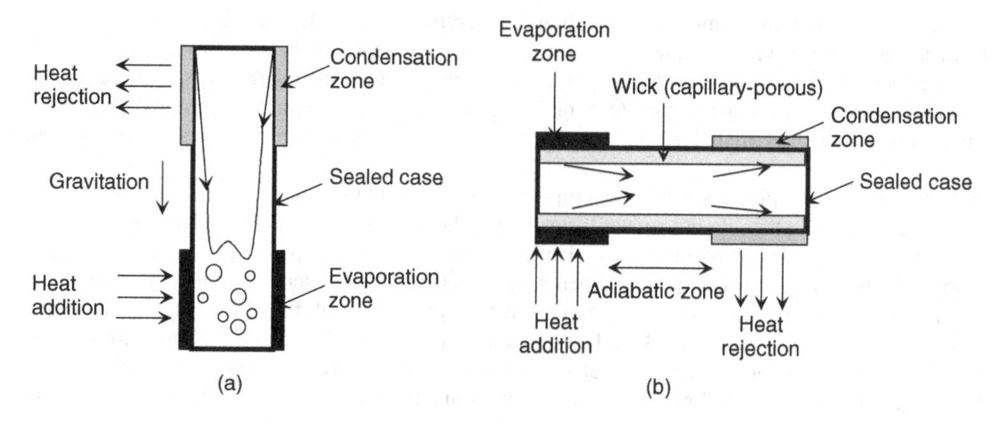

Figure 7.1 Two basic heat pipe configurations: (a) thermosiphon and (b) capillary driven.

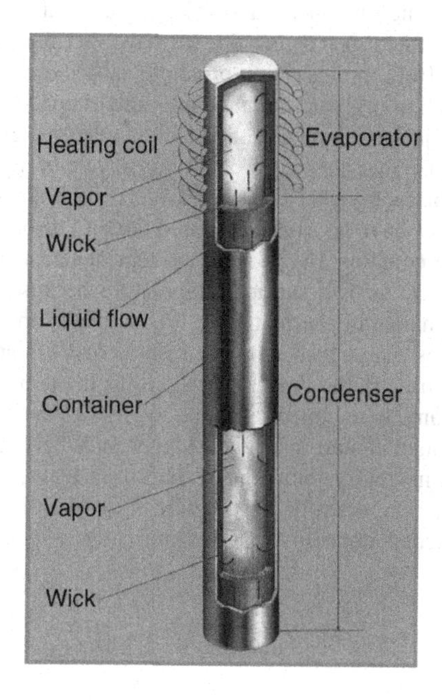

Figure 7.2 A cutaway view of a cylindrical heat pipe (*Courtesy of Los Alamos National Laboratory. Copyright © 1998–2002 The Regents of the University of California*).

A heat pipe is a synergistic engineering structure which, within certain limitations on the manner of use, is equivalent to a material having a thermal conductivity greatly exceeding that of any known metal. Shown in Figure 7.2 is a cutaway view of a cylindrical heat pipe with a homogeneous screen wick. Working fluid is vaporized in the evaporator section and flows toward the condenser section where it deposits its heat by condensation. Capillary forces in the porous wick return the condensed working fluid to the evaporator section. Heat transfer occurs through the capillary movement of fluids. The "pumping" action of surface tension forces may be sufficient to move liquids from a low temperature zone to a high temperature zone (with subsequent return in vapor form using as the driving force the difference in vapor pressure at the two temperatures). Such a closed system, requiring no external pumps, may be of particular interest in space reactors in moving heat from the reactor core to a radiating system. In the absence of gravity, the forces must only be such as to overcome the capillary and the drag of the returning vapor through its channels.

Note that in a heat pipe assembly, the coil supporting rod and the induction coil are assembled as one integral unit and they do not rotate. Instead, only the outer shell, or jacket, rotates on the heavy duty inner bearings mounted on each end of the nonrotating coil support rod. This construction eliminates the need for rotary joints. When an AC voltage of commercial frequency is supplied, the induction coil generates flux lines whose direction alternates with the power supply frequency. And, since the roll shell is mounted on the same axis as the induction coil, the shell functions as one complete turn of a secondary coil. Therefore, the coil, which receives the power, does not heat up; rather, the shell heats up, following Faraday's law. Thus, the roll shell itself is the heat source, not some remotely located heater or boiler. It is well known that the electromagnetic induction method is almost 100% efficient in converting electrical energy into heat. The shell has several gun-drilled holes running the full width of the roll, called *jacket chambers*, the number of which will vary with roll specifications. In each of the chambers, a small amount of thermal medium is placed, after which, each chamber is sealed and evacuated. So, we have thermal medium in a vacuum. When the roll is operating, the heat from the induction principle causes this thermal medium to vaporize. Since the pressure of vaporization is greater than the pressure of condensation, the vapor must move to any cooler area within the jacket chamber and it then condenses, giving off to the shell surface the latent heat of vaporization. Thus, there is a continuous cycle of vaporization and condensation taking place in the vacuum of each jacket chamber, which is the phenomenon known as the *heat pipe principle*. These heat pipes have an extremely rapid rate of heat transmission (almost the speed of sound) and each heat pipe contains a very large amount of latent heat. The heat pipe action is what maintains the highly accurate roll surface temperature because it responds so rapidly, and automatically, to any slight change of thermal load. So, with a temperature correction device, the accurate surface temperature is maintained not only in the cross direction but also in the machine direction. Since no oil flows through the journals of the rolls, the temperature of the journal, where the support bearing for the frame is mounted, is about one half of the roll surface temperature. This means that the external bearings should last much longer and that high temperature bearings are not always needed. So, with no rotary joints, no seals, no oil leaks, and cooler-running bearings, the maintenance of the rolls is noticeably and significantly less than that of conventional rolls. More importantly, environmental concerns that are normally associated with oil and heat rolls are eliminated.

7.2.1 Heat Pipe Use

In heat pipe utilization, there are three primary objectives that we mainly expect from heat pipes:

- **To act as a primary heat conductive path.** When a heat source and heat sink need to be placed apart, a heat pipe can be a very effective heat conduction path for transporting heat from the heat source to the heat sink.
- **To aid heat conduction of a solid.** Heat pipes can add to the efficiency and transport capacity of a thermal shunt.
- **To aid heat spreading of a plane.** Heat pipes can be used to increase the heat spreading across a large heat sink base, thereby effectively increasing the base thermal conductivity. The effect of this is the decrease of the temperature gradient across the base (increase the efficiency), thereby lowering the heat source temperature.

7.3 Heat Pipe Applications

Heat pipes are used for a wide variety of heat-transfer applications covering the complete spectrum of thermal applications. Heat pipes are ideal for any application where heat must be transferred with a minimum thermal gradient either to increase the size of heat sink, to relocate the sink to a remote location, or where isothermal surfaces are required. Some typical heat pipe applications include

- cooling of electronic devices and computers,
- cooling of high-heat-load optical components,
- cooling of milling machine spindles,
- cooling of injection molds,
- cryogenic systems,
- aircraft thermal control systems,
- cooling of engine components in conventional aircraft,
- spacecraft systems,
- heat exchangers,
- waste heat recovery systems,
- various industrial processes, for example, metallurgical, chemical, pharmaceutical, food, oil refining, power generation, transportation, communication, electronics, and so on, and
- solar energy conversion and power generation systems.

Heat pipes are used in a wide range of products like air conditioners, refrigerators, heat exchangers, transistors, capacitors, and so on. Heat pipes are also used in laptops to reduce the working temperature for better efficiency. Their application in the field of cryogenics is very significant, especially in the development of space technology, particularly for spacecraft temperature equalization, component cooling, temperature control, and radiator design in satellites, and moderator cooling, removal of heat from the reactor at emitter temperature, and elimination of troublesome thermal gradients along the emitter and collector in spacecraft.

Heat pipes are extremely effective in transferring heat from one location to another. A common spaceflight heat pipe has an effective thermal conductivity many thousands of times that of copper. Although there are many ground applications for heat pipes, in a space-borne environment radiation and conduction are the sole means of heat transfer, so heat pipes are a fundamental aspect of a satellite thermal and structural subsystem design.

Figure 7.3 shows two configurations of heat pipes in laptops: (a) the heat pipe connected to the keyboard setup and (b) the setup of the heat pipe connected to the back screen. In the past Xie *et al.* (2001) have conducted comprehensive case studies on the above configurations and their applications in notebook computers.

7.3.1 Heat Pipe Coolers

Although heat pipes have been known in their basic working principles since the nineteenth century, it is only since the 1960s that they have been extensively used in several industrial fields. Owing to their relatively simple structure and improved thermal characteristics, heat pipe coolers show substantial gain in weight and size reduction. In addition, they are totally maintenance free, with a proven operating reliability in excess of 30 years. Heat pipe coolers (Figure 7.4) can be designed and manufactured with an electrical insulation up to 12 kV, while the two parts of the thermal circuit, the condensing and evaporating sections, respectively, can be physically separated to avoid hazardous contacts and dust accumulation. The versatility of these systems allows great freedom in designing customized thermal solutions.

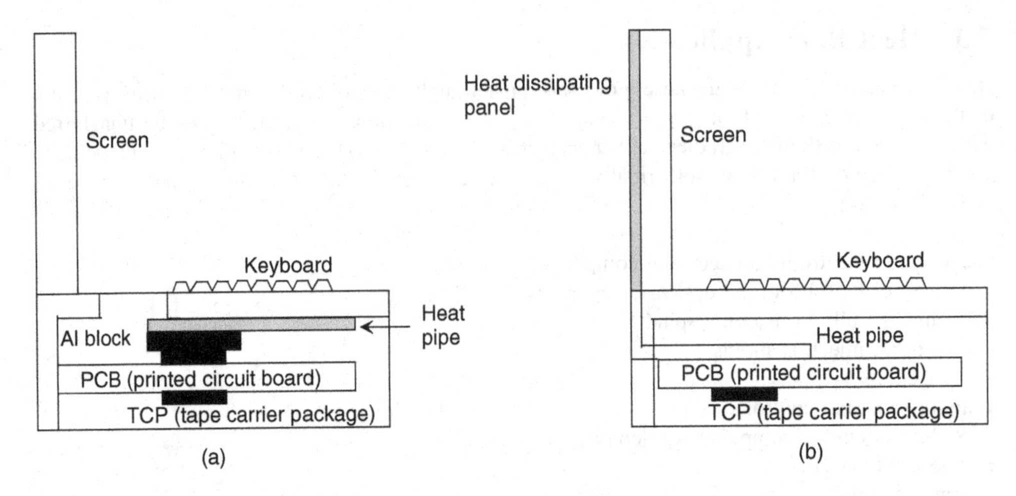

Figure 7.3 (a) The heat pipe connected to the keyboard setup. (b) The setup of the heat pipe connected to the back screen (Adapted from Xie *et al.*, 2001).

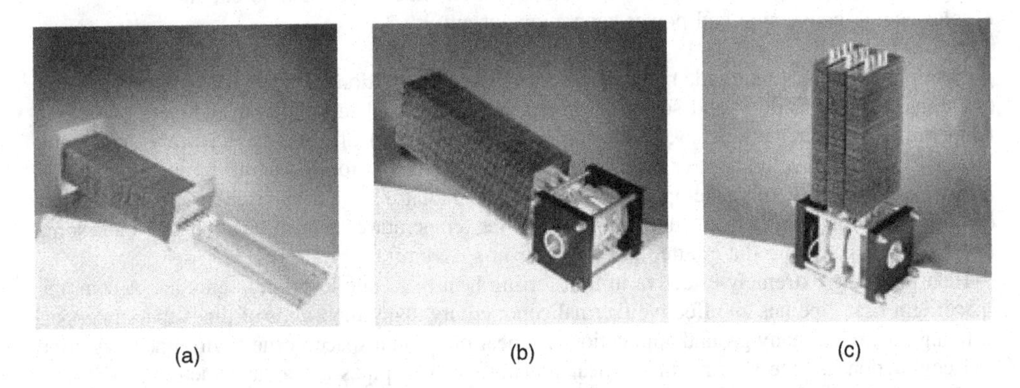

Figure 7.4 (a) A customized heat pipe unit, especially developed for cooling with natural air (for a total power of 1 kW, with 7.5 kV Al_2O_3 insulators, being used on onboard equipment in a subway train). (b) A stack of 7.5 kV insulated heat pipe coolers (with Al_2O_3 insulators) designed for natural air cooling of a pair of thyristors for onboard equipment in a subway train. Light construction with aluminum evaporator and fins. (c) A stack of noninsulated heat pipe sinks designed to cool a pair of thyristors with forced ventilation, mounted with special clamping equipment. The equipment is intended for use in conjunction with an AC mill drive (*Courtesy of Bosari Thermal Management s.r.l.*).

7.3.2 Insulated Water Coolers

Manufactured with the most advanced technologies, water cooled and insulated heat sinks provide electrical separation of the electronic component from the water circuit by means of high thermal conductivity ceramic components (Figure 7.5). They are further designed with oversized superficial discharged paths and feature very low total thermal resistance. They now offer a reliable, efficient, economic, and ecologically favorable alternative to the conventional cooling and treated water coolant systems.

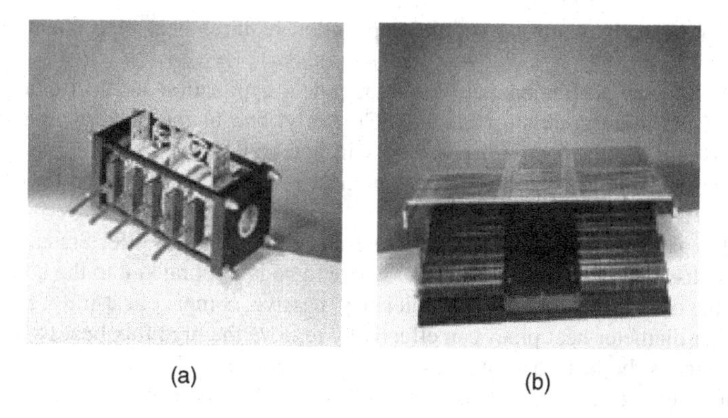

(a) (b)

Figure 7.5 (a) An insulated water cooler. (b) A heat exchanger designed for the natural air cooling of a sealed container (*Courtesy of Bosari Thermal Management s.r.l.*).

7.3.3 Heat Exchanger Coolers

This is considered a new solution for the cooling of sealed containers. On the basis of the heat pipe technology, the heat exchangers have wider radiating surfaces, thus saving weight and dimensions, when compared with the conventional cooling systems. In this regard, Figure 7.5b shows a heat exchanger designed for the natural air cooling of a sealed container, designed for onboard equipment of a tramway (dissipating surface $\sim 10\,m^2$ and total weight 25 kg).

7.4 Heat Pipes for Electronics Cooling

All electronic components, from microprocessors to high-end power converters, generate heat, and rejection of this heat is necessary for their optimum and reliable operation. As electronic design allows higher throughput in smaller packages, dissipating the heat load becomes a critical design factor. Many of today's electronic devices require cooling beyond the capability of standard metallic heat sinks. The heat pipe is meeting this need and is rapidly becoming a mainstream thermal management tool. In fact, heat pipes have been commercially available since the mid 1960s. Only in the past few years, however, has the electronics industry adopted heat pipes as reliable, cost-effective solutions for high-end cooling applications.

As mentioned earlier, a heat pipe is essentially a passive heat-transfer device with an extremely high effective thermal conductivity. The two-phase heat-transfer mechanism results in heat-transfer capabilities from one hundred to several thousand times that of an equivalent piece of copper. The heat pipe, in its simplest configuration, is a closed, evacuated, cylindrical vessel with the internal walls lined with a capillary structure or wick that is saturated with a working fluid. Since the heat pipe is evacuated and then charged with the working fluid prior to being sealed, the internal pressure is set by the vapor pressure of the fluid. As heat is input at the evaporator, fluid is vaporized, creating a pressure gradient in the pipe. This pressure gradient forces the vapor to flow along the pipe to a cooler section where it condenses, giving up its latent heat of vaporization. The working fluid is then returned to the evaporator by the capillary forces developed in the wick structure.

Heat pipes can be designed to operate over a very broad range of temperatures from cryogenic ($<-243\,°C$) applications utilizing titanium alloy/nitrogen heat pipes to high temperature applications ($>2000\,°C$) using tungsten/silver heat pipes. In electronic cooling applications where it is desirable to maintain junction temperatures below $125-150\,°C$, copper–water heat pipes are typically used.

Copper–methanol heat pipes are used if the application requires heat pipe operation below 0 °C (Garner, 1996).

Perhaps the best way to demonstrate the heat pipe's application to electronics cooling is to present a few of the more common examples. Currently, one of the highest volume applications for heat pipes is cooling the Pentium processors in notebook computers. Because of the limited space and power available in notebook computers, heat pipes are ideally suited for cooling the high power chips (see Figure 7.3).

Fan-assisted heat sinks require electrical power and reduce battery life. Standard metallic heat sinks capable of dissipating the heat load are too large to be incorporated into the notebook package. Heat pipes, on the other hand, offer a high-efficiency, passive, compact heat-transfer solution. Three or four millimeter diameter heat pipes can effectively remove the high flux heat from the processor. The heat pipe spreads the heat load over a relatively large-area heat sink, where the heat flux is so low that it can be effectively dissipated through the notebook case to ambient air. The heat sink can be the existing components of the notebook, from electromagnetic interference shielding under the key pad to metal structural components.

Typical thermal resistances for such applications at 6–8 W heat loads are 4–6 °C/W. High power mainframe, mini-mainframe, server, and workstation chips may also employ heat pipe heat sinks. High-end chips dissipating up to 100 W are outside the capabilities of conventional heat sinks. Heat pipes are used to transfer heat from the chip to a fin stack large enough to convect the heat to the supplied air stream. The heat pipe isothermalizes the fins, eliminating the large conductive losses associated with standard sinks. The heat pipe heat sinks dissipate loads in the 75–100 W range with resistances from 0.2 to 0.4 °C/W, depending on the available airflow.

In addition, other high-power electronics including silicon controlled rectifiers, insulated gate bipolar transistors, and thyristors often utilize heat pipe heat sinks. In fact, heat pipe heat sinks are capable of cooling several devices with total heat loads up to 5 kW. These heat sinks are also available in an electrically isolated version, where the fin stack can be at ground potential with the evaporator operating at device potentials of up to 10 kV. Typical thermal resistances for the high-power heat sinks range from 0.05 to 0.1 °C/W. Again, the resistance is predominately controlled by the available fin volume and airflow (for details, see Garner, 1996).

Example 7.1

Consider a heat pipe used as a heat sink for electronic cooling. The heat load is 16 W and the thermal resistance of the heat pipe is 4 °C/W. What is the temperature difference involved in the dissipation of this heat load?

Solution

The rate of heat dissipation is the temperature difference divided by the thermal resistance. Then,

$$\dot{Q} = \frac{\Delta T}{R} \longrightarrow \Delta T = \dot{Q}R = (16\ \text{W})(4\,^{\circ}\text{C/W}) = 64\,^{\circ}\text{C}$$

7.5 Types of Heat Pipes

Various innovative applications of heat pipes demand a complete and thorough understanding of the physical phenomena occurring in a heat pipe. In this regard, efforts have been directed toward more detailed numerical and analytical modeling of conventional heat pipes, thermosiphons, rotating heat pipes, and micro heat pipes as well as capillary pump loops (Faghri, 1996).

Since the 1960s, various types of heat pipe heat exchangers have been developed, which are as follows:

- capillary pumped loop heat pipes,
- gas loaded heat pipes,
- variable conductance heat pipes,
- micro and miniature heat pipes (particularly for microelectronics cooling),
- coaxial heat pipes,
- rotating heat pipes,
- pulsating heat pipes,
- osmotic heat pipes,
- chemical heat pipes,
- gravity-driven geothermal heat pumps,
- thermosiphon heat pipes (here, the word "thermosiphon" is used to describe both single-phase and evaporative gravity-assisted heat transport devices),
- low temperature and cryogenic heat pipes, and
- alkali metal heat pipes.

7.5.1 Micro Heat Pipes

Cotter (1984) first introduced the concept of micro heat pipes incorporated into semiconductor devices to provide more uniform temperature distribution and better heat transfer. The primary operating principles of micro heat pipes are essentially the same as those occurring in larger, more conventional heat pipes. Heat applied to one end of the heat pipe vaporizes the liquid in that region and forces it to move to the cooler end where it condenses and gives up the latent heat of vaporization. This vaporization and condensation process causes the liquid–vapor interface in the liquid arteries to change continually along the pipe, as illustrated in Figure 7.6, and results in a capillary pressure difference between the evaporator and condenser regions. This capillary pressure difference promotes the flow of the working fluid from the condenser back to the evaporator through the triangular-shaped corner regions. These corner regions serve as liquid arteries; thus, no wicking structure is required.

7.5.2 Cryogenic Heat Pipes

Cryogenic heat pipes operate between 4 and 200 K. Typical working fluids include helium, argon, oxygen, and krypton. The amount of heat that can be transferred for cryogenic heat pipes is quite low because of the small heats of vaporization, high viscosities, and small surface tensions of the working fluids.

7.6 Heat Pipe Components

In general, a traditional heat pipe structure is of a hollow cylindrical container filled with a vaporizable liquid as working fluid as shown in Figure 7.7.

A heat pipe typically consists of a sealed container lined with a wicking material. The container is evacuated and backfilled with just enough liquid to fully saturate the wick. Because heat pipes operate on a closed two-phase cycle and only pure liquid and vapor are present within the container, the working fluid will remain at saturation conditions as long as the operating temperature is between the triple point and the critical state. As illustrated in Figure 7.7, a heat pipe consists of three distinct

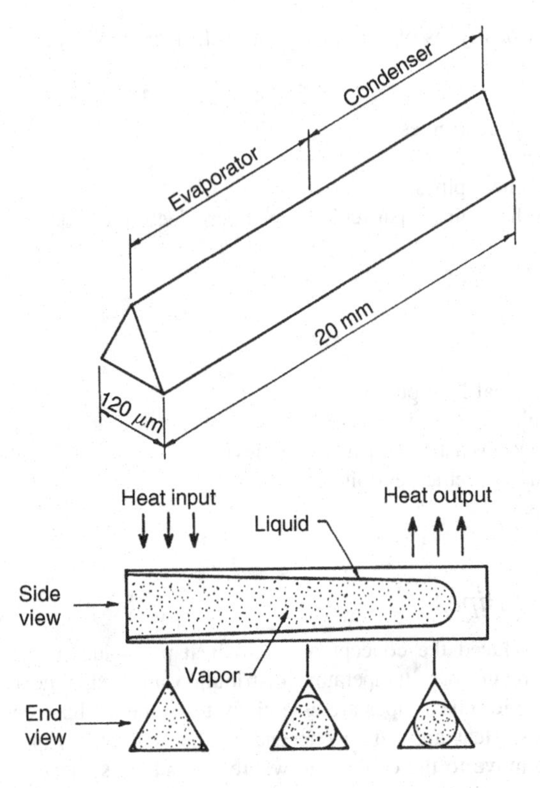

Figure 7.6 A micro heat pump (Peterson, 1994).

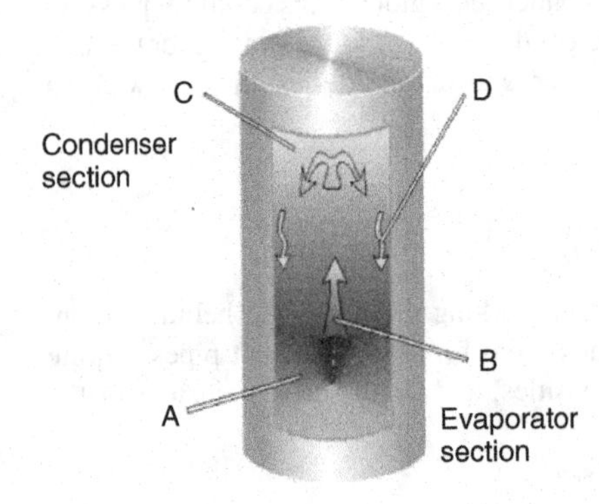

A. Heat is absorbed in the evaporating section.

B. Fluid boils to vapor phase.

C. Heat is released from the upper part of cylinder to the environment; vapor condenses to liquid phase.

D. Liquid returns by gravity to the lower part of cylinder (evaporating section).

Figure 7.7 A heat pipe structure (*Courtesy of Heat Pipe Technology, Inc.*).

regions: an evaporator or heat addition region, a condenser or heat rejection region, and an adiabatic or isothermal region. When heat is added to the evaporator region of the container, the working fluid present in the wicking structure is heated until it vaporizes. The high temperature and corresponding high pressure in this region cause the vapor to flow to the cooler condenser region, where the vapor condenses, giving up its latent heat of vaporization. The capillary forces existing in the wicking structure then pump the liquid back to the evaporator.

Basically a heat pipe consists of three major components, namely,

- the container, which can be constructed from glass, ceramics, or metals;
- a working fluid, which can vary from nitrogen or helium for low-temperature (cryogenic) heat pipes to lithium, potassium, or sodium for high-temperature (liquid metal) heat pipes; and
- a wicking structure or capillary structure, constructed from woven fiberglass, sintered metal powders, screens, wire meshes, or grooves.

Each of these three components is equally important, with careful consideration given to the material type, thermophysical properties, and compatibility. For example, the container material must be compatible with both the working fluid and the wicking structure, strong enough to withstand pressures associated with the saturation temperatures encountered during storage and normal operation, and must have a high thermal conductivity. In addition to these characteristics, which are primarily concerned with the internal effects, the container material must be resistant to corrosion resulting from interaction with the environment and must be malleable enough to be formed into the appropriate size and shape.

7.6.1 Container

Basic requirements of the heat pipe case include a container capable of maintaining a leak-proof seal and structural integrity throughout the entire pressure range to which the heat pipe will be exposed. Therefore, the function of the container is to isolate the working fluid from the outside environment. It has to therefore be leak proof, maintain the pressure differential across its walls, and enable transfer of heat to take place from and into the working fluid. Possible materials include pure metal alloys such as aluminum, stainless steel, or copper; composite materials, either metal or carbon composite; or for higher temperature applications, refractory materials or linings to prevent corrosion.

Careful consideration should be given to the selection of the container or case material for heat pipes. Various factors that should be considered include the following (Peterson, 1994):

- compatibility with the wicking structure and working fluid,
- operating temperature range of the proposed device,
- evaporator and condenser sizes and shapes,
- applicability,
- reliability,
- strength to weight ratio,
- internal operating pressure,
- thermal conductivity,
- ease of fabrication (including welding, machinability, and ductility),
- possibility of external corrosion,
- porosity, and
- wettability.

In addition to the compatibility problem associated with heat pipes, it must be remembered that heat pipes and thermosiphons are in fact "unfired pressure vessels" and as a result must be designed to meet the appropriate pressure vessel codes.

7.6.2 Working Fluid

A first consideration in the identification of a suitable working fluid is the operating vapor temperature range. Within the approximate temperature band, several possible working fluids may exist, and a variety of characteristics must be examined in order to determine the most acceptable of these fluids for the application considered. The prime requirements are as follows:

- compatibility with wick and wall materials,
- good thermal stability,
- wettability of wick and wall materials,
- vapor pressure not too high or low over the operating temperature range,
- high latent heat,
- high thermal conductivity,
- low liquid and vapor viscosities,
- high surface tension, and
- acceptable freezing or pour point.

The selection of the working fluid must also be based on thermodynamic considerations which are concerned with the various limitations to heat flow occurring within the heat pipe like, viscous, sonic, capillary, entrainment, and nucleate boiling levels. These will be explained later.

In heat pipe design, a high value of surface tension is desirable in order to enable the heat pipe to operate against gravity and to generate a high capillary driving force. In addition to high surface tension, it is necessary for the working fluid to wet the wick and the container material, that is, the contact angle should be zero or very small. The vapor pressure over the operating temperature range must be sufficiently high to avoid high vapor velocities, which tend to set up large temperature gradient and cause flow instabilities.

A high latent heat of vaporization is desirable in order to transfer large amounts of heat with minimum fluid flow, and hence to maintain low pressure drops within the heat pipe. The thermal conductivity of the working fluid should preferably be high in order to minimize the radial temperature gradient and to reduce the possibility of nucleate boiling at the wick or wall surface. The resistance to fluid flow will be minimized by choosing fluids with low values of vapor and liquid viscosities. Table 7.1 lists a few media with their useful temperature ranges.

Heat pipe working fluids range from helium and nitrogen for cryogenic temperatures, to liquid metals like sodium and potassium for high temperature applications. Some of the more common heat pipe fluids used for electronics cooling applications are ammonia, water, acetone, and methanol.

Although many working fluids are used in various heat pipe models, the most common are water and methanol. Water is thermodynamically superior to methanol under most conditions and becomes the fluid of choice where it is applicable. The useful range for water is generally 50–200 °C. The useful range for methanol is slightly lower, generally 20–120 °C. Since methanol freezes at a very low temperature, −97 °C, it is useful in gravity-aided, pool-boiling applications where water heat pipes would be subject to freezing.

Heat pipe working fluids including water maintain the normal freezing point. Properly designed heat pipes, however, will not be damaged by the freezing and thawing of the working fluid. Heat pipes will not operate until the temperature rises above the freezing temperature of the fluid.

Water at atmospheric pressure boils at 100 °C. Inside a heat pipe, the working fluid (e.g., water) is not at atmospheric pressure. The internal pressure of the heat pipe is the saturation pressure of the fluid at the corresponding fluid temperature. As such, the fluid in a heat pipe will boil at any temperature above its freezing point. Therefore, at room temperature (e.g., 20 °C), a water heat pipe is under partial vacuum, and the heat pipe will boil as soon as heat is input.

Owing to the vaporization and condensation of the working fluid that takes place in a heat pipe (as a basis for the operation), the selection of an appropriate working fluid is a crucial factor in the design and manufacture of heat pipes. It is important to ensure that the operating temperature range is adequate for the application. While most applications involving the use of heat pipes in the thermal control of electronic devices and systems require the use of a working fluid with

Table 7.1 Some heat pipe fluids and their temperature ranges.

Medium	Melting Point Temperature (°C)	Boiling Point Temperature (°C) (at Atmospheric Pressure)	Application Range (°C)
Helium	−271	−261	−271 to −269
Nitrogen	−210	−196	−203 to −160
Ammonia	−78	−33	−60 to 100
Acetone	−95	57	0 to 120
Methanol	−98	64	10 to 130
Flutec PP2	−50	76	10 to 160
Ethanol	−112	78	0 to 130
Water	0	100	30 to 200
Toluene	−95	110	50 to 200
Mercury	−39	361	250 to 650
Sodium	98	892	600 to 1200
Lithium	179	1340	1000 to 1800
Silver	960	2212	1800 to 2300

Source: Narayanan (2001).

boiling temperatures between 250 and 375 K both cryogenic heat pipes (operating in the 5 to 100 K temperature range) and liquid-metal heat pipes (operating in the 750 to 5000 K temperature range) have been developed and used. Figure 7.8 illustrates the possible working temperature ranges for some of the various heat pipe fluids. In addition to the thermophysical properties of the working fluid, consideration must be given to the ability of the working fluid, to the wettability of the working fluid, and to the wick and wall materials. Further criteria for the selection of the working fluids, including a number of other factors such as liquid and vapor pressure and compatibility of the materials, are considered.

7.6.3 Selection of Working Fluid

Because the basis for operation of a heat pipe is the vaporization and condensation of the working fluid, selection of a suitable working fluid is perhaps the most important aspect of the design and manufacture process. Factors affecting the selection of an appropriate working fluid include

- operating temperature range,
- vapor pressure,
- thermal conductivity,
- compatibility with the wick and case materials, the stability, and
- toxicity.

It should be noted that the theoretical operating temperature range for a given heat pipe is typically between the critical temperature and triple state of the working fluid. Above the critical temperature, the working fluid exists in a vapor state and no increase in pressure will force it to

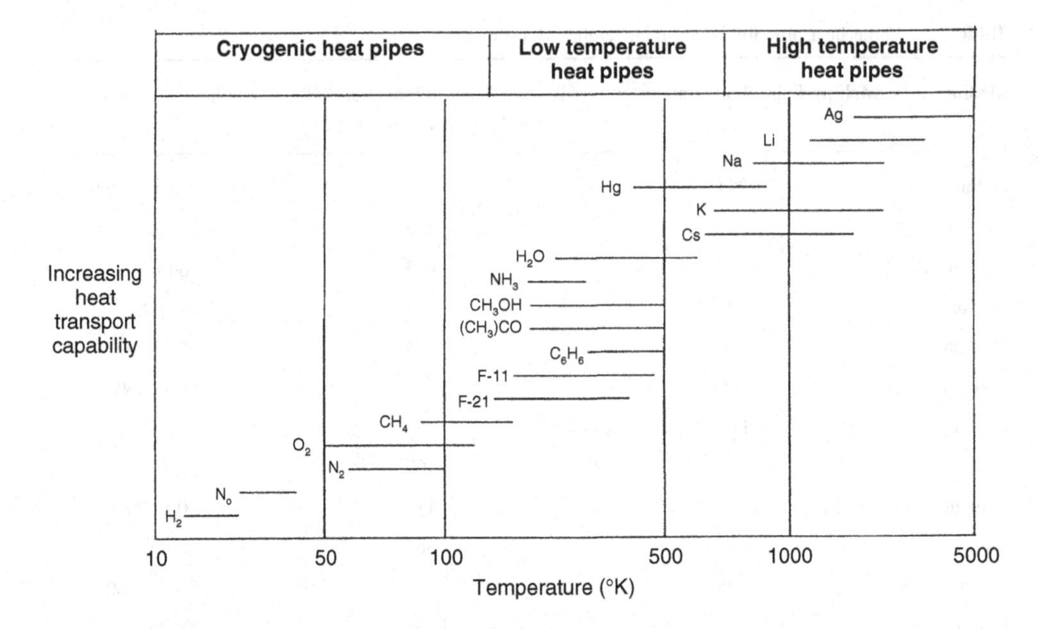

Figure 7.8 Temperature ranges of some heat pipe working fluids (Peterson, 1994).

return to a liquid state. As a result, when working fluids are above their critical temperature, the capillary pumping mechanism provided by the wicking structure ceases to function. Similarly, when the operating temperature is below the triple state, the working fluid exists in the solid and vapor states. While some heat transfer may occur due to sublimation, operation in this temperature range should be avoided (Peterson, 1994).

7.6.4 Wick or Capillary Structure

The concept of utilizing a wicking structure as part of a passive two-phase heat-transfer device capable of transferring large quantities of heat with a minimal temperature drop was first introduced by Gaugler (1944).

The wick or capillary structure is porous and made of materials like steel, aluminum, nickel, or copper in various ranges of pore sizes. They are fabricated using metal foams and, more particularly, felts, the latter being more frequently used. By varying the pressure on the felt during assembly, various pore sizes can be produced. By incorporating removable metal mandrels, an arterial structure can also be molded in the felt.

Fibrous materials, like ceramics, have also been used widely. They generally have smaller pores. The main disadvantage of ceramic fibers is that they have little stiffness and usually require a continuous support by a metal mesh. Thus, while the fiber itself may be chemically compatible with the working fluids, the supporting materials may cause problems. More recently, interest has turned to carbon fibers as a wick material. Carbon fiber filaments have many fine longitudinal grooves on their surface, have high capillary pressures, and are chemically stable. Many heat pipes constructed using carbon fiber wicks seem to show a greater heat transport capability.

The main goal of the wick is to generate capillary pressure to transport the working fluid from the condenser to the evaporator. It must also be able to distribute the liquid around the evaporator section to any area where heat is likely to be received by the heat pipe. Often, these two functions require wicks of different forms. The selection of the wick for a heat pipe depends on many factors, several of which are closely linked to the properties of the working fluid.

Note that the maximum capillary head generated by a wick increases with decrease in pore size. The wick permeability increases with increasing pore size. Another feature of the wick which must be optimized is its thickness. The heat transport capability of the heat pipe is raised by increasing the wick thickness. The overall thermal resistance at the evaporator also depends on the conductivity of the working fluid in the wick. There are some other necessary properties of the wick, namely, compatibility with the working fluid and wettability.

The two most important properties of a wick are the pore radius and the permeability. The pore radius determines the pumping pressure that the wick can develop. The permeability determines the frictional losses of the fluid as it flows through the wick. The most common types of wicks that are used are as follows (Narayanan, 2001):

- **Sintered powder metal.** This process will provide high power handling, low temperature gradients, and high capillary forces for antigravity applications. A complex sintered wick with several vapor channels and small arteries is used to increase the liquid flow rate. Very tight bends in the heat pipe can be achieved with this type of structure.
- **Grooved tube.** The small capillary driving force generated by the axial grooves is adequate for low-power heat pipes when operated horizontally or with gravity assistance. The tube can be readily bent. When used in conjunction with screen mesh, the performance can be considerably enhanced.
- **Screen mesh or cable or fiber.** This type of wick is used in the majority of the products and provides readily variable characteristics in terms of power transport and orientation sensitivity, according to the number of layers and mesh counts used.

Figure 7.9 shows several heat pipe wick structures. It is important to select the proper wick structure for your application. The above list is in order of decreasing permeability and decreasing pore radius. Grooved wicks have a large pore radius and a high permeability; as a result, the pressure losses are low and the pumping head is also low. Grooved wicks can transfer high heat loads in a horizontal or gravity-aided position, but cannot transfer large loads against gravity. The powder metal wicks at the opposite end of the list have small pore radii and relatively low permeability. They are limited by pressure drops in the horizontal position but can transfer large loads against gravity.

The wicking structure has two functions in heat pipe operation:

- providing the mechanism by which the working fluid is returned from the condenser to the evaporator and
- ensuring that the working fluid is evenly distributed over the evaporator surface.

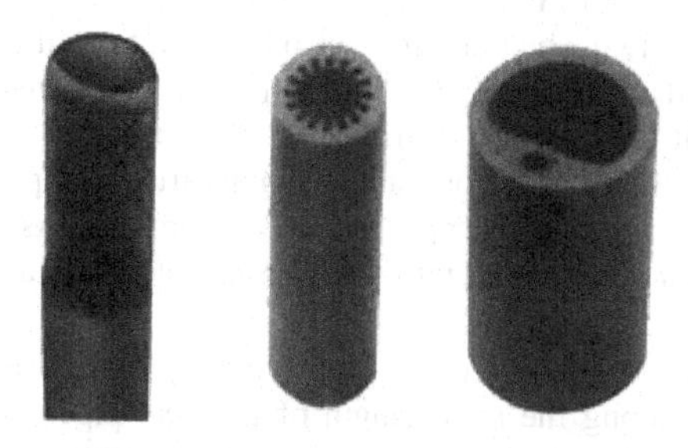

Figure 7.9 Wick structures (Garner, 1996) (*Reproduced by permission of Flomerics, Inc.*).

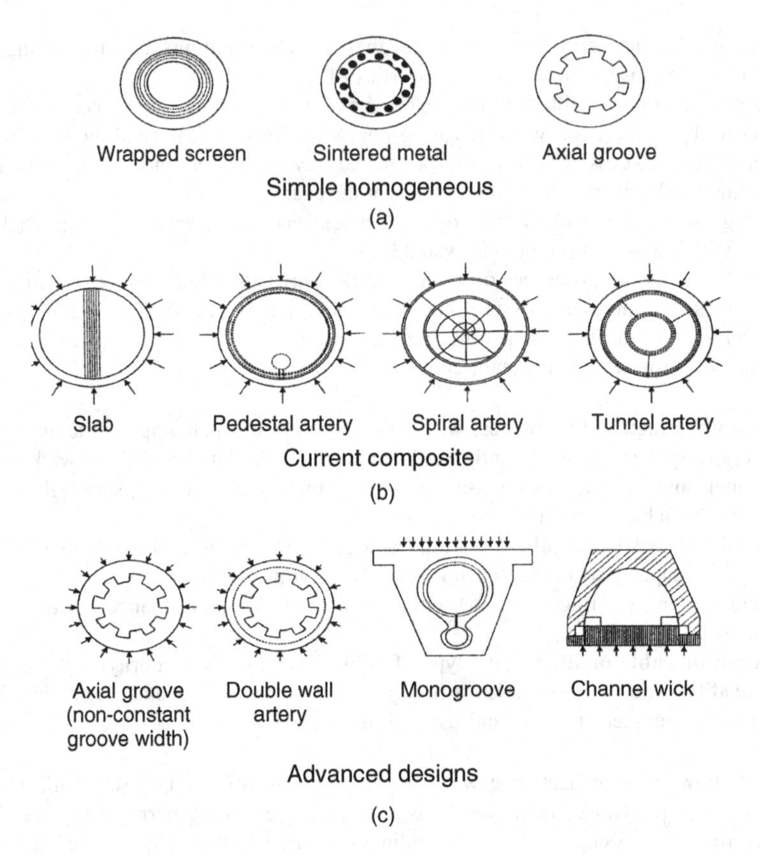

Figure 7.10 Some common heat pipe wicking configurations and their structures. (a) Simple homogeneous, (b) current composite, and (c) advanced designs (Peterson, 1994).

Figure 7.10 illustrates several common wicking structures presently in use, along with several more advanced concepts under development. In order to provide a flow path with low flow resistance through which the liquid can be returned from the condenser to the evaporator, an open porous structure with a high permeability is desirable. However, to increase the capillary pumping pressure, a small pore size is necessary. Solutions to this apparent dichotomy can be achieved through the use of a nonhomogeneous wick made of several different materials or through composite wicking structures similar to those shown in Figure 7.10b.

The wicking structure has two functions in heat pipe operation: it is both the vehicle and the mechanism through which the working fluid returns from the condenser to the evaporator and it ensures that the working fluid is evenly distributed circumferentially over the entire evaporator surface. Figure 7.10 illustrates several common wicking structures presently in use, along with several composite and high capacity concepts under development. As shown in Figure 7.10, the various wicking structures can be divided into three broad categories as follows:

- **Homogeneous structures.** Homogeneous wicks typically comprise of a single material and are distributed uniformly along the axial length of the heat pipe. The most common types of homogeneous wicks include wrapped screen, sintered metal, axially grooved and crescent as shown in Figure 7.10a.

- **Current composite.** In order to provide a flow path with low flow resistance through which the liquid can be returned from the condenser to the evaporator, an open porous structure with a high permeability is desirable. However, to increase the capillary pumping pressure, a small pore size is necessary. Solutions to this apparent dichotomy can be achieved through the use of a nonhomogeneous wick made of several different materials or through a composite wicking structure. Composite wicks are typically comprised of a combination of several types or porosities of materials and/or configurations. Examples of these types of wick structures are illustrated in Figure 7.10b.
- **Advanced designs.** Most of these are relatively new (Figure 7.10c) and consist of variations on the composite wicking structures. Again, the two functions of the wicking structure (i.e., circumferential distribution and axial fluid transport) are achieved by different segments of the capillary structure. The basic design of this advanced capacity configuration consists of two large axial channels, one for vapor flow and the other for liquid flow. In this type of heat pipe, several improvements result from the separation of the liquid and vapor channels. First, because the axial liquid transport can be handled independently from the circumferential distribution, a high heat transport capacity can be achieved. Second, by separating the two channels, the viscous pressure drop normally associated with heat pipes in which the liquid and vapor flows occur within the same channel can be greatly reduced. Third, with the majority of the fluid located in an external artery, heat transfer in the evaporator and condenser takes place across a relatively thin film of liquid in the circumferential wall grooves, thereby increasing the heat-transfer coefficient. While somewhat different in shape, the basic principle of operation of the other advanced designs is the same: Separate the circumferential distribution and axial liquid flow to maximize the capillary pumping and reduce the liquid pressure drop.

7.7 Operational Principles of Heat Pipes

Inside the container is a liquid under its own pressure that enters the pores of the capillary material, wetting all internal surfaces. Applying heat at any point along the surface of the heat pipe causes the liquid at that point to boil and enter a vapor state. When that happens, the liquid picks up the latent heat of vaporization. The gas, which then has a higher pressure, moves inside the sealed container to a colder location where it condenses. Thus, the gas gives up the latent heat of vaporization and moves heat from the input to the output end of the heat pipe. Heat pipes have an effective thermal conductivity many thousands of times that of copper. The heat transfer or transport capacity of a heat pipe is specified by its "Axial Power Rating (APR)." It is the energy moving axially along the pipe. The larger the heat pipe diameter, the greater is the APR. Similarly, the longer the heat pipe the lesser is the APR. Heat pipes can be built in almost any size and shape.

Heat pipes transfer heat by the evaporation and condensation of a working fluid. As stated earlier, a heat pipe is a vacuum-tight vessel which is evacuated and partially backfilled with a working fluid. As heat is input at the evaporator, fluid is vaporized, creating a pressure gradient in the pipe. This pressure gradient forces the vapor to flow along the pipe to the cooler section where it condenses, giving up its latent heat of vaporization. The working fluid is then returned to the evaporator by capillary forces developed in the porous wick structure or by gravity.

A heat pipe is said to be operating against gravity when the evaporator is located above the condenser. In this orientation, the working fluid must be pumped against gravity back to the evaporator. All heat pipes have wick structures that pump the working fluid back to the evaporator using the capillary pressure developed in the porous wick. The finer the pore radius of a wick structure, the higher the heat pipe can operate against gravity. A thermosiphon is similar to a heat pipe, but has no wick structure and will only operate gravity aided.

A heat pipe (Figure 7.7) consists of a vacuum-tight envelope, a wick structure, and a working fluid. The heat pipe is evacuated and then backfilled with a small quantity of working fluid, just enough to saturate the wick. The atmosphere inside the heat pipe is set by an equilibrium of liquid and vapor. As heat enters at the evaporator this equilibrium is upset, generating vapor at a slightly higher pressure. This higher pressure vapor travels to the condenser end where the slightly lower temperatures cause the vapor to condense, giving up its latent heat of vaporization. The condensed fluid is then pumped back to the evaporator by the capillary forces developed in the wick structure. This continuous cycle transfers large quantities of heat with very low thermal gradients. A heat pipe's operation is passive, being driven only by the heat that is transferred. This passive operation results in high reliability and long life.

Both heat pipes and thermosiphons operate on a closed two-phase cycle and utilize the latent heat of vaporization to transfer heat with very small temperature gradients. However, the operation of these two devices is significantly different. In a heat pipe, as illustrated earlier, heat added to the bottom portion of a thermosiphon vaporizes the working fluid. During this phase change process, the fluid picks up the heat associated with its latent heat of vaporization. Because the vapor in the evaporator region is at a higher temperature and hence at a higher pressure than the vapor in the condenser, the vapor rises and flows to the cooler condenser where it gives up the latent heat of vaporization (buoyancy forces assist this process). Gravitational forces then cause the condensate film to flow back down the inside of the heat pipe wall where it can again be vaporized. Although the inside surface of a thermosiphon may occasionally be lined with grooves or a porous structure to promote return of the condensate to the evaporator or increase the heat-transfer coefficient, thermosiphons principally rely upon the local gravitational acceleration for the return of the liquid from the evaporator to the condenser. By definition, then, for proper operation the evaporator of a thermosiphon must be located below the condenser or dryout of the evaporator will occur.

Alternatively, heat pipes utilize some sort of capillary wicking structure to promote the flow of liquid from the condenser to the evaporator and as a result can be used in a horizontal orientation, microgravity environments, or even applications where the capillary structure must "pump" the liquid against gravity from the evaporator to the condenser. It is this single characteristic – the dependence of the local gravitational field to promote the flow of the liquid from the condenser to the evaporator – that differentiates thermosiphons from heat pipes.

7.7.1 Heat Pipe Operating Predictions

Historically, the use of metallic heat sinks has been sufficient to provide the required thermal management for most electronic cooling applications. However, with the new breed of compact devices dissipating larger heat loads, the use of metallic heat sinks is sometimes limited because of the weight and physical size required. Accordingly the use of heat pipes is becoming a solution of choice.

The performance of natural convection heat sinks is directly dependent on the effective surface area; more effective surface area results in better performance. A heat pipe embedded into the base material of a standard aluminum extrusion can reduce the overall temperature difference along the base material, tending to isothermalize the base material. In essence, the localized heat source is spread equally along the length of the heat pipe, increasing the overall efficiency of the heat sink. Although an embedded heat pipe heat sink is slightly more expensive because of the added cost of the heat pipe, it is an easy method of improving the performance of a marginal extrusion.

A more elegant approach is to design a heat sink that fully utilizes the characteristics of a heat pipe. Typical extruded heat sinks have limited aspect ratios and thick fins, which result in lower surface area per length. The material thickness adds unnecessary weight, and, more importantly, obstructs the cooling air flow. To alleviate the extrusion limits, bonded fin heat sinks have been developed which allow the use of a tall, thin fin, which optimizes cooling flow. But bonded fin heat sinks can also be limited by the conduction losses in the base plate for concentrated heat sources.

A heat pipe used in conjunction with parallel plate fins provides more efficient surface area with minimum volume demands. This design application is useful when there is not enough physical volume or airflow above the device to use an extrusion, and allows the designer much latitude in component arrangement. The heat pipe can transport the heat to a "remote" parallel plate fin stack that has enough volume to dissipate the heat. Heat pipes can be designed into most electronic devices for various power levels, and may even allow the use of a natural convection heat sink (DeHoff and Grubb, 2000).

Predicting or developing an optimum heat pipe thermal solution requires use of theoretical and empirical relationships, wisdom and design experience, and knowledge of the application design parameters and system flexibilities. For design concepts and preliminary designs, it can be useful to have a guideline for heat pipe performance. These are general performance guidelines based on a "standard" powder metal wick structure. Alternative powders and production techniques are available that may increase the performance by upward of 500%. The above operating limitations can be summarized to predict heat pipe performances based on three orientation categories and the performance profiles for such cases are shown in Figure 7.11.

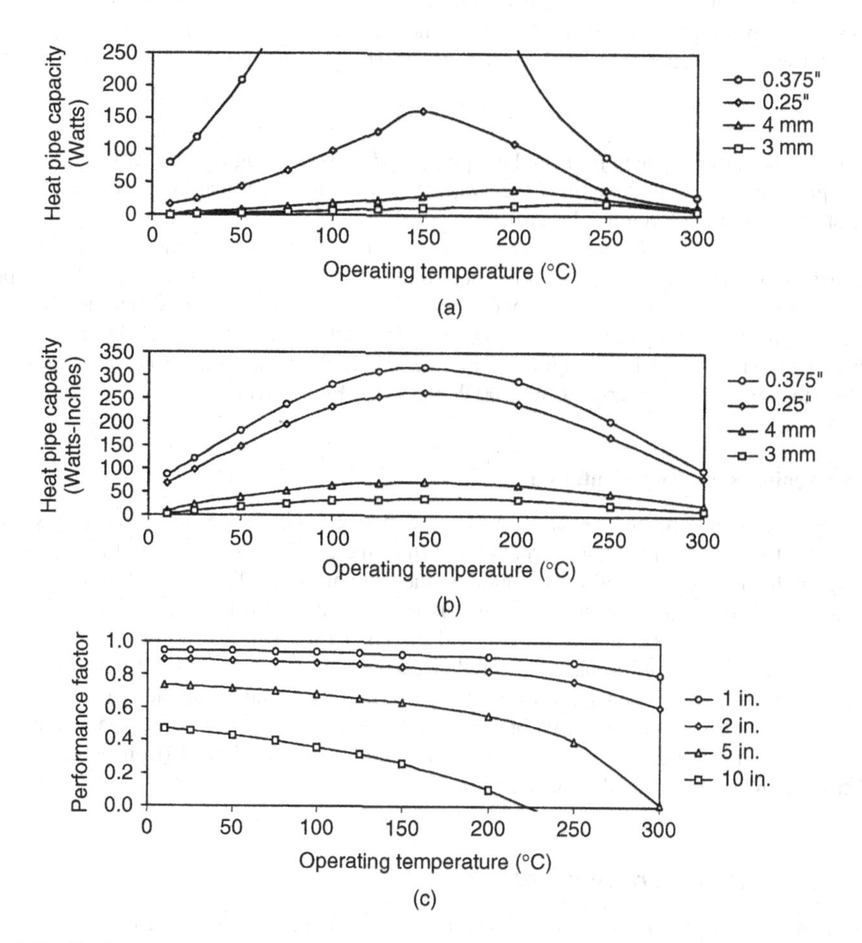

Figure 7.11 Performance curves: (a) for gravity-aided operation, (b) for horizontal operation, and (c) for various heights against gravity (*Courtesy of Thermacore International, Inc.*).

7.7.1.1 Gravity-Aided Orientation

The evaporator is at a lower elevation than the condenser. The gravity-aided orientation is the most efficient, since the heat pipe acts as a thermosiphon and gravity will return the condensed fluid to the evaporator. A sintered powder wick structure may still be needed to handle the heat flux in the evaporator. Heat pipe operation is typically limited by the *flooding limit* or the *boiling limit* (at elevated temperatures above 175 °C). These two limitations are greatly affected by the diameter of the heat pipe: a larger diameter heat pipe will carry more power. Figure 7.11a can be used as a guideline for the selection of a "standard" copper–water powder wick heat pipe in the gravity-aided orientation. The area below each curve is the allowable operating region. For the miniature heat pipes (3 and 4 mm) use the greater of the "gravity-aided" and "horizontal" curves.

7.7.1.2 Horizontal Orientation

The horizontal orientation relies on the wick structure to provide the capillary pressure to return the condensed fluid to the evaporator. The heat pipe operation is typically limited by the *capillary limit*. This limitation is greatly affected by the diameter of the heat pipe (a larger diameter heat pipe will carry more power) and the length of the heat pipe (a longer heat pipe will carry less power). A useful parameter is the *effective length* (DeHoff and Grubb, 2000):

$$L_{eff} = L_{adi} + 0.5(L_{eva} + L_{con}) \tag{7.1}$$

where L_{eff} is the effective length of the heat pipe, m; L_{adi} is the length of the adiabatic section of the heat pipe, m; L_{eva} is the length of the evaporator section of the heat pipe, m; and L_{con} is the length of the adiabatic section of the heat pipe, m.

In conjunction with this, Figure 7.11b can be used as a guideline for the selection of a "standard" copper–water powder wick heat pipe in the horizontal orientation. The capacity of a heat pipe can be determined by taking the appropriate value from the figure in "watt-inches" and dividing by the effective length. For example, a 0.25 in. OD heat pipe with a total length of 8 in., an evaporator length of 1 in., and a condenser length of 5 in. operated at 25 °C gives an effective length of 5 in. Therefore the heat pipe can carry 20 W [100 W-in. (from Figure 7.11b)/5 in.].

7.7.1.3 Against Gravity Orientation

The evaporator is at a higher elevation than the condenser. Heat pipe operation against gravity orientation relies solely on the wick structure to return the condensed fluid up to the higher evaporator. Again the heat pipe operation is limited by the capillary limit. This orientation is very similar to the horizontal orientation, except that the effects of gravity must be accounted for. A larger elevation difference between the evaporator and the condenser results in a lower power capacity. Figure 7.11c shows the performance factor that can be used as a guideline for the selection of a "standard" copper–water powder wick heat pipe in the horizontal orientation. The performance factor must be applied to the capacity obtained from Figure 7.11b. For example, the heat pipe from the previous example operated 5 in. against gravity would carry 14 W [0.7 (from Figure 7.11c) × 20 W from above] (DeHoff and Grubb, 2000).

7.7.2 Heat Pipe Arrangement

As mentioned above, the orientation and layout of a heat pipe design are critical. When the design allows, the heat source should be located below or at the same elevation as the cooling section for best performance. This orientation allows gravity to aid the capillary action, and results in a greater heat-carrying capability. If this orientation is unacceptable, then a sintered powder wick structure will be necessary. Additionally, heat pipes have the ability to adhere to the physical constraints of the system, and can be bent around obstructions. For a cylindrical heat pipe, the typical bend radius is three times the heat pipe diameter. Tighter bend radii are possible, but may reduce the heat-transfer capability. Since a bend in the heat pipe will have a small impact on performance, the number of bends should be limited. A good rule of thumb is to assume a 1 °C temperature loss for each 90° bend, if the heat pipe is not operated near one of the limitations. The heat pipe is also capable of being flattened (a 3-mm diameter heat pipe can be flattened to 2 mm) (DeHoff and Grubb, 2000). Again, flattening has a minimal effect on performance if the vapor space is not collapsed or the heat pipe is not operating near a limitation. The heat pipe can conform to the system space restrictions and can transport the heat from the source to the fin stack or heat sink, where the heat is effectively dissipated (DeHoff and Grubb, 2000).

7.8 Heat Pipe Performance

The above sections provided an overview of heat pipes and their performance. More important, though, is the proper use of the heat pipe in a heat sink and the increased heat sink capabilities that are provided by the utilization of heat pipes. Most applications use a remote fin stack design, which consists of an aluminum evaporator block (heat input section), the heat pipe (heat transport section), and aluminum fins (heat dissipation section).

Over the last 10 years, a host of computationally inclined heat pipe investigators in the United States have been busy modeling heat pipe transient operation. The difficulty of transient heat pipe modeling can be immense, especially if a simulation of the frozen start-up problem is attempted. Important mechanisms related to transient heat pipe operation include the transition from free molecule to continuum flow in the vapor space, the migration of the melt front in capillary structures, mass transfer between the liquid and vapor regions, compressibility effects and shock formation in the vapor flow, and the possibility of externally imposed body forces on the working fluid in its liquid phase. Performance-limiting mechanisms during power transitions in recently proposed heat pipe systems include evaporator entrainment, freeze-out of the working fluid inventory in the condenser, evaporator capillary limits, and nucleate boiling departure in the evaporator.

A thermal resistance network, analogous to electrical circuits, is the quickest way to predict the overall performance of a parallel plate/heat pipe heat sink. The thermal resistance network is considered a good approach to determine design feasibility (DeHoff and Grubb, 2000). The heat pipe heat sink can be represented by a resistance network, as shown in Figure 7.12. Although this network neglects the interface between the device and heat sink, it can easily be added.

Each of the above resistances can be solved to calculate an associated temperature drop using the Fourier's law of conduction equation as follows:

$$\dot{Q} = kA\frac{\Delta T}{L} \tag{7.2}$$

and with the thermal resistance it can be rewritten as

$$\dot{Q} = \frac{\Delta T}{R} \tag{7.3}$$

Therefore, the temperature drop in the evaporator section (block) can be calculated by the conduction heat transfer for the evaporator section (R''_{eva}):

$$\Delta T_{\text{eva}} = \frac{\dot{Q}L_{\text{eva}}}{k_{\text{eva}}A_{\text{eva}}} \tag{7.4}$$

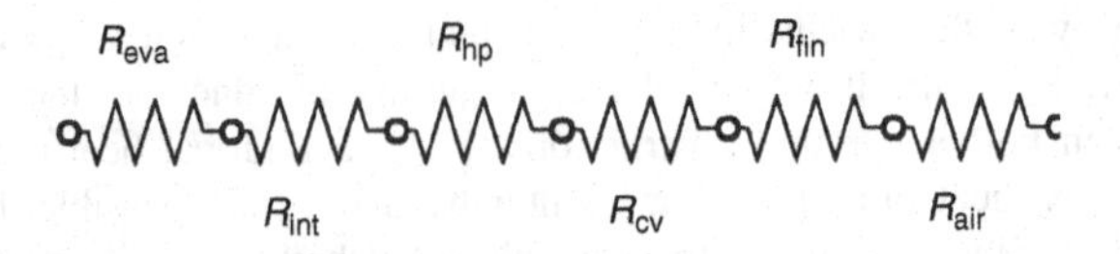

Figure 7.12 Thermal resistance network for a heat pipe (*Courtesy of Thermacore International, Inc.*).

The loss associated with the interface between the evaporator block and the heat pipe (R_{int}) can be calculated using the thermal resistance of the interface material, which is typically solder ($R''_{int} \approx 0.5\,°\text{C/W}\cdot\text{cm}^2$) or thermal epoxy ($R''_{int} \approx 1.0\,°\text{C/W}\cdot\text{cm}^2$), and the interface area.

$$\Delta T_{int} = \frac{\dot{Q}R''_{int}}{\pi D_{hp} L_{eva}} \tag{7.5}$$

The detailed analysis of heat pipe is rather complex. The total thermal resistance of a heat pipe is the sum of the resistances due to conduction through the evaporator section wall and wick, evaporation or boiling, axial vapor flow, condensation, and conduction losses back through the condenser section wick and wall. A rough guide for a copper/water heat pipe with a powder metal wick structure is to use $0.2\,°\text{C/W}\cdot\text{cm}^2$ for the thermal resistance at the evaporator and condenser (applied over the heat input/output areas) and $0.02\,°\text{C/W}\cdot\text{cm}^2$ for axial resistance (applied over the cross-sectional area of the vapor space) in the following equation:

$$\Delta T_{hp} = \frac{\dot{Q}R''_{eva}}{\pi D_{hp} L_{eva}} + \frac{\dot{Q}R''_{ax-hp}}{\pi D_{vs}^2/4} + \frac{\dot{Q}R''_{con}}{\pi D_{hp} L_{con}} \tag{7.6}$$

The resistance in transferring the heat from the fin to the air (R_{cv}) is calculated using the convection coefficient as follows:

$$\Delta T_{cv} = \frac{\dot{Q}}{h A_{fin}} \tag{7.7}$$

The conductive losses that are associated with the fin (R_{fin}) are governed by the fin efficiency which is defined as

$$\eta_{fin} = \frac{\tanh(m_{fin} L_{eff})}{m_{fin} L_{eff}} \tag{7.8}$$

where

$$m_{fin} = \sqrt{\frac{2h}{k_{fin} Z_{fin}}} \tag{7.9}$$

So, the temperature drop in the fin results in

$$\Delta T_{fin} = \Delta T_{cv}(1 - \eta_{fin}) \tag{7.10}$$

The temperature drop in the air by cooling can be written as

$$\Delta T_{air} = \frac{1}{2}\left(\frac{\dot{Q}}{\dot{m} c_p}\right) \tag{7.11}$$

The overall performance of the sink is the sum of the individual temperature drops as follows:

$$\Delta T_{total} = \Delta T_{eva} + \Delta T_{int} + \Delta T_{hp} + \Delta T_{cv} + \Delta T_{fin} + \Delta T_{air} \tag{7.12}$$

The total thermal resistance of the sink to the surroundings becomes

$$R_{\text{total}} = \frac{\Delta T_{\text{total}}}{\dot{Q}} \tag{7.13}$$

where

$$R_{\text{total}} = R_{\text{eva}} + R_{\text{int}} + R_{\text{hp}} + R_{\text{cv}} + R_{\text{fin}} + R_{\text{air}} \tag{7.14}$$

or with the flux thermal resistance:

$$R''_{\text{total}} = \frac{\Delta T_{\text{total}}}{\dot{q}} \tag{7.15}$$

where

$$R''_{\text{total}} = R''_{\text{eva}} + R''_{\text{int}} + R''_{\text{hp}} + R''_{\text{cv}} + R''_{\text{fin}} + R''_{\text{air}} \tag{7.16}$$

In summary, the above calculation should provide a reasonable estimate on the feasibility of a heat pipe heat sink.

7.8.1 Effective Heat Pipe Thermal Resistance

The other primary heat pipe design consideration is the effective heat pipe thermal resistance or overall heat pipe ΔT at a given design power. As the heat pipe is a two-phase heat-transfer device, a constant effective thermal resistance value cannot be assigned. The effective thermal resistance is not constant but a function of a large number of variables, such as heat pipe geometry, evaporator length, condenser length, wick structure, and working fluid.

The total thermal resistance of a heat pipe is the sum of the resistances due to conduction through the wall, conduction through the wick, evaporation or boiling, axial vapor flow, condensation, and conduction losses back through the condenser section wick and wall.

The evaporator and condenser resistances are based on the outer surface area of the heat pipe. The axial resistance is based on the cross-sectional area of the vapor space. This design guide is only useful for powers at or below the design power for the given heat pipe.

Example 7.2

Consider a 1.27 cm diameter copper/water heat pipe. It is 30.5 cm long with a 1 cm diameter vapor space. Assume that the heat pipe is dissipating 75 W with a 5 cm evaporator and a 5 cm condenser length (for details, see Garner, 1996). Determine the total temperature drop.

Solution

The evaporator heat flux equals the power divided by the heat input area:

$$\dot{q}_{\text{evap}} = \dot{q}_{\text{cond}} = \frac{\dot{Q}}{A_{\text{evap}}} = \frac{\dot{Q}}{\pi DL} = \frac{75\ \text{W}}{\pi(1.27\ \text{cm})(5\ \text{cm})} = 3.76\ \text{W/cm}^2$$

The axial heat flux equals the power divided by the cross-sectional area of the vapor space:

$$\dot{q}_{\text{axial}} = \frac{\dot{Q}}{A_{\text{axial}}} = \frac{\dot{Q}}{\pi D^2/4} = \frac{75\ \text{W}}{\pi(1\ \text{cm})^2/4} = 95.5\ \text{W/cm}^2$$

For a copper/water heat pipe with a powder metal wick structure, thermal resistance at the evaporator and condenser can be taken as $0.2\,°\mathrm{C/W\cdot cm^2}$ and the axial resistance for the vapor space can be taken as $0.02\,°\mathrm{C/W\cdot cm^2}$. Then, the total temperature drop may be determined based on Equation 7.15 as

$$\begin{aligned}\Delta T_{\mathrm{total}} &= \dot{q}_{\mathrm{evap}} R''_{\mathrm{evap}} + \dot{q}_{\mathrm{axial}} R''_{\mathrm{axial}} + \dot{q}_{\mathrm{cond}} R''_{\mathrm{cond}} \\ &= (3.76\ \mathrm{W/cm^2})(0.2\ °\mathrm{C/W\cdot cm^2}) + (95.5\ \mathrm{W/cm^2})(0.02\ °\mathrm{C/W\cdot cm^2}) \\ &\quad + (3.76\ \mathrm{W/cm^2})(0.2\ °\mathrm{C/W\cdot cm^2}) \\ &= \mathbf{3.4\,°C}\end{aligned}$$

7.9 Design and Manufacture of Heat Pipes

The design and manufacture of heat pipes is an extremely complex process, as shown in Figure 7.13, involving many different physical variables such as size, shape, weight, and volume; thermophysical properties such as working fluid, wicking structure, and case material properties; and other design aspects, such as thermal load, transport length, evaporator/condenser length, acceptable temperature drop, operating temperature range, gravitational environment, source–sink interfaces, fluid inventory, life/reliability, and safety (Peterson, 1994).

In addition to these specific areas, the design and manufacture of heat pipes is governed by three operational considerations: the effective operating temperature range, which is determined by the

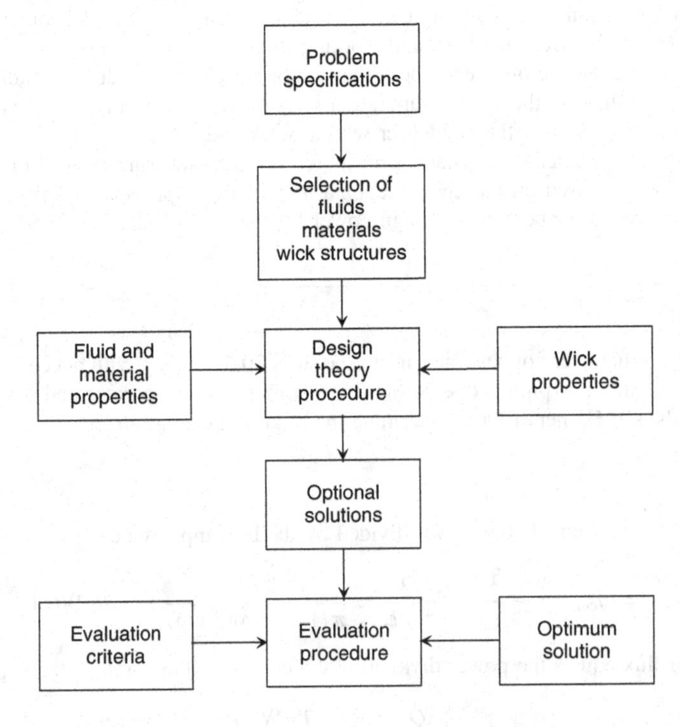

Figure 7.13 Heat pipe design flow chart (adapted from Peterson, 1994).

selection of the working fluid; the maximum power the heat pipe is capable of transporting, which is determined by the ultimate pumping capacity of the wick structure (for the capillary wicking limit); and the maximum evaporator heat flux, which is determined by the point at which nucleate boiling occurs.

Because, as illustrated in Figure 7.13, all three of these operational considerations must be included. The design process first requires that the design specifications for the problem under consideration be clearly identified. Once this has been accomplished, preliminary selection of the working fluid, wicking structure, and case materials can be performed. Finally, using an iterative process, various combinations of working fluids, evaporator and condenser sizes, and case/wick material combinations can be evaluated. While experience is extremely helpful, the new designer can, using the guidelines outlined above, develop an improved, if not optimal, design. Perhaps the most difficult part of the design process is determining how the various components utilized in heat pipe and thermosiphon construction affect the different design requirements. Table 7.2 presents a matrix which gives some indication of how each of the three primary components – the

Table 7.2 Heat pipe components and their effect on design requirements.

Design Requirements	Working Fluid	Wick Material	Case Material
Thermal performance			
• Transport capacity	SF	SF	WF
• Operating temperature range	SF	WF	WF
• Temperature drop	MF	WF	WF
Mechanical			
• Physical requirements (size, weight, etc.)	WF	WF	MF
• Wall thickness (internal pressure)	WF	NF	SF
• Sink-source interface	NF	NF	SF
• Dynamic/static loads	WF	SF	MF
Reliability and safety			
• Material compatibility	SF	SF	SF
• External corrosion	NF	NF	–
• Fabrication	MF	MF	MF
• Pressure containment/leakage	WF	MF	SF
• Toxicity	SF	WF	WF
Gravitational environment			
• >1 g	SF	SF	SF
• 1 g	MF	MF	WF
• <1 g	WF	MF	WF

SF: strong factor; MF: moderate factor; WF: weak factor; NF: negligible factor.
Source: Peterson (1994).

working fluid, wick material, and heat pipe case material – affect the various design requirements. As shown, no single component appears to be more important than any other, and very few design requirements are affected by only one of these three components.

There are many factors to consider when designing a heat pipe, including

- compatibility of materials,
- operating temperature range,
- diameter,
- power limitations,
- thermal resistances, and
- operating orientation.

However, the design issues are reduced to two major considerations by limiting the selection to copper/water heat pipes for cooling electronics. Considerations are the amount of power that the heat pipe is capable of carrying and its effective thermal resistance.

There are three properties of wicks that are important in heat pipe design (Faghri, 1995):

- **Minimum capillary radius.** This parameter should be small if a large capillary pressure difference is required, such as in terrestrial operation for a long heat pipe with the evaporator above the condenser or in cases where a high heat transport capability is needed.
- **Permeability.** Permeability is a measure of the wick resistance to axial liquid flow. This parameter should be large in order to have a small liquid pressure drop and therefore higher heat transport capability.
- **Effective thermal conductivity.** A large value for this parameter gives a small temperature drop across the wick, which is a favorable condition in heat pipe design.

A high thermal conductivity and permeability and a low minimum capillary radius are somewhat contradictory properties in most wick designs. For example, a homogeneous wick may have a small minimum capillary radius and a large effective thermal conductivity, but have a small permeability. Therefore, the designer must always make trade-offs between these competing factors to obtain an optimal wick design.

Some heat pipes material-related issues investigated during the past decade include liquid–solid wetting and surface phenomena, fluid container compatibility, wick development, and protection of heat pipes from the surrounding conditions.

The most important heat pipe design consideration is the amount of power the heat pipe is capable of transferring. Heat pipes can be designed to carry a few watts or several kilowatts, depending on the application. Heat pipes can transfer much higher powers for a given temperature gradient than even the best metallic conductors. If driven beyond its capacity, however, the effective thermal conductivity of the heat pipe will be significantly reduced. Therefore, it is important to assure that the heat pipe is designed to safely transport the required heat load.

The maximum heat transport capability of the heat pipe is governed by several limiting factors which must be addressed when designing a heat pipe. There are five primary heat pipe heat transport limitations. These heat transport limits, which are a function of the heat pipe operating temperature, include viscous, sonic, capillary pumping, entrainment or flooding, and boiling. Each heat transport limitation is summarized in Table 7.3.

7.9.1 The Thermal Conductivity of a Heat Pipe

Heat pipes do not have a set thermal conductivity like solid materials because of the two-phase heat transfer. Instead, the effective thermal conductivity improves with length. For example, a

Table 7.3 Heat pipe heat transport limitations.

Heat Transport Limit	Description	Cause	Potential Solution
Viscous	Viscous forces prevent vapor flow in the heat pipe	Heat pipe operating below recommended operating temperature	Increase heat pipe operating temperature or find alternative working fluid
Sonic	Vapor flow reaches sonic velocity when exiting heat pipe evaporator, resulting in a constant heat pipe transport power and large temperature gradients	Power–temperature combination, too much power at low operating temperature	This is typically only a problem at start-up. The heat pipe will carry a set power and the large ΔT will self-correct as the heat pipe warms up
Entrainment/ Flooding	High-velocity vapor flow prevents condensate from returning to evaporator	Heat pipe operating above designed power input or at too low an operating temperature	Increase vapor space diameter or operating temperature
Capillary	Sum of gravitational, liquid, and vapor flow pressure drops exceed the capillary pumping head of the heat pipe wick structure	Heat pipe input power exceeds the design heat transport capacity of the heat pipe	Modify heat pipe wick structure design or reduce power input
Boiling	Film boiling in heat pipe evaporator typically initiates at 5–10 W/cm^2 for screen wicks and 20–30 W/cm^2 for powder metal wicks	High radial heat flux causes film boiling resulting in heat pipe dry out and large thermal resistances	Use a wick with a higher heat flux capacity or spread out the heat load

Source: Garner (1996) (Reproduced by permission of Flomerics, Inc.).

10-cm-long heat pipe carrying 100 W will have close to the same thermal gradient as a 12-in. (30 cm)-long pipe carrying the same power. Thus, the 12-in. pipe would have a higher effective thermal conductivity. Unlike solid materials, a heat pipe's effective thermal conductivity will also change with the amount of power being transferred and with the evaporator and condenser sizes. For a well-designed heat pipe, effective thermal conductivity can range from 10 to 10,000 times the effective thermal conductivity of copper depending on the length of the heat pipe.

7.9.2 Common Heat Pipe Diameters and Lengths

Some heat pipes are intentionally made easy to bend. Many heat pipes are made of annealed copper. These units can be readily bent and formed to fit customer applications. The most common heat pipe diameters in current production are 2, 3, 4, and 6 mm and 1/4 (6.4 mm) and 5/8 in. (17.9 mm) in diameter. Lengths vary with the diameter from about 12 in. (30 mm) for 2-mm diameter units to about 4 ft (1.2 m) for 5/8-in. (17.9 mm) diameter units.

Heat pipes are manufactured in a multitude of sizes and shapes. Unusual application geometry can be easily accommodated by the heat pipe's versatility to be shaped as a heat transport device. If some range of motion is required, heat pipes can even be made of flexible material. Two of the most common heat pipes are constant temperature (the heat pipe maintains a constant temperature or temperature range) and diode (the heat pipe allows heat transfer in only one direction).

7.10 Heat-Transfer Limitations

There are various parameters that put limitations and constraints on the steady and transient operation of heat pipes. In other words, the rate of heat transport through a heat pipe is subject to a number of operating limits. The following physical phenomena might limit heat transport in heat pipes (Faghri, 1995):

- **Capillary limit.** For a given capillary wick structure and working fluid combination, the pumping ability of the capillary structure to provide the circulation for a given working medium is limited. This limit is usually called the *capillary* or *hydrodynamic limit*.
- **Sonic limit.** For some heat pipes, especially those operating with liquid metal working fluids, the vapor velocity may reach sonic or supersonic values during the start-up or steady state operation. This choked working condition is called the *sonic limit*.
- **Boiling limit.** If the radial heat flux or the heat pipe wall temperature becomes excessively high, boiling of the working fluid in the wick may severely affect the circulation of the working fluid and lead to the boiling limit.
- **Entrainment limit.** When the vapor velocity in the heat pipe is sufficiently high, the shear force existing at the liquid–vapor interface may tear the liquid from the wick surface and entrain it into the vapor flow stream. This phenomenon reduces the condensate return to the evaporator and limits the heat transport capability.
- **Frozen start-up limit.** During the start-up process from the frozen state, vapor from the evaporation zone may be refrozen in the adiabatic or condensation zones. This may deplete the working fluid from the evaporation zone and cause dryout of the evaporator.
- **Continuum vapor limit.** For small heat pipes, such as micro heat pipes, and for heat pipes with very low operating temperatures, the vapor flow in the heat pipe may be in the free molecular or rarefied condition. The heat transport capability under this condition is limited because the continuum vapor state has not been reached.
- **Vapor pressure limit (viscous limit).** When the viscous forces dominate the vapor flow, as for a liquid metal heat pipe, the vapor pressure at the condenser end may reduce to zero. Under this condition the heat transport of the heat pipe may be limited. A heat pipe operating at temperatures below its normal operating range can encounter this limit, which is also know as the vapor pressure limit.
- **Condenser heat-transfer limit.** The maximum heat rate capable of being transported by a heat pipe may be limited by the cooling ability of the condenser. The presence of noncondensible gases can reduce the effectiveness of the condenser.

The heat-transfer limitation can be any of the above limitations, depending on the size and shape of the pipe, working fluid, wick structure, and operating temperature. The lowest limit among the eight constraints defines the maximum heat transport limitation of a heat pipe at a given temperature.

Although heat pipes are very efficient heat-transfer devices, they are subject to a number of heat-transfer limitations. These limitations determine the maximum heat-transfer rate a particular heat pipe can achieve under certain working conditions. The type of limitation that restricts the operation of the heat pipe is determined by the limitation which has the lowest value at a specific heat pipe working temperature.

7.11 Heat Pipes in HVAC

The common practice in air conditioning was to design equipment basically to cool the air. However, the human body actually responds to additional factors other than temperature. To provide truly comfortable conditions, we must control humidity. Overlooking this aspect, in many instances we have produced environments which are cold, clammy, stuffy, and even unhealthy. The heat pipe in this regard appears to be an effective solution to overcome such problems.

In HVAC applications, a heat pipe is a simple device that can transfer heat from one point to another without having to use an external power supply. It is a sealed tube that has been partially filled with a vaporizable working fluid (such as alcohol or a freon, e.g., R-22) from which all air has been evacuated. The sealed refrigerant, which will boil under low-grade heat, absorbs heat from the warm return air such as in an air-conditioning system and vaporizes inside the tube. The vapor then travels to the other end of the heat pipe (the high end), which is placed in the stream of cold air that is produced by the air conditioner. In most cases, on the inside wall of the pipe, a wicking material transports the liquified fluid by capillary action. The heat that was absorbed from the warm air at the low end is now transferred from the refrigerant's vapor through the pipe's wall into the cool supply air. This loss of heat causes the vapor inside the tube to condense back into a fluid. The condensed refrigerant then travels by gravity to the low end of the heat pipe where it begins the cycle all over again.

Serious work on heat pipes began in the 1960s with applications of heat pipes in the space program. The largest amount of research was done in the United States by NASA, with heat pipes earning a place in the domain of aerospace applications. Only recently have heat pipes been applied to HVAC (Dinh, 1996). The application of heat pipes for heat recovery in cold climates is widely recognized. In northern Europe and Canada, heat recovery with heat pipes has proven itself to be very reliable and economical. With advancement of heat pipes with a low air pressure drop, made possible by loop configurations, heat recovery applications can now be extended to milder climates and still be economical. A new possibility is cooling recovery in summertime, which is now economical enough to be considered. The application of heat pipes to increase the dehumidification capacity of a conventional air conditioner is one of the most attractive applications. By using dehumidifier heat pipes, one can decrease the relative humidity in the conditioned space (typically by 10%) resulting in noticeably improved indoor air quality and reduced power demand. Heat pipes also promise to improve indoor air quality greatly and at the same time help conserve energy.

Heat pipes provide a large number of benefits in HVAC applications, including

- improved comfort level,
- reduced moisture and better humidity control,
- higher productivity,
- improved air quality,
- easy retrofitting of existing systems,
- no moving parts,
- no additional energy for operation,
- dramatic reduction in HVAC loads and hence energy and money savings,
- low payback time,
- less maintenance costs, and
- less installation costs.

7.11.1 Dehumidifier Heat Pipes

As mentioned earlier, heat pipes can dramatically improve the moisture removal capabilities of many air-conditioning systems. Air can be precooled by simply transferring heat from the warm incoming air to the cool supply air (Figure 7.14). This "bypassing" can be accomplished by placing the low end of a heat pipe in the return air and the high end in the supply air. Heat is removed from the warm upstream air and rerouted to the cool downstream air. This heat, in effect, bypasses the evaporator, although the air that contained the heat does indeed pass through the air-conditioning coil. The total amount of cooling required is slightly reduced and some of the air conditioner's

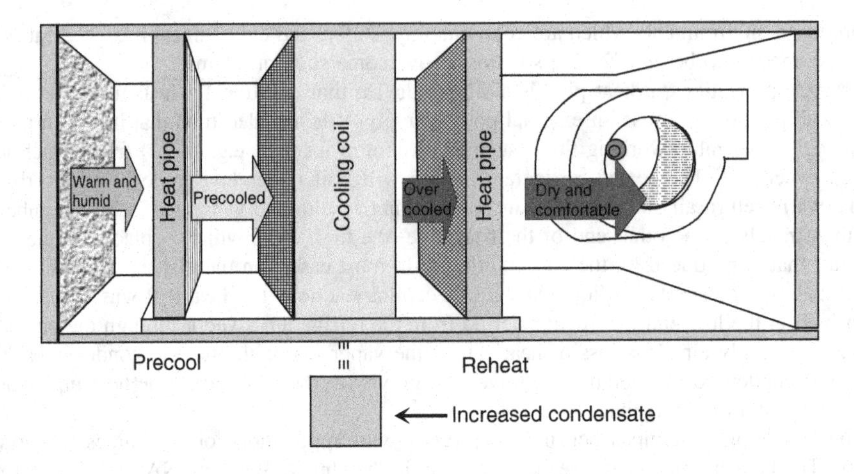

Figure 7.14 Heat pipe system for dehumidification (*Courtesy of Heat Pipe Technology, Inc.*).

sensible capacity is therefore exchanged for additional latent capacity. Now the unit can cope with high-moisture air more efficiently. To accomplish a heat transfer around a cooling coil through utilization of heat pipe technology, different configurations may be used. One method is to arrange several heat pipes in parallel banks with the evaporator coil separating the pipes' evaporator ends and condenser ends. Fins (much like those found in air conditioner coils) may then be attached to the outside surface of the heat pipes to improve the heat transfer between the tubes and the air.

Nowadays, libraries, restaurants, storage facilities, and supermarkets benefit the most from heat pipe technology that needs moisture-controlled air to preserve goods and products kept inside, to prevent the increased wear and tear associated with high humidity, or to increase occupant comfort. Any air-conditioning system that uses reheat, desiccants, or mechanical dehumidification is a good candidate for heat pipe assistance.

When reheat is used, the energy savings that can be accomplished through heat pipe dehumidification assistance can be substantial. While the percentage of energy savings may vary greatly from customer to customer because of the number of variables, one of the best examples reported so far involves a chain restaurant that was retrofitted with heat pipes. A restaurant was selected for the test because restaurants have traditionally been victimized by extremely humid interior conditions. High humidity causes interior fixtures and building materials to deteriorate at an accelerated rate because of water condensation. High humidity also results in increased energy and equipment repair/replacement costs. In addition to the geographic location, the elements that contribute to high humidity in restaurants include customer loads, cooking loads, and code requirements concerning the rate of movement of outside air to the building's interior. Analysis of data from the test site indicated that outside air requirements can be a key causative factor of extremely high interior humidity.

7.11.1.1 Working Principle

The working principle of the heat pipe for dehumidification in an air-conditioning system is quite simple. The heat pipe performs heat transfer between the warm return air from the room to the cold supply air from the air-conditioning coil. In the operation, heat pipes may be described as having two sections: precool and reheat (Figure 7.14). The first section is located in the incoming air stream. When warm air passes over the heat pipes, the refrigerant vaporizes, carrying heat to the second section of heat pipes, placed downstream. Because some heat has been removed from the air before encountering the evaporator coil, the incoming air stream section is called the *precool heat pipe*. Air passing through the evaporator coil is brought down to a lower temperature, resulting in greater condensate removal. The "overcooled" air is then reheated to a comfortable temperature by the reheat heat pipe section, using the heat transferred from the precool heat pipe. This entire process of precool and reheat is accomplished with no additional energy use. The result is an air-conditioning system with the ability to remove 50–100% more moisture than regular systems.

With advantages such as high effectiveness, low air drag, moderate cost, and no cross contamination, heat pipes offer a perfect solution to humidity-related internal air quality problems. When properly applied, heat pipes can be very economical, to reduce both initial cost and operating cost. Even in a case of retrofitting existing systems with custom built-to-fit heat pipes, the savings in avoiding reheat can pay for the heat pipe installation within a few years. From the mechanical integrity and maintenance standpoint heat pipes, having no moving parts, are expected to outlast other components of the HVAC system. As with any coil, periodic cleaning should keep heat pipes working at peak efficiency. Corrosive atmospheres can be handled by coating heat pipes with corrosion-prevention plastic coatings (Dinh, 1996).

7.11.1.2 Indoor Dehumidifier Heat Pipes

Indoor air quality is improved by using heat pipes in the air-conditioning system (Figure 7.15), creating a situation of enhanced comfort, greater health, elimination of mold and mildew, and reduction of building deterioration.

Energy saving through the use of heat pipes is achieved in the following ways:

- by the elimination of reheat and the additional air-conditioning load imposed by the reheat and
- by setting the thermostat a few degrees higher to achieve the same comfort level because of the lower relative humidity.

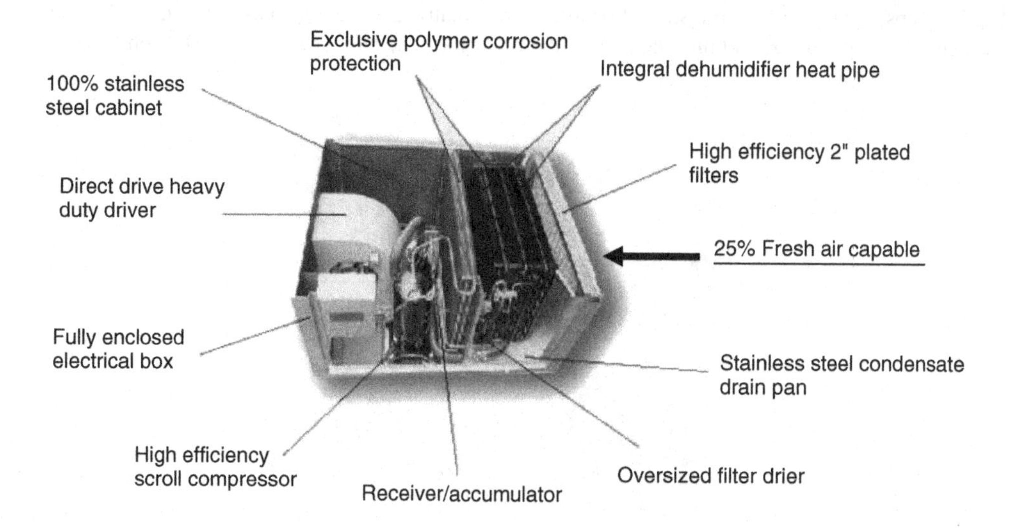

Figure 7.15 An indoor dehumidifier with heat pipes (*Courtesy of Heat Pipe Technology Inc.*).

7.11.2 Energy Recovery Heat Pipes

- Heat pipes offer a highly efficient way to recover energy from a building's exhaust air and reuse that same energy to precool or preheat incoming fresh air without having to use an external power supply. Heat pipes are most commonly installed to control humidity in hot, humid climates and as air preheaters in steam boilers, air dryers, waste heat recovery from exhaust steam, and waste heat recovery from conditioned air.
- Heat pipes are most cost-effective when the air streams are adjacent. Within the heat pipe, a charge of refrigeration continuously evaporates, condenses, and migrates by capillary action through the porous wick. Since the only thing that moves is the refrigerant, and it is self-contained, low maintenance and long life are obtained. These self-contained, no-moving-part devices have many applications. The example shown here is sensible heat transfer between adjacent fresh-air-intake and stale exhaust air streams. When the two air streams are farther apart, a set of individual coils and circulating heat-transfer fluid connecting them provides simple heat transfer, with no restrictions on exhaust and intake location. Energy transfer wheels go even further than run-around coils and heat pipes, in that they can control both temperature and humidity. In winter, they recover both sensible and latent heat from exhaust air; in summer, they can both cool and dehumidify the incoming fresh air. Seals and laminar flow of air through the wheels reduce the mixing of exhaust air and incoming air. A further precautionary step in the process purges each sector of the wheel briefly, using fresh air to blow away any unpleasant residual effects of the exhaust air on the wheel surfaces. These systems are much more maintenance prone and may not be as cost-effective as the heat pipe.

In conjunction with the above introductory information, energy (heat) recovery heat pipes (e.g., Figures 7.16 and 7.17) provide economical and reliable recovery of both heat and cooling. Here, the heat pipe is assembled into arrays (bundles of tubes). The bundle of finned heat pipes extends through the wall separating the inlet and exhaust ducts in a pattern that resembles conventional finned coil heat exchangers. Each of the separate pipes, however, is a sealed element consisting of an annular wick on the inside of the full length of the tube, in which an appropriate heat-transfer fluid is absorbed. The heat transferred from the hot exhaust gases evaporates the fluid in the wick, causing the vapor to expand into the core of the heat pipe. The latent heat of vaporization is carried

Figure 7.16 An energy recovery heat pipe (*Courtesy of Heat Pipe Technology, Inc.*).

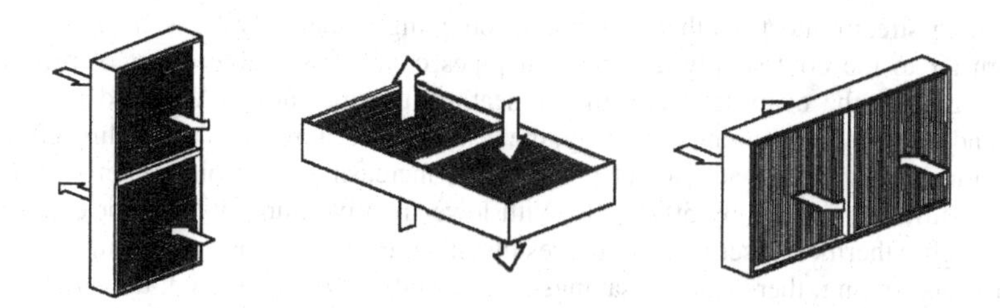

Figure 7.17 Three configurations of energy recovery heat pipes: (a) Over and under horizontal air streams with heat pipes in a vertical plane. (b) Side by side vertical air streams with heat pipe in a horizontal plane. (c) Side by side horizontal air streams with heat pipe in a vertical plane *(Courtesy of Heat Pipe Technology, Inc.)*.

with the vapor to the cold end of the tube, where it is removed by transference to the cold gas as the vapor condenses. The condensate is then carried back in the wick to the hot end of the tube by capillary action and by gravity (if the tube is tilted from the horizontal), where it is recycled.

The heat pipe is compact and efficient because (i) the finned-tube bundle is inherently a good configuration for convective heat transfer in both ducts and (ii) the evaporative–condensing cycle within the heat tubes is a highly efficient method of transferring heat internally. It is also free of cross-contamination.

The heat pipe sits level in a vertical plane and exchanges heat/cooling in both directions. Energy recovery heat pipes with over- and under-horizontal airflows provide an excellent one-directional energy transfer with the warmer air stream at the bottom. Some of their benefits can be underlined as follows:

- made of high quality copper tube for reliability and longevity,
- highest heat-transfer effectiveness,
- bidirectional heat transfer (except for vertical module),
- no tilting necessary,
- no mechanical seasonal changeover necessary,
- high thermal conductivity and great heat transfer,
- better thermal response than any metal,
- equality of thermal distribution,
- small size and lightweight,
- flexible design,
- cost effectiveness,
- reliability,
- no electrical or mechanical power required; maintenance free,
- both flat and tube series available,
- flexible shape and length,
- easy combination with properly designed spreader,
- less cost,
- no need for reheat,
- no mechanical or electrical input required,
- maintenance free,
- lower operating and maintenance costs,
- longer operational time, and
- environmentally benign.

In an air-conditioning system, the colder the air that passes over the cooling coil (evaporator), more is the moisture that is condensed out. The heat pipe is designed to have one section in the warm incoming stream and the other in the cold outgoing stream. By transferring heat from the warm return air to the cold supply air, the heat pipes create the double effect of precooling the air before it goes to the evaporator and then reheating it immediately. Activated by temperature difference and therefore consuming no energy, the heat pipe, because of its precooling effect, allows the evaporator coil to operate at a lower temperature, increasing the moisture removal capability of the air-conditioning system by 50–100%. With lower relative humidity, indoor comfort can be achieved at higher thermostat settings, which results in net energy savings. Generally, for each 1 °C rise in thermostat setting, there is a 7% savings in electricity cost. In addition, the precooling effect of the heat pipe allows the use of a smaller compressor.

Design flexibility is achieved because of the heat pipe's ability to transfer heat efficiently. While traditional heat sinks must be located on the heat source, heat pipes transfer heat away to areas where dissipation is more convenient or airflow is greater. Heat pipes can be bent and formed into a variety of configurations while maintaining their heat-transfer properties.

Heat pipes can be used efficiently by libraries, restaurants, cold storage facilities, supermarkets, applications requiring controlled/reduced humidity, and applications where reheat or desiccants are necessary.

Hill and Lau (1993) studied supermarket air-conditioning systems equipped with heat pipe heat exchangers. Operation in four different climates was considered. The heat pipe heat exchangers were used to save refrigeration energy by reducing the humidity of the refrigerated spaces. Rosenfeld and North (1995) discussed the use of heat pipes in "porous media heat exchangers" for the cooling of high-heat-load optical components.

7.12 Concluding Remarks

This chapter deals with heat pipes particularly for thermal applications, and discusses related matters from the structure, features, technical aspects, operational aspects, technical details, heat pipe fluids, design and manufacturing aspects, heat-transfer limits, energy savings, types and applications points of view. Some examples are given to highlight the importance of the topic and show the benefits of the technology for some specific thermal applications and for refrigeration at large. It is clearly indicated that micro heat pipes are now really essential equipment for electronics cooling applications.

Nomenclature

A	cross-sectional area, m^2; surface area, m^2
A_{eva}	heat-transfer area of evaporator section, m^2
A_{fin}	total surface area of fins, m^2
c_p	specific heat, J/kg · °C
C	volumetric heat capacity, J/m^3 · °C
D	diameter, m
h	convection heat-transfer coefficient, W/m^2 · °C
k	thermal conductivity, W/m · °C
L	length, m
L_{con}	length of condenser section, m
L_{eff}	effective fin length, m
L_{eva}	length of evaporator section, m
m_{fin}	fin factor for uniform cross-sectional area
$\dot{m}$	mass flow rate of cooling air, kg/s
M	moisture content, kg/kg

q	specific heat transfer, kJ/kg
$\dot{q}$	heat flux, W/m^2
$\dot{Q}$	heat-transfer rate, kW
R	thermal resistance, °C/W
R_{air}	thermal resistance of cooling air flow
R_{axial}	thermal resistance along heat pipe
R_{con}	thermal resistance of the heat pipe condenser
R_{cv}	thermal resistance due to convection
R_{eva}	thermal resistance of the heat pipe evaporator block
R_{fin}	thermal resistance due to fin efficiency
R_{hp}	thermal resistance of heat pipe
R_{int}	thermal resistance of evaporator block heat pipe interface
R_{total}	thermal resistance of heat sink to ambient
R''	flux thermal resistance, °C/W · m^2
T	temperature, °C
ΔT_{air}	temperature change in cooling air flow, °C
ΔT_{eva}	temperature change in evaporator section (block), °C
ΔT_{cv}	temperature change due to convection, °C
ΔT_{fin}	temperature change due to fin efficiency, °C
ΔT_{hp}	temperature change in heat pipe, °C
ΔT_{int}	temperature change in the interface of evaporator block and heat pipe, °C
Z_{evap}	thickness of the evaporator block, m
Z_{fin}	thickness of the fin, m
η_{fin}	fin efficiency

Study Problems

Heat Pipes

7.1 What is a heat pipe? Why is it also referred as a superconductor?

7.2 Describe the geometry and operation of a heat pipe.

7.3 What is called as the heart of a heat pipe and why?

7.4 Provide a list of heat pipe liquids used.

7.5 Provide a list of heat pipe wall and wick materials used.

7.6 What are the basic characteristics of a heat pipe? What is the most attractive feature of heat pipes.

7.7 What are the three main objectives we expect from heat pipes?

7.8 List some of the typical heat pipe applications.

Types of Heat Pipes

7.9 Give a list of the types of heat pipes.

7.10 What are the characteristics of cryogenic heat pipes.

Heat Pipe Components

7.11 What characteristics should the container material of a heat pipe have?

7.12 What factors should be considered in the selection of the container material of a heat pipe?

7.13 What characteristics should the working fluid of a heat pipe have?

7.14 Why is a high value of surface tension desirable for the working fluid of a heat pipe?

7.15 Why is a high value of latent heat of vaporization desirable for the working fluid of a heat pipe?

7.16 What are the most commonly used working fluids in heat pipes? What are the useful temperature ranges for these fluids? Which working fluid is more suitable for subfreezing temperature application?

7.17 What are the two most important properties of a wick and why are they important?

7.18 What are the most common types of wicks that are used in heat pipes?

7.19 What are the two functions of the wicking structure in heat pipe operation?

7.20 What are the three broad categories of wicking structures? Describe each briefly.

7.21 What are the three improvements accomplished in advanced wicking structure designs?

Heat Pipe Performance

7.22 What does the total thermal resistance of a heat pipe consist of?

7.23 Consider a heat pipe with a heat rejection of 100 W. The effective thermal resistance of the heat pipe is 0.4 °C/W. What is the temperature difference involved in the dissipation of this heat load?

7.24 Consider a heat pipe with a heat rejection of 35 W with a temperature change of 28 °C. What is the effective thermal resistance of the heat pipe?

7.25 A heat pipe is used as a heat sink for cooling of an electronic component. The temperature difference across the heat sink is 30 °C and the effective thermal resistance of the heat sink is 1.8 °C/W. What is the rate of heat rejected?

7.26 Consider a copper/water heat pipe. The evaporator and condenser heat flux is 4.9 W/cm^2 and the axial heat flux of the vapor space is 124 W/cm^2. What is the total temperature drop?

7.27 Consider a 0.8 cm diameter copper/water heat pipe. It is 22 cm long with a 0.55 cm diameter vapor space. The heat pipe is dissipating 58 W with a 6 cm evaporator and a 7 cm condenser length. Determine the total temperature drop.

Design and Manufacture of Heat Pipes

7.28 What are the three operational considerations that govern the design and manufacture of heat pipes?

7.29 List some of the factors to consider when designing a heat pipe.

7.30 What are the three properties of wicks that are important in heat pipe design?

7.31 What are the common heat pipe diameters and lengths?

Heat-Transfer Limitations

7.32 What are the physical phenomena that might limit heat transport in heat pipes?

Heat Pipes in HVAC

7.33 What are the benefits of heat pipes in HVAC applications?

7.34 Describe the working principle of the heat pipe used for dehumidification in an air-conditioning system with the help of a schematic of the system.

7.35 Explain how an indoor dehumidifier heat pump provides energy savings.

7.36 List some of the benefits of energy recovery heat pipes.

References

Cotter, T.P. (1984) *Principles and Prospects of Micro Heat Pipes*. Proceedings of the 5th International Heat Pipe Conference. Tsukuba, Japan, pp. 328–337.

DeHoff, R. and Grubb, K. (2000) *Heat Pipe Application Guidelines*, Thermacore Inc., Lancaster, PA.

Dincer, I. (1997) *Heat Transfer in Food Cooling Applications*, Taylor & Francis, Washington, DC.

Dincer, I. (2003) *Refrigeration Systems and Applications*, Wiley, 1st ed, London.

Dinh, K. (1996) Heat pipes in HVAC, in *Heat Pipe Technology* (eds J. Andrews *et al.*), Pergamon, London, pp. 357–363.

Faghri, A. (1995) *Heat Pipe Science and Technology*, Taylor & Francis, Washington, DC.

Faghri, A. (1996) Heat pipe simulation: from promise to reality, in *Heat Pipe Technology* (eds J. Andrews *et al.*), Pergamon, London, pp. 1–21.

Garner, S.D. (1996) Heat pipes for electronics cooling applications. *Electronics Cooling*, **September**, 1–9.

Gaugler, R.S. (1944) *Heat Transfer Devices*. US Patent 2, 350,348.

Hill, J.M. and Lau, A.S. (1993) Performance of supermarket air-conditioning systems equipped with heat pipe heat exchangers. *ASHRAE Transactions*, **99** (1), 1315–1319.

Narayanan, S. (2001) *What Is a Heat Pipe? The Chemical Engineering Resource Page*, http://www.cheresources.com/htpipes.shtml Last Accessed 11 September 2009.

Peterson, G.P. (1994) *An Introduction to Heat Pipes*, John Wiley & Sons, Ltd., New York.

Rosenfeld, J.H. and North, M.T. (1995) Porous media heat exchangers for cooling of high-power optical components. *Optical Engineering*, **34** (2), 335–342.

Xie, H., Aghazadeh, M. and Toth, J. (2001) *The Use of Heat Pipes in the Cooling of Portables with High Power Packages – A Case Study with the Pentium. Processor-Based Notebooks and Sub-notebooks*, Intel Corporation, Chandler, AZ and Thermacore International Inc., Lancaster, PA.

7.32 [illegible] of pipes [illegible] heat transport [illegible].

7.33 [illegible] heat pipes [illegible]

7.34 [illegible] air conditioning [illegible] of the system.

7.35 [illegible] pump [illegible] energy saving.

7.36 [illegible]

References

[illegible] *Principles and Practice* [illegible] Conference, [illegible].

[illegible] (2000) *Heat Pipe* [illegible].

[illegible] (1990) [illegible] *Applications*, [illegible], Washington, DC.

[illegible] London.

[illegible] Heat pipes [illegible] London.

[illegible]

[illegible] (2000) [illegible]

[illegible]

[illegible] Committee [illegible]

[illegible] *Heat Pipe* [illegible]

Hill, [illegible] Performance of [illegible] system [illegible] equipped with heat pipe [illegible] ASHRAE [illegible].

[illegible] *Heat Pipes* [illegible]

[illegible] Wiley & Sons [illegible]

[illegible] (1995) [illegible] heat exchange [illegible] high power [illegible].

[illegible] Alamdari [illegible] *Heat Pipes* [illegible] Conference [illegible].

Appendix A

Conversion Factors

Refrigeration Systems and Applications İbrahim Dinçer and Mehmet Kanoğlu

Table A.1 Conversion factors for commonly used quantities.

Quantity	SI to English	English to SI
Area	1 m^2 = 10.764 ft^2 = 1550.0 in.^{2}	1 ft^2 = 0.00929 m^2 1 in.^{2} = 6.452 × 10^{-4} m^2
Density	1 kg/m^3 = 0.06243 lb$_m$/ft^3	1 lb$_m$/ft^3 = 16.018 kg/m^3 1 slug/ft^3 = 515.379 kg/m^3
Energy	1 J = 9.4787 × 10^{-4} Btu	1 Btu = 1055.056 J 1 cal = 4.1868 J 1 lb$_f$·ft = 1.3558 J 1 hp·h = 2.685 × 10^6 J
Energy per unit mass	1 J/kg = 4.2995 × 10^{-4} Btu/lb$_m$	1 Btu/lb$_m$ = 2326 J/kg
Force	1 N = 0.22481 lb$_f$	1 lb$_f$ = 4.448 N 1 pdl = 0.1382 N
Gravitation	g = 9.80665 m/s^2	g = 32.17405 ft/s^2
Heat flux	1 W/m^2 = 0.3171 Btu/h·ft^2	1 Btu/h·ft^2 = 3.1525 W/m^2 1 kcal/h·m^2 = 1.163 W/m^2 1 cal/s·cm^2 = 41870.0 W/m^2
Heat generation (volume)	1 W/m^3 = 0.09665 Btu/h·ft^3	1 Btu/h·ft^3 = 10.343 W/m^3
Heat transfer coefficient	1 W/m^2·K = 0.1761 Btu/h·ft^2·°F	1 Btu/h·ft^2·°F = 5.678 W/m^2·K 1 kcal/h·m^2·°C = 1.163 W/m^2·K 1 cal/s·m^2·°C = 41870.0W/m^2·K
Heat transfer rate	1 W = 3.4123 Btu/h	1 Btu/h = 0.2931 W
Length	1 m = 3.2808 ft = 39.370 in 1 km = 0.621371 mi	1 ft = 0.3048 m 1 in. = 2.54 cm = 0.0254 m 1 mi = 1.609344 km 1 yd = 0.9144 m
Mass	1 kg = 2.2046 lb$_m$ 1 ton (metric) = 1000.0 kg 1 grain = 6.47989 × 10^{-5} kg	1 lb$_m$ = 0.4536 kg 1 slug = 14.594 kg
Mass flow rate	1 kg/s = 7936.6 lb$_m$/h = 2.2046 lb$_m$/s	1 lb$_m$/h = 0.000126 kg/s 1 lb$_m$/s = 0.4536 kg/s
Power	1 W = 1 J/s = 3.4123 Btu/h = 0.737562 lb$_f$·ft/s 1 hp (metric) = 0.735499 kW 1 ton of refrig. = 3.51685 kW	1 Btu/h = 0.2931 W 1 Btu/s = 1055.1 W 1 lb$_f$·ft/s = 1.3558 W 1 hp(UK) = 745.7 W
Pressure and stress (Pa = N/m^2)	1 Pa = 0.020886 lb$_f$/ft^2 = 1.4504 × 10^{-4} lb$_f$/in.^{2} = 4.015 × 10^{-3} in water = 2.953 × 10^{-4} in Hg	1 lb$_f$/ft^2 = 47.88 Pa 1 lb$_f$/in^2 = 1 psi = 6894.8 Pa 1 stand. atm. = 1.0133 × 10^5 Pa 1 bar = 1 × 10^5 Pa
Specific heat	1 J/kg·K = 2.3886 × 10^{-4} Btu/lb$_m$·°F	1 Btu/lb$_m$·°F = 4187.0 J/kg·K

Table A.1 (*continued*)

Quantity	SI to English	English to SI
Surface tension	1 N/m = 0.06852 lb_f/ft	1 lb_f/ft = 14.594 N/m 1 dyn/cm = 1 × 10^{-3} N/m
Temperature	T(K) = T(°C) + 273.15 = T(°R)/1.8 = [T(°F) + 459.67]/1.8 T(°C) = [T(°F) – 32.0]/1.8	T(°R) = 1.8T(K) = T(°F) + 459.67 = 1.8T(°C) + 32.0 = 1.8[T(K) – 273.15] + 32
Temperature difference	1 K = 1 °C = 1.8 °R = 1.8 °F	1 °R = 1 °F = 1 K/1.8 = 1 °C/1.8
Thermal conductivity	1 W/m·K = 0.57782 Btu/h·ft·°F	1 Btu/h·ft·°F = 1.731 W/m·K 1 kcal/h·m·°C = 1.163 W/m·K 1 cal/s·cm·°C = 418.7 W/m·K
Thermal diffusivity	1 m^2/s = 10.7639 ft^2/s	1 ft^2/s = 0.0929 m^2/s 1 ft^2/h = 2.581 × 10^{-5} m^2/s
Thermal resistance	1 K/W = 0.52750 °F·h/Btu	1 °F·h/Btu = 1.8958 K/W
Velocity	1 m/s = 3.2808 ft/s 1 km/s = 0.62137 mi/h	1 ft/s = 0.3048 m/s 1 ft/min = 5.08 × 10^{-3} m/s
Viscosity (dynamic) (kg/m·s = N·s/m^2)	1 kg/m·s = 0.672 lb_m/ft·s = 2419.1 lb_m/fh·h	1 lb_m/ft·s = 1.4881 kg/m·s 1 lb_m/ft·h = 4.133 × 10^{-4} kg/m·s 1 centipoise (cP) = 10^{-2} poise = 1 × 10^{-3} kg/m·s
Viscosity (kinematic)	1 m^2/s = 10.7639 ft^2/s = 1 × 10^4 stokes	1 ft^2/s = 0.0929 m^2/s 1 ft^2/h = 2.581 × 10^{-5} m^2/s 1 stoke = 1 cm^2/s
Volume	1 m^3 = 35.3134 ft^3 1 L = 1 dm^3 = 0.001 m^3	1 ft^3 = 0.02832 m^3 1 in^3 = 1.6387 × 10^{-5} m^3 1 gal (US) = 0.003785 m^3 1 gal (UK) = 0.004546 m^3
Volumetric flow rate	1 m^3/s = 35.3134 ft^3/s = 1.2713 × 10^5 ft^3/h	1 ft^3/s = 2.8317 × 10^{-2} m^3/s 1 ft^3/min = 4.72 × 10^{-4} m^3/s 1 ft^3/h = 7.8658 × 10^{-6} m^3/s 1 gal (US)/min = 6.309 × 10^{-5} m^3/s

Appendix B

Thermophysical Properties

Refrigeration Systems and Applications İbrahim Dinçer and Mehmet Kanoğlu

Table B.1 Ideal-gas specific heats of various common gases at 300 K.

Gas	Formula	Gas constant, R kJ/kg · K	c_p kJ/kg · K	c_v kJ/kg · K	k
Air	–	0.2870	1.005	0.718	1.400
Argon	Ar	0.2081	0.5203	0.3122	1.667
Butane	C_4H_{10}	0.1433	1.7164	1.5734	1.091
Carbon dioxide	CO_2	0.1889	0.846	0.657	1.289
Carbon monoxide	CO	0.2968	1.040	0.744	1.400
Ethane	C_2H_6	0.2765	1.7662	1.4897	1.186
Ethylene	C_2H_4	0.2964	1.5482	1.2518	1.237
Helium	He	2.0769	5.1926	3.1156	1.667
Hydrogen	H_2	4.1240	14.307	10.183	1.405
Methane	CH_4	0.5182	2.2537	1.7354	1.299
Neon	Ne	0.4119	1.0299	0.6179	1.667
Nitrogen	N_2	0.2968	1.039	0.743	1.400
Octane	C_8H_{18}	0.0729	1.7113	1.6385	1.044
Oxygen	O_2	0.2598	0.918	0.658	1.395
Propane	C_3H_8	0.1885	1.6794	1.4909	1.126
Steam	H_2O	0.4615	1.8723	1.4108	1.327

Note: The unit kJ/kg · K is equivalent to kJ/kg · °C.
Source: This table is taken from *Thermodynamics: An Engineering Approach* by Cengel, Y.A. and Boles, M.A., 6th ed., 2008, McGraw Hill, New York. Reproduced with permission of The McGraw-Hill Companies.

Table B.2 Properties of common liquids.

Substance	*Boiling data at 1 atm*		*Freezing data*		*Liquid properties*		
	Normal boiling point, °C	Latent heat of vaporization h_{fg}, kJ/kg	Freezing point, °C	Latent heat of fusion h_{if}, kJ/kg	Temperature, °C	Density ρ, kg/m^3	Specific heat c_p, kJ/kg · K
Ammonia	−33.3	1357	−77.7	322.4	−33.3	682	4.43
					−20	665	4.52
					0	639	4.60
					25	602	4.80
Argon	−185.9	161.6	−189.3	28	−185.6	1394	1.14
Benzene	80.2	394	5.5	126	20	879	1.72
Brine (20% sodium chloride by mass)	103.9	–	−17.4	–	20	1150	3.11

Table B.2 *(continued)*

Substance	Boiling data at 1 atm: Normal boiling point, °C	Boiling data at 1 atm: Latent heat of vaporization h_{fg}, kJ/kg	Freezing data: Freezing point, °C	Freezing data: Latent heat of fusion h_{if}, kJ/kg	Liquid properties: Temperature, °C	Liquid properties: Density ρ, kg/m^3	Liquid properties: Specific heat c_p, kJ/kg · K
n-Butane	−0.5	385.2	−138.5	80.3	−0.5	601	2.31
Carbon dioxide	−78.4*	230.5 (at 0 °C)	−56.6		0	298	0.59
Ethanol	78.2	838.3	−114.2	109	25	783	2.46
Ethyl alcohol	78.6	855	−156	108	20	789	2.84
Ethylene glycol	198.1	800.1	−10.8	181.1	20	1109	2.84
Glycerine	179.9	974	18.9	200.6	20	1261	2.32
Helium	−268.9	22.8	–	–	−268.9	146.2	22.8
Hydrogen	−252.8	445.7	−259.2	59.5	−252.8	70.7	10.0
Isobutane	−11.7	367.1	−160	105.7	−11.7	593.8	2.28
Kerosene	204–293	251	−24.9	–	20	820	2.00
Mercury	356.7	294.7	−38.9	11.4	25	13,560	0.139
Methane	−161.5	510.4	−182.2	58.4	−161.5	423	3.49
					−100	301	5.79
Methanol	64.5	1100	−97.7	99.2	25	787	2.55
Nitrogen	−195.8	198.6	−210	25.3	−195.8	809	2.06
					−160	596	2.97
Octane	124.8	306.3	−57.5	180.7	20	703	2.10
Oil (light)					25	910	1.80
Oxygen	−183	212.7	−218.8	13.7	−183	1141	1.71
Petroleum	–	230–384			20	640	2.0
Propane	−42.1	427.8	−187.7	80.0	−42.1	581	2.25
					0	529	2.53
					50	449	3.13
Refrigerant-134a	−26.1	217.0	−96.6	–	−50	1443	1.23
					−26.1	1374	1.27
					0	1295	1.34
					25	1207	1.43
Water	100	2257	0.0	333.7	0	1000	4.22
					25	997	4.18
					50	988	4.18
					75	975	4.19
					100	958	4.22

*Sublimation temperature. (At pressures below the triple-point pressure of 518 kPa, carbon dioxide exists as a solid or gas. Also, the freezing-point temperature of carbon dioxide is the triple-point temperature of −56.5°C.)

Source: This table is taken from *Thermodynamics: An Engineering Approach by Cengel*, Y.A. and Boles, M.A., 6th ed., 2008, McGraw Hill, New York. Reproduced with permission of The McGraw-Hill Companies.

Table B.3 Saturated refrigerant-134a – Temperature table.

		Specific volume, m^3/kg		*Internal energy*, kJ/kg			*Enthalpy*, kJ/kg			*Entropy*, kJ/kg · K		
Temp., T °C	Sat. press., P_{sat} kPa	Sat. liquid, v_f	Sat. vapor, v_g	Sat. liquid, u_f	Evap., u_{fg}	Sat. vapor, u_g	Sat. liquid h_f	Evap., h_{fg}	Sat. vapor, h_g	Sat. liquid, s_f	Evap., s_{fg}	Sat. vapor, s_g
−40	51.25	0.0007054	0.36081	−0.036	207.40	207.37	0.000	225.86	225.86	0.00000	0.96866	0.96866
−38	56.86	0.0007083	0.32732	2.475	206.04	208.51	2.515	224.61	227.12	0.01072	0.95511	0.96584
−36	62.95	0.0007112	0.29751	4.992	204.67	209.66	5.037	223.35	228.39	0.02138	0.94176	0.96315
−34	69.56	0.0007142	0.27090	7.517	203.29	210.81	7.566	222.09	229.65	0.03199	0.92859	0.96058
−32	76.71	0.0007172	0.24711	10.05	201.91	211.96	10.10	220.81	230.91	0.04253	0.91560	0.95813
−30	84.43	0.0007203	0.22580	12.59	200.52	213.11	12.65	219.52	232.17	0.05301	0.90278	0.95579
−28	92.76	0.0007234	0.20666	15.13	199.12	214.25	15.20	218.22	233.43	0.06344	0.89012	0.95356
−26	101.73	0.0007265	0.18946	17.69	197.72	215.40	17.76	216.92	234.68	0.07382	0.87762	0.95144
−24	111.37	0.0007297	0.17395	20.25	196.30	216.55	20.33	215.59	235.92	0.08414	0.86527	0.94941
−22	121.72	0.0007329	0.15995	22.82	194.88	217.70	22.91	214.26	s237.17	0.09441	0.85307	0.94748
−20	132.82	0.0007362	0.14729	25.39	193.45	218.84	25.49	212.91	238.41	0.10463	0.84101	0.94564
−18	144.69	0.0007396	0.13583	27.98	192.01	219.98	28.09	211.55	239.64	0.11481	0.82908	0.94389
−16	157.38	0.0007430	0.12542	30.57	190.56	221.13	30.69	210.18	240.87	0.12493	0.81729	0.94222
−14	170.93	0.0007464	0.11597	33.17	189.09	222.27	33.30	208.79	242.09	0.13501	0.80561	0.94063
−12	185.37	0.0007499	0.10736	35.78	187.62	223.40	35.92	207.38	243.30	0.14504	0.79406	0.93911
−10	200.74	0.0007535	0.099516	38.40	186.14	224.54	38.55	205.96	244.51	0.15504	0.78263	0.93766
−8	217.08	0.0007571	0.092352	41.03	184.64	225.67	41.19	204.52	245.72	0.16498	0.77130	0.93629
−6	234.44	0.0007608	0.085802	43.66	183.13	226.80	43.84	203.07	246.91	0.17489	0.76008	0.93497
−4	252.85	0.0007646	0.079804	46.31	181.61	227.92	46.50	201.60	248.10	0.18476	0.74896	0.93372
−2	272.36	0.0007684	0.074304	48.96	180.08	229.04	49.17	200.11	249.28	0.19459	0.73794	0.93253
0	293.01	0.0007723	0.069255	51.63	178.53	230.16	51.86	198.60	250.45	0.20439	0.72701	0.93139
2	314.84	0.0007763	0.064612	54.30	176.97	231.27	54.55	197.07	251.61	0.21415	0.71616	0.93031
4	337.90	0.0007804	0.060338	56.99	175.39	232.38	57.25	195.51	252.77	0.22387	0.70540	0.92927
6	362.23	0.0007845	0.056398	59.68	173.80	233.48	59.97	193.94	253.91	0.23356	0.69471	0.92828
8	387.88	0.0007887	0.052762	62.39	172.19	234.58	62.69	192.35	255.04	0.24323	0.68410	0.92733
10	414.89	0.0007930	0.049403	65.10	170.56	235.67	65.43	190.73	256.16	0.25286	0.67356	0.92641
12	443.31	0.0007975	0.046295	67.83	168.92	236.75	68.18	189.09	257.27	0.26246	0.66308	0.92554
14	473.19	0.0008020	0.043417	70.57	167.26	237.83	70.95	187.42	258.37	0.27204	0.65266	0.92470
16	504.58	0.0008066	0.040748	73.32	165.58	238.90	73.73	185.73	259.46	0.28159	0.64230	0.92389
18	537.52	0.0008113	0.038271	76.08	163.88	239.96	76.52	184.01	260.53	0.29112	0.63198	0.92310
20	572.07	0.0008161	0.035969	78.86	162.16	241.02	79.32	182.27	261.59	0.30063	0.62172	0.92234
22	608.27	0.0008210	0.033828	81.64	160.42	242.06	82.14	180.49	262.64	0.31011	0.61149	0.92160
24	646.18	0.0008261	0.031834	84.44	158.65	243.10	84.98	178.69	263.67	0.31958	0.60130	0.92088
26	685.84	0.0008313	0.029976	87.26	156.87	244.12	87.83	176.85	264.68	0.32903	0.59115	0.92018
28	727.31	0.0008366	0.028242	90.09	155.05	245.14	90.69	174.99	265.68	0.33846	0.58102	0.91948
30	770.64	0.0008421	0.026622	92.93	153.22	246.14	93.58	173.08	266.66	0.34789	0.57091	0.91879
32	815.89	0.0008478	0.025108	95.79	151.35	247.14	96.48	171.14	267.62	0.35730	0.56082	0.91811
34	863.11	0.0008536	0.023691	98.66	149.46	248.12	99.40	169.17	268.57	0.36670	0.55074	0.91743
36	912.35	0.0008595	0.022364	101.55	147.54	249.08	102.33	167.16	269.49	0.37609	0.54066	0.91675
38	963.68	0.0008657	0.021119	104.45	145.58	250.04	105.29	165.10	270.39	0.38548	0.53058	0.91606
40	1017.1	0.0008720	0.019952	107.38	143.60	250.97	108.26	163.00	271.27	0.39486	0.52049	0.91536
42	1072.8	0.0008786	0.018855	110.32	141.58	251.89	111.26	160.86	272.12	0.40425	0.51039	0.91464
44	1130.7	0.0008854	0.017824	113.28	139.52	252.80	114.28	158.67	272.95	0.41363	0.50027	0.91391
46	1191.0	0.0008924	0.016853	116.26	137.42	253.68	117.32	156.43	273.75	0.42302	0.49012	0.91315
48	1253.6	0.0008996	0.015939	119.26	135.29	254.55	120.39	154.14	274.53	0.43242	0.47993	0.91236

Table B.3 (*continued*)

		Specific volume, m^3/kg		*Internal energy*, kJ/kg			*Enthalpy*, kJ/kg			*Entropy*, kJ/kg · K		
Temp., T °C	Sat. press., P_{sat} kPa	Sat. liquid, v_f	Sat. vapor, v_g	Sat. liquid, u_f	Evap., u_{fg}	Sat. vapor, u_g	Sat. liquid h_f	Evap., h_{fg}	Sat. vapor, h_g	Sat. liquid, s_f	Evap., s_{fg}	Sat. vapor, s_g
52	1386.2	0.0009150	0.014265	125.33	130.88	256.21	126.59	149.39	275.98	0.45126	0.45941	0.91067
56	1529.1	0.0009317	0.012771	131.49	126.28	257.77	132.91	144.38	277.30	0.47018	0.43863	0.90880
60	1682.8	0.0009498	0.011434	137.76	121.46	259.22	139.36	139.10	278.46	0.48920	0.41749	0.90669
65	1891.0	0.0009750	0.009950	145.77	115.05	260.82	147.62	132.02	279.64	0.51320	0.39039	0.90359
70	2118.2	0.0010037	0.008642	154.01	108.14	262.15	156.13	124.32	280.46	0.53755	0.36227	0.89982
75	2365.8	0.0010372	0.007480	162.53	100.60	263.13	164.98	115.85	280.82	0.56241	0.33272	0.89512
80	2635.3	0.0010772	0.006436	171.40	92.23	263.63	174.24	106.35	280.59	0.58800	0.30111	0.88912
85	2928.2	0.0011270	0.005486	180.77	82.67	263.44	184.07	95.44	279.51	0.61473	0.26644	0.88117
90	3246.9	0.0011932	0.004599	190.89	71.29	262.18	194.76	82.35	277.11	0.64336	0.22674	0.87010
95	3594.1	0.0012933	0.003726	202.40	56.47	258.87	207.05	65.21	272.26	0.67578	0.17711	0.85289
100	3975.1	0.0015269	0.002630	218.72	29.19	247.91	224.79	33.58	258.37	0.72217	0.08999	0.81215

Source: Tables B.3 through B.8 are taken from *Thermodynamics: An Engineering Approach by Cengel*, Y.A. and Boles, M.A., 6th ed., 2008, McGraw Hill, New York. Reproduced with permission of The McGraw-Hill Companies.

Table B.4 Saturated refrigerant-134a – Pressure table.

		Specific volume, m^3/kg		*Internal energy*, kJ/kg			*Enthalpy*, kJ/kg			*Entropy*, kJ/kg · K		
Press., P kPa	Sat. temp., T_{sat} °C	Sat. liquid, v_f	Sat. vapor, v_g	Sat. liquid, u_f	Evap., u_{fg}	Sat. vapor, u_g	Sat. liquid h_f	Evap., h_{fg}	Sat. vapor, h_g	Sat. liquid, s_f	Evap., s_{fg}	Sat. vapor, s_g
60	−36.95	0.0007098	0.31121	3.798	205.32	209.12	3.841	223.95	227.79	0.01634	0.94807	0.96441
70	−33.87	0.0007144	0.26929	7.680	203.20	210.88	7.730	222.00	229.73	0.03267	0.92775	0.96042
80	−31.13	0.0007185	0.23753	11.15	201.30	212.46	11.21	220.25	231.46	0.04711	0.90999	0.95710
90	−28.65	0.0007223	0.21263	14.31	199.57	213.88	14.37	218.65	233.02	0.06008	0.89419	0.95427
100	−26.37	0.0007259	0.19254	17.21	197.98	215.19	17.28	217.16	234.44	0.07188	0.87995	0.95183
120	−22.32	0.0007324	0.16212	22.40	195.11	217.51	22.49	214.48	236.97	0.09275	0.85503	0.94779
140	−18.77	0.0007383	0.14014	26.98	192.57	219.54	27.08	212.08	239.16	0.11087	0.83368	0.94456
160	−15.60	0.0007437	0.12348	31.09	190.27	221.35	31.21	209.90	241.11	0.12693	0.81496	0.94190
180	−12.73	0.0007487	0.11041	34.83	188.16	222.99	34.97	207.90	242.86	0.14139	0.79826	0.93965
200	−10.09	0.0007533	0.099867	38.28	186.21	224.48	38.43	206.03	244.46	0.15457	0.78316	0.93773
240	−5.38	0.0007620	0.083897	44.48	182.67	227.14	44.66	202.62	247.28	0.17794	0.75664	0.93458
280	−1.25	0.0007699	0.072352	49.97	179.50	229.46	50.18	199.54	249.72	0.19829	0.73381	0.93210
320	2.46	0.0007772	0.063604	54.92	176.61	231.52	55.16	196.71	251.88	0.21637	0.71369	0.93006
360	5.82	0.0007841	0.056738	59.44	173.94	233.38	59.72	194.08	253.81	0.23270	0.69566	0.92836
400	8.91	0.0007907	0.051201	63.62	171.45	235.07	63.94	191.62	255.55	0.24761	0.67929	0.92691
450	12.46	0.0007985	0.045619	68.45	168.54	237.00	68.81	188.71	257.53	0.26465	0.66069	0.92535
500	15.71	0.0008059	0.041118	72.93	165.82	238.75	73.33	185.98	259.30	0.28023	0.64377	0.92400
550	18.73	0.0008130	0.037408	77.10	163.25	240.35	77.54	183.38	260.92	0.29461	0.62821	0.92282
600	21.55	0.0008199	0.034295	81.02	160.81	241.83	81.51	180.90	262.40	0.30799	0.61378	0.92177
650	24.20	0.0008266	0.031646	84.72	158.48	243.20	85.26	178.51	263.77	0.32051	0.60030	0.92081

(*continued overleaf*)

Table B.4 *(continued)*

Press., P kPa	Sat. temp., T_{sat} °C	*Specific volume,* m^3/kg Sat. liquid, v_f	Sat. vapor, v_g	*Internal energy,* kJ/kg Sat. liquid, u_f	Evap., u_{fg}	Sat. vapor, u_g	*Enthalpy,* kJ/kg Sat. liquid h_f	Evap., h_{fg}	Sat. vapor, h_g	*Entropy,* kJ/kg · K Sat. liquid, s_f	Evap., s_{fg}	Sat. vapor, s_g
700	26.69	0.0008331	0.029361	88.24	156.24	244.48	88.82	176.21	265.03	0.33230	0.58763	0.91994
750	29.06	0.0008395	0.027371	91.59	154.08	245.67	92.22	173.98	266.20	0.34345	0.57567	0.91912
800	31.31	0.0008458	0.025621	94.79	152.00	246.79	95.47	171.82	267.29	0.35404	0.56431	0.91835
850	33.45	0.0008520	0.024069	97.87	149.98	247.85	98.60	169.71	268.31	0.36413	0.55349	0.91762
900	35.51	0.0008580	0.022683	100.83	148.01	248.85	101.61	167.66	269.26	0.37377	0.54315	0.91692
950	37.48	0.0008641	0.021438	103.69	146.10	249.79	104.51	165.64	270.15	0.38301	0.53323	0.91624
1000	39.37	0.0008700	0.020313	106.45	144.23	250.68	107.32	163.67	270.99	0.39189	0.52368	0.91558
1200	46.29	0.0008934	0.016715	116.70	137.11	253.81	117.77	156.10	273.87	0.42441	0.48863	0.91303
1400	52.40	0.0009166	0.014107	125.94	130.43	256.37	127.22	148.90	276.12	0.45315	0.45734	0.91050
1600	57.88	0.0009400	0.012123	134.43	124.04	258.47	135.93	141.93	277.86	0.47911	0.42873	0.90784
1800	62.87	0.0009639	0.010559	142.33	117.83	260.17	144.07	135.11	279.17	0.50294	0.40204	0.90498
2000	67.45	0.0009886	0.009288	149.78	111.73	261.51	151.76	128.33	280.09	0.52509	0.37675	0.90184
2500	77.54	0.0010566	0.006936	166.99	96.47	263.45	169.63	111.16	280.79	0.57531	0.31695	0.89226
3000	86.16	0.0011406	0.005275	183.04	80.22	263.26	186.46	92.63	279.09	0.62118	0.25776	0.87894

Table B.5 Superheated refrigerant-134a.

T °C	v m^3/kg	u kJ/kg	h kJ/kg	s kJ/kg · K	v m^3/kg	u kJ/kg	h kJ/kg	s kJ/kg · K	v m^3/kg	u kJ/kg	h kJ/kg	s kJ/kg · K
	$P = 0.06$ MPa ($T_{sat} = -36.95$°C)				$P = 0.10$ MPa ($T_{sat} = -26.37$°C)				$P = 0.14$ MPa ($T_{sat} = -18.77$°C)			
Sat.	0.31121	209.12	227.79	0.9644	0.19254	215.19	234.44	0.9518	0.14014	219.54	239.16	0.9446
−20	0.33608	220.60	240.76	1.0174	0.19841	219.66	239.50	0.9721				
−10	0.35048	227.55	248.58	1.0477	0.20743	226.75	247.49	1.0030	0.14605	225.91	246.36	0.9724
0	0.36476	234.66	256.54	1.0774	0.21630	233.95	255.58	1.0332	0.15263	233.23	254.60	1.0031
10	0.37893	241.92	264.66	1.1066	0.22506	241.30	263.81	1.0628	0.15908	240.66	262.93	1.0331
20	0.39302	249.35	272.94	1.1353	0.23373	248.79	272.17	1.0918	0.16544	248.22	271.38	1.0624
30	0.40705	256.95	281.37	1.1636	0.24233	256.44	280.68	1.1203	0.17172	255.93	279.97	1.0912
40	0.42102	264.71	289.97	1.1915	0.25088	264.25	289.34	1.1484	0.17794	263.79	288.70	1.1195
50	0.43495	272.64	298.74	1.2191	0.25937	272.22	298.16	1.1762	0.18412	271.79	297.57	1.1474
60	0.44883	280.73	307.66	1.2463	0.26783	280.35	307.13	1.2035	0.19025	279.96	306.59	1.1749
70	0.46269	288.99	316.75	1.2732	0.27626	288.64	316.26	1.2305	0.19635	288.28	315.77	1.2020
80	0.47651	297.41	326.00	1.2997	0.28465	297.08	325.55	1.2572	0.20242	296.75	325.09	1.2288
90	0.49032	306.00	335.42	1.3260	0.29303	305.69	334.99	1.2836	0.20847	305.38	334.57	1.2553
100	0.50410	314.74	344.99	1.3520	0.30138	314.46	344.60	1.3096	0.21449	314.17	344.20	1.2814
	$P = 0.18$ MPa ($T_{sat} = -12.73$°C)				$P = 0.20$ MPa ($T_{sat} = -10.09$°C)				$P = 0.24$ MPa ($T_{sat} = -5.38$°C)			
Sat.	0.11041	222.99	242.86	0.9397	0.09987	224.48	244.46	0.9377	0.08390	227.14	247.28	0.9346
−10	0.11189	225.02	245.16	0.9484	0.09991	224.55	244.54	0.9380				
0	0.11722	232.48	253.58	0.9798	0.10481	232.09	253.05	0.9698	0.08617	231.29	251.97	0.9519
10	0.12240	240.00	262.04	1.0102	0.10955	239.67	261.58	1.0004	0.09026	238.98	260.65	0.9831
20	0.12748	247.64	270.59	1.0399	0.11418	247.35	270.18	1.0303	0.09423	246.74	269.36	1.0134
30	0.13248	255.41	279.25	1.0690	0.11874	255.14	278.89	1.0595	0.09812	254.61	278.16	1.0429
40	0.13741	263.31	288.05	1.0975	0.12322	263.08	287.72	1.0882	0.10193	262.59	287.06	1.0718

Table B.5 *(continued)*

T °C	v m^3/kg	u kJ/kg	h kJ/kg	s kJ/kg · K	v m^3/kg	u kJ/kg	h kJ/kg	s kJ/kg · K	v m^3/kg	u kJ/kg	h kJ/kg	s kJ/kg · K
50	0.14230	271.36	296.98	1.1256	0.12766	271.15	296.68	1.1163	0.10570	270.71	296.08	1.1001
60	0.14715	279.56	306.05	1.1532	0.13206	279.37	305.78	1.1441	0.10942	278.97	305.23	1.1280
70	0.15196	287.91	315.27	1.1805	0.13641	287.73	315.01	1.1714	0.11310	287.36	314.51	1.1554
80	0.15673	296.42	324.63	1.2074	0.14074	296.25	324.40	1.1983	0.11675	295.91	323.93	1.1825
90	0.16149	305.07	334.14	1.2339	0.14504	304.92	333.93	1.2249	0.12038	304.60	333.49	1.2092
100	0.16622	313.88	343.80	1.2602	0.14933	313.74	343.60	1.2512	0.12398	313.44	343.20	1.2356
	$P = 0.28$ MPa ($T_{sat} = -1.25$°C)				$P = 0.32$ MPa ($T_{sat} = 2.46$°C)				$P = 0.40$ MPa ($T_{sat} = 8.91$°C)			
Sat.	0.07235	229.46	249.72	0.9321	0.06360	231.52	251.88	0.9301	0.051201	235.07	255.55	0.9269
0	0.07282	230.44	250.83	0.9362								
10	0.07646	238.27	259.68	0.9680	0.06609	237.54	258.69	0.9544	0.051506	235.97	256.58	0.9305
20	0.07997	246.13	268.52	0.9987	0.06925	245.50	267.66	0.9856	0.054213	244.18	265.86	0.9628
30	0.08338	254.06	277.41	1.0285	0.07231	253.50	276.65	1.0157	0.056796	252.36	275.07	0.9937
40	0.08672	262.10	286.38	1.0576	0.07530	261.60	285.70	1.0451	0.059292	260.58	284.30	1.0236
50	0.09000	270.27	295.47	1.0862	0.07823	269.82	294.85	1.0739	0.061724	268.90	293.59	1.0528
60	0.09324	278.56	304.67	1.1142	0.08111	278.15	304.11	1.1021	0.064104	277.32	302.96	1.0814
70	0.09644	286.99	314.00	1.1418	0.08395	286.62	313.48	1.1298	0.066443	285.86	312.44	1.1094
80	0.09961	295.57	323.46	1.1690	0.08675	295.22	322.98	1.1571	0.068747	294.53	322.02	1.1369
90	0.10275	304.29	333.06	1.1958	0.08953	303.97	332.62	1.1840	0.071023	303.32	331.73	1.1640
100	0.10587	313.15	342.80	1.2222	0.09229	312.86	342.39	1.2105	0.073274	312.26	341.57	1.1907
110	0.10897	322.16	352.68	1.2483	0.09503	321.89	352.30	1.2367	0.075504	321.33	351.53	1.2171
120	0.11205	331.32	362.70	1.2742	0.09775	331.07	362.35	1.2626	0.077717	330.55	361.63	1.2431
130	0.11512	340.63	372.87	1.2997	0.10045	340.39	372.54	1.2882	0.079913	339.90	371.87	1.2688
140	0.11818	350.09	383.18	1.3250	0.10314	349.86	382.87	1.3135	0.082096	349.41	382.24	1.2942
	$P = 0.50$ MPa ($T_{sat} = 15.71$°C)				$P = 0.60$ MPa ($T_{sat} = 21.55$°C)				$P = 0.70$ MPa ($T_{sat} = 26.69$°C)			
Sat.	0.041118	238.75	259.30	0.9240	0.034295	241.83	262.40	0.9218	0.029361	244.48	265.03	0.9199
20	0.042115	242.40	263.46	0.9383								
30	0.044338	250.84	273.01	0.9703	0.035984	249.22	270.81	0.9499	0.029966	247.48	268.45	0.9313
40	0.046456	259.26	282.48	1.0011	0.037865	257.86	280.58	0.9816	0.031696	256.39	278.57	0.9641
50	0.048499	267.72	291.96	1.0309	0.039659	266.48	290.28	1.0121	0.033322	265.20	288.53	0.9954
60	0.050485	276.25	301.50	1.0599	0.041389	275.15	299.98	1.0417	0.034875	274.01	298.42	1.0256
70	0.052427	284.89	311.10	1.0883	0.043069	283.89	309.73	1.0705	0.036373	282.87	308.33	1.0549
80	0.054331	293.64	320.80	1.1162	0.044710	292.73	319.55	1.0987	0.037829	291.80	318.28	1.0835
90	0.056205	302.51	330.61	1.1436	0.046318	301.67	329.46	1.1264	0.039250	300.82	328.29	1.1114
100	0.058053	311.50	340.53	1.1705	0.047900	310.73	339.47	1.1536	0.040642	309.95	338.40	1.1389
110	0.059880	320.63	350.57	1.1971	0.049458	319.91	349.59	1.1803	0.042010	319.19	348.60	1.1658
120	0.061687	329.89	360.73	1.2233	0.050997	329.23	359.82	1.2067	0.043358	328.55	358.90	1.1924
130	0.063479	339.29	371.03	1.2491	0.052519	338.67	370.18	1.2327	0.044688	338.04	369.32	1.2186
140	0.065256	348.83	381.46	1.2747	0.054027	348.25	380.66	1.2584	0.046004	347.66	379.86	1.2444
150	0.067021	358.51	392.02	1.2999	0.055522	357.96	391.27	1.2838	0.047306	357.41	390.52	1.2699
160	0.068775	368.33	402.72	1.3249	0.057006	367.81	402.01	1.3088	0.048597	367.29	401.31	1.2951
	$P = 0.80$ MPa ($T_{sat} = 31.31$°C)				$P = 0.90$ MPa ($T_{sat} = 35.51$°C)				$P = 1.00$ MPa ($T_{sat} = 39.37$°C)			
Sat.	0.025621	246.79	267.29	0.9183	0.022683	248.85	269.26	0.9169	0.020313	250.68	270.99	0.9156
40	0.027035	254.82	276.45	0.9480	0.023375	253.13	274.17	0.9327	0.020406	251.30	271.71	0.9179
50	0.028547	263.86	286.69	0.9802	0.024809	262.44	284.77	0.9660	0.021796	260.94	282.74	0.9525
60	0.029973	272.83	296.81	1.0110	0.026146	271.60	295.13	0.9976	0.023068	270.32	293.38	0.9850
70	0.031340	281.81	306.88	1.0408	0.027413	280.72	305.39	1.0280	0.024261	279.59	303.85	1.0160

(*continued overleaf*)

Table B.5 *(continued)*

T °C	v m³/kg	u kJ/kg	h kJ/kg	s kJ/kg · K	v m³/kg	u kJ/kg	h kJ/kg	s kJ/kg · K	v m³/kg	u kJ/kg	h kJ/kg	s kJ/kg · K
80	0.032659	290.84	316.97	1.0698	0.028630	289.86	315.63	1.0574	0.025398	288.86	314.25	1.0458
90	0.033941	299.95	327.10	1.0981	0.029806	299.06	325.89	1.0860	0.026492	298.15	324.64	1.0748
100	0.035193	309.15	337.30	1.1258	0.030951	308.34	336.19	1.1140	0.027552	307.51	335.06	1.1031
110	0.036420	318.45	347.59	1.1530	0.032068	317.70	346.56	1.1414	0.028584	316.94	345.53	1.1308
120	0.037625	327.87	357.97	1.1798	0.033164	327.18	357.02	1.1684	0.029592	326.47	356.06	1.1580
130	0.038813	337.40	368.45	1.2061	0.034241	336.76	367.58	1.1949	0.030581	336.11	366.69	1.1846
140	0.039985	347.06	379.05	1.2321	0.035302	346.46	378.23	1.2210	0.031554	345.85	377.40	1.2109
150	0.041143	356.85	389.76	1.2577	0.036349	356.28	389.00	1.2467	0.032512	355.71	388.22	1.2368
160	0.042290	366.76	400.59	1.2830	0.037384	366.23	399.88	1.2721	0.033457	365.70	399.15	1.2623
170	0.043427	376.81	411.55	1.3080	0.038408	376.31	410.88	1.2972	0.034392	375.81	410.20	1.2875
180	0.044554	386.99	422.64	1.3327	0.039423	386.52	422.00	1.3221	0.035317	386.04	421.36	1.3124
	P = 1.20 MPa (T_{sat} = 46.29°C)				P = 1.40 MPa (T_{sat} = 52.40°C)				P = 1.60 MPa (T_{sat} = 57.88°C)			
Sat.	0.016715	253.81	273.87	0.9130	0.014107	256.37	276.12	0.9105	0.012123	258.47	277.86	0.9078
50	0.017201	257.63	278.27	0.9267								
60	0.018404	267.56	289.64	0.9614	0.015005	264.46	285.47	0.9389	0.012372	260.89	280.69	0.9163
70	0.019502	277.21	300.61	0.9938	0.016060	274.62	297.10	0.9733	0.013430	271.76	293.25	0.9535
80	0.020529	286.75	311.39	1.0248	0.017023	284.51	308.34	1.0056	0.014362	282.09	305.07	0.9875
90	0.021506	296.26	322.07	1.0546	0.017923	294.28	319.37	1.0364	0.015215	292.17	316.52	1.0194
100	0.022442	305.80	332.73	1.0836	0.018778	304.01	330.30	1.0661	0.016014	302.14	327.76	1.0500
110	0.023348	315.38	343.40	1.1118	0.019597	313.76	341.19	1.0949	0.016773	312.07	338.91	1.0795
120	0.024228	325.03	354.11	1.1394	0.020388	323.55	352.09	1.1230	0.017500	322.02	350.02	1.1081
130	0.025086	334.77	364.88	1.1664	0.021155	333.41	363.02	1.1504	0.018201	332.00	361.12	1.1360
140	0.025927	344.61	375.72	1.1930	0.021904	343.34	374.01	1.1773	0.018882	342.05	372.26	1.1632
150	0.026753	354.56	386.66	1.2192	0.022636	353.37	385.07	1.2038	0.019545	352.17	383.44	1.1900
160	0.027566	364.61	397.69	1.2449	0.023355	363.51	396.20	1.2298	0.020194	362.38	394.69	1.2163
170	0.028367	374.78	408.82	1.2703	0.024061	373.75	407.43	1.2554	0.020830	372.69	406.02	1.2421
180	0.029158	385.08	420.07	1.2954	0.024757	384.10	418.76	1.2807	0.021456	383.11	417.44	1.2676

Table B.6 Saturated refrigerant-134a – Temperature table.

Temp., T °F	Sat. press., P_{sat} psia	*Specific volume,* ft³/lbm: Sat. liquid, v_f	Sat. vapor, v_g	*Internal energy,* Btu/lbm: Sat. liquid, u_f	Evap., u_{fg}	Sat. vapor, u_g	*Enthalpy,* Btu/lbm: Sat. liquid h_f	Evap., h_{fg}	Sat. vapor, h_g	*Entropy,* Btu/lbm · R: Sat. liquid, s_f	Evap., s_{fg}	Sat. vapor, s_g
−40	7.432	0.01130	5.7796	−0.016	89.167	89.15	0.000	97.100	97.10	0.00000	0.23135	0.23135
−35	8.581	0.01136	5.0509	1.484	88.352	89.84	1.502	96.354	97.86	0.00355	0.22687	0.23043
−30	9.869	0.01143	4.4300	2.990	87.532	90.52	3.011	95.601	98.61	0.00708	0.22248	0.22956
−25	11.306	0.01150	3.8988	4.502	86.706	91.21	4.526	94.839	99.36	0.01058	0.21817	0.22875
−20	12.906	0.01156	3.4426	6.019	85.874	91.89	6.047	94.068	100.12	0.01405	0.21394	0.22798
−15	14.680	0.01163	3.0494	7.543	85.036	92.58	7.574	93.288	100.86	0.01749	0.20978	0.22727
−10	16.642	0.01171	2.7091	9.073	84.191	93.26	9.109	92.498	101.61	0.02092	0.20569	0.22660

Table B.6 (*continued*)

Temp., T °F	Sat. press., P_{sat} psia	Specific volume, ft^3/lbm: Sat. liquid, v_f	Sat. vapor, v_g	Internal energy, Btu/lbm: Sat. liquid, u_f	Evap., u_{fg}	Sat. vapor, u_g	Enthalpy, Btu/lbm: Sat. liquid h_f	Evap., h_{fg}	Sat. vapor, h_g	Entropy, Btu/lbm·R: Sat. liquid, s_f	Evap., s_{fg}	Sat. vapor, s_g
−5	18.806	0.01178	2.4137	10.609	83.339	93.95	10.650	91.698	102.35	0.02431	0.20166	0.22598
0	21.185	0.01185	2.1564	12.152	82.479	94.63	12.199	90.886	103.08	0.02769	0.19770	0.22539
5	23.793	0.01193	1.9316	13.702	81.610	95.31	13.755	90.062	103.82	0.03104	0.19380	0.22485
10	26.646	0.01201	1.7345	15.259	80.733	95.99	15.318	89.226	104.54	0.03438	0.18996	0.22434
15	29.759	0.01209	1.5612	16.823	79.846	96.67	16.889	88.377	105.27	0.03769	0.18617	0.22386
20	33.147	0.01217	1.4084	18.394	78.950	97.34	18.469	87.514	105.98	0.04098	0.18243	0.22341
25	36.826	0.01225	1.2732	19.973	78.043	98.02	20.056	86.636	106.69	0.04426	0.17874	0.22300
30	40.813	0.01234	1.1534	21.560	77.124	98.68	21.653	85.742	107.40	0.04752	0.17509	0.22260
35	45.124	0.01242	1.0470	23.154	76.195	99.35	23.258	84.833	108.09	0.05076	0.17148	0.22224
40	49.776	0.01251	0.95205	24.757	75.253	100.01	24.873	83.907	108.78	0.05398	0.16791	0.22189
45	54.787	0.01261	0.86727	26.369	74.298	100.67	26.497	82.963	109.46	0.05720	0.16437	0.22157
50	60.175	0.01270	0.79136	27.990	73.329	101.32	28.131	82.000	110.13	0.06039	0.16087	0.22127
55	65.957	0.01280	0.72323	29.619	72.346	101.97	29.775	81.017	110.79	0.06358	0.15740	0.22098
60	72.152	0.01290	0.66195	31.258	71.347	102.61	31.431	80.013	111.44	0.06675	0.15396	0.22070
65	78.780	0.01301	0.60671	32.908	70.333	103.24	33.097	78.988	112.09	0.06991	0.15053	0.22044
70	85.858	0.01312	0.55681	34.567	69.301	103.87	34.776	77.939	112.71	0.07306	0.14713	0.22019
75	93.408	0.01323	0.51165	36.237	68.251	104.49	36.466	76.866	113.33	0.07620	0.14375	0.21995
80	101.45	0.01334	0.47069	37.919	67.181	105.10	38.169	75.767	113.94	0.07934	0.14038	0.21972
85	110.00	0.01347	0.43348	39.612	66.091	105.70	39.886	74.641	114.53	0.08246	0.13703	0.21949
90	119.08	0.01359	0.39959	41.317	64.979	106.30	41.617	73.485	115.10	0.08559	0.13368	0.21926
95	128.72	0.01372	0.36869	43.036	63.844	106.88	43.363	72.299	115.66	0.08870	0.13033	0.21904
100	138.93	0.01386	0.34045	44.768	62.683	107.45	45.124	71.080	116.20	0.09182	0.12699	0.21881
105	149.73	0.01400	0.31460	46.514	61.496	108.01	46.902	69.825	116.73	0.09493	0.12365	0.21858
110	161.16	0.01415	0.29090	48.276	60.279	108.56	48.698	68.533	117.23	0.09804	0.12029	0.21834
115	173.23	0.01430	0.26913	50.054	59.031	109.08	50.512	67.200	117.71	0.10116	0.11693	0.21809
120	185.96	0.01446	0.24909	51.849	57.749	109.60	52.346	65.823	118.17	0.10428	0.11354	0.21782
130	213.53	0.01482	0.21356	55.495	55.071	110.57	56.080	62.924	119.00	0.11054	0.10670	0.21724
140	244.06	0.01521	0.18315	59.226	52.216	111.44	59.913	59.801	119.71	0.11684	0.09971	0.21655
150	277.79	0.01567	0.15692	63.059	49.144	112.20	63.864	56.405	120.27	0.12321	0.09251	0.21572
160	314.94	0.01619	0.13410	67.014	45.799	112.81	67.958	52.671	120.63	0.12970	0.08499	0.21469
170	355.80	0.01681	0.11405	71.126	42.097	113.22	72.233	48.499	120.73	0.13634	0.07701	0.21335
180	400.66	0.01759	0.09618	75.448	37.899	113.35	76.752	43.726	120.48	0.14323	0.06835	0.21158
190	449.90	0.01860	0.07990	80.082	32.950	113.03	81.631	38.053	119.68	0.15055	0.05857	0.20911
200	504.00	0.02009	0.06441	85.267	26.651	111.92	87.140	30.785	117.93	0.15867	0.04666	0.20533
210	563.76	0.02309	0.04722	91.986	16.498	108.48	94.395	19.015	113.41	0.16922	0.02839	0.19761

Table B.7 Saturated refrigerant-134a – Pressure table.

		Specific volume, ft^3/lbm		*Internal energy*, Btu/lbm			*Enthalpy*, Btu/lbm			*Entropy*, Btu/lbm · R		
Press., P psia	Sat. temp., T °F	Sat. liquid, v_f	Sat. vapor, v_g	Sat. liquid, u_f	Evap., u_{fg}	Sat. vapor, u_g	Sat. liquid h_f	Evap., h_{fg}	Sat. vapor, h_g	Sat. liquid, s_f	Evap., s_{fg}	Sat. vapor, s_g
5	−53.09	0.01113	8.3785	−3.918	91.280	87.36	−3.907	99.022	95.11	−0.00945	0.24353	0.23408
10	−29.52	0.01144	4.3753	3.135	87.453	90.59	3.156	95.528	98.68	0.00742	0.22206	0.22948
15	−14.15	0.01165	2.9880	7.803	84.893	92.70	7.835	93.155	100.99	0.01808	0.20908	0.22715
20	−2.43	0.01182	2.2772	11.401	82.898	94.30	11.445	91.282	102.73	0.02605	0.19962	0.22567
25	7.17	0.01196	1.8429	14.377	81.231	95.61	14.432	89.701	104.13	0.03249	0.19213	0.22462
30	15.37	0.01209	1.5492	16.939	79.780	96.72	17.006	88.313	105.32	0.03793	0.18589	0.22383
35	22.57	0.01221	1.3369	19.205	78.485	97.69	19.284	87.064	106.35	0.04267	0.18053	0.22319
40	29.01	0.01232	1.1760	21.246	77.307	98.55	21.337	85.920	107.26	0.04688	0.17580	0.22268
45	34.86	0.01242	1.0497	23.110	76.221	99.33	23.214	84.858	108.07	0.05067	0.17158	0.22225
50	40.23	0.01252	0.94791	24.832	75.209	100.04	24.948	83.863	108.81	0.05413	0.16774	0.22188
55	45.20	0.01261	0.86400	26.435	74.258	100.69	26.564	82.924	109.49	0.05733	0.16423	0.22156
60	49.84	0.01270	0.79361	27.939	73.360	101.30	28.080	82.030	110.11	0.06029	0.16098	0.22127
65	54.20	0.01279	0.73370	29.357	72.505	101.86	29.510	81.176	110.69	0.06307	0.15796	0.22102
70	58.30	0.01287	0.68205	30.700	71.688	102.39	30.867	80.357	111.22	0.06567	0.15512	0.22080
75	62.19	0.01295	0.63706	31.979	70.905	102.88	32.159	79.567	111.73	0.06813	0.15245	0.22059
80	65.89	0.01303	0.59750	33.201	70.151	103.35	33.394	78.804	112.20	0.07047	0.14993	0.22040
85	69.41	0.01310	0.56244	34.371	69.424	103.79	34.577	78.064	112.64	0.07269	0.14753	0.22022
90	72.78	0.01318	0.53113	35.495	68.719	104.21	35.715	77.345	113.06	0.07481	0.14525	0.22006
95	76.02	0.01325	0.50301	36.578	68.035	104.61	36.811	76.645	113.46	0.07684	0.14307	0.21991
100	79.12	0.01332	0.47760	37.623	67.371	104.99	37.869	75.962	113.83	0.07879	0.14097	0.21976
110	85.00	0.01347	0.43347	39.612	66.091	105.70	39.886	74.641	114.53	0.08246	0.13703	0.21949
120	90.49	0.01360	0.39644	41.485	64.869	106.35	41.787	73.371	115.16	0.08589	0.13335	0.21924
130	95.64	0.01374	0.36491	43.258	63.696	106.95	43.589	72.144	115.73	0.08911	0.12990	0.21901
140	100.51	0.01387	0.33771	44.945	62.564	107.51	45.304	70.954	116.26	0.09214	0.12665	0.21879
150	105.12	0.01400	0.31401	46.556	61.467	108.02	46.945	69.795	116.74	0.09501	0.12357	0.21857
160	109.50	0.01413	0.29316	48.101	60.401	108.50	48.519	68.662	117.18	0.09774	0.12062	0.21836
170	113.69	0.01426	0.27466	49.586	59.362	108.95	50.035	67.553	117.59	0.10034	0.11781	0.21815
180	117.69	0.01439	0.25813	51.018	58.345	109.36	51.497	66.464	117.96	0.10284	0.11511	0.21795
190	121.53	0.01452	0.24327	52.402	57.349	109.75	52.912	65.392	118.30	0.10524	0.11250	0.21774
200	125.22	0.01464	0.22983	53.743	56.371	110.11	54.285	64.335	118.62	0.10754	0.10998	0.21753
220	132.21	0.01490	0.20645	56.310	54.458	110.77	56.917	62.256	119.17	0.11192	0.10517	0.21710
240	138.73	0.01516	0.18677	58.746	52.591	111.34	59.419	60.213	119.63	0.11603	0.10061	0.21665
260	144.85	0.01543	0.16996	61.071	50.757	111.83	61.813	58.192	120.00	0.11992	0.09625	0.21617
280	150.62	0.01570	0.15541	63.301	48.945	112.25	64.115	56.184	120.30	0.12362	0.09205	0.21567
300	156.09	0.01598	0.14266	65.452	47.143	112.60	66.339	54.176	120.52	0.12715	0.08797	0.21512
350	168.64	0.01672	0.11664	70.554	42.627	113.18	71.638	49.099	120.74	0.13542	0.07814	0.21356
400	179.86	0.01757	0.09642	75.385	37.963	113.35	76.686	43.798	120.48	0.14314	0.06848	0.21161
450	190.02	0.01860	0.07987	80.092	32.939	113.03	81.641	38.041	119.68	0.15056	0.05854	0.20911
500	199.29	0.01995	0.06551	84.871	27.168	112.04	86.718	31.382	118.10	0.15805	0.04762	0.20566

Table B.8 Superheated refrigerant-134a.

T °F	v ft^3/lbm	u Btu/lbm	h Btu/lbm	s Btu/lbm·R	v ft^3/lbm	u Btu/lbm	h Btu/lbm	s Btu/lbm·R	v ft^3/lbm	u Btu/lbm	h Btu/lbm	s Btu/lbm·R
	$P = 10$ psia ($T_{sat} = -29.52$°F)				$P = 15$ psia ($T_{sat} = -14.15$°F)				$P = 20$ psia ($T_{sat} = -2.43$°F)			
Sat.	4.3753	90.59	98.68	0.22948	2.9880	92.70	100.99	0.22715	2.2772	94.30	102.73	0.22567
−20	4.4856	92.13	100.43	0.23350								
0	4.7135	95.41	104.14	0.24174	3.1001	95.08	103.68	0.23310	2.2922	94.72	103.20	0.22671
20	4.9380	98.77	107.91	0.24976	3.2551	98.48	107.52	0.24127	2.4130	98.19	107.12	0.23504
40	5.1600	102.20	111.75	0.25761	3.4074	101.95	111.41	0.24922	2.5306	101.70	111.07	0.24311
60	5.3802	105.72	115.67	0.26531	3.5577	105.50	115.38	0.25700	2.6461	105.28	115.07	0.25097
080	5.5989	109.32	119.68	0.27288	3.7064	109.13	119.42	0.26463	2.7600	108.93	119.15	0.25866
100	5.8165	113.01	123.78	0.28033	3.8540	112.84	123.54	0.27212	2.8726	112.66	123.29	0.26621
120	6.0331	116.79	127.96	0.28767	4.0006	116.63	127.74	0.27950	2.9842	116.47	127.52	0.27363
140	6.2490	120.66	132.22	0.29490	4.1464	120.51	132.02	0.28677	3.0950	120.37	131.82	0.28093
160	6.4642	124.61	136.57	0.30203	4.2915	124.48	136.39	0.29393	3.2051	124.35	136.21	0.28812
180	6.6789	128.65	141.01	0.30908	4.4361	128.53	140.84	0.30100	3.3146	128.41	140.67	0.29521
200	6.8930	132.77	145.53	0.31604	4.5802	132.66	145.37	0.30798	3.4237	132.55	145.22	0.30221
220	7.1068	136.98	150.13	0.32292	4.7239	136.88	149.99	0.31487	3.5324	136.78	149.85	0.30912
	$P = 30$ psia ($T_{sat} = 15.37$°F)				$P = 40$ psia ($T_{sat} = 29.01$°F)				$P = 50$ psia ($T_{sat} = 40.23$°F)			
Sat.	1.5492	96.72	105.32	0.22383	1.1760	98.55	107.26	0.22268	0.9479	100.04	108.81	0.22188
20	1.5691	97.56	106.27	0.22581								
40	1.6528	101.17	110.35	0.23414	1.2126	100.61	109.58	0.22738				
60	1.7338	104.82	114.45	0.24219	1.2768	104.34	113.79	0.23565	1.0019	103.84	113.11	0.23031
80	1.8130	108.53	118.59	0.25002	1.3389	108.11	118.02	0.24363	1.0540	107.68	117.43	0.23847
100	1.8908	112.30	122.80	0.25767	1.3995	111.93	122.29	0.25140	1.1043	111.55	121.77	0.24637
120	1.9675	116.15	127.07	0.26517	1.4588	115.82	126.62	0.25900	1.1534	115.48	126.16	0.25406
140	2.0434	120.08	131.42	0.27254	1.5173	119.78	131.01	0.26644	1.2015	119.47	130.59	0.26159
160	2.1185	124.08	135.84	0.27979	1.5750	123.81	135.47	0.27375	1.2488	123.53	135.09	0.26896
180	2.1931	128.16	140.34	0.28693	1.6321	127.91	140.00	0.28095	1.2955	127.66	139.65	0.27621
200	2.2671	132.32	144.91	0.29398	1.6887	132.10	144.60	0.28803	1.3416	131.87	144.28	0.28333
220	2.3408	136.57	149.56	0.30092	1.7449	136.36	149.27	0.29501	1.3873	136.15	148.98	0.29036
240	2.4141	140.89	154.29	0.30778	1.8007	140.70	154.03	0.30190	1.4326	140.50	153.76	0.29728
260	2.4871	145.30	159.10	0.31456	1.8562	145.12	158.86	0.30871	1.4776	144.93	158.60	0.30411
280	2.5598	149.78	163.99	0.32126	1.9114	149.61	163.76	0.31543	1.5223	149.44	163.53	0.31086
	$P = 60$ psia ($T_{sat} = 49.84$°F)				$P = 70$ psia ($T_{sat} = 58.30$°F)				$P = 80$ psia ($T_{sat} = 65.89$°F)			
Sat.	0.7936	101.30	110.11	0.22127	0.6821	102.39	111.22	0.22080	0.59750	103.35	112.20	0.22040
60	0.8179	103.31	112.39	0.22570	0.6857	102.73	111.62	0.22155				
80	0.8636	107.23	116.82	0.23407	0.7271	106.76	116.18	0.23016	0.62430	106.26	115.51	0.22661
100	0.9072	111.16	121.24	0.24211	0.7662	110.76	120.68	0.23836	0.66009	110.34	120.11	0.23499
120	0.9495	115.14	125.68	0.24991	0.8037	114.78	125.19	0.24628	0.69415	114.42	124.69	0.24304
140	0.9908	119.16	130.16	0.25751	0.8401	118.85	129.73	0.25398	0.72698	118.52	129.29	0.25083
160	1.0312	123.25	134.70	0.26496	0.8756	122.97	134.31	0.26149	0.75888	122.68	133.91	0.25841
180	1.0709	127.41	139.30	0.27226	0.9105	127.15	138.94	0.26885	0.79003	126.89	138.58	0.26583
200	1.1101	131.63	143.96	0.27943	0.9447	131.40	143.63	0.27607	0.82059	131.16	143.31	0.27310
220	1.1489	135.93	148.69	0.28649	0.9785	135.71	148.39	0.28317	0.85065	135.49	148.09	0.28024
240	1.1872	140.30	153.48	0.29344	1.0118	140.10	153.21	0.29015	0.88030	139.90	152.93	0.28726
260	1.2252	144.75	158.35	0.30030	1.0449	144.56	158.10	0.29704	0.90961	144.37	157.84	0.29418
280	1.2629	149.27	163.29	0.30707	1.0776	149.10	163.06	0.30384	0.93861	148.92	162.82	0.30100
300	1.3004	153.87	168.31	0.31376	1.1101	153.71	168.09	0.31055	0.96737	153.54	167.86	0.30773
320	1.3377	158.54	173.39	0.32037	1.1424	158.39	173.19	0.31718	0.99590	158.24	172.98	0.31438

(continued overleaf)

Table B.8 *(continued)*

T °F	v ft³/lbm	u Btu/lbm	h Btu/lbm	s Btu/lbm·R	v ft³/lbm	u Btu/lbm	h Btu/lbm	s Btu/lbm·R	v ft³/lbm	u Btu/lbm	h Btu/lbm	s Btu/lbm·R
	$P = 90$ psia ($T_{sat} = 72.78$°F)				$P = 100$ psia ($T_{sat} = 79.12$°F)				$P = 120$ psia ($T_{sat} = 90.49$°F)			
Sat.	0.53113	104.21	113.06	0.22006	0.47760	104.99	113.83	0.21976	0.39644	106.35	115.16	0.21924
080	0.54388	105.74	114.80	0.22330	0.47906	105.18	114.05	0.22016				
100	0.57729	109.91	119.52	0.23189	0.51076	109.45	118.90	0.22900	0.41013	108.48	117.59	0.22362
120	0.60874	114.04	124.18	0.24008	0.54022	113.66	123.65	0.23733	0.43692	112.84	122.54	0.23232
140	0.63885	118.19	128.83	0.24797	0.56821	117.86	128.37	0.24534	0.46190	117.15	127.41	0.24058
160	0.66796	122.38	133.51	0.25563	0.59513	122.08	133.09	0.25309	0.48563	121.46	132.25	0.24851
180	0.69629	126.62	138.22	0.26311	0.62122	126.35	137.85	0.26063	0.50844	125.79	137.09	0.25619
200	0.72399	130.92	142.97	0.27043	0.64667	130.67	142.64	0.26801	0.53054	130.17	141.95	0.26368
220	0.75119	135.27	147.78	0.27762	0.67158	135.05	147.47	0.27523	0.55206	134.59	146.85	0.27100
240	0.77796	139.69	152.65	0.28468	0.69605	139.49	152.37	0.28233	0.57312	139.07	151.80	0.27817
260	0.80437	144.19	157.58	0.29162	0.72016	143.99	157.32	0.28931	0.59379	143.61	156.79	0.28521
280	0.83048	148.75	162.58	0.29847	0.74396	148.57	162.34	0.29618	0.61413	148.21	161.85	0.29214
300	0.85633	153.38	167.64	0.30522	0.76749	153.21	167.42	0.30296	0.63420	152.88	166.96	0.29896
320	0.88195	158.08	172.77	0.31189	0.79079	157.93	172.56	0.30964	0.65402	157.62	172.14	0.30569
	$P = 140$ psia ($T_{sat} = 100.50$°F)				$P = 160$ psia ($T_{sat} = 109.50$°F)				$P = 180$ psia ($T_{sat} = 117.69$°F)			
Sat.	0.33771	107.51	116.26	0.21879	0.29316	108.50	117.18	0.21836	0.25813	109.36	117.96	0.21795
120	0.36243	111.96	121.35	0.22773	0.30578	111.01	120.06	0.22337	0.26083	109.94	118.63	0.21910
140	0.38551	116.41	126.40	0.23628	0.32774	115.62	125.32	0.23230	0.28231	114.77	124.17	0.22850
160	0.40711	120.81	131.36	0.24443	0.34790	120.13	130.43	0.24069	0.30154	119.42	129.46	0.23718
180	0.42766	125.22	136.30	0.25227	0.36686	124.62	135.49	0.24871	0.31936	124.00	134.64	0.24540
200	0.44743	129.65	141.24	0.25988	0.38494	129.12	140.52	0.25645	0.33619	128.57	139.77	0.25330
220	0.46657	134.12	146.21	0.26730	0.40234	133.64	145.55	0.26397	0.35228	133.15	144.88	0.26094
240	0.48522	138.64	151.21	0.27455	0.41921	138.20	150.62	0.27131	0.36779	137.76	150.01	0.26837
260	0.50345	143.21	156.26	0.28166	0.43564	142.81	155.71	0.27849	0.38284	142.40	155.16	0.27562
280	0.52134	147.85	161.35	0.28864	0.45171	147.48	160.85	0.28554	0.39751	147.10	160.34	0.28273
300	0.53895	152.54	166.50	0.29551	0.46748	152.20	166.04	0.29246	0.41186	151.85	165.57	0.28970
320	0.55630	157.30	171.71	0.30228	0.48299	156.98	171.28	0.29927	0.42594	156.66	170.85	0.29656
340	0.57345	162.13	176.98	0.30896	0.49828	161.83	176.58	0.30598	0.43980	161.53	176.18	0.30331
360	0.59041	167.02	182.32	0.31555	0.51338	166.74	181.94	0.31260	0.45347	166.46	181.56	0.30996
	$P = 200$ psia ($T_{sat} = 125.22$°F)				$P = 300$ psia ($T_{sat} = 156.09$°F)				$P = 400$ psia ($T_{sat} = 179.86$°F)			
Sat.	0.22983	110.11	118.62	0.21753	0.14266	112.60	120.52	0.21512	0.09642	113.35	120.48	0.21161
140	0.24541	113.85	122.93	0.22481								
160	0.26412	118.66	128.44	0.23384	0.14656	113.82	121.95	0.21745				
180	0.28115	123.35	133.76	0.24229	0.16355	119.52	128.60	0.22802	0.09658	113.41	120.56	0.21173
200	0.29704	128.00	138.99	0.25035	0.17776	124.78	134.65	0.23733	0.11440	120.52	128.99	0.22471
220	0.31212	132.64	144.19	0.25812	0.19044	129.85	140.42	0.24594	0.12746	126.44	135.88	0.23500
240	0.32658	137.30	149.38	0.26565	0.20211	134.83	146.05	0.25410	0.13853	131.95	142.20	0.24418
260	0.34054	141.99	154.59	0.27298	0.21306	139.77	151.59	0.26192	0.14844	137.26	148.25	0.25270
280	0.35410	146.72	159.82	0.28015	0.22347	144.70	157.11	0.26947	0.15756	142.48	154.14	0.26077
300	0.36733	151.50	165.09	0.28718	0.23346	149.65	162.61	0.27681	0.16611	147.65	159.94	0.26851
320	0.38029	156.33	170.40	0.29408	0.24310	154.63	168.12	0.28398	0.17423	152.80	165.70	0.27599
340	0.39300	161.22	175.77	0.30087	0.25246	159.64	173.66	0.29098	0.18201	157.97	171.44	0.28326
360	0.40552	166.17	181.18	0.30756	0.26159	164.70	179.22	0.29786	0.18951	163.15	177.18	0.29035

Table B.9 Thermophysical properties of pure water at atmospheric pressure.

T (°C)	ρ (kg/m^3)	$\mu \times 10^3$ (kg/m·s)	$\nu \times 10^6$ (m^2/s)	k (W/m·K)	$\beta \times 10^5$ (K^{-1})	c_p (J/kg·K)	Pr
0	999.84	1.7531	1.7533	0.5687	−6.8140	4209.3	12.976
5	999.96	1.5012	1.5013	0.5780	1.5980	4201.0	10.911
10	999.70	1.2995	1.2999	0.5869	8.7900	4194.1	9.2860
15	999.10	1.1360	1.1370	0.5953	15.073	4188.5	7.9910
20	998.20	1.0017	1.0035	0.6034	20.661	4184.1	6.9460
25	997.07	0.8904	0.8930	0.6110	20.570	4180.9	6.0930
30	995.65	0.7972	0.8007	0.6182	30.314	4178.8	5.3880
35	994.30	0.7185	0.7228	0.6251	34.571	4177.7	4.8020
40	992.21	0.6517	0.6565	0.6351	38.530	4177.6	4.3090
45	990.22	0.5939	0.5997	0.6376	42.260	4178.3	3.8920
50	988.04	0.5442	0.5507	0.6432	45.780	4179.7	3.5350
60	983.19	0.4631	0.4710	0.6535	52.330	4184.8	2.9650
70	977.76	0.4004	0.4095	0.6623	58.400	4192.0	2.5340
80	971.79	0.3509	0.3611	0.6698	64.130	4200.1	2.2010
90	965.31	0.3113	0.3225	0.6759	69.620	4210.7	1.9390
100	958.35	0.2789	0.2911	0.6807	75.000	4221.0	1.7290

Source: D.J. Kukulka (1981) *Thermodynamic and Transport Properties of Pure and Saline Water*, MSc Thesis, State University of New York at Buffalo.

Table B.10 Thermophysical properties of air at atmospheric pressure.

T (K)	ρ (kg/m^3)	c_p (kJ/kg·K)	$\mu \times 10^7$ (kg/m·s)	$\nu \times 10^6$ (m^2/s)	$k \times 10^3$ (W/m·K)	$a \times 10^6$ (m^2/s)	Pr
200	1.7458	1.007	132.5	7.59	18.10	10.30	0.737
250	1.3947	1.006	159.6	11.44	22.30	15.90	0.720
300	1.1614	1.007	184.6	15.89	26.30	22.50	0.707
350	0.9950	1.009	208.2	20.92	30.00	29.90	0.700
400	0.8711	1.014	230.1	26.41	33.80	38.30	0.690
450	0.7740	1.021	250.7	32.39	37.30	47.20	0.686
500	0.6964	1.030	270.1	38.79	40.70	56.70	0.684
550	0.6329	1.040	288.4	45.57	43.90	66.70	0.683
600	0.5804	1.051	305.8	52.69	46.90	76.90	0.685
650	0.5356	1.063	322.5	60.21	49.70	87.30	0.690
700	0.4975	1.075	338.8	68.10	52.40	98.00	0.695
750	0.4643	1.087	354.6	76.37	54.90	109.00	0.702
800	0.4354	1.099	369.8	84.93	57.30	120.00	0.709
850	0.4097	1.110	384.3	93.80	59.60	131.00	0.716
900	0.3868	1.121	398.1	102.90	62.00	143.00	0.720
950	0.3666	1.131	411.3	112.20	64.30	155.00	0.723

Source: I. Dinçer (1997) *Heat Transfer in Food Cooling Applications*, Taylor & Francis, Washington, DC; and C. Borgnakke and R.E. Sonntag (1997) *Thermodynamic and Transport Properties*, Wiley, New York.

Table B.11 Thermophysical properties of ammonia (NH_3) gas at atmospheric pressure.

T (K)	ρ (kg/m^3)	c_p (kJ/kg·K)	$\mu \times 10^7$ (kg/m·s)	$\nu \times 10^6$ (m^2/s)	$k \times 10^3$ (W/m·K)	$a \times 10^6$ (m^2/s)	Pr
300	0.6994	2.158	101.5	14.70	24.70	16.66	0.887
320	0.6468	2.170	109.0	16.90	27.20	19.40	0.870
340	0.6059	2.192	116.5	19.20	29.30	22.10	0.872
360	0.5716	2.221	124.0	21.70	31.60	24.90	0.870
380	0.5410	2.254	131.0	24.20	34.00	27.90	0.869
400	0.5136	2.287	138.0	26.90	37.00	31.50	0.853
420	0.4888	2.322	145.0	29.70	40.40	35.60	0.833
440	0.4664	2.357	152.5	32.70	43.50	39.60	0.826
460	0.4460	2.393	159.0	35.70	46.30	43.40	0.822
480	0.4273	2.430	166.5	39.00	49.20	47.40	0.822
500	0.4101	2.467	173.0	42.20	52.50	51.90	0.813
520	0.3942	2.504	180.0	45.70	54.50	55.20	0.827
540	0.3795	2.540	186.5	49.10	57.50	59.70	0.824
560	0.3708	2.577	193.5	52.00	60.60	63.40	0.827
580	0.3533	2.613	199.5	56.50	63.68	69.10	0.817

Source: I. Dinçer (1997) *Heat Transfer in Food Cooling Applications*, Taylor & Francis, Washington, DC; and C. Borgnakke and R.E. Sonntag (1997) *Thermodynamic and Transport Properties*, Wiley, New York.

Table B.12 Thermophysical properties of carbon dioxide (CO_2) gas at atmospheric pressure.

T (K)	ρ (kg/m^3)	c_p (kJ/kg·K)	$\mu \times 10^7$ (kg/m·s)	$\nu \times 10^6$ (m^2/s)	$k \times 10^3$ (W/m·K)	$a \times 10^6$ (m^2/s)	Pr
280	1.9022	0.830	140.0	7.36	15.20	9.63	0.765
300	1.7730	0.851	149.0	8.40	16.55	11.00	0.766
320	1.6609	0.872	156.0	9.39	18.05	12.50	0.754
340	1.5618	0.891	165.0	10.60	19.70	14.20	0.746
360	1.4743	0.908	173.0	11.70	21.20	15.80	0.741
380	1.3961	0.926	181.0	13.00	22.75	17.60	0.737
400	1.3257	0.942	190.0	14.30	24.30	19.50	0.737
450	1.1782	0.981	210.0	17.80	28.20	24.50	0.728
500	1.0594	1.020	231.0	21.80	32.50	30.10	0.725
550	0.9625	1.050	251.0	26.10	36.60	36.20	0.721
600	0.8826	1.080	270.0	30.60	40.70	42.70	0.717
650	0.8143	1.100	288.0	35.40	44.50	49.70	0.712
700	0.7564	1.130	305.0	40.30	48.10	56.30	0.717
750	0.7057	1.150	321.0	45.50	51.70	63.70	0.714
800	0.6614	1.170	337.0	51.00	55.10	71.20	0.716

Source: I. Dinçer (1997) *Heat Transfer in Food Cooling Applications*, Taylor & Francis, Washington, DC; and C. Borgnakke and R.E. Sonntag (1997) *Thermodynamic and Transport Properties*, Wiley, New York.

Table B.13 Thermophysical properties of hydrogen (H_2) gas at atmospheric pressure.

T (K)	ρ (kg/m^3)	c_p (kJ/kg·K)	$\mu \times 10^7$ (kg/m·s)	$\nu \times 10^6$ (m^2/s)	$k \times 10^3$ (W/m·K)	$a \times 10^6$ (m^2/s)	Pr
100	0.2425	11.23	42.1	17.40	67.00	24.60	0.707
150	0.1615	12.60	56.0	34.70	101.00	49.60	0.699
200	0.1211	13.54	68.1	56.20	131.00	79.90	0.704
250	0.0969	14.06	78.9	81.40	157.00	115.00	0.707
300	0.0808	14.31	89.6	111.00	183.00	158.00	0.701
350	0.0692	14.43	98.8	143.00	204.00	204.00	0.700
400	0.0606	14.48	108.2	179.00	226.00	258.00	0.695
450	0.0538	14.50	117.2	218.00	247.00	316.00	0.689
500	0.0485	14.52	126.4	261.00	266.00	378.00	0.691
550	0.0440	14.53	134.3	305.00	285.00	445.00	0.685
600	0.0404	14.55	142.4	352.00	305.00	519.00	0.678
700	0.0346	14.61	157.8	456.00	342.00	676.00	0.675
800	0.0303	14.70	172.4	569.00	378.00	849.00	0.670
900	0.0269	14.83	186.5	692.00	412.00	1030.00	0.671

Source: I. Dinçer (1997) *Heat Transfer in Food Cooling Applications*, Taylor & Francis, Washington, DC; and C. Borgnakke and R.E. Sonntag (1997) *Thermodynamic and Transport Properties*, Wiley, New York.

Table B.14 Thermophysical properties of oxygen (O_2) gas at atmospheric pressure.

T (K)	ρ (kg/m^3)	c_p (kJ/kg·K)	$\mu \times 10^7$ (kg/m·s)	$\nu \times 10^6$ (m^2/s)	$k \times 10^3$ (W/m·K)	$a \times 10^6$ (m^2/s)	Pr
100	3.9450	0.962	76.4	1.94	9.25	2.44	0.796
150	2.5850	0.921	114.8	4.44	13.80	5.80	0.766
200	1.9300	0.915	147.5	7.64	18.30	10.40	0.737
250	1.5420	0.915	178.6	11.58	22.60	16.00	0.723
300	1.2840	0.920	207.2	16.14	26.80	22.70	0.711
350	1.1000	0.929	233.5	21.23	29.60	29.00	0.733
400	0.9620	0.942	258.2	26.84	33.00	36.40	0.737
450	0.8554	0.956	281.4	32.90	36.30	44.40	0.741
500	0.7698	0.972	303.3	39.40	41.20	55.10	0.716
550	0.6998	0.988	324.0	46.30	44.10	63.80	0.726
600	0.6414	1.003	343.7	53.59	47.30	73.50	0.729
700	0.5498	1.031	380.8	69.26	52.80	93.10	0.744
800	0.4810	1.054	415.2	86.32	58.90	116.00	0.743
900	0.4275	1.074	447.2	104.60	64.90	141.00	0.740

Source: I. Dinçer (1997) *Heat Transfer in Food Cooling Applications*, Taylor & Francis, Washington, DC; and C. Borgnakke and R.E. Sonntag (1997) *Thermodynamic and Transport Properties*, Wiley, New York.

Table B.15 Thermophysical properties of water vapor (steam) gas at atmospheric pressure.

T (K)	ρ (kg/m^3)	c_p (kJ/kg·K)	$\mu \times 10^7$ (kg/m·s)	$\nu \times 10^6$ (m^2/s)	$k \times 10^3$ (W/m·K)	$a \times 10^6$ (m^2/s)	Pr
380	0.5863	2.060	127.1	21.68	24.60	20.40	1.060
400	0.5542	2.014	134.4	24.25	26.10	23.40	1.040
450	0.4902	1.980	152.5	31.11	29.90	30.80	1.010
500	0.4405	1.985	170.4	38.68	33.90	38.80	0.998
550	0.4005	1.997	188.4	47.04	37.90	47.40	0.993
600	0.3652	2.026	206.7	56.60	42.20	57.00	0.993
650	0.3380	2.056	224.7	66.48	46.40	66.80	0.996
700	0.3140	2.085	242.6	77.26	50.50	77.10	1.000
750	0.2931	2.119	260.4	88.84	54.90	88.40	1.000
800	0.2739	2.152	278.6	101.70	59.20	100.00	1.010
850	0.2579	2.186	296.9	115.10	63.70	113.00	1.020

Source: I. Dinçer (1997) *Heat Transfer in Food Cooling Applications*, Taylor & Francis, Washington, DC; and C. Borgnakke and R.E. Sonntag (1997) *Thermodynamic and Transport Properties*, Wiley, New York.

Table B.16 Thermophysical properties of some solid materials.

Composition	T (K)	ρ (kg/m^3)	k (W/m·K)	c_p (J/kg·K)
Aluminum	273–673	2,720	204.0–250.0	895
Asphalt	300	2,115	0.0662	920
Bakelite	300	1,300	1.4	1,465
Brass (70% Cu + 30% Zn)	373–573	8,520	104.0–147.0	380
Carborundum	872	–	18.5	–
Chrome brick	473	3,010	2.3	835
	823	–	2.5	–
Diatomaceous silica, fired	478	–	0.25	–
Fire clay brick	478	2,645	1.0	960
	922	–	1.5	–
Bronze (75% Cu + 25% Sn)	273–373	8,670	26.0	340
Clay	300	1,460	1.3	880
Coal (anthracite)	300	1,350	0.26	1,260
Concrete (stone mix)	300	2,300	1.4	880
Constantan (60% Cu + 40% Ni)	273–373	8,920	22.0–26.0	420
Copper	273–873	8,950	385.0–350.0	380
Cotton	300	80	0.06	1,300
Glass				
Plate (soda lime)	300	2,500	1.4	750
Pyrex	300	2,225	1.4	835
Ice	253	–	2.03	1,945
	273	920	1.88	2,040

Table B.16 *(continued)*

Composition	T (K)	ρ (kg/m^3)	k (W/m·K)	c_p (J/kg·K)
Iron (C ≈ 4% cast)	273–1,273	7,260	52.0–35.0	420
Iron (C ≈ 0.5% wrought)	273–1,273	7,850	59.0–35.0	460
Lead	273–573	–	–	–
Leather (sole)	300	998	0.159	–
Magnesium	273–573	1,750	171.0–157.0	1,010
Mercury	273–573	13,400	8.0–10.0	125
Molybdenum	273–1,273	10,220	125.0–99.0	251
Nickel	273–673	8,900	93.0–59.0	450
Paper	300	930	0.18	1,340
Paraffin	300	900	0.24	2,890
Platinum	273–1,273	21,400	70.0–75.0	240
Rock				
Granite, Barre	300	2,630	2.79	775
Limestone, Salem	300	2,320	2.15	810
Marble, Halston	300	2,680	2.80	830
Rubber, vulcanized				
Soft	300	1,100	0.13	2,010
Sandstone, Berea	300	2,150	2.90	745
Hard	300	1,190	0.16	–
Sand	300	1,515	0.27	800
Silver	273–673	10,520	410.0–360.0	230
Soil	300	2,050	0.52	1,840
Steel (C ≈ 1%)	273–1,273	7,800	43.0–28.0	470
Steel (Cr ≈ 1%)	273–1,273	7,860	62.0–33.0	460
Steel (18% Cr + 8% Ni)	273–1,273	7,810	16.0–26.0	460
Snow	273	110	0.049	–
Teflon	300	2,200	0.35	–
Tin	273–473	7,300	65.0–57.0	230
Tissue, human				
Skin	300	–	0.37	–
Fat layer (adipose)	300	–	0.2	–
Muscle	300	–	0.41	–
Tungsten	273–1,273	19,350	166.0–76.0	130
Wood, cross grain				
Fir	300	415	0.11	2,720
Oak	300	545	0.17	2,385
Yellow pine	300	640	0.15	2,805
White pine	300	435	0.11	–
Wood, radial				
Fir	300	420	0.14	2,720
Oak	300	545	0.19	2,385
Zinc	273–673	7,140	112.0–93.0	380

Source: I. Dinçer (1997) *Heat Transfer in Food Cooling Applications*, Taylor & Francis, Washington, DC; and F.P. Incropera and D.P. DeWitt (1998) *Fundamentals of Heat and Mass Transfer*, Wiley, New York.

Appendix C

Food Refrigeration Data

Refrigeration Systems and Applications İbrahim Dinçer and Mehmet Kanoğlu

Table C.1 Data on storage temperatures, relative humidities, freezing temperatures, and storage periods of several food commodities.

Product	Storage Temperature (°C)	Relative Humidity (%)	Frozen Temperature (°C)	Storage Period
• Bread	−18.0–12.0	–	–	12 m
• Cake – fruit	0.6 and below	80–85	–	8–10 m
• Candy				
Chocolate covered	0.0–15.6	50–60	–	3–5 m
Hard	0.0–27.0	70–80	–	2–4 m
• Canned foods	1.7	–	–	8–12 m
• Cereal foods	1.7	75–80	–	2–5 m
• Cheese				
American	0.0–0.6	80–90	−8.5	12–20 m
Cheddar	−1.1–1.1	70–75	–	3–6 m
Camembert	0.0–2.2	85–90	–	2 m
Cream	0.0–1.1	80–85	–	12–20 m
Gorgonzola	−1.1–1.1	80–85	–	3–6 m
Gruyere	10.6	80–85	–	2–3 m
Frozen	0.0	–	–	2–12 m
Limburger	0.0–1.1	85–90	−7.3	2–3 m
Roquefort	0.0–0.6	80–85	−16.2	2–3 m
Swiss	1.1–1.7	85–90	−9.6	8–12 m
• Chocolate	0.0–10.0	50–60	–	3–6 m
Frozen	<−18.0	–	–	2–12 m
• Cider	0.0	80–90	–	6–8 m
• Coffee				
Green	−1.1–0.0	70–75	–	3–6 m
Roasted	−1.1–0.0	85	–	2–4 m
• Cream				
Frozen	0.0 and below	–	–	8–12 m
Unfrozen	0.0–1.7	85	–	10–12 d
• Dried fruits	0.0–4.4	65–70	–	12 m
• Eggs				
Dried	0.0	65–75	–	6–10 m
Liquid frozen	−18.0–(−12.2)	–	–	12–24 m
Shell	−0.6–0.0	85–90	−2.5	8–10 m

Table C.1 (*continued*)

Product	Storage Temperature (°C)	Relative Humidity (%)	Frozen Temperature (°C)	Storage Period
• Fats				
Butter	0.0–1.7	80	–	4–5 w
Frozen	<0.0	–	–	4–12 m
Margarine	0.0–1.7	80	–	6–8 m
• Ferns				
Dagger	4.4–7.2	80	−4.6	2–16 w
Asparagus	4.4–1.2	80	−2.7	7–10 d
• Fish (chilled)				
White fish in ice	−1.0–0.0	–	–	12–18 d
Large fish in ice	−1.0–0.0	–	–	21–22 d
High-fat fish in ice	−1.0–0.0	–	–	4–5 d
Shellfish in ice	−1.0–0.0	–	–	6–10 d
Smoked fish	0.0–1.0	–	–	–
Packaged fish	1.0–2.0	–	–	3–7 d
Packaged shellfish	1.0–2.0	–	–	3–7 d
Smoked haddock	0.0–1.0	–	–	8–10 d
Unwrapped kippers	0.0–1.0	–	–	10–14 d
Smoked salmon	0.0–1.0	–	–	10 d
• Flour	−0.6–0.0	75–80	–	2–5 m
Fruits and Vegetables				
• Apples				
Golden	1.5–3.0	85–90	–	4 m
Jonagold	0.0–0.5	85–90	–	4 m
Jonathan	3.0–4.0	85–90	–	3–5 m
Lord Derby	3.0–4.0	90–95	–	3–4 m
McIntosh	2.0–3.0	85–90	–	4–6 m
Rome beauty	1.0–2.0	85–90	–	5–6 m
Yellow Newton	1.0–1.5	85–90	–	5–6 m
York imperial	−0.5	85–90	–	5.6 m
• Apricots	−1.1–1.1	85–90	−2.1	2–4 w
• Artichokes				
Globe	−0.6–0.0	85–95	−2.5	1–3 w
Jeruselam	−0.6–0.0	85–95	−2.5	2–5 m

(*continued overleaf*)

Table C.1 (*continued*)

Product	Storage Temperature (°C)	Relative Humidity (%)	Frozen Temperature (°C)	Storage Period
• Avocados				
Mexican	0.0–1.1	85–90	−1.8	2–6 w
Californian	4.4–7.2	85–90	−2.6	2–6 w
• Bananas	11.7–12.8	90–95	−16.0	1–2 w
• Beans	0.0–1.1	85–95	−1.2	1–4 w
• Beets				
Bunch	0.0–1.1	85–90	–	1–2 w
Topped	0.0–1.1	90–95	–	1–3 m
• Blackberries	0.0–1.1	80–85	−1.7	7–10 d
• Blueberries	0.0–1.1	80–85	−1.7	7–10 d
• Broccoli	−1.1–1.1	85–90	−1.5	10–20 d
• Brussels sprouts	0.0–1.1	85–95	−1.1	10–20 d
• Cabbage	0.0–1.1	90–95	−0.8	3–5 m
• Cantaloupes	0.0–1.1	80–85	−1.1	7–10 d
• Carrots				
Topped	0.0–1.1	90–95	−1.3	4–6 m
Bunch	0.0–1.1	90–95	−1.3	7–10 d
• Cauliflower	0.0–1.1	85–90	−1.0	2–3 w
• Celery	0.0–1.1	90–95	−1.2	3–4 m
• Cherries				
Sour	0.0–1.1	80–85	−2.2	10–20 d
Sweet	0.0–1.1	80–85	−4.1	10–20 d
• Chicory	0.0–1.1	85–90	−1.1	10–30 d
• Citron	0.0–1.1	75–80	–	2–4 m
• Cucumbers	4.4	80–90	−0.8	10 d
• Currants	0.0–1.1	80–85	−1.1	10 d
• Corn – green	0.0–1.1	85–90	−1.7	1–2 w
• Cranberries	0.0–1.1	85–90	−2.6	1–4 m
• Dates	−2.2–0.0	70–75	–	10–12 m
• Dewberries	0.0–1.1	80–85	−1.7	7–10 d
• Elderberries	0.0–1.1	85–90	−1.1	7–10 d
• Endive	0.0–1.1	90–95	−0.6	2–3 w

Table C.1 (*continued*)

Product	Storage Temperature (°C)	Relative Humidity (%)	Frozen Temperature (°C)	Storage Period
• Figs	−2.2–0.0	85–90	−2.2	2–6 w
• Garlic – cured	−1.1–0.6	70–75	−3.6	6–8 m
• Gooseberries	0.0–1.1	85–90	−1.7	2–3 w
• Grapefruit	0.0–10.0	85–90	−2.0	3–8 w
• Grapes				
American	0.0–1.1	80–85	−2.5	3–4 w
Vinifer	0.0–1.1	85–90	−3.9	4–6 m
• Honeydews	1.1–3.3	75–85	−1.7	2 w
• Horseradish	−2.2–0.0	90–95	−3.1	10–12 m
• Kale	0.0–1.1	85–90	−1.1	10–15 d
• Leeks – green	0.0–1.1	85–90	−1.5	1–3 m
• Lemons	10.0–12.8	85–90	−2.1	3–5 m
• Lettuce	0.0–1.1	90–95	−0.4	1–3 w
• Limes	4.4–14.4	85–90	−1.5	6–8 w
• Loganberries	0.0–1.1	85–90	−1.3	1–2 w
• Mandarins	1.7–4.4	80–85	−2.0	4–6 w
• Mangoes	0.0–10.0	85–90	−0.6	1–7 w
• Mushrooms	0.0–1.1	Dry	−0.6	3–10 d
Spawn	1.1–2.2	75	–	1 m
• Nectarines	0.0–1.1	75–85	−2.0	1–2 m
• Olives – fresh	7.2–10.0	85–90	−1.9	4–6 w
• Onions	0.0–1.1	70–75	−1.0	5–8 m
• Oranges	3.3–4.4	85–90	−2.5	8–10 w
• Parsley	0.0–1.1	80–85	−1.1	7–10 d
• Parsnips	0.0–1.1	85–90	−1.7	2–6 m
• Peaches	−1.1–1.1	85–90	−1.4	1–5 w
• Pears				
Anjou	−1.0	95	–	4–6 m
Beurre Hardy	−1.0–(−0.5)	90–95	–	2–3 m
Bosc	−1.0	90–95	–	3–4 m
Keiffler	−1.0	90–95	–	2–3 m
Williams	−1.0–(−0.5)	95	–	1–3 m

(*continued overleaf*)

Table C.1 (*continued*)

Product	Storage Temperature (°C)	Relative Humidity (%)	Frozen Temperature (°C)	Storage Period
• Peas – green	0.0–1.1	85–90	−1.1	1–3 w
• Peppers				
Chilli	0.0–1.1	85–90	−1.1	2 w
Green	0.0–1.1	85–90	−0.6	1 m
Sweet	0.0–1.1	85–90	−1.0	4–6 w
• Pineapples	4.4–7.2	85–90	−1.6	2–4 w
• Pomegranates	1.1–1.7	80–85	−2.2	2–4 m
• Plums	0.0–1.1	80–85	−2.2	2–3 w
• Popcorn	0.0–1.1	70–85	–	4–6 m
• Potatoes	4.4–7.2	85–90	−1.7	6–8 m
Cured	4.4	85–90	–	6–8 m
Sweet	10.0–12.8	80–90	−1.9	4–6 m
• Prunes	0.0–1.1	80–85	−2.2	6 m
• Pumpkins	10.0–12.8	70–75	−1.0	2–6 m
• Quinces	0.0–1.1	80–85	−2.1	3–4 m
• Radishes				
Black	0.0–1.1	90–95	−0.6	2–4 m
Red	0.0–1.1	90–95	−0.6	3–4 m
• Raspberries	0.0–1.1	80–85	−1.1	7–14 d
• Rhubarb	0.0–1.1	85–90	−2.0	2–4 w
• Sauerkraut	0.0–1.1	80–85	–	3–6 m
• Spinach	0.0–1.1	90–95	−0.9	7–10 d
• Strawberries	0.0–1.1	80–85	−1.1	7–10 d
• Tomatoes				
Green	10.0–15.6	85–90	−0.8	1–6 w
Ripe	0.0–10.0	85–90	−0.8	1–2 w
• Tangerines	0.0–1.1	80–85	−2.2	1–3 m
• Turnips	0.0–1.1	90–95	−0.8	2–4 m
• Vegetables – frozen	−18.0 and below	–	–	12 m
• Watermelons	4.4–7.2	75–85	−1.7	2–3 w
• Gelatin	6.1–10.0	–	–	–
• Glucose	−1.1–0.6	80–90	−2.2	3–5 m

Table C.1 *(continued)*

Product	Storage Temperature (°C)	Relative Humidity (%)	Frozen Temperature (°C)	Storage Period
• Herbs	0.0–1.1	75	–	2 y
• Honey				
In comb	4.4–7.5	85	–	1–3 m
Strained	4.4–7.5	83–88	–	3–5 m
• Ice	−2.2	–	–	–
• Ice cream				
Hardening	−21.0 and below	–	–	–
Storage	−21.0 and below	–	–	3–12 m
• Juices – fruit	2.2	–	–	6–12 m
• Macaroni	0.0–1.1	60–70	–	2–4 m
• Malt	0.6–1.1	80–85	–	3–5 m
Meat				
• Bacon				
Dry cured	0.0–4.5	75–80	–	1 m
Wiltshire (uw)	−2.0–0.0	75–80	–	3–5 w
Wiltshire (vp)	−2.0–0.0	75–80	–	3–5 w
Frozen	−18.0 and below	–	–	3 m
• Beef				
Carcass (uw)	4.0	90	–	1–2 w
Carcass (uw)	−1.0–0.0	90	–	3–4 w
Carcass package	−10.0	–	–	4–8 m
Carcass package	−18.0	–	–	12 m
Boneless joints (vp)	−1.0–0.0	–	–	12 w
Retail cuts (wr)	4.0	–	–	1–4 d
Retail cuts (vp)	4.0	–	–	2 w
Minced (wr)	4.0	–	–	1 d
Minced (wp)	4.0	–	–	1–2 w
• Frog's leg	−18.0 and below	–	–	4–12 m
• Game	0.0–1.5	90	–	2 w
• Hams				
Chilled	−1.7	75–80	–	2 w
Frozen	−18.0 and below	–	–	3–12 m

(*continued overleaf*)

Table C.1 *(continued)*

Product	Storage Temperature (°C)	Relative Humidity (%)	Frozen Temperature (°C)	Storage Period
• Lamb (uw)	4.0	90	–	1–2 w
• Lamb (uw)	−1.0–0.0	90	–	2–3 w
• Lamb (vp)	−1.0–0.0	–	–	10 w
Frozen carcass	−18.0–(−10.0)	–	–	3–10 m
• Lard				
In paper	−1.0–0.0	80–95	–	4–8 m
Boxes	−1.0–0.0	–	–	8 m
Tierces	4.5	80–95	–	6–8 m
Offal (uw)	−1.0–0.0	90	–	7 d
Offal (vp)	−18.0	–	–	12 m
• Pigeons – chill	−1.1–1.7	75–80	–	2 w
• Pork				
Carcass	−1.0–0.0	90	–	2 w
Joints (vp)	−1.0–0.0	–	–	3 w
Retail cuts (wr)	4.0	–	–	3 d
Minced	4.0	–	–	1 d
Frozen carcass, cuts	−18.0–(−10.0)	–	–	2–6 m
• Poultry (wr)	4.0	90	–	1 w
Chicken (wr)	−1.0–0.0	90	–	2 w
Chicken (in ice)	0.0	–	–	7–10 d
Chicken (frozen)	−18.0	–	–	6–8 m
Chicken (vp)	−18.0	–	–	12 m
• Rabbits				
Chill	−0.5–1.5	90	–	2 w
Frozen	−18.0	–	–	6–8 m
• Sausage				
Chill	−1.0–1.0	90	–	2 w
Frozen	−18.0	–	–	6 m
• Tripe				
Chill	2.2	75–80	–	2 w
Frozen	−18.0 and below	–	–	6–8 w
• Veal (uw)	4.0	90	–	6–8 d
• Veal (uw)	−1.0–0.0	–	–	3 w
Frozen carcass, cuts	−18.0	–	–	9 m

Table C.1 *(continued)*

Product	Storage Temperature (°C)	Relative Humidity (%)	Frozen Temperature (°C)	Storage Period
• Venison	−18.0	–	–	6–12 w
• Milk				
Condensed	1.7–4.4	70–75	–	6–8 m
Evaporated	1.7–4.4	70–75	−1.3	4–8 m
Powdered	−1.1–1.7	50–60	–	3–6 m
• Mincemeat	−2.2–0.0	70–80	–	3–5 m
• Molasses	4.4–10.0	75–80	–	–
• Nuts				
Brazil	0.0–1.7	80–85	−3.6	3 m
Chestnuts	0.0–1.7	80–85	−3.6	3 m
Coconuts	0.0–1.7	80–85	−3.6	3 m
Peanuts	0.0–1.7	80–85	–	3 m
Walnuts	0.0–1.7	80–85	–	3 m
• Oatmeal	0.0–1.1	75–80	–	6–12 m
• Oils				
Coconut	0.6–1.7	85–86	–	3–5 m
Cottonseed	1.7–4.4	85–86	–	3–5 m
Olco	10.0–12.8	65–70	–	4–6 m
Olive	2.2–4.4	81–87	–	3–5 m
Palm	1.7–2.8	81–87	–	3–5 m
Peanut	1.1–2.2	81–82	–	3–5 m
• Spaghetti	0.0–0.6	81–82	–	3–5 m
• Sugar	4.4	75–80	–	1–5 m
• Syrups	4.4	80–85	–	3–5 m
• Tallow – edible	1.7–4.4	80–85	–	6–8 m
• Tapioca	0.0–0.6	69–72	–	2–4 m
• Tobacco	4.4–10.0	75–80	–	–
• Wheat	0.0	80	–	2–3 m
• Wines	10.0–12.8	75–80	–	3–5 m
• Yeast	0.0–1.1	84–86	–	1–2 w

as, added sugar; d, day; m, month; un, unwrapped; vp, vacuum packed; w, week; wr, wrapped; y, year.
Source: Cambridge Refrigeration Technology, UK.

Table C.2 Transport temperatures and conditions of several food commodities.

Commodity	Carrying Temperature (°C)	Temperature Limits (°C)	Freezing Temperature (°C)	Ventilation	Storage Days
Fruits					
• Apples	0.0	−0.5–(+2.0)	−1.5	H	VD
• Apricot	−0.5	−0.5–0.0	−1.5	H	20
• Avocado	7.0	4.5–13.0	−0.5	H	30
• Banana					
Lacatan	14.0	14.0–15.0	−1.0	MP	24
Other varieties	12.0	12.0–13.5	−1.0	MP	24
• Cherry	−0.5	−1.0–0.0	−1.5	L	20
• Kiwi fruit	−0.5	−0.5–(+0.5)	−2.0	H	50/75
• Grape	−0.5	−1.0–(+0.5)	−1.5	L	50/100
• Grapefruit	10.0	4.5–16.0	−1.0	MP (or 1% CO_2)	40
• Lemon	10.0	5.0–16.0	−1.5	MP (or 1% CO_2)	80
• Melon					
Honeydew	10.0	10.0–21.0	–	M	90
Cantaloupe	3.0	2.0–4.5	–	M	15
Water	10.0	4.5–10.0	–	L	15
• Nectarines	−0.5	−0.5–(+0.5)	−1	M/H	30
• Orange	4.5	3.0–7.0	−1.0–(−0.5)	MP (or 1% CO_2)	40/50
• Peach	−0.5	−1.0–(−0.5)	−1.5	M/H	30
• Pear	−0.5	−1.0–(+0.5)	−1.5	3% CO_2	60/150
• Pineapple	8.5	7.0–10.0	−1.0	L	30
• Plantain	12.0	12.0–13.5	−1.0	MP	24
• Plum	−0.5	−0.5–(+0.5)	−1.0	H	20/35
• Tangerine orange	4.5	3.0–7.0	−1.5	MP (or 1% CO_2)	40
Vegetables					
• Artichoke					
Globe	0.0	−0.5–(+4.0)	−1.0	L	14/20
Jerusalem	0.0	−0.5–(+4.0)	–	L	60
• Asparagus	0.0	0.0–1.1	−0.5	M	20
• Beans (French)	0.0	0.0–7.0	−0.5	M/H	20
• Beetroot	0.0	0.0–1.0	−0.5	L	60/90
• Broccoli					
Sprouting	0.0	0.0–1.0	−0.5	H	10
W. cauliflower	0.0	0.0–1.0	−0.5	H	30
• Brussels sprout	0.0	0.0–1.0	−0.5	H	30
• Cabbage	0.0	0.0–1.0	−0.5	H	20
• Carrots	0.0	−0.5–(+0.5)	−1.0	L	70

Table C.2 *(continued)*

Commodity	Carrying Temperature (°C)	Temperature Limits (°C)	Freezing Temperature (°C)	Ventilation	Storage Days
• Cauliflower	0.0	0.0–1.0	−0.5	H	30
• Celery	0.0	0.0–1.0	−0.3	H	60/90
• Chicory	0.0	0.0–1.0	−0.5	H	14/20
• Cucumber	7.0	7.0–10.0	−0.3	H	14
• Eggplant	7.0	7.0–10.0	−0.5	L	14
• Garlic	0.0	0.0–1.0	−0.5	L	150
• Ginger	12.0	10.0–13.0	–	L	150
• Leek	0.0	0.0–1.0	−0.5	M	60
• Lettuce					
Iceberg	0.0	0.0–1.0	−0.5	H	40
Other varieties	0.0	0.0–1.0	0.0	H	20
• Onions	0.0	0.0–1.0	−0.5	M	30/120
• Peas in pod	0.0	0.0–1.0	−0.5	M	7/20
• Peppers (sweet)	7.0	7.0–10.0	−0.5	M	20
• Potatoes					
Ware	7.0	4.5–10.0	−0.5	M	60+
Seed	4.5	2.0–7.0	−0.5	M	150+
Sweet	13.0	13.0–16.0	−1.0	L	120
• Pumpkin	10.0	10.0–13.0	−0.5	L	60/90
• Rhubarb	0.0	0.0–1.0	−0.5	L	15/30
• Salsify	0.0	0.0–1.0	−1.0	L	–
• Squash					
Winter	10.0	7.0–13.0	−0.5	L	60/90
Summer	7.0	7.0–10.0	−0.3	M	60
• Tomato					
Green	13.0	10.0–16.0	−0.5	H	20
Firm ripe	7.0	7.0–10.0	−0.5	H	14
Meat, Dairy Produce, and Fish					
• Bacon	−1.0	−2.0–(+4.5)	–	None	30
• Beef packaged	−1.5	−1.5–0.0	–	None	70
• Butter	0.0	−1.0–(+4.5)	–	None	30
• Cheese	2.0	0.0–10.0	–	As required	–
• Cream	0.0	–	−1.0– (+0.5)	None	10
• Eggs	0.0	–	−1.0– (+0.5)	H	180

(continued overleaf)

Table C.2 *(continued)*

Commodity	Carrying Temperature (°C)	Temperature Limits (°C)	Freezing Temperature (°C)	Ventilation	Storage Days
• Fats	0.0	–	−1.0– (+0.5)	None	–
• Fish	0.0	–	−1.0– (+0.5)	None	–
Iced	−0.5	–	−2.0–0.0	None	–
Salt smoked	−0.5	−2.0–(+4.4)	–	H	–
• Game	0.0	−1.5–0.0	–	–	14
• Ham					
Fresh cured	−0.5	−1.5–(+0.5)	–	None	21
Canned	4.5	0.0–10.0	–	None	–
• Lamb and mutton	−1.5	−1.5–0.0	–	None	30
Packaged	−1.5	−1.5–0.0	–	None	70
• Lard	0.0	−1.5–(+4.5)	–	None	180
• Margarine	0.0	−1.5–(+0.5)	–	None	–
• Meat products	−0.5	−1.5–(+0.5)	–	None	–
• Milk					
Pasteurized	0.0	−1.5–(+1.0)	–	None	14
Sterilized	0.0	−1.5–(+1.0)	–	None	30
Concentrated	0.0	−1.5–(+1.0)	–	None	–
• Pork	−1.5	−1.5–0.0	–	None	14
• Poultry	−1.0	−1.5–(+1.5)	–	None	14
Miscellaneous					
• Beer	2.0	0.5–3.0	–	None	120
• Chocolate	7.0	4.5–13.0	–	None	150
• Confectionary	7.0	4.5–13.0	–	None	150
• Flowers					
Cut	0.0	−0.5–(+4.5)	−0.5	H	–
Florists, greens	0.0	−0.5–(+4.5)	−0.5	H	<30
• Hops	4.5	−2.0–(+10.0)	–	H	100
• Nuts					
Chestnuts	0.0	−1.0–(+1.5)	–	L	180
Others	0.0	−1.0–(+10.0)	–	L	180
• Yeast					
Active	0.0	−0.5–(+1.0)	–	None	14
Died	0.0	0.0–10.0	–	None	–

L, one time per hour; M, two to three times per hour; H, more than four times per hour; MP: maximum possible.
Source: Cambridge Refrigeration Technology, UK.

Table C.3 Data on storage temperatures, relative humidities, freezing temperatures, and storage periods of several food commodities.

Product	Moisture Content (%)	Initial Freezing Point (°C)	Specific Heat above Freezing (J/kg·K)	Specific Heat below Freezing (J/kg·K)	Thermal Conductivity (W/m·K)
Vegetables					
Artichokes, globe	84.9	−1.2	3680.3010	1903.3440	0.5666
Artichokes, Jerusalem	78.0	−2.5	3449.2200	1816.6800	0.5325
Asparagus	92.4	−0.6	3931.4760	1997.5440	0.6035
Beans, snap	90.3	−0.7	3861.1470	1971.1680	0.5932
Beans, Lima	70.2	−0.6	3187.9980	1718.7120	0.4941
Beets	87.6	−1.1	3770.7240	1937.2560	0.5799
Broccoli	90.7	−0.6	3874.5430	1976.1920	0.5952
Brussels sprouts	86.0	−0.8	3717.1400	1917.1600	0.5720
Cabbage	92.2	−0.9	3924.7780	1995.0320	0.6025
Carrots	87.8	−1.4	3777.4220	1939.7680	0.5809
Cauliflower	91.9	−0.8	3914.7310	1991.2640	0.6011
Celeriac	88.0	−0.9	3784.1200	1942.2800	0.5818
Celery	94.6	−0.5	4005.1540	2025.1760	0.6144
Collards	90.6	−0.8	3871.1940	1974.9360	0.5947
Corn, sweet, yellow	76.0	−0.6	3382.2400	1791.5600	0.5227
Cucumbers	96.0	−0.5	4052.0400	2042.7600	0.6213
Eggplant	92.0	−0.8	3918.0800	1992.5200	0.6016
Endive	93.8	−0.1	3978.3620	2015.1280	0.6104
Garlic	58.6	−0.8	2799.5140	1573.0160	0.4369
Ginger, root	81.7	–	3573.1330	1863.1520	0.5508
Horseradish	78.7	−1.8	3472.6630	1825.4720	0.5360
Kale	84.5	−0.5	3666.9050	1898.3200	0.5646
Kohlrabi	91.0	−1.0	3884.5900	1979.9600	0.5966
Leeks	83.0	−0.7	3616.6700	1879.4800	0.5572
Lettuce, iceberg	95.9	−0.2	4048.6910	2041.5040	0.6208
Mushrooms	91.8	−0.9	3911.3820	1990.0080	0.6006
Okra	89.6	−1.8	3837.7040	1962.3760	0.5897
Onions	89.7	−0.9	3841.0530	1963.6320	0.5902
Onions, dried flakes	3.9	–	967.6110	885.9840	0.1672
Parsley	87.7	−1.1	3774.0730	1938.5120	0.5804
Parsnips	79.5	−0.9	3499.4550	1835.5200	0.5399
Peas, green	78.9	−0.6	3479.3610	1827.9840	0.5370
Peppers, freeze dried	2.0	–	903.9800	862.1200	0.1579
Peppers, sweet, green	92.2	−0.7	3924.7780	1995.0320	0.6025

(*continued overleaf*)

Table C.3 (*continued*)

Product	Moisture Content (%)	Initial Freezing Point (°C)	Specific Heat above Freezing (J/kg·K)	Specific Heat below Freezing (J/kg·K)	Thermal Conductivity (W/m·K)
Potatoes, main crop	79.0	−0.6	3482.7100	1829.2400	0.5375
Potatoes, sweet	72.8	−1.3	3275.0720	1751.3680	0.5069
Pumpkins	91.6	−0.8	3904.6840	1987.4960	0.5996
Radishes	94.8	−0.7	4011.8520	2027.6880	0.6154
Rhubarb	93.6	−0.9	3971.6640	2012.6160	0.6094
Rutabaga	89.7	−1.1	3841.0530	1963.6320	0.5902
Salsify (veg. oyster)	77.0	−1.1	3415.7300	1804.1200	0.5276
Spinach	91.6	−0.3	3904.6840	1987.4960	0.5996
Squash, Summer	94.2	−0.5	3991.7580	2020.1520	0.6124
Squash, Winter	87.8	−0.8	3777.4220	1939.7680	0.5809
Tomatoes, mature green	93.0	−0.6	3951.5700	2005.0800	0.6065
Tomatoes, ripe	93.8	−0.5	3978.3620	2015.1280	0.6104
Turnip greens	91.1	−0.2	3887.9390	1981.2160	0.5971
Turnip	91.9	−1.1	3914.7310	1991.2640	0.6011
Watercress	95.1	−0.3	4021.8990	2031.4560	0.6168
Yams	69.6	–	3167.9040	1711.1760	0.4911
Fruits					
Apples, fresh	83.9	−1.1	3646.8110	1890.7840	0.5616
Apples, dried	31.8	–	1901.9820	1236.4080	0.3048
Apricots	86.3	−1.1	3727.1870	1920.9280	0.5735
Avocados	74.3	−0.3	3325.3070	1770.2080	0.5143
Bananas	74.3	−0.8	3325.3070	1770.2080	0.5143
Blackberries	85.6	−0.8	3703.7440	1912.1360	0.5700
Blueberries	84.6	−1.6	3670.2540	1899.5760	0.5651
Cantaloupes	89.8	−1.2	3844.4020	1964.8880	0.5907
Cherries, sour	86.1	−1.7	3720.4890	1918.4160	0.5725
Cherries, sweet	80.8	−1.8	3542.9920	1851.8480	0.5463
Cranberries	86.5	−0.9	3733.8850	1923.4400	0.5744
Currants, black	82.0	−1.0	3583.1800	1866.9200	0.5523
Currants, red & white	84.0	−1.0	3650.1600	1892.0400	0.5621
Dates, cured	22.5	−15.7	1590.5250	1119.6000	0.2589
Figs, fresh	79.1	−2.4	3486.0590	1830.4960	0.5380
Figs, dried	28.4	–	1788.1160	1193.7040	0.2880
Gooseberries	87.9	−1.1	3780.7710	1941.0240	0.5813
Grapefruit	90.9	−1.1	3881.2410	1978.7040	0.5961

Table C.3 (*continued*)

Product	Moisture Content (%)	Initial Freezing Point (°C)	Specific Heat above Freezing (J/kg·K)	Specific Heat below Freezing (J/kg·K)	Thermal Conductivity (W/m·K)
Grapes, American	81.3	−1.6	3559.7370	1858.1280	0.5488
Grapes, European type	80.6	−2.1	3536.2940	1849.3360	0.5454
Lemons	87.4	−1.4	3764.0260	1934.7440	0.5789
Limes	88.3	−1.6	3794.1670	1946.0480	0.5833
Mangoes	81.7	−0.9	3573.1330	1863.1520	0.5508
Melons, casaba	92.0	−1.1	3918.0800	1992.5200	0.6016
Melons, honeydew	89.7	−0.9	3841.0530	1963.6320	0.5902
Melons, watermelon	91.5	−0.4	3901.3350	1986.2400	0.5991
Nectarines	86.3	−0.9	3727.1870	1920.9280	0.5735
Olives	80.0	−1.4	3516.2000	1841.8000	0.5424
Oranges	82.3	−0.8	3593.2270	1870.6880	0.5537
Peaches, fresh	87.7	−0.9	3774.0730	1938.5120	0.5804
Peaches, dried	31.8	–	1901.9820	1236.4080	0.3048
Pears	83.8	−1.6	3643.4620	1889.5280	0.5611
Persimmons	64.4	−2.2	2993.7560	1645.8640	0.4655
Pineapples	86.5	−1.0	3733.8850	1923.4400	0.5744
Plums	85.2	−0.8	3690.3480	1907.1120	0.5680
Pomegranates	81.0	−3.0	3549.6900	1854.3600	0.5473
Prunes, dried	32.4	–	1922.0760	1243.9440	0.3077
Quinces	83.8	−2.0	3643.4620	1889.5280	0.5611
Raisins, seedless	15.4	–	1352.7460	1030.4240	0.2239
Raspberries	86.6	−0.6	3737.2340	1924.6960	0.5749
Strawberries	91.6	−0.8	3904.6840	1987.4960	0.5996
Tangerines	87.6	−1.1	3770.7240	1937.2560	0.5799
Whole Fish					
Cod	81.2	−2.2	3556.3880	1856.8720	0.5022
Haddock	79.9	−2.2	3512.8510	1840.5440	0.4955
Halibut	77.9	−2.2	3445.8710	1815.4240	0.4851
Herring, kippered	59.7	−2.2	2836.3530	1586.8320	0.3904
Mackerel. Atlantic	63.6	−2.2	2966.9640	1635.8160	0.4107
Perch	78.7	−2.2	3472.6630	1825.4720	0.4892
Pollock, Atlantic	78.2	−2.2	3455.9180	1819.1920	0.4866
Salmon, pin	76.4	−2.2	3395.6360	1796.5840	0.4773
Tuna, bluefin	68.1	−2.2	3117.6690	1692.3360	0.4341
Whiting	80.3	−2.2	3526.2470	1845.5680	0.4976

(*continued overleaf*)

Table C.3 *(continued)*

Product	Moisture Content (%)	Initial Freezing Point (°C)	Specific Heat above Freezing (J/kg·K)	Specific Heat below Freezing (J/kg·K)	Thermal Conductivity (W/m·K)
Shellfish					
Clams	81.8	−2.2	3576.4820	1864.4080	0.5054
Lobster, American	76.8	−2.2	3409.0320	1801.6080	0.4794
Oysters	85.2	−2.2	3690.3480	1907.1120	0.5230
Scallop, meat	78.6	−2.2	3469.3140	1824.2160	0.4887
Shrimp	75.9	−2.2	3378.8910	1790.3040	0.4747
Beef					
Brisket	55.2	–	2685.6480	1530.3120	0.3670
Carcass, choice	57.3	−2.2	2755.9770	1556.6880	0.3780
Carcass, select	58.3	−1.7	2789.4670	1569.2480	0.3832
Liver	69.0	−1.7	3147.8100	1703.6400	0.4388
Ribs, whole	54.5	–	2662.2050	1521.5200	0.3634
Round, full cut, lean/fat	64.8	–	3007.1520	1650.8880	0.4170
Round, full cut, lean	70.8	–	3208.0920	1726.2480	0.4482
Sirloin, lean	71.7	−1.7	3238.2330	1737.5520	0.4528
Short loin, steak, lean	69.6	–	3167.9040	1711.1760	0.4419
Short loin, T-bone steak	69.7	–	3171.2530	1712.4320	0.4424
Tenderloin, lean	68.4	–	3127.7160	1696.1040	0.4357
Veal, lean	75.9	–	3378.8910	1790.3040	0.4747
Lamb					
Cuts, lean	73.4	−1.9	3295.1660	1758.9040	0.4617
Leg, whole, lean	74.1	–	3318.6090	1767.6960	0.4653
Pork					
Backfat	7.7	–	1094.8730	933.7120	0.1200
Bacon	31.6	–	1895.2840	1233.8960	0.2443
Belly	36.7	–	2066.0830	1297.9520	0.2708
Carcass	49.8	–	2504.8020	1462.4880	0.3390
Ham, whole lean	68.3	–	3124.3670	1694.8480	0.4352
Ham, cured lean	55.9	–	2709.0910	1539.1040	0.3707
Shoulder, whole lean	72.6	−2.2	3268.3740	1748.8560	0.4575
Sausage					
Braunschweiger	48.0	–	2444.5200	1439.8800	0.3296
Frankfurter	53.9	−1.9	2642.1110	1513.9840	0.3603
Italian	51.1	–	2548.3390	1478.8160	0.3457
Polish	53.2	–	2618.6680	1505.1920	0.3566

Table C.3 *(continued)*

Product	Moisture Content (%)	Initial Freezing Point (°C)	Specific Heat above Freezing (J/kg·K)	Specific Heat below Freezing (J/kg·K)	Thermal Conductivity (W/m·K)
Pork	44.5	–	2327.3050	1395.9200	0.3114
Smoked links	39.3	–	2153.1570	1330.6080	0.2844
Poultry Products					
Chicken	66.0	−2.8	3047.3400	1665.9600	0.4232
Duck	48.5	–	2461.2650	1446.1600	0.3322
Turkey	70.4	–	3194.6960	1721.2240	0.4461
Egg					
White	87.8	−0.6	3777.4220	1939.7680	0.5538
White, dried	14.6	–	1325.9540	1020.3760	0.1388
Whole	75.3	−0.6	3358.7970	1782.7680	0.4830
Whole, dried	3.1	–	940.8190	875.9360	0.0736
Yolk	48.8	−0.6	2471.3120	1449.9280	0.3327
Yolk, salted	50.8	−17.2	2538.2920	1475.0480	0.3440
Yolk, sugared	51.2	−3.9	2551.6880	1480.0720	0.3463
Dairy Products					
Butter	17.9	–	1436.4710	1061.8240	0.1575
Cheese					
Camembert	51.8	–	2571.7820	1487.6080	0.3497
Cheddar	36.8	−12.8	2069.432	1299.208	0.2647
Cottage, uncreamed	79.8	−1.2	3509.502	1839.288	0.5087
Cream	53.8	–	2638.7620	1512.7280	0.3610
Gouda	41.5	–	2226.8350	1358.2400	0.2913
Limburger	48.4	−7.4	2457.9160	1444.9040	0.3304
Mozzarella	54.1	–	2648.8090	1516.4960	0.3627
Parmesan, hard	29.2	–	1814.9080	1203.7520	0.2216
Roquefort	39.4	−16.3	2156.5060	1331.8640	0.2794
Swiss	37.2	−10.0	2082.8280	1304.2320	0.2669
Processed American	39.2	−6.9	2149.8080	1329.3520	0.2783
Cream					
Half and half	80.6	–	3536.2940	1849.3360	0.5130
Table	73.8	−2.2	3308.5620	1763.9280	0.4744
Heavy whipping	57.7	–	2769.3730	1561.7120	0.3832
Ice Cream					
Chocolate	55.7	−5.6	2702.3930	1536.5920	0.3718
Strawberry	60.0	−5.6	2846.4000	1590.6000	0.3962
Vanilla	61.0	−5.6	2879.8900	1603.1600	0.4019

(continued overleaf)

Table C.3 *(continued)*

Product	Moisture Content (%)	Initial Freezing Point (°C)	Specific Heat above Freezing (J/kg·K)	Specific Heat below Freezing (J/kg·K)	Thermal Conductivity (W/m·K)
Milk					
Canned, condensed	27.2	−15.0	1747.9280	1178.6320	0.2102
Evaporated	74.0	−1.4	3315.2600	1766.4400	0.4756
Skim	90.8	–	3877.8920	1977.4480	0.5708
Skim, dried	3.2	–	944.1680	877.1920	0.0741
Whole	87.7	−0.6	3774.0730	1938.5120	0.5533
Whole, dried	2.5	–	920.7250	868.4000	0.0702
Whey, acid, dried	3.5	–	954.2150	880.9600	0.0758
Whey, sweet, dried	3.2	–	944.1680	877.1920	0.0741
Nuts, Shelled					
Almonds	4.4	–	984.3560	892.2640	0.0809
Filberts	5.4	–	1017.8460	904.8240	0.0866
Peanuts, raw	6.5	–	1054.6850	918.6400	0.0929
Peanuts, salted and roasted	1.6	–	890.5840	857.0960	0.0651
Pecans	4.8	–	997.7520	897.2880	0.0832
Walnuts, english	3.6	–	957.5640	882.2160	0.0764
Candy	–	–	–	–	–
Fudge, vanilla	10.9	–	1202.0410	973.9040	0.1178
Marshmallows	16.4	–	1386.2360	1042.9840	0.1490
Milk chocolate	1.3	–	880.5370	853.3280	0.0634
Peanut brittle	1.8	–	897.2820	859.6080	0.0662
Juice and Beverages					
Apple juice, Unsweetened	87.9	–	3780.7710	1941.0240	0.5092
Grapefruit juice, sweetened	87.4	–	3764.0260	1934.7440	0.5071
Grape juice, unsweetened	84.1	–	3653.5090	1893.2960	0.4932
Lemon juice	92.5	–	3934.8250	1998.8000	0.5285
Lime juice, unsweetened	92.5	–	3934.8250	1998.8000	0.5285
Orange juice	89.0	−0.4	3817.6100	1954.8400	0.5138
Pineapple juice, unsweetened	85.5	–	3702.0695	1911.5080	0.4993
Prune juice	81.2	–	3556.3880	1856.8720	0.4810
Tomato juice	93.9	–	3981.7110	2016.3840	0.5344
Cranberry-apple juice	82.8	–	3609.9720	1876.9680	0.4878
Cranberry-grape juice	85.6	–	3703.7440	1912.1360	0.4995
Fruit punch drink	88.0	–	3784.1200	1942.2800	0.5096
Club soda	99.9	–	4182.6510	2091.7440	0.5596

Table C.3 *(continued)*

Product	Moisture Content (%)	Initial Freezing Point (°C)	Specific Heat above Freezing (J/kg·K)	Specific Heat below Freezing (J/kg·K)	Thermal Conductivity (W/m·K)
Cola	89.4	–	3831.0060	1959.8640	0.5155
Cream soda	86.7	–	3740.5830	1925.9520	0.5041
Ginger ale	91.2	–	3891.2880	1982.4720	0.5230
Grape soda	88.8	–	3810.9120	1952.3280	0.5130
Lemon-lime soda	89.5	–	3834.3550	1961.1200	0.5159
Orange soda	87.6	–	3770.7240	1937.2560	0.5079
Root beer	89.3	–	3827.6570	1958.6080	0.5151
Chocolate milk, 2% fat	83.6	–	3636.7640	1887.0160	0.4911
Miscellaneous					
Honey	17.1	–	1409.6790	1051.7760	0.2118
Maple syrup	32.0	–	1908.6800	1238.9200	0.2744
Popcorn, air-popped	4.1	–	974.3090	888.4960	0.1572
Popcorn, oil-popped	2.8	–	930.7720	872.1680	0.1518
Yeast, Baker's, compressed	69.0	–	3147.8100	1703.6400	0.4298

Note: Moisture content and initial freezing data from ASHRAE Handbook of Refrigeration (1998) (*Reproduced by permission of ASHRAE*); other specific heats and thermal conductivities were calculated using the correlations presented in Chapter 7.

Table C.4 Thermal diffusivity data of some food products.

Products	Water Content (% by mass)	Temperature (°C)	Apparent Density (kg/m3)	Thermal Diffusivity (m2/s)
Fruits and Vegetables				
Apple, red whole[a]	85	0–30	840	0.14 × 10–6
Apple, dried	42	23	856	0.096 × 10–6
Apple sauce	37	5	–	0.11 × 10–6
	37	65	–	0.11 × 10–6
	80	5	–	0.12 × 10–6
	80	65	–	0.14 × 10–6
Apricots, dried	44	23	1323	0.11 × 10–6
Bananas, flesh	76	5	–	0.12 × 10–6
	76	65	–	0.14 × 10–6
Cherries, flesh[b]	–	0–30	1050	0.13 × 10–6
Dates	35	23	1319	0.10 × 10–6
Figs	40	23	1241	0.096 × 10–6
Jam, strawberry	41	20	1310	0.12 × 10–6

(continued overleaf)

Table C.4 (*continued*)

Products	Water Content (% by mass)	Temperature (°C)	Apparent Density (kg/m3)	Thermal Diffusivity (m2/s)
Jelly, grape	42	20	1320	0.12 × 10–6
Peaches[b]	–	2–32	960	0.14 × 10–6
Peaches, dried	43	23	1259	0.12 × 10–6
Potatoes, whole	–	0–70	1040–1070	0.13 × 10–6
Potatoes, mashed	78	5	–	0.12 × 10–6
	78	65	–	0.15 × 10–6
Prunes	43	23	1219	0.12 × 10–6
Raisins	32	23	1380	0.11 × 10–6
Strawberries, flesh	92	5	–	0.13 × 10–6
Sugar beets	–	0–60	–	0.13 × 10–6
Fish and Meat				
Codfish	81	5	–	0.12 × 10–6
	81	65	–	0.14 × 10–6
Halibut[c]	76	40–65	1070	0.15 × 10–6
Beef, chuck[d]	66	40–65	1060	0.12 × 10–6
Beef, round[d]	71	40–65	1090	0.13 × 10–6
Beef, tongue[d]	68	40–65	1060	0.13 × 10–6
Beefstick	37	20	1050	0.11 × 10–6
Bologna	65	20	1000	0.13 × 10–6
Corned beef	65	5	–	0.11 × 10–6
	65	65	–	0.13 × 10–6
Ham, country	72	20	1030	0.14 × 10–6
Ham, smoked	64	5	–	0.12 × 10–6
Ham, smoked[d]	64	40–65	1090	0.13 × 10–6
Pepperoni	32	20	1060	0.093 × 10–6
Salami	36	20	960	0.13 × 10–6
Cakes				
Angel food	36	23	147	0.26 × 10–6
Applesauce	24	23	300	0.12 × 10–6
Carrot	22	23	320	0.12 × 10–6
Chocolate	32	23	340	0.12 × 10–6
Pound	23	23	480	0.12 × 10–6
Yellow	25	23	300	0.12 × 10–6
White	32	23	446	0.10 × 10–6

[a]Data are applicable only to raw whole apple.
[b]Freshly harvested.
[c]Stored frozen and thawed prior to test.
[d]Data are applicable only where the juices exuded during heating remain in the food samples.
Source: ASHRAE Handbook of Refrigeration (1998) (*Reproduced by permission of ASHRAE*).

Subject Index